AF817141

Pathogenesis of Invertebrate Microbial Diseases

This study, covering the present state of knowledge of viral, bacterial, and fungal diseases of invertebrates, attempts to provide a single source of basic information that will be useful to new investigators and established scientists. It addresses the molecular and cellular interactions between microbial pathogens and their invertebrate hosts, including the physiological and behavioral responses of the host. The book is also intended to serve as a resource for pathologists working with other organisms, revealing the similarities as well as differences among disciplines, and as stimulus to further research by systematically identifying gaps in present knowledge within each area of specialization.

Pathogenesis of Invertebrate Microbial Diseases

edited by

ELIZABETH W. DAVIDSON

ALLANHELD, OSMUN Publishers

DEDICATION

Dedicated to the memory of Arthur M. Heimpel (1923-1979).

Dr. Heimpel was a pioneer in insect pathology and a founding member of the Society for Invertebrate Pathology. Together with Dr. Thomas Angus of the Sault Ste. Marie, Ontario Insect Pathology Research Institute, Dr. Heimpel contributed the fundamental research on the site and mode of action of the *Bacillus thuringiensis* toxins. As head of the U.S. Department of Agriculture Insect Pathology Laboratory at Beltsville, Maryland, he was influential in the establishment of Invertebrate Pathology as a science. Above all, he will be remembered for his kindness and encouragement toward young scientists.

ALLANHELD, OSMUN & CO. PUBLISHERS, INC.

Published in the United States of America in 1981
by Allanheld, Osmun & Co. Publishers, Inc.
(A Division of Littlefield, Adams & Company)
81 Adams Drive, Totowa, New Jersey 07512

Copyright 1981 by Elizabeth W. Davidson

All rights reserved. No part of this publication may be reproduced, stored in a retrieval system, or transmitted in any form or by any means, electronic, mechanical, photocopying, recording, or otherwise, without the prior permission of the publisher.

The following chapters were produced by U.S. Government employees during working hours and are not subject to copyright: Chapter 5, Couch; Chapter 7, Kawanishi; Chapter 12, Sparks.

LIBRARY OF CONGRESS CATALOGING IN PUBLICATION DATA
Main entry under title:

Pathogenesis of invertebrate microbial diseases.

Based on symposia held at the Annual Meeting of the Society for Invertebrate Pathology in East Lansing, Mich. (1977) and at the International Colloquim of Invertebrate Pathology in Prague, Czechoslovakia (1978).
Includes index.
1. Invertebrates--Diseases--Congresses. 2. Microorganisms, Pathogenic--Congresses. I. Davidson, Elizabeth W.
QR325.P37 592'.02'3 81-65007
ISBN 0-86598-014-4 AACR2

81 82 83 84 / 10 9 8 7 6 5 4 3 2 1

Printed in the United States of America

TABLE OF CONTENTS

SECTION II - BACTERIA

SECTION IV - RESISTANCE

"Pathogenicity is a highly skilled trade."

L. Thomas, 1978. Science, 200, 1459-1562.[1]

PREFACE

This book grew out of symposia held at the Annual Meeting of the Society of Invertebrate Pathology in East Lansing, Michigan (1977) and at the International Colloquium of Invertebrate Pathology in Prague, Czechoslovakia (1978), and is published with the permission of the Publication Board of the Society for Invertebrate Pathology. Scientists from eight countries have contributed to this volume; some, but not all, participated in the original symposia. In assessing the current information available on pathogenesis of invertebrate microbial diseases, the convenor and participants in the symposia recognized the need for a single source of basic information which would be available to new investigators and established scientists. We also felt that knowledge of the pathology of a microbial disease is basic to most studies utilizing the microorganisms. We therefore hope that this volume will prove useful for more applied studies. Another, equally important goal of this volume is to stimulate further research. Each author has indicated areas in his/her specialty for which further research is needed. Finally, we hope that this volume will serve as a resource for pathologists working with other organisms, revealing the similarities, as well as the differences, among our disciplines.

It will be immediately apparent to the reader that the depth of information on pathogenesis varies enormously among groups of microorganisms and hosts. In some areas, much current effort is directed toward basic identification of microorganisms associated with diseases of the host, while in

[1]Copyright 1978 by the American Association for the Advancement of Science; reprinted by permission.

other areas studies are elucidating molecular interactions between host and pathogen. It is our hope that obvious gaps in the knowledge will serve as stimuli for further research, and especially as intriguing problems for new scientists entering the field.

As editor, I am particularly grateful to Becky Payne, my typist. I am also grateful to Robert Faust, who provided information for the Dedication, to Caryl Romine, who proofread the final copy, and to my husband and son who were patient and supportive during the editing process. Above all, I am grateful to the authors for the very high quality of manuscripts submitted.

Elizabeth W. Davidson
Tempe, Arizona
March 17, 1981

SECTION I

VIRUSES

Chapter 1

BACULOVIRUS

by Peter Faulkner*

INTRODUCTION

Diseases of insects such as the honey bee and the silkworm have been known for centuries, and were the subject of speculation in the oldest literature of Asia and Europe. Bergold (1953) reports that a description of a disease of the silkworm characterized by the presence in tissues of crystalline occlusion bodies (OB's) was given in 1856 by Maestri and Cornalia. These workers established a correlation between the presence of OB's and the collapse of a population due to disease. At first the etiologic agent of the silkworm disease was considered to be a bacterium or a protozoan (Steinhaus, 1946) but general acceptance that the agent was viral in nature appears to have followed demonstrations by Von Prowazek (1913) and Acqua (1918-19) that extracts of tissue infected with silkworm jaundice retained infectivity after passing through an ultrafilter. Early in this century several forest insects were shown to be susceptible to similar infections called nuclear polyhedrosis diseases which were characterized by the formation of polyhedron-shaped OB's (Steinhaus, 1946, 1949), but it was not until the 1940's that their role in infection and their composition was elucidated by Bergold. He showed that polyhedra purified from *Bombyx mori*, *Lymantria monarcha*, and *L.* (*Porthetria*) *dispar* infected with nuclear polyhedrosis diseases could be dissolved in weak alkaline solutions (6-8 mM Na_2CO_3 in 0.05M NaCl) at about pH 10.8 to solubilize a noninfectious protein (now called polyhedrin) and to release rodshaped infectious virus particles (Bergold, 1947). Subsequently the virions were shown to be enveloped and to contain DNA (Bergold, 1953). The nuclear

*Department of Microbiology and Immunology, Queen's University, Kingston, Ontario, K7L 3N6, Canada

PATHOGENESIS OF INVERTEBRATE MICROBIAL DISEASES. Davidson, ed. © Allanheld, Osmun (Totowa, N.J.)

polyhedrosis viruses (NPV) and granulosis viruses (GV) (Family Baculoviridae) are two of four classes of insect virus diseases in which proteinaceous occlusion bodies protect the virions during natural transmission. The other two classes are entomopox viruses (Family Poxviridae) in which brick-shaped pox virus particles are occluded and cytoplasmic polyhedrosis viruses (CPV) (Family Reoviridae) which contain spherical virions possessing segmented double stranded RNA (Matthews, 1979).

Lists containing the names of several hundred inclusion-type viruses of insects have been published (Bergold, 1963; Smith, 1973; van der Geest and van der Laan, 1971; Martignoni and Iwai, 1977; Martignoni, 1981). Early observations of baculoviruses were made with acutely diseased insects and the viruses were usually named after the insect species from which they were derived. It is now recognized that many insects collected in the field are chronically infected with less virulent strains of disease-causing microorganisms; thus the pattern seen in invertebrates is similar to that in higher animals where several potential microbial pathogens and physiological mechanisms may interact to produce morbidity and sometimes mortality.

Over the past 5 years, probably as a consequence of international symposia which have focused on safety aspects of baculoviruses (Summers et al., 1975; Summers and Kawanishi, 1978; Mittenberger, 1980) there has been a concentrated effort by investigators to characterize baculovirus species and strains, to investigate the consequences of virus-cell interaction and to consider the genetic potential. In this chapter an attempt will be made to interpret observations on baculovirus pathogenesis in the light of current information on their biochemical properties and replication patterns. Reviews on the history of baculovirus infections are in books authored and edited by Steinhaus (1946, 1949, 1963) and Smith (1967). More recent studies on the course of infection with baculoviruses in insects and cell cultures are reviewed by Harrap (1973), Granados (1976, 1980), Summers (1977), Tinsley and Harrap (1978) and Stoltz and Vinson (1979).

THE GENUS *BACULOVIRUS*

Description

Three families of viruses (the Poxviridae, Reoviridae, and the Baculoviridae) contain genera which are characterized by the formation of crystalline occlusion bodies (OB's) in cells of disease insects (Matthews, 1979). The OB's are infectious *per os* and represent the form in which the virus is transmitted under natural conditions. The Baculoviridae contain a single genus, *Baculovirus*, which is divided into three sub groups, A,

B, and C based on differing arrangements for the packaging of the virion.

In sub group A, the nuclear polyhedrosis viruses (NPV), virions are present in a polyhedron-shaped inclusion body (PIB). PIB's (reported diameter up to 15 micrometers) encapsulate the elongated virions (260-280 nm in length by 60-70 nm in diameter) in two conformations. NPV nucleocapsids may be enveloped either singly (S-NPV) or in multiples (M-NPV) of up to 20 or more per bundle surrounded by a common envelope (Summers et al., 1975; Harrap and Payne, 1979). Based on chemical analysis (Bergold, 1963) the M-NPV of Bombyx mori contains about 350 nucleocapsids in bundles of two to nine and comprising about 5% of the mass of the OB.

Sub group B of the genus consists of the granulosis viruses (GV). Each inclusion body, which is often elipsoidal in section (about 200 x 400 nm), occludes a single enveloped virion (Huger, 1963; Summers et al., 1975).

Proposed sub group C (Matthews, 1979) has been diverse but the main characteristics are that the viruses are rod shaped enveloped nucleocapsids but no inclusions are produced. The type species is Oryctes virus derived from the Rhinoceros beetle. The virus particles are rather shorter and wider than those baculoviruses and may possess a "tail" structure associated with the nucleocapsid (Payne et al., 1977). Other well characterized members of this group have been derived from parasitic braconid wasps (Stoltz and Vinson, 1979). The reader is referred to Chapter 2 and to reviews by Stoltz and Vinson (1979) and Harrap and Payne (1979) for further information on this subgroup.

Tinsley (1978) states that more than 600 viruses have been isolated from dead or moribund insects and a WHO technical report (Smith, 1973) lists more than 200 species of insects and mites of importance in agriculture and forestry from which baculoviruses have been isolated. Martignoni and Iwai (1977) report 337 named isolates of baculoviruses from insects, 280 from the order Lepidoptera and the remainder from Hymenoptera, Coleoptera, Diptera, Neuroptera, Orthoptera, and Trichoptera. All of these fall into sub groups A and B (the NPV and GV). Until recently it was believed that baculoviruses were confined to the class Insecta, but typical baculovirus inclusion bodies have now been described for the pink shrimp Penaeus duorarum (Couch, 1974) (see Chapter 5). A description has been given of a possible baculovirus derived from the insect-parasitic fungus Strongwellsea magna (Federici and Humber, 1977). No data on the chemistry of the virion and its transmission are available but if the baculovirus nature is confirmed sub group C may include viruses that infect the plant kingdom.

The taxonomy of baculoviruses has resulted in much confusion in the literature as present practice continues to be that of naming the virus by coupling it to the Latin binomial of the host insect, e.g. Autographa californica NPV, Tricoplusia ni NPV, T. ni GV, Pieris rapae GV. This procedure has obvious drawbacks especially if more than one baculovirus is derived from the same insect as occurred for example with the cabbage looper, T. ni NPV before the M-NPV and S-NPV forms were recognized. Another source of taxonomic confusion is in the naming a virus which has been re-isolated in a second species of insect as could occur when the host range is extended. Fortunately newer techniques are now available to characterize baculoviruses (Harrap and Payne, 1979). For example, the pattern obtained after digesting baculovirus DNA with restriction endonucleases can distinguish species of baculovirus and even characterize variants within species (Smith and Summers, 1978; Lee and Miller, 1978; Vlak and Odink, 1979). Other systems of naming baculoviruses have been suggested (for a brief review see Tinsley and Harrap, 1978), but the present practice is to continue to use the Latin binomial of the host from which the virus was isolated and to cite, where possible, data on biological and biochemical properties.

The occlusion bodies

Thin sections of OB from NPV and GV revealed that they consist of a paracrystalline matrix of protein which surrounds the enveloped occluded viruses (Morgan et al., 1955; Bergold, 1962; Harrap, 1972a; Huger, 1963; Hughes, 1978). Several ultrastructure investigations have demonstrated that OB's are bounded by a membrane (Summers and Arnott, 1969; Harrap, 1972a; Hess and Falcon, 1978) which is contiguous with the lattice structure (Chung et al., 1980) and is rich in carbohydrate (Minion et al., 1979).

The terms polyhedrin and granulin are used to describe the protein solution obtained when baculovirus OB's are dissolved in dilute alkali and subsequently freed of virions by centrifugation (Summers et al., 1975). Analytical ultracentrifugation of polyhedrin and granulin showed that the principle component sedimented at 11-13 S (molecular weight approximately 300,000 daltons), but split components called sixths and eighteenths (molecular weight 47,000 and 15,000) were observed by Bergold (1947) with L. dispar NPV. Other NPV and GV also showed similar major components (Bergold, 1963; Rohrmann, 1977; Scharnhorst and Weaver, 1980). Alkali solubilized polyhedrin is now considered to consist of 12 S aggregates of 27,000 to 30,000 MW subunit. Other workers have found that the predominating peptide in detergent-mercaptoethanol treated polyhedrin and granulin is a peptide of MW 28,000 and there are reports that one or more minor peptides of higher or lower molecular weights are present (Cibulsky et al., 1977a; Padhi et al., 1974; McCarthy and Lui, 1976). Early studies on polyhedrin and granulin are difficult to interpret since it was not recognized until 1975 that inclusion bodies produced in vivo carry an endogenous proteolytic enzyme (Summers and Smith, 1975;

Eppstein and Thoma, 1975). The enzyme may be inactivated by heating PIB's at 70°C and by certain protolytic inhibitors (Summers and Smith, 1975). Polyhedrins have been characterized as phenol soluble phosphorylated protein (Summers and Smith, 1975, 1976) but the phosphorous content appears variable (Dobos and Cochran, 1980) and may vary due to differences in post translational modification during polyhedrin synthesis. Thus Scharnhorst and Weaver (1980) could separate four polyhedrin entities by isoelectric focusing and Carstens et al. (1979) in studies on biosynthesis of baculovirus peptides, provide evidence for post translational modification of polyhedrin.

Peptide mapping has been carried out on granulins and polyhedrins as a means of possible characterization of baculoviruses (Summers and Smith, 1975, 1976; Chibulski et al., 1977a). Peptide maps show differences among most isolates in number of acidic and basic peptide spots but there are few differences in neutral peptides. There were clear similarities between NPV's and GV's from the noctuid subfamilies studied. Major differences were found between M-NPV and S-NPV types from the same insect species. An important finding of Cibulski et al. (1977a) was that A. californica M-NPV can be distinguished from T. ni M-NPV polyhedrin on the basis of the tryptic peptide pattern. Subsequent studies have shown that the two viruses have different DNA restriction endonuclease patterns (Miller and Dawes, 1978b). Investigations by workers who measured in vivo host range and virulence had previously led to the conclusion that the viruses were identical.

Until recently it was considered that polyhedrin may not be a virus instructed protein but may arise as a host derived product as a response to virus infection. Several lines of evidence now clearly indicate that polyhedrin is indeed coded for by the virus genome. Maruniak et al. (1979) analyzed polyhedrins of A. californica NPV that had been passed through several species of insects and also had been propagated in several insect culture lines. Analysis of tryptic digests of these polyhedrins as well as amino acid analysis and N-terminal analysis all indicated a common composition. Van de Beek et al. (1980) isolated viral specific RNA from A. californica NPV infected cell cultures by binding the RNA to sheared virion DNA that has been covalently linked to cellulose. Eluted RNA was translated in an in vitro system and polyhedrin was immunoprecipitated by reaction with a specific serum. A further line of evidence that polyhedrin is virus coded is the isolation of temperature sensitive (ts) mutants of A. californica NPV by Brown et al. (1980) and Lee and Miller (1979). Many mutants failed to produce polyhedrin at the restrictive temperature.

The virus

Three sources of virus particles have been described: i. occluded virus (OV) obtained by dissolution of polyhedra in dilute alkali (Bergold, 1947) or in gut juice (Vail et al. 1979); ii. non-occluded virus (NOV) released from tissue culture cells by budding at the cell membrane (Faulkner and Henderson, 1972; Henderson et al., 1974), or present within cells (Volkman et al., 1976); iii. virus in hemolymph of infected insects either in the form of nucleocapsids with fragile envelopes or as naked nucleocapsids (Kawarabata, 1974; Stairs and Ellis, 1971; Summers and Volkman, 1976). Each of these forms of virus has low infectivity *per os* (Dougherty et al., 1975; Kawarabata and Aratake, 1978; Stairs, 1980) but all are infectious when injected into susceptible insects. Since OV are formed within the nucleus and are enveloped therein whereas NOV appear to pick up their envelopes outside the nucleus or at the cell membrane, it is not surprising that differences in infectivity and serological properties for these forms of virus have been described (Volkman et al., 1976; Volkman and Summers, 1977).

Virus DNA

Early analysis indicated that approximately 11% of the mass of baculovirus OV was DNA (Bergold, 1963) and recent intensive studies now indicate that a single nucleic acid component ranging from about MW 65-100 x 10^6 is extracted from the virions. When the DNA is centrifuged to equilibrium in ethidium bromide-cesium chloride solutions two bands usually separate having densities of 1.58 gm/ml and 1.54 gm/ml (Summers and Anderson, 1973; Vlak and Odink, 1979; Scharnhorst and Weaver, 1979). These bands were infectious *in vivo* (Strokovskaya et al., 1977). EM examination of the bands indicated that the densest band consisted of Form I closed circular DNA with highly twisted circles and the lighter band consisted of Form II relaxed open circles. Several workers have estimated the molecular weight of NPV DNA by measuring the length of the relaxed circle. For example, Burgess (1977) found a range from 58-93.9 x 10^6 for a series of NPV's from seven insects. Molecular weights of NPV DNA have also been estimated by sedimentation in neutral and alkaline gradients (for a review see Harrap and Payne, 1979). There appears to be little or no reiteration of DNA sequences as measured by reassociation kinetics of melted DNA (Kelly 1977; Scharnhorst and Weaver, 1979).

Recently the size of baculovirus DNA's has been calculated following cleavage of the DNA with restriction endonucleases (REN) and summation of the molecular weights of the fragments after separation in agarose gels (Rohrmann and Beaudreau, 1977; Miller and Dawes, 1978a; Smith and Summers, 1978; Vlak and

Odink, 1979). The values obtained by this method for A. californica NPV ranged from 72-84 x 10^6 and are similar to those calculated using other techniques. REN analysis has also proved to be a powerful tool for identification and characterization of baculoviruses. The patterns are characteristic not only for a species of virus but are capable of picking up strains which have been isolated by plaquing techniques (Smith and Summers, 1978; Lee and Miller, 1978). Physical maps of baculovirus genomes have been constructed by REN analysis (Smith and Summers, 1979, 1980; Miller and Dawes, 1979) and DNA sequence homology between viruses has also been studied. Thus, Jewell and Miller (1980) nick translated DNA's from several NPV's and using Southern blot hybridization studied sequence homology to separated A. californica NPV fragments. NPV's derived from such disparate insects as the tussock moth, the gypsy moth, and the corn earworm shared homologous regions equivalent to about 1% of the genome. DNA restriction patterns of recombinant baculovirus produced by dual infection of G. mellonella larvae with A. californica and G. mellonella NPV's (Croizier et al., 1980) or which occur naturally (Smith and Summers, 1980) have been studied. An analysis of the segregation pattern of three virus proteins and polyhedrin among the recombinants identified regions of DNA sequences which coded for the peptides on the physical map (Summers et al., 1980).

Virus proteins

Purified OV and NOV have been examined by SDS-PAGE for viral structural proteins and a review of the findings is given in tabular form by Harrap and Payne (1979). Investigators have found 11 to 25 polypeptides ranging in MW from 9-160,000. Most but not all NPV's are reported to contain at least two major polypeptides, one 10-20,000 and another between 30-40,000 (Harrap et al. 1977; Summers and Smith, 1978). Two dimensional methods of separation (O'Farrell, 1975) have not yet been applied to baculovirus structural protein, but when this is done it is likely that they will lead to further resolution of structural peptides. When the viral envelope is removed by treatment with detergent (most often NP40) the isolated nucleocapside or capsid fractions contain fewer polypeptides but the two major fractions already referred to appear to remain associated with the nucleocapsid and it appears that many of the higher molecular weight components are envelope-associated glycopeptides. Viruses which are derived from closely related hosts, e.g. M-NPV's from S. frugiperda, S. littoralis, and S. exempta, were readily distinguished by major differences in the gel profiles of structural polypeptides (Harrap et al., 1977). M-NPV's have also been distinguished from S-NPV's isolated from the same insect species (Cibulsky et al., 1977b). Recently Tweeten et al. (1980) found that a highly basic core protein, of MW 12,000, could be extracted with dilute acid from purified nucleocapsids from several NPV and GV OV's. This protein, VP-12, was virus specific because it was only found in nucleocapsids and virus infected cells but not in profiles from uninfected cells. EM observations indicated that the basic protein was associated with viral in a DNA-protein complex.

Molecular genetics and biochemical aspects of replication of Baculoviruses

Studies of the biochemical events that accompany genome expression in baculoviruses are just beginning to be published. Parental viral DNA was detected in cells infected at a high multiplicity of infection one hour post-infection (Tjia et al., 1979). Host cell DNA synthesis is suppressed within a few hours of virus infection (Knudson and Tinsley, 1978; Brown et al., 1979) and by 6-8 hours post infection viral DNA synthesis is detected by filter hybridization (Knudson and Tinsley, 1978; Brown et al., 1979; Tjia et al., 1979).

The kinetics of virus protein synthesis have been investigated by observing the appearance of viral specific peptides (Carstens et al., 1979; Dobos and Cochran, 1980; Wood, 1980a) or viral antigens (Summers et al., 1978). Thirty or more viral specific polypeptides have been detected in infected cultures. With high multiplicity of infection the first virus induced polypeptides appeared 3 hours post infection and most could be detected by 10 hours after infection when nucleocapsids were also seen in cells (Carstens et al., 1979). Polyhedrin appeared as a late protein 24 hours or more post infection. By this time extracellular and intracellular virus particle morphogenesis was at a maximum (Volkman and Summers, 1977). Pulse-chase experiments may indicate that a precursor to polyhedrin underwent post-translational modification during synthesis (Carstens et al., 1979).

Temperature sensitive mutants of *A. californica* NPV have been isolated and characterized (Brown et al., 1979; Lee and Miller, 1979). They have been sorted into 5 or more complementation groups and classified so far by their phenotypic response at the restricted temperature, e.g. alteration of cytopathic effect, plaque size, restriction in NOV production, or DNA synthesis. Brown and Faulkner (1980) produced a partial genetic map for temperature sensitive NOV mutants and found that mutants affecting NOV production appeared to be clustered and separated on the map from mutants affecting OB production. With the availability of physical maps of the *A. californica* NPV genome based on REN analysis it is anticipated that marker rescue techniques will be used to localize the lesion in the temperature sensitive mutants and by inference to identify genes that control specific gene functions. These studies will be aided by recent reports that viral DNA may be used in transfection experiments (Burand et al., 1980; Carstens et al., 1980). An investigation of a morphology mutant (m-5) of A. californica which produces a single large crystal in infected cells (Brown et al., 1979) has revealed that the m-5 polyhedrin has an altered tryptic peptide pattern compared to the prototype (Brown et al., 1980) and investigations of the structure of the genome by restriction enzyme mapping have indicated that the alteration is in the region of the putative gene for polyhedrin synthesis (E. Carstens, personal communication).

Serological relationship between baculoviruses

The topic is reviewed by Mazzone and Tignor (1976), DiCapua and Norton (1976), and by Harrap and Payne (1979). Many comparisons between baculoviruses were made using immunodiffusion techniques but some studies have used complement fixation, hemagglutination and virus neutralization; in the past few years baculovirus antigens have been compared using Radioimmune Assay (RIA) and Enzyme Linked Immunoabsorbent Assay (ELISA). It is difficult to correlate the results obtained by different workers as, until recently, relatively crude techniques were used to prepare antigen and many different inoculation routes and schedules were used for antisera production. It was not until 1977 that it was shown that careful purification of antigens could be used to eliminate cross reactions between OV and polyhedrin or granulin (Bell and Orlob, 1977), a finding confirmed by McCarthy and Lambiase (1980) who measured serological reactions of nine Plusinae baculovirus components including nucleocapsids, solubilized envelope proteins as well as polyhedrin and granulin. Nowadays it is common to use methods such as affinity chromatography and SDS-PAGE (Crawford et al., 1978; Summers and Hoops, 1980) for stringent purification of antigen. In general the following conclusions can be drawn: i. Most workers find no cross reaction between insect host antigen and inclusion bodies, virions, polyhedrin and granulin. ii. Cross reactions are not usually observed between polyhedrin or granulin and the virion of the hemologous virus when the polyhedrin has been highly purified before being used to raise antibody. This observation (Crawford et al., 1978) made it possible to investigate polyhedrin synthesis in cultures infected with temperature sensitive mutant of A. californica NPV. It was possible to distinguish those mutant strains which produced NOV but were restricted in polyhedrin synthesis (Brown et al., 1979). iii. Polyhedrins have been found to cross react with other polyhedrins and also with granulins. This cross reactivity was considered to be broadly based in early literature (reviewed by Mazzone and

Tignor, 1976) but in evaluating it nowadays one can never be sure that rigorous purification methods were used to separate the virus components from host material. However recent investigations using RIA and ELISA have also confirmed cross reactivity between polyhedrins (Crawford et al., 1978; Crook and Payne, 1980). In an investigation of four variants of the RIA technique, Crawford et al. (1978) were able to alter the sensitivity of the methods for discriminating between NPV's.

Virus neutralization in tissue culture is a powerful method to distinguish between antigens related to infectivity in tissue culture, thus Volkman et al. (1976) were able to distinguish between OV and NOV of *A. californica* NPV. Neutralization studies have also been carried out *in vivo* to distinguish between the tussock moth and *A. californica* NPV (Martignoni et al., 1980).

Notwithstanding the difficulties associated with a specific method there is no doubt that serological techniques are highly effective in detecting antigens in contaminated samples such as larval extracts or bird droppings (Crawford et al., 1978; Kelly et al., 1978; Crook and Payne, 1980) and have a role to play in virus identification and epidemiological studies.

Host range of Baculoviruses

Insects, plants and animals: NPV infect insects other than their homologous hosts (Aizawa, 1963; Ignoffo, 1973a) and the consensus of opinion is that cross transmission is often demonstrated within insect taxonomic families but is limited between families and rarely occurs between orders. Even this generalization is suspect however, because as pointed out by Knudson (1980) the data on cross transmission were mostly obtained in the laboratory and there is none on virus host range in nature. The GV display a high degree of host specificity. Huger (1963) published a table which shows that 14 GV from several Lepidopteran families failed to replicate in a group of up to 15 heterologous insect larvae from either the same or closely related families. In an investigation of the host spectrum of the codling moth GV it was demonstrated that the virus was infectious only to some species within the same sub family of the Tortricidae. In the same study the NPV of *Choristoneura muriana* was infectious to four insect species from two subfamilies (Huber, 1980).

The host ranges of NPV from *A. californica* and *H. zea* have been carefully investigated as these viruses are used as microbial insecticides. *Autographa californica* NPV is reported to infect 28 insect species (Van der Beek, 1980). *Heliothis zea* NPV was tested in many species of vertebrates and plants with no harmful effects. These tests were part of an elaborate program required for licensing of the virus for general distribution as a microbial insecticide and included rigorous acute and chronic toxicity testing in small animals, primates, fish, birds, and tests for replication in plants (Ignoffo, 1973a, b). The lack of toxicity of NPV for higher animals and man is not surprising since it is well established that NPV's are present in the soil and often on marketed products such as vegetables (Jaques, 1975; Thomas et al., 1972; Heimpel et al., 1973).

Invertebrate and vertebrate cell cultures: Among the baculoviruses only NPV (Granados, 1976) and the sub-group C *Oryctes* virus (Kelly, 1975) have been reported to grow in insect tissue culture. The number of invertebrate cell lines that are susceptible to insect viruses is small, possibly because the lines were selected for optimal growth and not the ability to support virus replication. Where tissue cultures are available, the homologous M-NPV derived from the same insect species as the cell line often grows in culture (see tables in Granados, 1976; Knudson and Buckley, 1977). It is noteworthy that *A. californica* NPV which has a broad host range *in vivo* also grows in seven Lepidopteran cell lines (Knudson, 1980). Host cell modification of the virus replication pattern has been demonstrated in tissue culture. *Autographa californica* NPV failed to form OB's in a *B. mori* cell line although NOV production was unaffected (Summers et al., 1978). Altered patterns of viral growth have also been seen when temperature sensitive or morphology mutants of *A. californica* NPV are grown in cell culture (Brown et al., 1979; Lee and Miller, 1979).

Studies have been reported on attempts to infect vertebrate cells in culture with baculoviruses. Himeno et al. (1967) reported that infectious DNA of *B. mori* NPV induced the production of infectious polyhedra in a small proportion of cultured human amnion cells twelve days post infection. In contrast, *H. zea* NPV OV did not produce cytopathic effect either immediately or after 3 blind passages in human primary embryo kidney, human diploid lung, African green monkey kidney cells, or in the Hela cell line. Neither did the baculovirus interfere with the replication of ECHO 11 virus in these cells (Ignoffo and Rafajko, 1973). McIntosh and Maramorosch (1973) reported no cytopathic effect in five human cell lines following exposure to *H. zea* NPV. Cell associated infectivity could be recovered from 3 of 5 cultures four weeks after inoculation. Extensive tests have also been carried out using *A. californica* NPV and *T. ni* NPV or GV in primate cell lines. Again, the virus was shown to persist for several weeks and after several passages in culture (McIntosh and Shamy, 1975).

Autographa californica NPV infects a poikilothermic viper cell line (McIntosh and Shamy, 1975). Immunofluorescent tests confirmed the presence of viral antigens in inoculated cultures, but complete viral replication was not demonstrated. EM studies by Granados (1980) indicate that enveloped nucleocapsids of A. californica NPV are internalized by viropexis into cytoplasmic vacuoles one to twenty-four hours post infection of Hela and poikilothermic fat head minnow cell line. Viral infectivity persisted in the culture for at least six weeks at 28° C but there was no replication of infectious progeny. Studies by McIntosh and Shamy (1980) indicate that A. californica NPV replicates in a Chinese hamster ovary (CHO) cell line. The evidence is based on the production of a tritiated thymidine labelled entity which banded at the density 1.24 to 1.25 in sucrose gradients similar to the banding pattern observed with A. californica NPV NOV. Material from the gradient could be immunoprecipitated with anti-A. californica NPV antiserum and could also produce a typical cytopathic effect when used to infect a T. ni cell line. In these experiments careful consideration was given to checking the authenticity of the vertebrate cell line and the purity of the virus preparation and the results strongly suggest that the baculovirus genome can be expressed in a mammalian cell line to produce infectious material under laboratory conditions.

PATHOGENESIS IN VIVO AND IN VITRO

An overview

In nature the infectious entities of the baculoviruses are the occlusion bodies of the nuclear polyhedrosis viruses (PIB) and the granulosis viruses (capsules) and the non-occluded virus particles of the subgroup C viruses. The virulence in laboratory feeding tests of OB's is variable and covers a broad range, e.g. the LD_{50} for T. ni NPV is less than 50 OB's per insect (Faulkner and Henderson, 1972) and for the spruce budworm NPV is greater than 3,000 OB's per 6th instar larva (F. T. Bird, personal communication). In tissue culture four forms of infectious particles have been described: i. occluded virus (OV) obtained by dissolution of PIB in dilute alkali (Knudson and Tinsley, 1974); ii. non-occluded virus (NOV) released from cells by budding at the cell membrane (Henderson et al., 1974; MacKinnon et al., 1974); iii. intracellular enveloped virus (Volkman et al., 1976; Volkman and Summers, 1977); iv. virus in hemolymph of infected insects (Faulkner and Henderson, 1972). These forms of virus particle have low infectivity per os (Dougherty et al., 1975) but all may be transmitted by injection into susceptible insects or into tissue culture. Structural components, infectivity, in vivo pathogenesis and in vitro morphogenesis of each of these viral entities have been investigated often in great detail.

Ultrastructural investigations reveal that many aspects of virus morphogenesis are similar in insects and in cell culture after the virus nucleocapsid reaches the cell nucleus. However, in insect larvae a complex sequence of changes is observed between the ingestion of OB's and the commencement of virus replication in the insect cell. The replication of NPV and GV occurs in two stages in most species of susceptible insects: (1) OB's are dissolved in the high alkaline midgut region to release enveloped viral particles (OV). These infect adjacent midgut epithelial cells and replicate in the nuclei. (2) Progeny nuclocapsids and possibly some nucleocapsids derived from the infecting virus then bud through a basement membrane and enveloped virus particles (NOV) are released into the hemocoel. Dissemination of NOV in the hemolymph occurs and leads to infection of secondary target organs such as epithelial cells lining the trachea of various organs, the blood cells, epidermis and the fat body. OB's are rarely seen in the gut cells of Lepidoptera cells but PIB and capsules invariably form in the nuclei of most other infected tissues. Tinsley (1979) has pointed out that this pathway of infection of baculoviruses is of fundamental importance and relevance to the whole infectious process and that the host response in terms of virus susceptibility and resistance to infection may hinge on a detailed understanding of the process. Tissue culture studies have helped to elucidate intermediate stages in the virus morphogenesis and some investigations have thrown light on cellular factors that can modify the outcome of infection. It seems appropriate therefore to review *in vivo* pathogenesis and tissue culture morphogenesis in separate sections.

Pathogenesis in insects

Gross pathology: The development of external symptoms of baculovirus infection is often paralleled by striking differences in insect behavior (Huger, 1963; Smith, 1967). Most leaf eating insects continue feeding as the infection proceeds but after several days they often become swollen and sluggish, finally feeding ceases and the insects are often found hanging by their prolegs to branches or stems. A commonly noted behavioral pattern is that diseased insects seek the highest point of a cage, plant, or tree and as the disease progresses they are found hanging head down (Steinhaus, 1949; Smith, 1967).

In the Lepidoptera the first cytopathological changes are seen in the nuclei of tissues of larvae several days after infection. The nucleus swells and a Feulgen-positive (DNA enriched) virogenic stroma (Xeros, 1956) is observed. Infected hemocytes observed by dark field microscopy exhibit a scintillating zone at the periphery of the swollen nucleus which is apparently due to Brownian movement of particles and precedes the appearance of OB's in the nucleoplasm.

A comprehensive description of the NPV affecting the alpine tent caterpillar, Malacosoma alpicola (Benz, 1963), indicated that the following tissues progressively showed cytopathology characteristic of the disease: (1) fat body, (2) hypodermis, (3) tracheal matrix, (4) the sheathes lining muscles and nerve fibers, (5) muscles, (6) ganglia, and (7) pericardial cells. In baculovirus infections of the Hymenoptera the infection may be restricted to epithelial cells of the midgut where the OB's form. Dissection of sawflies late in the infection revealed a milky white midgut section packed with a suspension of OB's. They were not observed in hemocytes, fat body, and trachea (Balch and Bird, 1944; Bird and Whalen, 1953).

The patterns of pathogenesis of NPV and GV in susceptible insects have revealed many elements in common when viewed in terms of gross pathology and also when observations have been made at the cellular level. Differences have been noted in such parameters as the specific tissues where replication of the virus occurs and recent reviews have pointed up differences in the location of virus particle morphogenesis within gut cells (Granados, 1978, 1980).

Routes of infection

The commonest route of infection with NPV and GV OB's is considered to be by ingestion during feeding of larvae. Although OB's have been found in pupae and adult Lepidoptera and Hymenoptera (Bergold, 1953), the insects were considered to have acquired the infection as larvae. Other modes of entry of polyhedra as a result of trauma or through the spiracles which would bring the OB's directly in contact with the hemolymph can be discounted as major pathways of transmission because of their low frequency of occurrence and because OB's injected directly into the hemolymph are not infectious (Bergold, 1953).

The literature also contains references to transmission of baculoviruses either by contamination of the surface of the ova during oviposition (transovum transmission) or by transovarial transmission where virus is contained within the ovum. The two conditions can be distinguished by surface sterilization of eggs. The topic has been reviewed by Longworth (1973) and discussed in detail by Etzel and Falcon (1976). Longworth cites only two instances where transovarian transmission seems to have been demonstrated in NPV infections of Attacus atlas and Prodenia litura. In a study of GV infection of the codling moth (Laspeyresia pomonella) virus was absent from fertile eggs obtained from insects that had been fed or injected with the GV. Virus was not observed in F^1 and F^2 generations derived from larvae that had been force-fed virus (Etzel and Falcon, 1976). Transovum transmission has been demonstrated by Hamm and Young (1974) who investigated the

transmission of an NPV from H. zea. Both methods of vertical transmission of the virus described above infer that the virus is transmitted as OB's or virus particles but investigations of survival and transmission of other latent forms have not yet been adequately studied. The early literature abounds with references to latent viral infections with baculoviruses (Aruga, 1963). Probably many of these reports can be attributed to low grade chronic infection of insect hosts but the possibility of presence of viral genomes in an extra-chromosomal repressed form or possibly integrated as a provirus into the insect chromosomes cannot be discounted without definitive investigations using modern methods of probing for nucleic acid sequences. An examination of cell extracts from A. californica NPV-infected S. frugiperda cells failed to detect integrated viral DNA sequences (Tjia et al., 1979) but this observation may have little bearing on viral persistence in the insect host where many cell types occur.

Dissolution of OB

The NPV and GV are unique among animal viruses because the virions are embedded in OB's and the infectious particles can only be released from them in susceptible hosts in regions of the gut which are of sufficiently high alkalinity. EM observations have indicated that rapid dissolution of OB's occurs within the gut possibly within a matter of minutes (Granados, 1980). The dissolution and release of virus from OB is apparently a non-specific process requiring only that the inclusions are present within a sufficiently alkaline region of the midgut. Thus dissolution takes place in larvae that are not considered susceptible for some baculoviruses. Paschke and Summers (1975) observed that T. ni GV and Rachiplusia ou NPV were degraded in the gut lumen of first and second star non-susceptible mosquito larvae.

Most EM studies of the early events in midgut cell infection with baculoviruses report that intact enveloped nucleocapsids are observed adjacent to microvilli prior to virus-cell fusion (Granados, 1980). They probably lie in a hostile environment. Virus obtained by dissolving OB from A. californica NPV in digestive juice (pH 10.4-10.6) of Estigmene acreae, a susceptible insect, was infectious in tissue culture (Vail et al., 1979). Infectivity was at a maximum after two minutes exposure at 26° C and could not be detected after 10 minutes. Virus released in alkali from purified OB is stable for long periods at high pH (Bergold, 1953), thus, inactivation by gut juice may be by consequence of the presence of degrading enzymes as well as toxic substances (Paschke and Summers, 1975).

The role if any of an endogenous protease present in occlusions of NPV (Summers and Smith, 1975; Eppstein and Thoma, 1975) and GV (Tweeten et al., 1978) is unclear. Presence of the enzyme in OB does not appear to be essential for infectivity. Protease was not detected in tissue culture derived T. ni OB (Zummer and Faulkner, 1979) or A. californica OB (Wood, 1980b) and the LD_{50} of tissue culture produced OB is similar to that of insect derived OB in the case of T. ni M-NPV (Faulkner and Henderson, 1972) and A. californica M-NPV (Vail et al., 1973). In addition the efficacy of A. californica NPV propagated in vivo and in vitro and used in field tests on crops was similar (Ignoffo et al., 1974). The biochemical properties of A. californica NPV occluded virus derived from OB's where protease has been inactivated by heating differ from those where the OB's possessed active protease. Protease activity was required for complete dissociation of polyhedrin from the viral membrane when OB's were dissolved in alkali (Wood, 1980b). Tweeten et al. (1978) reported that the rate of solubilization of Plodia interpunctella GV decreased if the protease was inactivated and they speculated that the enzyme, activated by the insect gut alkalinity, functions in the initiation of infection by removing protein from the viral envelope.

A proteinaceous synergistic factor that enhances the infection of an NPV in the armyworm, Pseudaletia unipuncta, was isolated from granulin of a GV that infects the armyworm. The factor displayed enzymatic activity and hydrolyzed p-nitrophenyl esters of fatty acids at rates depending on the chain length (Hara et al., 1976). A further analysis by Yamamoto and Tanada (1978a), showed that the synergistic factor is composed of polypeptides and phospholipids and that when treated with phospholipase C the enhancing activity of the factor was lost. In a previous study it was shown that there was an increased uptake of nucleocapsids into midgut cells of insect larvae when they were fed mixtures of NPV with the factor (Tanada et al., 1975). It is postulated that the phospholipid component may play a major role in virus attachment (Yamamoto and Tanada, 1978b) and that the site of action is the cell membrane of the microvillus (Tanada et al., 1980).

Gut juice enzymes which degrade carbohydrates may also have a role in the initial phases of dissolution of OB. Minion et al. (1979) showed that the envelope of OB's from H. zea NPV is rich in carbohydrate. They found that "ghosts" isolated by prolonged digestion of OB in alkali consisted entirely of carbohydrate moieties.

There has been conjecture that the peritrophic membrane may be a barrier to infection in the insect midgut (Paschke and Summers, 1975; Granados, 1980). This tissue appears thin and unstructured as concentric lamelli in the EM (Tanada et al. 1975), but it may consist of closely interwoven chitin micro-fibres (Smith, 1968). The PM of the uninfected Douglas fir tussock moth (Orygia pseudotsugata) was seen as a one micrometer-thick membrane that possessed no pores or other discontinuities (Brandt et al., 1978). It was suggested by

Harrap (1970) that holes in the PM allow transport of the virions across the barrier in lepidopteran larvae. Micrographs also support the suggestion that disruption of the PM occurs in infection since Tanada et al. (1975) show dissolving OB's on both sides of the membrane.

Virus interaction with cells of the midgut

Observations made on larvae infected with *Aglas urticae* NPV (Harrap and Robertson, 1968; Harrap, 1970) and with *T. ni* GV (Summers, 1969, 1971) were the first to demonstrate that baculoviruses enter midgut columnar epithelial cells by fusion, replicate within nuclei and subsequently give rise to progeny virus which are released through a basement membrane into the hemocoel. Previously most tissues of larvae had been implicated as sites of OB formation but the key role of the midgut epithelial cells in initiating a round of replication had not been recognized. Gross histological changes were not seen by light microscopy in midgut cells of insects infected with *A. urticae* NPV although by 72 hours post infection cytopathic effects were observed in fat body cells (Harrap and Robertson, 1968). However examination of gut tissues in the electron microscope 24 hours after infection revealed virus replication in columnar epithelial but not goblet cells of the midgut. Virus particles and bodies resembling OB in their crystalline structure, but lacking virus particles were also seen in nuclei of infected cells. By 48 hours post infection virus was seen to replicate in tracheolar cells associated with the basement membrane and in the cytoplasm of epithelial cells. Both Harrap and Robertson (1968) and Summers (1969) found that the process of cell infection occurred by fusion of enveloped viral nucleocapsids with the plasma membrane of a microvillus. Microvilli are extensions into the gut lumen of the columnar epithelial cells. They occur as a regular array within the gut and make up a brush border (Anderson and Harvey, 1966). Goblet cells which contain a large vacuole, comprise the second principal cell type of the midgut. In sections they are always seen surrounded by epithelial cells and they do not appear to be susceptible to baculovirus infection.

The process of virus entry to epithelial cells by fusion has been confirmed in other studies of GV and NPV infections in larvae and some excellent micrographs of the process have been published (Granados, 1978; Tanada et al., 1975; Kawanishi et al., 1972; Adams et al., 1977). The dissolution of OB occurs rapidly and enveloped particles have been seen 1-4 hours post infection of larvae (Tanada et al., 1975; Granados, 1978). The presence of a PM does not seem to impede the dissolution process since dissolving OB's were observed on both sides of the PM (Tanada et al., 1975), and enveloped nucleocapsids are regularly seen adjacent to microvilli. Most investigators comment that unenveloped extracellular nucleocapsids are only rarely seen in sections of the midgut (Granados, 1980). Tanada et al. (1975) working with *P. unipunctella* M-NPV observed both singles and enveloped bundles of nucleocapsids in the process of fusion with microvilli and noted that a state of "interdigitation" with the membrane of the microvilli accompanies the fusion process. Only unenveloped nucleocapsids have been observed within microvilli in most studies where the infection has been investigated in detail (Summers, 1969, 1971; Tanada et al., 1975; Granados, 1978). There is a single report (Tanada and Leutenegger, 1970) that *T. ni* GV entered cells by viropexis since vesicles containing enveloped nucleocapsids were observed in epithelial cells. A recent study by Granados (1980) of early events of virus invasion for *E. acrea* GV found that entry into larval midgut cells was by membrane fusion and not by pinocytosis.

A second mechanism by which OV released from OB may enter the hemocoel was suggested by Tanada and Leutenegger (1970). On the basis of observations that OV was present in intercellular spaces of the gut, they suggested that virus particles entering the junction between two columnar cells moved along the intercellular space into the basement membrane from whence they passed to the hemocoel. This observation has not been confirmed by other workers. Tinsley (1979) points out that the region of the gut where columnar cells and the basal lamellae lie is convoluted and it is not easy to determine whether enveloped particles or baculoviruses seen there are actually inside or outside the cell membrane.

A few nucleocapsids entering a microvillus may distintegrate (Harrap, 1970) but most were considered to traverse the cytoplasm to the vicinity of the nuclear membrane. Bundles of nucleocapsids that enter a microvillus by fusion apparently soon separate as micrographs indicate they travel separately to the nucleus (Tanada et al., 1975).

In recent studies with *A. californica* NPV, Granados and Lawler (1980) reported that after entry into epithelial cells by membrane fusion some nucleocapsids passed directly through the cytoplasm to the basement membrane where enveloping of nucleocapsids was detected within four hours of infection. In these experiments large doses of OB's were fed and virus infectivity could be detected in hemolymph by plaque assay within 30 minutes. The observation that the virus could be detected in tissue culture is of significance because the assay is hundreds of times more sensitive to virus budded from plasma membranes than it is to virus released by dissolution of OB's (Volkman and Summers, 1977) and thus tends to confirm the existence of a hitherto unknown mechanism by which NPV can invade the insect gut and release virus to the hemolymph.

Virus morphogenesis in the midgut

Observations of Harrap and Robertson (1968) and Summers (1969, 1971) provided the first evidence that virus replication occurs within the nucleus of columnar epithelial cells of the midgut of susceptible larvae. Nucleocapsids of GV and NPV may uncoat (i.e. lose the capsid to release the genomic nucleoprotein complex) by different mechanisms (Granados, 1980). Summers (1971) noted many instances of end-on association between nucleocapsids of *T. ni* GV and nuclear membrane pores and postulated that release of the inner components of the virion may occur directly through the nuclear pore. Granados (1980) provided further visual evidence for uncoating of nucleocapsids of *E. acrea* GV at nuclear pores of midgut cells. Tanada and Hess (1976) working with *P. unipuncta* NPV also noted the attachment of empty capsids at nuclear pores of midgut cells and considered that uncoating was similar to that described for GV. Recently however, Granados (1978) has published micrographs showing the uncoating of *H. zea* NPV within nuclei of midgut cells at 2 to 4 hours post inoculation. In view of the few studies made on uncoating it is too early to decide whether different mechanisms obtain with NPV and GV (Granados, 1980).

The complex nature of the replication pattern of baculoviruses in infected cells is documented in several recent reviews (Tinsley and Harrap, 1978; Granados, 1976, 1980). In gut cells enlargement of the nucleus was observed soon after infection and patches of virogenic stroma, an electron dense chromatin network involved in nucleocapsid assembly, were seen (Tanada and Hess, 1976; Granados, 1980). Soon afterward nucleocapsids were seen associated with the virogenic stroma, usually aligned with one end attached to it. In some sections Tanada and Hess (1976) observed profiles of the same dimensions as nucleocapsids in stages of increasing electron density. The gut cells of Lepidoptera differ from those of other infected tissues because OB do not form in them. Thus nucleocapsids become enveloped and exit via the basal membrane to the hemocoel. *In vivo* studies (Summers, 1971; Tanada and Hess, 1976; Granados, 1980; Adams et al., 1977) have shown that nucleocapsids may acquire an envelope (a) within the nucleus following *de novo* membrane synthesis (b) at the nuclear membrane either by budding into the perinuclear space or by becoming embedded in the completed double nuclear membrane (c) by budding at the plasma membrane. Each of these mechanisms obviously represents a highly complex process and details are not understood. Similar mechanisms for acquiring envelopes have been reported in *in vitro* studies (MacKinnon et al., 1974; Raghow and Grace, 1974; Hirumi et al., 1975). At the present time the inescapable conclusion is that particle morphogenesis in midgut and other cells of insects, and in tissue culture leads to a heterogeneous population of NOV particles composed of nucleocapsids enveloped by membranes of diverse origin. Although the EM observations cited above indicate that NOV entering the hemocoel are enveloped, the particles may also be fragile. Kawarabata (1974) isolated infectious particles by sucrose gradient centrifugation from the hemolymph of silkworm infected with an NPV and found only unenveloped nucleocapsids, yet enveloped virus is always considered to be released from plasma membranes. Spike-like projections (peplomers) have been seen at one end of the rod shaped virions which bud into the hemocoel from gut cells, and it has been suggested that these particles are responsible for secondary and subsequent infections of host tissues (Adams et al., 1977; Summers and Volkman, 1976; Hess and Falcon, 1977; Granados, 1980). Adams et al. (1977) consider that acquisition of a viral envelope at the plasma membrane of an epithelial cell occurs by a process similar to that described for animal viruses, and they present micrographs which show portions of the cell membrane transformed to peplomer morphology at loci where nucleocapsids accumulate.

Secondary infection of other larval tissues

Adams et al. (1977) suggest that there are two forms of baculovirus virions: i. enveloped nucleocapsids (OV) which are occluded within the nucleus in OB. These viruses are released by dissolution and infect larval midgut cells. ii. A spreading or hemocoelic form of enveloped nucleocapsids which bud from the midgut and other tissues and which infect tissues in secondary and subsequent stages infection. Tracheolar cells associated with the midgut basal lamina are probably the first to be infected by virus particles that bud from the columnar epithelial cells (Harrap and Robertson, 1968). *Autographa californica* NPV OB's were detected in hemocytes 16 hours post-infection suggesting that they had been infected early (Granados, 1980). Adams et al. (1977) suggested that infection of tracheolar cells which surround virus susceptible tissues such as the fat body, hypodermis, Malpighian tubules, silk glands and epithelial linings of the reproductive tissues and the nerve tracts are the principal targets of secondary infection.

The nucleus is the site for the initial events of the replication of both NPV and GV. As indicated above, an early indicator of baculovirus infection is the swelling of the nucleus and the development of the virogenic stroma. Further ultrastructural changes within the NPV leading to the formation of OB are believed to be confined to the nucleus in NPV infections but it appears that capsules (OB's) may form in both nucleus and cytoplasm in GV infections, possibly due to some disintegration of the nuclear membrane (Smith, 1967). A table presented by Huger (1963) lists 10 studies where the site of development of OB in granulosis infections was reported. Development of GV capsules was seen in the nucleus (4), in the cytoplasm (2), and nucleus and cytoplasm (4 instances). Falcon and Hess (1977) made a comparative EM study of the replication of A. californica NPV and Spodoptera NPV in the insect Spodoptera exigua and found that replication of Spodoptera NPV was confined to the nucleus whereas all stages in the replication of the A. californica virus were observed in cytoplasmic areas. In commenting on this observation Granados (1980) pointed out that the tissues were fixed 5 days p.i. at which times many cycles of replication may have taken place, and aberrant patterns of replication may have occurred in the later cycles.

Virus morphogenesis in cell culture

Details of NPV morphogenesis have been revealed by electron microscopy of thin sections in larvae and in cell cultures. An advantage in using cell cultures to study virus replication is that the experimental conditions can be manipulated so that all cells are infected at about the same time (high multiplicity of infection), and the sequence of events can be monitored more closely. Some caution has to be exercised in generalizing from cell culture studies as it must be borne in mind that cells in culture represent a select group that can be maintained in a defined medium and are not under the constraints of cells within the living host. Even so, many similar structures have been identified in infected tissues from insects and in cell culture (Summers and Arnott, 1969; Harrap, 1972b; Falcon and Hess, 1977; Raghow and Grace, 1974; MacKinnon et al., 1974; Knudson, and Harrap, 1976; Hirumi et al., 1975). In cell cultures it has proven possible to monitor the replication cycle by measuring the appearance of progeny virus by an infectivity test (Raghow and Grace, 1974; Knudson and Harrap, 1976). Working with S. frugiperda NPV in S. frugiperda cells Knudson and Harrap (1976) found that virus entered the cells by 1 hour p.i. and was uncoated. There was an eclipse period of 9 hours during which no virus was detected and a latent period of 12 hours from the time of infection until progeny virus particles appeared in the culture medium. Polyhedron formation was first detected by 18 hours p.i. and final OB morphogenesis was considered complete by 48 hours p.i. Measurement of the kinetics of virus production in the cell culture have indicated that there is a shutdown in release of plasma membrane-budded virus (NOV) at about the time that OB morphogenesis occurs in the nucleus (Volkman et al., 1976).

Virus entry into cell cultures is by attachment of the NOV to the plasma membrane followed by invagination which draws in the particle by phagocytosis (Raghow and Grace, 1974; Hirumi et al., 1975; Knudson and Harrap, 1976). Particles were seen within cytoplasmic vacuoles in partially ruptured vacuoles in the middle region of the cytoplasm. Later nucleocapsids were often observed in end-on orientation to nuclear pores similar to the conformation observed with *in vivo* infections with NPV (Hirumi et al., 1975; Raghow and Grace, 1974). In tissue culture as with gut cells *in vivo* uncoating of the nucleocapsid occurs within the nucleus since intact nucleocapsids were observed within nuclei at the early stages of infection with *A. californica* NPV (Hirumi et al., 1975).

It is considered that during the initial stage of virus morphogenesis nucleocapsids form within the nucleus and pass to the plasma membrane where envelopment occurs and NOV are liberated from the cell surface. EM studies of the early stages of viral infection show that as nuclear swelling takes place an electron dense material, the virogenic stroma, is laid down in patches. Soon after, full length rod-shaped nucleocapsids are observed laying at the surface of the virogenic stroma and apparently contigous with its substance. It is considered that mature nucleocapsids produced early in the infectious cycle pass to the region of the plasma membrane. Enveloping of the nucleocapsids has also been regularly observed as they pass from the nucleus to the plasma membrane and, as has been observed *in vivo*, the membranes are of diverse origin (MacKinnon et al., 1974). Nucleocapsids have been seen lying in bulges of the double nuclear membrane as it extends into the cytoplasm, and it has been suggested that these buds could pinch off the nucleocapsid and form a vacuole in which it may traverse the cytoplasm. The vacuole then disintegrates at the vicinity of the plasma membrane where budding of NOV occurs (Adams et al., 1977; Hirumi et al., 1975). Virions were also seen in vacuoles in the cytoplasm of the cell (Knudson and Harrap, 1976) and in enveloped structures within the perinuclear space (MacKinnon et al., 1974).

As infection progresses accumulations of short, open-ended membrane profiles have been seen within nuclei both *in vivo* (Stoltz et al., 1973) and *in vitro* (MacKinnon et al., 1974; Knudson and Harrap, 1976; Hurumi et al., 1975). Stoltz et al., (1973) suggested that these membranes could have arisen by *de novo* morphogenesis since serial sectioning of tissue did not reveal points of contact between the intranuclear membrane and membranes of the nuclear envelope. Soon after membrane proliferation was observed nucleocapsids were seen to associate in an end-on configuration with short pieces of membrane, and sometimes several nucleocapsids were seen lined up on a membrane (MacKinnon et al., 1974). Single enveloped nucleocapsids and several nucleocapsids within a bundle were observed in sections *in vivo* (Summers and Arnott, 1969) and *in vitro* (MacKinnon et al., 1974; Raghow and Grace, 1974; Knudson and Harrap, 1976; Hirumi et al., 1975).

At about the time that envelopment of nucleocapsids is in progress OB morphogenesis commences with the appearance of a fibrous material associated with developing occlusion bodies (Summers and Arnott, 1969; MacKinnon et al., 1974). Investigations by Chung et al. (1980) have revealed the fibrous substance was contiguous with the lattice of crystallizing occlusion bodies and it was suggested that accumulation of the fibrous masses seen at the periphery of the occlusion body condensed to form the polyhedron membrane. Thread-like structures seen within the fibrous material (designated membrane spacers by Summers and Arnott, 1969; spacers by MacKinnon et al., 1974; condensed fibrous sheets by Chung et al., 1980) are considered to form by condensation of the fibrous material and they are structurally related to the occlusion body envelope (Chung et al., 1980).

Only enveloped nucleocapsids are found within OB but crystallization of OB can occur without occlusion of a virus (Summers and Arnott, 1969; Hirumi et al., 1975; Chung et al., 1980). Aberrant occlusion bodies have also been noted in tissues from *Trichoplusia ni* larvae (Summers and Arnott, 1969). Certain morphology mutants of *A. californica* NPV do not contain virus particles (Brown et al., 1980). It is not known what factors stimulate crystallization in cells. Prolonged serial passage of *T. ni* and *A. californica* NPV in tissue culture apparently favors the production of baculovirus variants that exhibit atypical patterns of morphogenesis in cell culture (MacKinnon et al., 1974; Hirumi et al., 1975; Zherebtsova, 1978). These often take the form of "empty" OB's with low *in vivo* infectivity (Potter et al., 1976, 1978).

CONCLUSIONS

Recent studies on the baculoviruses have characterized their biochemical and biophysical properties in order to describe the nature of the virion and inclusion bodies (Harrap and Payne, 1979). In addition there is a renewed interest in investigating details of virus infection and morphogenesis (Granados, 1980). With those viruses that grow in tissue culture (the NPV) it has been possible to investigate replicative pathways. The stimulus for this work has been twofold. First, the lists of viruses already referred to have until very recently described virus derived from the various insect species and there has been little attempt to compare viruses even from closely related insect species; thus some of the viruses may represent species with extended host ranges or there may only be minor genetic variations between some of them (variants). It is impossible to describe viruses solely in terms of their ultrastructure and their virulence, thus more detailed information on the biochemical properties and serological relationship has been sought (Harrap and Payne,

1979). Second, some baculoviruses are effective microbial insecticides and three of them, the NPV's of the corn earworm, Heliothis zea, the Douglas fir tussock moth, Orgyia psuedotsugata, and the gypsy moth, L. dispar are licensed for general distribution in the U.S.. A fourth virus, the NPV of Autographa californica which has a broad host range of susceptible insects particularly those on brassicae crops, may be licensed shortly (Summers and Kawanishi, 1978).

Further studies with baculovirus will likely be directed towards continued dissection of the viral genome in order to relate gene structure to expression of such biological properties as host range and virulence. Cytopathological investigations have already pointed up the key elements in the mechanisms of pathogenesis and these should provide guidance in planning experiments to relate them to molecular genetic expression. These studies will lead to greater understanding of the baculoviruses and will hopefully contribute to meaningful decisions on the role these viruses should play in microbial control.

ACKNOWLEDGEMENT

Supported in part by a grant from the Medical Research Council of Canada. The author gratefully acknowledges discussions with Dr. R. R. Granados concerning the structure of this chapter and thanks Mrs. Shannon Perry for assembling the manuscript.

REFERENCES

Acqua, C. 1918-19. Richeche sulla malattia del Rend. 1 st. bacol. Scuola super. agr. Portici, 3, 243. Cited in Bergold (1953).

Adams, J. R., Goodwin, R. H. and Wilcox T. A. 1977. Electron microscope investigation on invasion and replication of insect baculoviruses in vivo and in vitro. Biol. Cellulaire, 28, 261-268.

Aizawa, K. 1963. The nature of infections caused by nuclear-polyhedrosis viruses. In Insect Pathology, An Advanced Treatise, E. A. Steinhaus, ed., Vol. 1. Academic Press, N.Y. pp. 382-412.

Anderson, E. and Harvey, W. R. 1966. Active transport by the cecropia midgut. II. Fine structure of the midgut epithelium. J. Cell. Biol., 31, 107-134.

Aruga, H. 1963. Induction of virus infections. In Insect Pathology, An Advanced Treatise, E. A. Steinhaus, ed., Vol. 1. Academic Press, N.Y. pp. 499-520.

Balch, R. E. and Bird, F. T. 1944. A disease of the European spruce sawfly, Gilpinia hercyniae (Mtg.) and its place in natural control. Sci. Agr., 25, 65-80.

Bell, C. D. and Orlob, G. B. 1977. Serological studies on virions and polyhedron protein of a nuclear polyhedrosis virus of the cabbage looper, Trichoplusia ni. Virol., 78, 162-172.

Benz, G. 1963. A nuclear polyhedrosis of Malacosoma alpicola (Staudinger) J. Insect Pathol., 5, 215-241.

Bergold, G. H. 1947. Die isolierung des polyder-virus und die natur der polyeder. Z. Naturforsch., 2b., 122-143.

Bergold, G. H. 1953. Insect viruses. Adv. Virus Res., 1, 91-139.

Bergold, G. H. 1962. The molecular structure of some insect virus inclusion bodies. J. Ultrastruct. Res., 8, 360-378.

Bergold, G. H. 1963. The nature of nuclear-polyhedrosis viruses. In Insect Pathology, An Advanced Treatise, E. A. Steinhaus, ed., Vol. 1. Academic Press, N.Y. pp. 413-456.

Bird, F. T. and Whalen, M. M. 1953. A virus disease of the European pine sawfly, Neodiprion sertifer (Geoffr.). Can. Ent., 85, 433-437.

Brandt, C. R., Adang, M. J. and Spence, K. D. 1978. The peritrophic membrane: Ultrastructure analysis and functions as a mechanical barrier to microbial infection in Orygia pseudotsugata. J. Invertebr. Pathol., 32, 12-24.

Brown, M., Crawford, A. M. and Faulkner, P. 1979. Genetic analysis of a baculovirus, Autographa californica nuclear polyhedrosis virus. I. Isolation of temperature sensitive mutants and assortment into complementation groups. J. Virol., 31, 335-387.

Brown, M. and Faulkner, P. 1980. A partial genetic map of the baculovirus, Autographa californica nuclear polyhedrosis virus, based on recombination studies with ts mutants. J. Gen. Virol., 48, 247-251.

Brown, M., Faulkner, P., Cochran, M. A. and Chung, K. L. 1980. Characterization of two morphology mutants of Autographa californica nuclear polyhedrosis virus with large cuboidal inclusion bodies. J. Gen. Virol., 50, 309-316.

Burand, J. P., Summers, M. D. and Smith, G. E. 1980. Transfection with baculovirus DNA. Virol., 101, 286-290.

Burgess, S. 1977. Molecular weights of lepidopteran baculovirus DNAs: Derivation by electron microscope. J. Gen. Virol., 37, 501-510.

Burges, H. D. and Hussey, N. W. 1971. Microbial Control of Insects and Mites. Academic Press, N.Y. 861 pp.

Carstens, E. B., Tjia, S. T. and Doerfler, W. 1979. Infection of Spodoptera frugiperda cells with Autographa californica nuclear polyhedrosis. Virol., 99, 386-398.

Carstens, E. B., Tjia, S. T. and Doerfler, W. 1980. Infectious DNA from Autographa californica nuclear polyhedrosis virus. Virol., 101, 311-314.

Chung, K. L., Brown, M. and Faulkner, P. 1980. Studies on the morphogenesis of polyhedral inclusion bodies of a baculovirus Autographa californica NPV. J. Gen. Virol., 46, 335-347.

Cibulsky, R. J., Harper, J. D. and Gudausakas, R. T. 1977a. Biochemical comparison of polyhedral proten from five nuclear polyhedrosis viruses infecting plusine larvae (Lepidoptera: Noctuidae). J. Invertebr. Pathol., 29, 182-191.

Cibulsky, R. J., Harper, J. D. and Gudauskas, R. T. 1977b. Biochemical comparison of virion protens from five nuclear polyhedrosis viruses infecting plusine larvae (Lepidoptera: Noctuidae). J. Invertebr. Pathol., 30, 303-313.

Couch, J. A. 1974. Free and occluded virus, similar to baculovirus, in hepatopancreas of pink shrimp. Nature, 247, 229-231.

Crawford, A. M., Faulkner, P. and Kalmakoff, J. 1978. Comparison of solid-phase radio-immunoassays for baculoviruses. App. Env. Micro., 36, 18-24.

Croizier, G., Godse, D. and Flak, J. 1980. Selection de types viraux dans les infections doubles a Baculovirus chez les larves de Lepidoptere. C.R. Acad. Sci. Paris,Ser. D, 290, 579-582.

Crook, N. E. and Payne, C. C. 1980. Comparison of three methods of ELISA for baculovirus. J. Gen. Virology, 46, 29-37.

DiCapua, R. A. and Norton, P. W. 1976. In Invertebrate Tissue Culture-Applications in Medicine, Biology and Agriculture, K. Maramorosch and E. Kurstak, eds. Academic Press, N.Y. pp. 317-330.

Dobos, P. and Cochran, M. A. 1980. Protein synthesis in cells infected by Autographa californica nuclear polyhedrosis virus (Ac-NPV): The effect of cytosine arabinoside. Virol., 103, 446-464.

Dougherty, E. M., Vaughn, J. L. and Reichelderfer, C. F. 1975. Characteristics of the non-occluded form of a nuclear polyhedrosis virus. Intervirol., 5, 109-121.

Eppstein, D. A. and Thoma, J. A. 1975. Alkaline protease associated with the matrix protein of a virus infecting the cabbage looper. Biochem. Biophys. Res. Comm., 62, 478-484.

Etzel, L. K. and Falcon, L. A. 1976. Studies on transovum and trans-statial transmission of a granulosis virus of the codling moth. J. Invertebr. Pathol., 27, 13-26.

Falcon, L. A. and Hess, R. T. 1977. Electron microscope study on the replication of Autographa nuclear polyhedrosis virus in Spodoptera exigua. J. Invertebr. Pathol. 29, 36-43.

Faulkner, P. and Hendersen, J. F. 1972. Serial passage of a nuclear polyhedrosis disease virus of the cabbage looper (Trichoplusia ni) in a continuous tissue culture cell line. Virol., 50, 920-924.

Federici, B. A. and Humber, R. A. 1977. A possible baculovirus in the insect-parasitic fungus Strongwellsea magna. J. Gen. Virol., 35, 387-391.

Granados, R. R. 1976. Infection and replication of insect pathogenic viruses in tissue cultures. Adv. Vir. Res., 20, 189.

Granados, R. R. 1978. Early events in the infection of Heliothis zea midgut cells by a baculovirus. Virol., 90, 170-174.

Granados, R. R. 1980. Infectivity and mode of action of baculoviruses. Biotechnol. Bioeng., 23, 1377-1405.

Granados, R. R. and Lawler, K. A. 1981. In vivo pathway of Autographa californica baculovirus invasion and infection. Virol., 108, 297-308.

Hamm, J. J. and Young, J. R. 1974. Mode of transmission of nuclear polyhedrosis virus to progeny of adult Heliothis zea. J. Invertebr. Pathol., 24, 70-81.

Hara, S., Tanada, Y. and Omi, E. M. 1976. Isolation and characterization of a synergistic enzyme from the capsule of a granulosis virus of the army worm Pseudaletia unipuncta. J. Invertebr. Pathol., 27, 115-124.

Harrap, K. A. 1970. Cell infection by a nuclear polyhedrosis virus. Virol., 42, 311-318.

Harrap, K. A. 1972a. The structure of nuclear polyhedrosis viruses. I. The inclusion body. Virol., 50, 114-123.

Harrap, K. A. 1972b. The structure of nuclear polyhedrosis viruses. III. Virus assembly. Virol., 50, 135-139.

Harrap, K. A. 1973. Virus infections in invertebrates, In Virus and Invertebrates, A. J. Gibbs, ed., North Holland, Amsterdam. pp. 271-299.

Harrap, K. A. and Payne, C. C. 1979. The structural properties and identification of insect viruses. Adv. Virus Res., 25, 273-355.

Harrap, K. A., Payne, C. C. and Robertson, J. S. 1977. The properties of three baculoviruses from closely related hosts. Virol., 79, 14-31.

Harrap, K. A. and Robertson, J. S. 1968. A possible infection pathway in the development of a nuclear polyhedrosis virus. J. Gen. Virol., 3, 221-225.

Heimpel, A. M., Thomas, E. D., Adams, J. R. and Smith, L. J. 1973. The presence of nuclear polyhedrosis viruses of Trichoplusia ni on cabbage from the market shelf. Env. Ent., 2, 72-75.

Henderson, J. F., Faulkner, P. and Mackinnon, E. A. 1974. Some biophysical properties of virus present in tissue cultures infected with the nuclear polyhedrosis virus of Trichoplusia ni. J. Gen. Virol., 22, 143-146.

Hess, R. T. and Falcon, L. A. 1977. Observations on the interaction of baculoviruses with the plasma membrane. J. Gen. Virol., 36, 525-530.

Hess, R. T. and Falcon, L. A. 1978. Electron microscope observations of the membrane surrounding polyhedral inclusion bodies of insects. Arch. Virol., 56, 169-176.

Himeno, H., Sakai, F. and Onodera, K. 1967. Formation of nuclear polyhedral bodies and nuclear polyhedrosis virus of silkworm in mammalian cells infected with viral DNA. J. Ultrastruct. Res., 33, 507-512.

Hirumi, H., Hirumi, K. and McIntosh, A. H. 1975. Morphogenesis of a nuclear polyhedrosis virus of the alphalfa looper in a continuous cabbage looper cell line. Ann. N.Y. Acad. Sci., 266, 302-326.

Huber, J. 1980. About the host spectrum of the codling moth granulosis virus. Symposium Proceedings: Safety Aspects of Baculoviruses as Biological Insecticides, H. G. Mittenburger, ed. Bundesministerium fur Forschung und Technologie, Bonn pp. 75-80.

Huger, A. 1963. Granulosis of Insects. In Insect Pathology, An Advanced Treatise, E. A. Steinhaus, ed., Vol. 1. Academic Press, N.Y. pp. 531-575.

Hughes, K. M. 1978. The macromolecular lattice of polyhedra. J. Invertebr. Pathol., 22, 168-174.

Ignoffo, C. M. 1973a. Effects of entopathogens on vertebrates. Ann. N.Y. Acad. Sci., 217, 141-164.

Ignoffo, C. M. 1973b. Development of a viral insecticide: Concept to commercialization. Expt. Parasitol., 33, 380-406.

Ignoffo, C. M., Hostetter, D. L. and Shapiro, M. 1974. Efficacy of insect viruses propagated in vivo and in vitro

J. Invertebr. Pathol., 24, 184.
Ignoffo, C. M. and Rafajko, R. F. 1973. In vitro attempts to infect primate cell lines with the nucleopolyhedrosis virus of Heliothis. J. Invertebr. Pathol., 20, 321-325.
Jewell, J. E. and Miller, L. K. 1980. DNA sequence homology relationships among six Lepidopteran nuclear polyhedrosis viruses. J. Gen. Virol., 48, 161-175.
Jaques, R. P. 1975. Persistence, accumulation and denaturation of nuclear polyhedrosis and granulosis viruses. In Baculoviruses for Insect Pest Control: Safety Considerations, M. D. Summers, R. Engler, L. A. Falcon and P. Vail, eds.. Amer. Soc. Microbiol., Washington, D.C. pp. 90-99.
Kawanishi, C. Y., Summers, M. D., Stoltz, D. B. and Arnott, H. J. 1972. Entry of an insect virus in vivo by fusion of viral envelope and microvillus membrane. J. Invertebr. Pathol., 20, 104-108.
Kawarabata, T. 1974. Highly infectious free virions in the hemolymph of the silkworm (Bombyx mori) infected with a nuclear polyhedrosis virus. J. Invertebr. Pathol., 24, 196-200.
Kawarabata, T. and Aratake, Y. 1978. Functional differences between occluded and non-occluded viruses of a nuclear polyhedrosis of the silkworm Bombyx mori, J. Invertebr. Pathol., 20, 321-325.
Kelly, D.C. 1975. "Oryctes" virus replication: Electron microscopic observations on infected moth and mosquito cells. Virol., 69, 596-606.
Kelly, D. C. 1977. The DNA contained by nuclear polyhedrosis viruses isolated from four Spodoptera sp. (Lepidoptera: Noctuidae): Genome size and homology assessed by DNA replication kinetics. Virol., 76, 468-471.
Kelly, D. C., Edwards, M. L., Evans, J. F. and Robertson, J. S. 1978. The use of an enzyme-linked immunosorbent assay to detect a nuclear polyhedrosis virus in Heliothis armigera larvae. J. Gen. Virol., 40, 465-469.
Knudson, D. L. 1980. Specificity of Baculoviruses. Symposium Proceedings: Safety Aspects of Baculoviruses as Biological Insecticides, H. G. Mittenburger, ed. Bundesministerium fur Forschung und Technologie, Bonn. pp. 241-256.
Knudson, D. L. and Buckley, S. M. 1977. Invertebrate cell culture methods for the study of invertebrate associated animal viruses. In Methods in Virology, K. Maramorosch and H. Koprowski, eds., Vol. 6. Academic Pr., N.Y. pp. 323-391.
Knudson, D. L. and Harrap, K. A. 1976. Replication of a nuclear polyhedrosis virus in a continuous cell culture of Spodoptera frugiperda: Microscopy study of the sequence of events of the virus infection. J. Virol., 17, 254-268.

Knudson, D. L. and Tinsley, T. W. 1974. Replication of a nuclear polyhedrosis virus in a continuous cell culture of Spodoptera frugiperda: Purification, assay of infectivity and growth characteristics of the virus. J. Virol., 14, 934-944.

Knudson, D. L. and Tinsley, T. W. 1978. Replication of a nuclear polyhedrosis virus in a continuous cell line of Spodoptera frugiperda: Partial characterization of the viral DNA, comparative DNA-DNA hybridization, and patterns of DNA synthesis. Virol., 87, 42-47.

Lee, H. H. and Miller, L. K. 1978. Isolation of genotypic variants of Autographa californica nuclear polyhedrosis virus. J. Virol., 27, 754-767.

Lee, H. H. and Miller, L. K. 1979. Isolation, complementation, and initial characterization of temperature-sensitive mutants of the baculovirus Autographa californica nuclear polyhedrosis virus. J. Virol., 31, 240.

Longworth, J. F. 1973. Viruses and Lepidoptera. In Viruses and Invertebrates, A. J. Gibbs, ed. American Elsevier, Amsterdam. pp. 428-441.

McIntosh, A. H. and Maramorosch, K. 1973. Retention of insect virus infectivity in mammalian cell culture. J. N.Y. Entomol. Soc., 81, 175-182.

McIntosh, A. H. and Shamy, R. 1975. Effects of the nuclear polyhedrosis virus (NPV) of Autographa californica on a vertebrate viper cell line. Ann. N.Y. Acad. Sci., 266, 327-331.

McIntosh, A. H. and Shamy, R. 1980. Biological studies of a baculovirus in a mammalian cell line. Intervirol., 13, 331-341.

MacKinnon, E. A., Henderson, J. F., Stoltz, D. B. and Faulkner, P. 1974. Morphologenesis of nuclear polyhedrosis virus under conditions of prolonged passage in vitro. J. Ultrastruct. Res., 9, 419-435.

Martignoni, M. E., Iwai, P. J. and Rohrmann, G. F. 1980. Serum neutralization of nuclear polyhedrosis viruses (Baculovirus, Subgroup A) pathogenic for Orgyia pseudotsugata. J. Invertebr. Pathol., 36, 12-20.

Martignoni, M. E. 1981. A catalog of viral diseases of insects, mites and ticks. In Microbial Control of Pests and Plant Diseases 1970-1980, H. D. Burges, ed. Appendix 2. Academic Pr., N.Y.

Maruniak, J. E., Summers, M. D., Falcon, L. A. and Smith, G. E. 1979. Autographa californica nuclear polyhedrosis virus structural proteins compared from in vivo and in vitro sources. Intervirol., 11, 82-88.

Matthews, R. E. F. 1979. Classification and nomenclature of viruses. Intervirol., 12, 129-296.

Mazzone, H. M. and Tignor, G. H. 1976. Insect viruses: Serological relationships. Adv. Virus Res., 20, 237-270.

McCarthy, W. J. and Lui, S. Y. 1976. Electrophoretic and serological characterization of Porthetria dispar polyhedron protein. J. Invertebr. Pathol. 28, 57-65.

McCarthy, W. J. and Lambiase, J. T. 1980. Serological relationships among Plusinae Baculoviruses. J. Invertebr. Pathol., 34, 170-177.

Miller, L. K. and Dawes, K. P. 1978a. Restriction endonuclease analysis for the identification of baculovirus pesticides. Appl. Environ. Microbiol., 35, 411-421.

Miller, L. K. and Dawes, K. P. 1978b. Restriction endonuclease analysis to distinguish two closely related nuclear polyhedrosis viruses: Autographa californica MNPV and Trichoplusia ni MNPV. Appl. Environ. Microbiol., 35, 1206.

Miller, L. K. and Dawes, K. P. 1979. Physical map of Autographa californica nuclear polyhedrosis virus. J. Virol., 29, 1044.

Minion, F. C., Coons, L. B. and Broome, J. R. 1979. Characterization of the polyhedral envelope of the nuclear polyhedrosis virus of Heliothis virescens. J. Invertebr. Pathol., 34, 303-307.

Mittenberger, H. G., ed. 1980. Safety aspects of baculoviruses as biological insecticides. Symposium Proceedings, Bundesministerium fur Forschung und Technologie, Bonn. p. 301.

Morgan, C., Moore, D. H. and Rose, H. M. 1955. The macromolecular paracrystalline lattice of insect virus polyhedral bodies demonstrated in ultrathin sections examined in electron microscope. J. Biophys. Biochem. Cytol., 1: 187-190.

O'Farrel, P. H. 1975. High resolution two-dimensional electrophoresis of proteins. J. Biol. Chem., 250, 4007-4021.

Padhi, S. B., Eikenberry, E. F. and Chase, T. 1974. Electrophoresis of the proteins of the nuclear polyhedrosis virus of Porthetria dispar. Intervirol., 4, 333-345.

Paschke, J. D. and Summers, M. D. 1975. Early events in the infection of the arthropod gut by pathogenic insect viruses. In Invertebrate Immunity, K. I. Maramorosch and R. E. Shope, eds. Academic Pr., N.Y. pp. 75-114.

Payne, C. C., Compson, D. and de Looze, S. M. 1977. Properties of the nucleocapsids of a virus isolated from Oryctes rhinoceros. Virol., 77, 269.

Potter, K. N., Faulkner, P. and Mackinnon E. A. 1976. Strain selection during serial passage of Trichoplusia ni nuclear polyhedrosis virus. J. Virol., 18, 1040-1050.

Potter, K. N., Jaques, R. P. and Faulkner, P. 1978. Modification of Trichoplusia ni nuclear polyhedrosis virus passaged in vivo. Intervirol, 9, 76-85.

Raghow, R. and Grace, T. D. C. 1974. Studies on a nuclear polyhedrosis virus in Bombyx mori cells in vitro. I. Multiplication kinetics and ultrastructural studies. J. Ultrastruct. Res., 47, 384.

Rohrmann, G. F. 1977. Characterization of N-polyhedrin of two Baculovirus strains pathogenic for Orgyia pseudotsugata. Biochem., 16, 1631-1634.

Rohrmann, G. F. and Beaudreau, G. S. 1977. Characterization of DNA from polyhedral inclusion bodies of the nucleopolyhedrosis single-rod virus pathogenic to Orgyia pseudotsugata. Virol., 83, 474-478.

Scharnhorst, D. W. and Weaver, R. F. 1979. Characterization of the DNA of the nuclear polyhdrosis virus of the cotton bollworm, Heliothis zea. J. Gen. Virol., 42, 633-636.

Scharnhorst, D. W. and Weaver, R. F. 1980. Structural analysis of the matrix protein from the NPV of Heliothis zea. Virol., 102, 468-472.

Smith, D. S. 1968. Insect Cells: Their Structure and Function. Oliver and Boyd, Edinburgh, 372 pgs.

Smith, K. N. 1967. Insect Virology. Academic Pr. N.Y. 256 pp.

Smith, G. E. 1973. The use of viruses for the control of insect pests and diseases vectors. World Health Org. Tech. Rep. Ser., Geneva, 48 pg.

Smith, G. E., and Summers, M. D. 1978. Analysis of baculovirus genomes with restriction endonucleases. Virol., 89, 517-527.

Smith, G. E. and Summers, M. D. 1979. Restriction maps of five Autographa californica MNPV variants, Trichoplusia ni MNPV and Galleria mellonella MNPV DNAs with the endonucleases SmaI, KpnI, BamHI, XhoI and EcoRI. J. Virol., 30, 828.

Smith, G. E. and Summers, M. D. 1980. Restriction map of Rachiplusia ou and Rachiplusia ou-Autographa californica baculovirus recombinants. J. Virol., 33, 311-319.

Stairs, G. R. 1980. Comparative infectivity of non-occluded viruses, polyhedra and viruses released from polyhedra for larvae of Galleria mellonella. J. Invertebr. Pathol., 36, 281-282.

Stairs, G. R. and Ellis, B. J. 1971. Electron microscope and microfilter studies of infectious nuclear polyhedrosis virus in Galleria mellonella larvae. J. Invertebr. Pathol., 17, 350-353.

Steinhaus, E. A. 1946. Insect Microbiology. Comstock. Publ. Co. Ithaca, N.Y.

Steinhaus, E. A. 1949. Principles of Insect Pathology. McGraw-Hill, N.Y.

Steinhaus, E. A. 1963. Insect Pathology, An Advanced Treatise, Vol. 1. Academic Pr., N.Y.

Stoltz, D. B., Pavan, C. and da Cunha, A. B. 1973. Nuclear polyhedrosis virus: A possible example of de novo intranuclear membrane morphogenesis. J. Gen. Virol., 19, 145-150.

Stoltz, D. B. and Vinson, S. B. 1979. Viruses and parasitism in insects. Adv. Virus Res., 24, 125-171.

Strokovskaya, I. N., Strokovskaya, L. I., Zherebtsova, E. N. and Gudz-Gorban, A. P. 1977. Supercoiled DNA of nuclear polyhedrosis virus of Galleria mellonella L. Arch. Virol., 53, 79-86.

Summers, M. D. 1969. Apparent in vivo pathway of granulosis virus invasion and infection. J. Virol., 4, 188-190.

Summers, M. D. 1971. Electron microscope observations on granulosis virus entry, uncoating and replication processes during infection of the mid-gut cells of Trichoplusia ni. J. Ultrastruct. Res., 35, 606-620.

Summers, M. D. 1977. Baculoviruses (Baculoviridae). In The Atlas of Insect and Plant Viruses, K. Maramorosch, ed. Academic Pr. N.Y. pp. 3-28.

Summers, M. D. and Anderson, D. L. 1973. Characterization of nuclear polyhedrosis virus DNAs. J. Virol., 12, 1336-1346.

Summers, M. D. and Arnott, H. J. 1969. Ultrastructural studies on inclusion formation and virus occlusion in nuclear polyhedrosis and granulosis virus infected cells of Trichoplusia ni (Hubner). J. Ultrastrut. Res., 28, 462-480.

Summers, M. D., Engler, R., Falcon, L. A. and Vail, P. eds. 1975. Insect Baculoviruses for Insect Pest Control: Safety Considerations. Amer. Soc. Microbiol., Washington, D.C.

Summers, M. D. and Hoops, P. 1980. Radioimmuneassay analysis of baculovirus granulins and polyhedrosis. Virol., 103, 89-98.

Summers, M. D. and Kawanishi, C. Y., eds. 1978. Viral Pesticides: Present Knowledge and Potential Effects on Public and Environmental Health. Symposium Proceedings, U.S. Environ. Protection Agency, Research Triangle Park, NC.

Summers, M. D. and Smith, G. E. 1975. Trichoplusia ni granulosis virus granulin: A phenol soluable phosphorylated protein. J. Virol., 16, 1108-1116.

Summers, M. D. and Smith, G. E. 1976. Comparative studies of Baculovirus granulins and polyhedrins. Intervirol., 6, 168-180.

Summers, M. D. and Smith, G. E. 1978. Baculovirus structural polypeptides. Virol., 84, 390-402.

Summers, M. D., Volkman, L. E. and Hsieh, C-H. 1978. Immunoperoxidase detection of baculovirus antigens in insect cells. J. Gen. Virol., 40, 545-557.

Summers, M. D., Smith, G. E., Knell, J. D. and Burand, J. P. 1980. Physical maps of Autographa californica and Rachiplusia ou nuclear polyhedrosis virus recombinants. J. Virol., 34, 693-703.

Summers, M. D. and Volkman, L. E. 1976. Comparison of biophysical and morphological properties of occluded and non-occluded baculoviruses from in vivo and in vitro host systems. J. Virol., 17, 962-972.

Tanada, Y. and Leutenegger, R. 1970. Multiplication of a granulosis virus in larval midgut cells of Trichoplusia ni and possible pathways of invasion into the hemocoel. J. Ultrastruct. Res., 30, 589-600.

Tanada, Y. and Hess, R. T. 1976. Development of a nuclear polyhedrosis virus in midgut cells and penetration of the virus into the hemocoel of the armyworm. J. Invertebr. Pathol., 23, 325-332.

Tanada, Y., Hess, R. T. and Omi, E. M. 1975. Invasion of a nuclear polyhedrosis virus in midgut of the armyworm, Pseudaletia unipuncta, and the enhancement of a synergistic enzyme. J. Invertebr. Pathol., 36, 99-104.

Tanada, Y., Inoue, H., Hess, R. T. and Omi, E. M. 1980. Site of action of a synergistic factor of the granulosis virus of the army worm, Pseudaletia unipuncta. J. Invertebr. Pathol., 34, 249-255.

Thomas, E. D., Reichelderfer, C. F. and Heimpel, A. M. 1972. Accumulation and persistence of a nuclear polyhedrosis virus of the cabbage looper in the field. J. Invertebr. Pathol., 20, 157-164.

Tinsley, T. W. 1978. Use of insect pathogenic viruses as pesticidal agents. Perspectives in Virol., 10, 199-209.

Tinsley, T. W. 1979. The potential in insect pathogenic viruses as pesticidal agents. Ann. Rev. Entomol., 24, 63-87.

Tinsley, T. W. and Harrap, K. A. 1978. Viruses of invertebrates. In Comprehensive Virology, H. Fraenkel-Conrat and R. R. Wagner, eds. 12, 1-101.

Tjia, S. T., Carstens, E. B. and Doerfler, W. 1979. Infection of Spodoptera frugiperda cells with Autographa californica nuclear polyhedrosis virus. Virol., 99, 399-409.

Tweeten, K. A., Bulla, L. A. and Consigli, R. A. 1978. Characterization of an alkaline protease associated with granulosis virus of Plodia interpunctella. J. Virol., 26, 702-711.

Tweeten, K. A., Bulla, L. A. and Consigli, R. A. 1980. Characterization of an extremely basic protein derived from granulosis virus nucleocapsids. J. Virol., 33, 877-886.

Vail, P. V., Jay, D. L. and Hink, W. F. 1973. Replication and infectivity of the nuclear polyhedrosis virus of the alfalfa looper, Autographa californica produced in cells grown in vitro. J. Invertebr. Pathol., 22, 231-237.

Vail, P. V., Romine, C. L. and Vaughan, J. L. 1979. Infectivity of nuclear polyhedrosis virus extracted with digestive juices. J. Invertebr. Pathol., 33, 328-330.

Van der Beek, C. P. 1980. The origin of the polyhedral proteins of the occluded polyhedrosis virus of Autographa californica. Meded Landbouwhogeschool Wageningen, 80-85.

Van der Beek, C. P., Saaijer-Riep, J. D. and Vlak, J. M. 1980. On the origin of the polyhedral protein of Autographa californica nuclear polyhedrosis virus. Isolation, characterization and translation of viral messenger RNA. Virol., 100, 326-333.

Van der Geest, L. P. S. and Van der Laan, P. A. 1971. Insect pathogens available for distribution. In Microbial Control of Insects and Mites, H. D. Burges and N. K. Hussey, eds. Appendix 5. Academic Pr. N.Y.

Von Prowazek, S. 1913. Untersuchungen uber die Gelbsucht des Seidenraupen. Zentr. Bakteriol. Parasitenk. Infect. I. Orig., 67, 268-284.

Vlak, J. M. and Odink, K. G. 1979. Characterization of Autographa californica nuclear polyhedrosis virus deoxyribonucleic acid. J. Gen. Virol., 44, 333-347.

Volkman, L. E. and Summers, M. D. 1977. Autographa californica nuclear polyhedrosis virus: Comparative infectivity of the occluded, alkali-liberated and nonoccluded forms. J. Invertebr. Pathol., 30, 102-103.

Volkman, L. E., Summers, M. D. and Hsieh, C. 1976. Occluded and non-occluded nuclear polyhedrosis virus grown in Trichoplusia ni: Comparative neutralization, comparative infectivity, and in vitro growth studies. J. Virol., 19, 820-832.

Wood, H. A. 1980a. Autographa californica nuclear polyhedrosis virus induced proteins in tissue culture. Virol., 102, 21-27.

Wood, H. A. 1980b. Protease degradation of Autographa californica nuclear polyhedrosis virus proteins. Virol., 103, 392-399.

Xeros, N. 1956. The virogenic stroma in nuclear and cytoplasmic polyhedroses. Nature, 178, 412-413.

Yamamoto, T. and Tanada, Y. 1978a. Phospholipid, an enhancing component in the synergistic factor of a granulosis virus of the army worm Pseudaletia unipuncta. J. Invertebr. Pathol., 31, 48-56.

Yamamoto, T. and Tanada, Y. 1978b. Biochemical properties of viral envelopes of insect baculoviruses and their role in infectivity. J. Invertebr. Pathol., 32, 202-211.

Zherebtsova, E. N. 1978. Replication of Baculoviruses in established insect cell lines: Phenomena of attenuation. Arch. Virol., 54, 283-290.

Zummer, M. and Faulkner, P. 1979. Absence of protease in Baculovirus polyhedral bodies propagated in vitro. J. Invertebr. Pathol., 33, 383-384.

Chapter 2

NON-OCCLUDED VIRUSES

by David C. Kelly*

INTRODUCTION

Non-occluded viruses, those invertebrate pathogenic viruses not packaged within protein inclusions or polyhedra late in infection, comprise a large heterogenous group of viruses. These viruses include iridescent viruses (large and small), densonucleosis viruses, rhabdoviruses, picorna viruses, _Tellina_ viruses, together with non-occluded baculoviruses. In addition retroviruses and reoviruses have been described in invertebrate cells in culture though as yet they have not been associated specifically with disease in invertebrates. A comprehensive review of the pathogenesis of this miscellany of viruses cannot yet be attempted since most viruses remain poorly studied. The general principals of pathogenesis of non-occluded viruses will be illustrated mainly by three groups of viruses whose pathogenesis is best understood - the iridescent viruses, densonucleosis viruses, and picorna viruses - though it must be stressed that the pathogenesis of none of the non-occluded viruses is even reasonably well elucidated.

This review will consider pathogenesis to be the course or cycle of events of a disease from its beginning to the production of characteristic symptoms (Jacobs et al., 1957).

The pathogenesis of non-occluded viruses in cells in culture is a useful adjunct to considering the pathogenesis of viruses in whole animals. Few viruses have been shown to replicate in cells in culture though some of the iridescent viruses and picorna viruses are notable exceptions.

*Natural Environment Research Council, Institute of Virology, Oxford, England.

THE VIRUSES

The biophysical and biochemical parameters of non-occluded viruses have been extensively reviewed in recent years (Tinsley and Harrap, 1978; Harrap and Payne, 1979; Payne and Kelly, 1980) and so the salient features of the structural and biochemical properties are presented in the first four tables to illustrate the essential characteristics of the viruses. The diversity of the viruses is shown by the fact that viruses with double stranded or single-stranded nucleic acid (DNA or RNA) are featured in the lists of viruses. As such it is not surprising that the strategy of viral replication and consequently the pathogenesis of the viruses are markedly different.

Large iridescent viruses

These spherical viruses are the largest insect viruses recorded, some being 180-200 nm in diameter. They contain a linear double stranded DNA genome about 2.8 x 10^8 daltons in size (Wagner and Paschke, 1977) which means that they contain sufficient genetic material to code for some 250 gene products; this approaches the size of a bacterial genome (1.2 x 10^9 daltons). Surprisingly despite this potentially large genetic repertory, the viruses have a narrow host range and have mainly been isolated from mosquitoes (Hasan et al., 1971). The viruses occur in low incidence, in natural populations infecting some 1 to 3% of the larvae (Hasan et al., 1971; Tinsley et al., 1971). The source of infection is probably debris of previously infected mosquito larvae, although a number of lower fungi and algae contain morphologically similar viruses (Kelly and Robertson, 1973) and it is possible that these act as a reservoir for the virus. The virus can infect larvae by oral ingestion but Hasan et al. (1971) have shown the virus is more infectious when injected into the hemocoel. This is consistent with the observations of Stoltz and Summers (1971) who showed that virus ingested by first and second instar larvae appear to be degraded shortly before entering the midgut. The virus was unable to penetrate the peritrophic membrane and consequently no virus was observed inside, or in contact with, midgut cells. It is probable that the infection pathway is similar to that of the small iridescent viruses considered next. However the virus gains entry into larvae, the virus in *Aedes taeniorhynchus* (the black salt marsh mosquito) is capable of infecting a variety of tissues within its host. Cells of the fat body, tracheal epithelium, imaginal discs, and epidermis are the primary sites of viral replication. Extensive destruction of the fat body results in the death of most infected mosquitoes before they reach the adult stage. To a lesser extent the hemocytes,

TABLE 2-1. PHYSICAL DIMENSIONS AND FORM OF NON-OCCLUDED DNA INVERTEBRATE VIRUSES

Virus	Dimensions	Form	Particle Weight
Large iridescent virus	180 nm	Icosahedral	2.1×10^9
Small iridescent virus	130 nm	Icosahedral	5.5×10^8
Densonucleosis virus	22 nm	Icosahedral	5.7×10^6
Baculovirus	80 x 220 nm	Bacilliform	4.6×10^8
Retrovirus	100 nm	Spherical	-

TABLE 2-2. BIOCHEMICAL PROPERTIES OF NON-OCCLUDED RNA INVERTEBRATE VIRUSES

Virus	Nucleic Acid		Proteins	
	Form	Size	No.	Size Range
Large iridescent virus	ds[1] linear	$2.4 - 2.8 \times 10^8$	9	15 to 97.5 $\times 10^3$
Small iridescent virus	ds linear	$1.2 - 1.4 \times 10^8$	22	11.7 to 150 $\times 10^3$
Densonucleosis virus	ss[2] linear	1.85×10^6	4	46 to 106 $\times 10^3$
Baculovirus	ds circular	9.2×10^7	12	9.7 to 75.7 $\times 10^3$
Retrovirus	ds linear	---	---	----

[1]Double stranded

[2]Single stranded

TABLE 2-3. PHYSICAL DIMENSIONS AND FORM OF NON-OCCLUDED RNA INVERTEBRATE VIRUSES

Virus	Dimensions	Form
Rhabdovirus	70 x 160 nm	Bacilliform
Tellina virus	70 nm	Icosahedral
Reovirus	57 nm	Icosahedral
Picorna virus[1]	≃35 nm	Icosahedral

[1]See extended data in Harrap and Payne (1978) for details of 7 groups of picorna viruses.

TABLE 2-4. BIOCHEMICAL PROPERTIES OF NON-OCCLUDED RNA INVERTEBRATE VIRUSES

Virus	Nucleic Acids		Proteins	
	Form	Size	No.	Size Range
Rhabdovirus	ss linear	-	-	-
Tellina virus	ds segmented linear	2.3 and 2.5 x 10^6	4	2.8 to 9.0 x 10^3
Reovirus	ds segmented linear	0.55 to 2.75 x 10^6	-	
Picornavirus[1]	ss linear	1.5 to 2.3 x 10^6	1 to 4	

[1] See extended data in Harrap and Payne (1979) for details of 7 groups of picorna viruses.

esophagus nerve, muscle and ovaries are infected (Hall and Anthony, 1971). The virus replicates in the cytoplasm of cells and large inclusions are observed in the cell cytoplasm. The aggregates of virus cause the larvae to develop a lime green coloration at an advanced stage of infection (Tinsley et al., 1971) and the larvae become sluggish and die. It is claimed that this virus can be transovarially transmitted, and although the evidence has not been obtained by stringent experimentation, it is probable that this form of transmission is important in the vertical spread of the virus (Fukada and Clark, 1975).

The virus has not been shown to fully replicate in mosquito cells in culture (Webb et al., 1974; 1975) and so it is difficult to assess the pathology of the virus in these cells. Nevertheless, it appears that the virus is rapidly taken up by viropexis and intact virions are released into the cytoplasm within an hour of infection. Subsequently progeny virus is assembled in cytoplasmic factories within 48 hours and then leaves the cell by budding. No doubt egress of virus from cells by budding at the plasma membrane occurs in insects but this has not yet been demonstrated.

Small iridescent viruses

The pathogenesis of small iridescent viruses is probably one of the best studied largely due to the independent efforts of K. M. Smith and J. B. Carter. The small iridescent viruses are icosahedral and are about 135 nm in diameter (Kelly and Robertson, 1973). The viruses contain linear double stranded DNA of molecular weight 120-140 x 10^6 (Kelly and Every, 1974; Hibbin and Kelly, 1980). The viruses are found ubiquitously in nature, occuring in Diptera, Coleoptera, Lepidoptera, Hymenoptera, Hemiptera, Mollusca and Crustacea (Carey et al., 1978; Kelly and Robertson, 1973). With two exceptions, the thirty isolates so far recorded are serologically related (Kelly et al., 1978; Kelly et al., 1979; D. C. Kelly, unpublished observations).

The pathogenesis of small iridescent viruses will be illustrated by the example of iridescent virus Type 1, *Tipula* iridescent virus from *Tipula paludosa* (Xeros, 1954). The virus is not infectious *per os* and presumably enters larvae through wounds or is introduced by parasites.

The small iridescent viruses readily replicate in a number of insect cell lines including *Spodoptera frugiperda*, *Trichoplusia ni* and *Aedes albopictus* cells (Kelly, 1976a; Brown et al., 1978). A plaque assay is available to titrate the viruses, and clones of the virus from *Simulium* spp. (iridescent virus Type 22) have been produced (J. A. Hibbin and D. C. Kelly, unpublished observations). The cultured cell pathology of the virus mimics that of the virus in whole animals. The virus replicates in the cell cytoplasm and autoradiographic studies showed that viral DNA synthesis also occurs in the cell cytoplasm (Bellett and Mercer, 1964). Iridescent virus protein synthesis in S. frugiperda cells occurs in a three phase cascade fashion similar to that of the morphologically similar frog virus 3 (Elliott and Kelly, 1980; D. C. Kelly and R. M. Elliott, unpublished observations). The synthesis of the third phase of proteins requires viral DNA synthesis.

In insect larvae iridescent virus infections result in systemic infection. The source of infection appears to be cadavers of infected larvae rather than feces in which the concentration of virus is too low (Carter, 1973a). The virus is far more infective when injected into the hemocoel of the insect than when ingested. Embryos do not become infected when eggs are laid in agar containing virus, and there is no evidence of infection via the spiracles or of transovarial transmission. It is likely that the virus is transmitted principally by cannibalism of infected larvae. Carter (1973b) proposes that new generations of Tipula oleraceae become infected by first instar larvae cannabalistically killing and eating fourth instar larvae. Once the virus enters the insect the virus replicates unabated and can eventually amount to greater than 30% of the body weight of the larva. Larvae become blue, iridescent and flaccid about 7 to 10 days after infection (Smith, 1967; Kelly et al., 1978). Autoradiographic studies by Morris (1970) showed that altered DNA and RNA metabolism accompanied a massive increase in size of fat body cells. The viral DNA appeared to replicate in the cell cytoplasm. Younghusband and Lee (1969) demonstrated that this virus was taken up by larval hemocytes within 8 h of inoculation and within 48 h progeny virus budded from the hemocytic plasma membrane. This released virus presumably serves to infect other larval tissues. Subsequently these authors showed by cytochemistry and autoradiography that the virus is taken up by hemocytes into lysosomes and subsequently viral DNA synthesis is associated with the virogenic stroma (Younghusband and Lee, 1970). In a further study, Yule and Lee (1973) using immunological and cytological methods showed that in hemocytes the virus structural protein was initially located outside the cytoplasmic viroplasm and then within. Viral DNA was located exclusively within the viroplasm. These authors also proposed a developmental sequence for the assembly of the virion in hemocytes. A comparison of the morphogenesis of iridescent virus Type 22 is given in Figs. 2-1 and 2-2 where the morphogenesis of the virus in its natural black fly host (Fig. 2-1) and in S. frugiperda cells are shown.

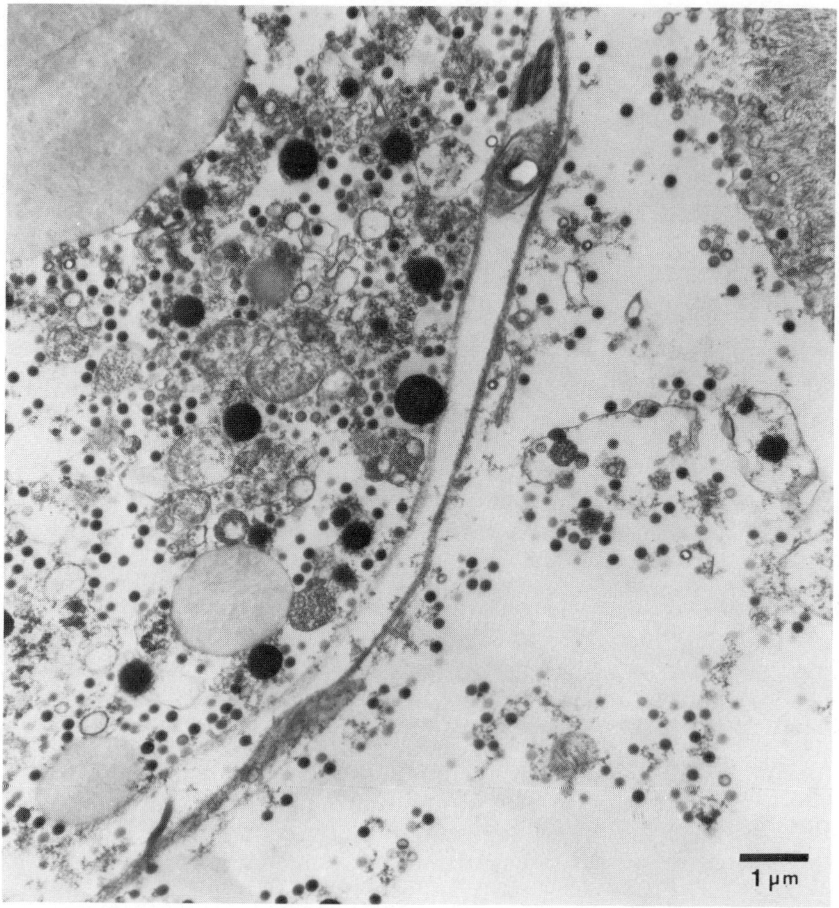

Figure 2-1. Iridescent virus Type 22, a small iridescent virus in the cytoplasm of the fat body of a Simulium larva.

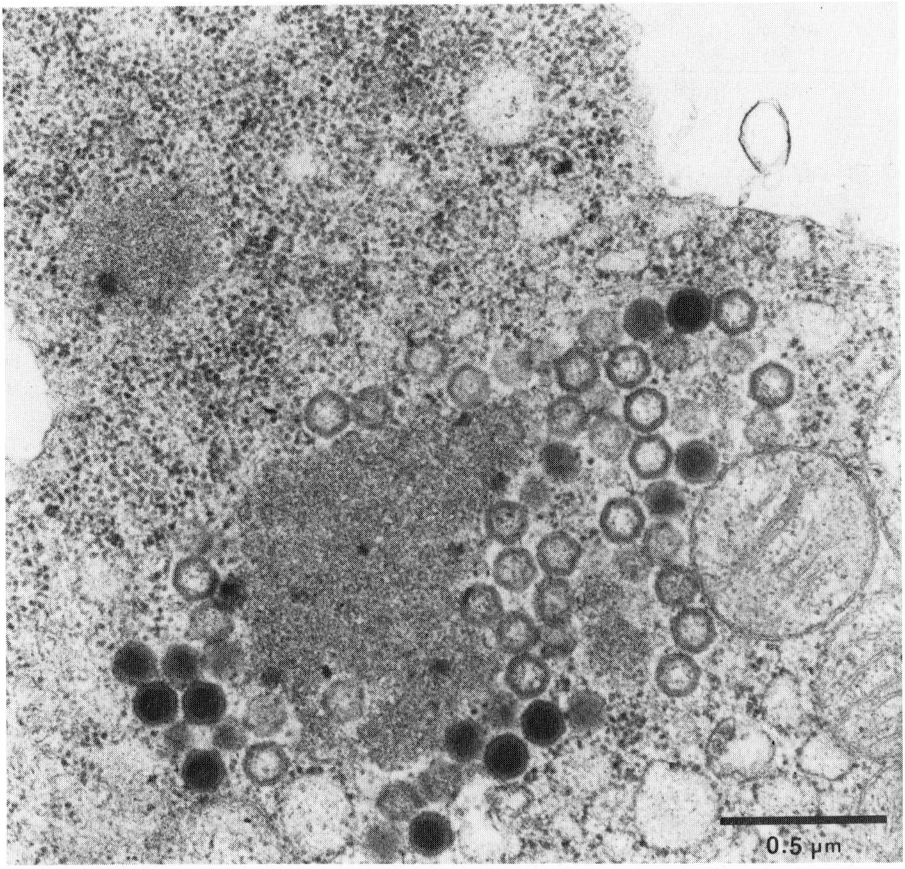

Figure 2-2. Iridescent virus Type 22 focus of infection in the cytoplasm of a Spodoptera frugiperda cell 22 hours after infection.

Densonucleosis viruses

These are the smallest invertebrate pathogenic viruses known, around 22 nm in diameter and containing linear single stranded DNA approximately 1.85 x 10^6 daltons in size (Kelly et al., 1977; 1980c; Kurstak, 1972). The best studied virus is the one isolated from Galleria mellonella (Armagier et al., 1965) although other densonucleosis viruses have been isolated from Junonia coenia (Rivers and Longworth, 1972), Agraulis vanillae (Kelly et al., 1980a), Sibine fusca (Meynadier et al., 1977b), Periplaneta fuliginosa (Suto, 1980), Simulium vittatum (Federici, 1976) and Bombyx mori (Nakagaki and Kawase, 1980).

The replication of the viruses in cells in culture is controversial. No one has successfully propagated the virus in a continuous cell line although the virus from J. coenia replicates in a primary B. mori cell line (Quiot et al., 1973) and the virus from G. mellonella grows in a primary G. mellonella ovarian cell culture (Vago and Luciani, 1965). It has also been claimed that the G. mellonella virus replicates in or transforms mouse cells (reviewed by Kurstak, 1972) although this observation awaits confirmation. Nevertheless the observations made in cells in part reflects the gross cellular pathology of the virus in larvae. The virus is highly infectious per os and presumably enters the midgut by direct penetration through the peritrophic membrane. The infection is characterized by the formation of dense, DNA positive inclusions in the hypertrophied nuclei of infected cells. This occurs in most tissues with the exception of the midgut (Garzon and Kurstak, 1976). This surprising observation may mean that the midgut has too high a pH to permit DNV replication (Garzon and Kurstak, 1976), or the virus circumvented this organ, because these authors injected virus into their experimental animal, or abnormal replication occurs and the virus enters the midgut subsequently to replicate throughout the insect. It appears that the virus can pass very quickly from the gut to other tissues. The resulting systemic infection of the larvae eventually destroys the internal organs of the larvae resulting in a flaccid cadaver. The general appearance of a densonucleosis virus infected cell is shown in Fig. 2-3.

Baculoviruses

The gross pathology of "conventional" baculoviruses is featured in Chapter 1. It is debatable whether non-occluded or occluded viruses are more common in nature and one suspects that the occluded viruses are more frequently recorded because of their more obvious pathology created by the presence of occlusions within the cell. This is also true of baculoviruses. Only one non-occluded baculovirus has been seriously studied -

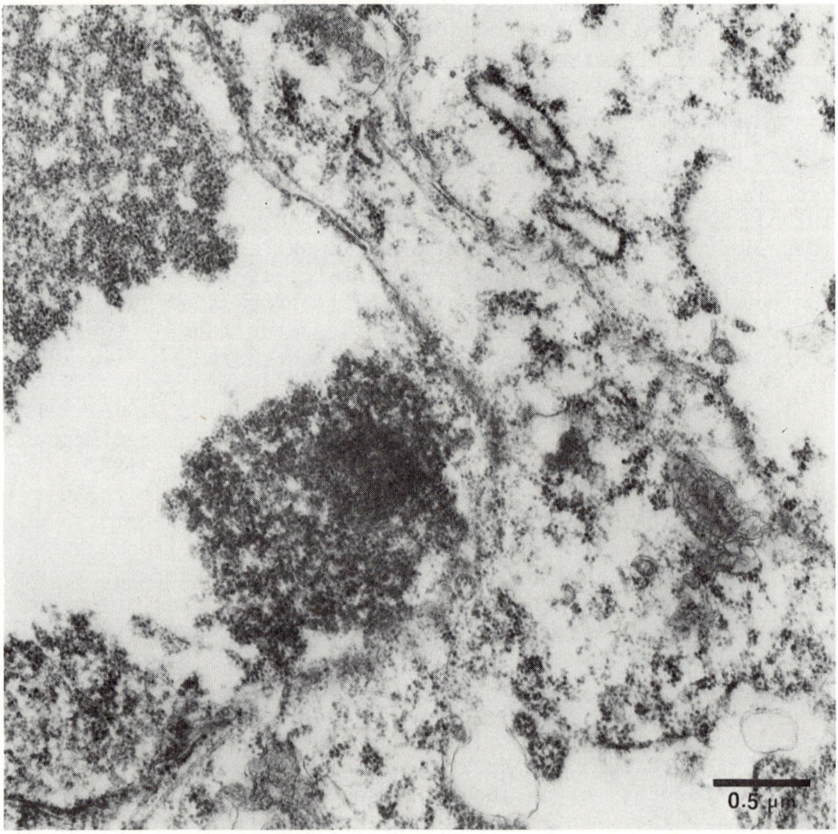

Figure 2-3. Densonucleosis virus Type 2 infection of *Aglais urticae* larvae. The degenerating nucleus (right) and cytoplasm both contain the virus particles.

that of Oryctes rhinoceros (Huger, 1966) - although there are reports of other non-occluded baculoviruses morphologically similar to Oryctes virus such as the virus in the crab Carcinus maenus (Bazin et al., 1974), and possibly the virus persistently infecting a Heliothis zea cell line (Granados et al., 1978; Kelly et al., 1980b), and other viruses morphologically dissimilar such as viruses from Tenebrio molitor (Thomas and Gouranton, 1975), Gyrinus natator (Gouranton, 1972) and Scolytus scolytus (Arnold and Barson, 1977). In addition non-occluded baculoviruses have been isolated from braconid and ichneumonid wasps which are a third morphological variant of non-occluded baculoviruses (Stoltz and Vinson, 1979).

The pathogenesis of these viruses is poorly understood. Oryctes virus is infectious per os and presumably entry of the virus into the larva or adult beetle is similar to that of occluded baculoviruses with the enveloped particles being taken up by fusion of the virus envelope with microvilli of the midgut and subsequent development similar to that of other baculoviruses. Virus replication in the tissues occurs in cell nuclei (Payne, 1974; Monsarrat et al., 1974). The virogenic stroma is not particularly pronounced and the singly enveloped particles accumulate in the nucleus. Virus is not found extensively in the cell cytoplasm nor at the plasma membrane and presumably cell to cell spread occurs by cell lysis. Larvae become infected systemically and most tissues become infected, particularly the fat body. Typically the larvae show a prolapsed rectum 15 days after infection (Bedford, 1976) and subsequently die. The infection of adults appears to be restricted to the midgut. Active virus production occurs in the regenerative crypts of the adult midgut and these cells subsequently become hypertrophied. The adults remain viable for many days after infection and show no obvious signs of infection and in fact act as flying reservoirs of the virus (Bedford, 1976). The virus is excreted from the adults in massive quantities (about 1 mg per day) and the excreta act as a source of inoculum, although the virus is viable in the excreta for only a short time.

Oryctes virus replicates in cells in culture where its morphogenesis in S. frugiperda cells resemble that which occurs in the insect (Kelly, 1976b). The appearance of Oryctes virus is shown in Fig. 2-4.

The baculoviruses of braconid wasps occur in large amounts in cell nuclei of the calyx or in the lumen of calyces and oviducts. They appear not to occur in any other tissue nor do they cause any disease condition in the adult wasp (Stoltz and Vinson, 1979). The particles are present in all females and it appears that an efficient mechanism must exist for maintenance and transmission of the virus. The hosts of these parasitic wasps, such as Orygia leucostigma, appear to act as a reservoir; this is fully discussed in the review by Stoltz and Vinson (1979).

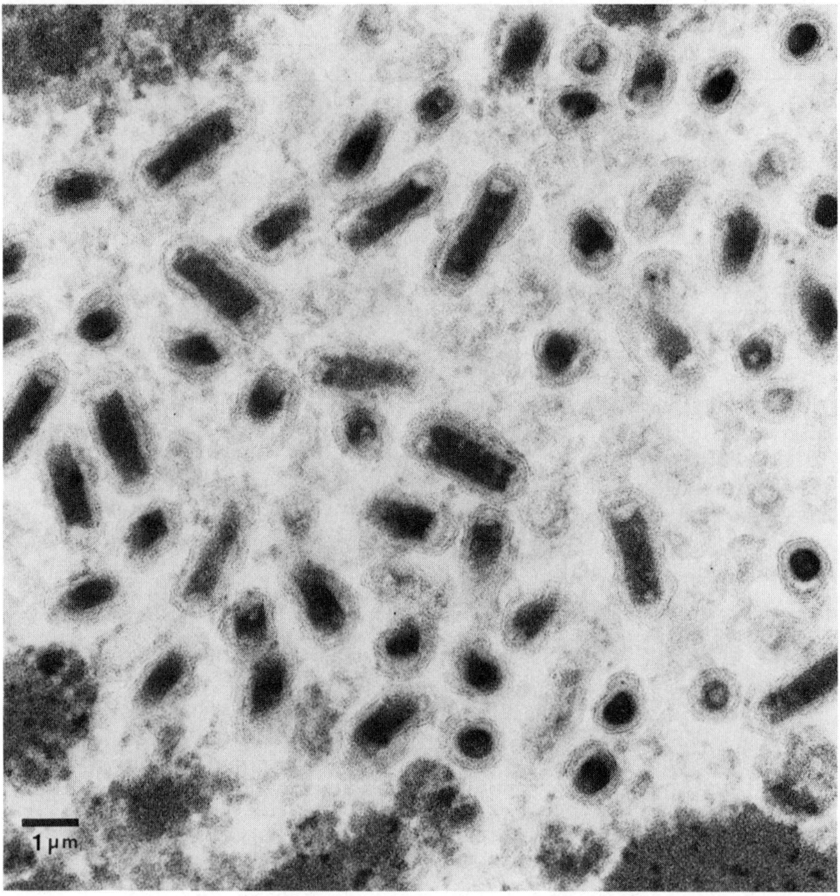

Figure 2-4. The singly enveloped non-occluded baculovirus from Oryctes rhinoceros in the nucleus of a midgut cell from a larva.

Studies on the pathogenesis of the other non-occluded viruses of invertebrates are restricted to electron microscope observations. A virus has been reported in the midgut of stick insects, Bacillus rossius, which is an apparently symptomless disease (Scali et al., 1980).

Picorna viruses

As pointed out by Longworth (1978) in his review of small isometric viruses of invertebrates, there has been little attention paid to the pathology of many small RNA viruses. The picorna viruses invariably develop in the cytoplasm of infected cells but there is no information on the replication and pathogenesis of the viruses in infected insects. Most picorna viruses are infectious per os and the gut is the primary site of replication. Subsequent pathogenesis depends on the invertebrate infected and whether the adult or immature forms are infected. When the gut is the major site of replication the symptoms of infection are vomiting and copious anal discharge (Longworth, 1978). In adults the brain and nervous system can be infected creating paralysis, especially in bees and crickets.

Bailey (1976) has written an excellent review on viruses attacking the honey bee. Most bee picorna viruses induce paralysis although a notable exception is sacbrood virus. This disease affects larvae and the symptoms are specific. Larvae with sacbrood fail to pupate and lie stretched on their backs with their heads to the cell capping. An accumulation of fluid occurs between the body and its unshed skin. The larva becomes pale yellow instead of white, because it is unable to shed its skin due to inhibition of the formation of chitinase. Eventually the larva dies and on drying it becomes brown (Bailey, 1976). Attempts to infect adult bees with sacbrood have failed (Bailey, 1967).

Chronic bee paralysis virus symptoms occur in adults and are typified by a bloated abdomen, trembling wings and body paralysis. Before death the bees crawl from the colony or are ejected by healthy bees. The virus is infectious per os, although virus injected into the adult hemolymph is considerably more infectious. The higher the temperature of the colony the more rapidly the bees die. The brain, nerve ganglia, hind-gut and other tissues develop basophilic cytoplasmic granules.

Acute bee paralysis causes adult bees to become flightless and death occurs quickly (within 5 to 6 days), in contrast to chronic bee paralysis where death occurs after a prolonged period of trembling. The spread of the virus probably occurs as an inapparent infection, probably via salivary gland secretions of adult bees which are added to pollen. Acute bee paralysis virus probably occurs in bumble bees and is the only honey bee pathogen for which an alternative host is known.

Picorna viruses of other insects are also infectious per os and cause paralysis in adults, e.g. cricket paralysis virus (Reinganum et al., 1970). In Teleogryllus oceanicus and T. commodus the virus causes a rapid and fatal paralysis. Early to mid-instar nymphs are most susceptible. Adults show some degree of resistance to the virus.

Picorna viruses with different symptoms include the P and Iota viruses of Drosophila melanogaster. Injection of P virus into adult flies induces sterility and shortens the life span of the flies. Just before death the flies become turgid with liquid.

A small divided genome virus from the black beetle Heteronychus arator induces paralysis and develops in the cytoplasm of gut and fat body tissues of larvae (Longworth and Carey, 1976).

Tellina viruses

Recently Dobos et al. (1980) proposed that a related group of bisegmented double stranded RNA viruses exist which are unlike reovirus and cytoplasmic polyhedrosis viruses. This group is as yet unnamed, but for convenience in this review they are called Tellina viruses after the virus isolated from a sand dwelling marine bivalve mollusc Tellina tenuis (Buchanan, 1973). Similar viruses have been isolated from D. melanogaster cell cultures (Teninges et al., 1979), and from an oyster (Hill, 1976). The three invertebrate viruses are similar in that they contain two RNA segments of similar size as well as proteins. The viruses were distinct as judged by serological tests and peptide mapping of the viral structural proteins although the oyster virus and the Tellina virus were related. Accompanying a high mortality among young T. tenuis were the symptoms of thin and chalky shells. The digestive gland which manufactures the shell precursor material was pale yellow in contrast to the normal dark brown, and was found to be necrotic (Buchanan, 1973).

Reoviruses

No reoviruses have been isolated directly from insects although recently Haars et al. (1980) described a reovirus in a D. melanogaster blood cell line. No evidence has yet been shown that the virus is truly an endogenous Drosophila virus and that it is not a contaminant.

Retroviruses

Recently Heine et al. (1980) described a retrovirus from *Drosophila* cells in culture. No evidence of retroviruses in insects has been obtained. It is possible that the virus is derived from transposons recently identified in *Drosophila* (Heine, Avery and Kelly, unpublished observations).

Rhabdoviruses

One such virus is known in invertebrates, the sigma virus of *D. melanogaster*. The virus is frequently found in natural populations of *Drosophila*. The virus confers on infected flies sensitivity to CO_2 (Printz, 1973). Otherwise the virus causes no obvious symptoms. Susceptible flies suffer from paralysis and die shortly after receiving sufficient CO_2 to induce narcosis. No physiological basis for the CO_2 sensitivity is known. The virus is systemically induced in flies, but the natural route of infection is not known.

CONCLUSIONS

Most non-occluded viruses of invertebrates appear to cause fatal disease. The mode of entry of the virus is either through the midgut or a wound. The cellular histopathology, the molecular interaction, and physiological effects of these viruses in the hosts have not yet been well studied, and there is an obvious need for future investigations of this type.

REFERENCES

Armagier, A., Vago, C. and Meynadier, G. 1965. Etude histopathologique d'un noveau type de virus mis en evidence chez le Lepidoptera, Galleria mellonella. Arch. Ges. Virusforsch., 15, 659-667.

Arnold, M. K. and Barson, G. 1977. Occurrence of virus-like particles in midgut epithelial cells of the large elm bark beetle, Scolytus scolytus. J. Invertebr. Pathol., 29, 373-381.

Bailey, L. 1967. The incidence of virus diseases in the honey bee. Ann. Appl. Biol., 60, 43-48.

Bailey, L. 1976. Viruses attacking the honey bee. Adv. Virus Res., 20, 271-304.

Bazin, F., Monsarrat, P., Bonami, J. R., Croizier, G., Meynadier, G., Quiot, J. M. and VAgo, C. 1974. Particles virales de type baculovirus observees chez le crabe Carcinus maenus. Rev. Trav. Inst. Peches Marit., 38, 205-208.

Bedford, G. O. 1976. Use of a virus against the coconut palm rhinoceros beetle in Fiji. Pans, 22, 11-16.

Bellett, A. J. D. and Mercer, E. H. 1964. The multiplication of Sericesthis iridescent virus in cell cultures from Antherea eucalytpi Scott. I. Qualitative experiments. Virology, 25, 22-32.

Brown, D. A., Lescott, T., Harrap, K. A. and Kelly, D. C. 1978. The replication and titration of iridescent virus Type 22 in Spodoptera friugperda cells. J. Gen. Virol., 38, 175-178.

Buchanan, J. S. 1973. Electron microscopic observations of virus-like particles in the digestive gland of a marine bivalve, Tellina tenis. Proc. Int'l. Colloq. Invertebr. Pathol., Oxford, U.K., p. 28.

Carey, G. P., Lescott, T., Robertson, J. S., Spencer, L. K. and Kelly, D. C. 1978. Three African isolates of small iridescent viruses. Virology, 85, 307-309.

Carter, J. B. 1973a. The mode of transmission of Tipula iridescent virus. I. Source of infection. J. Invertebr. Pathol., 21, 123-130.

Carter, J. B. 1973b. The mode of transmission of Tipula iridescent virus. II. Route of infection. J. Invertebr. Pathol., 21, 136-143.

Dobos, P., Hill, B. J., Hallett, R., Kells, D. T. C., Becht, T. and Teninges, D. 1979. Biophysical and biochemical characterization of five animal viruses with bisegmented double stranded RNA genomes. J. Virol., 32, 593-605.

Elliott, R. M. and Kelly, D. C. 1980. Frog virus 3 replication: Induction and intracellular distribution of polypeptides in infected cells. J. Virol., 33, 28-51.

Federici, B. A. 1976. Pathology and histochemistry of a densonucleosis virus in larvae of the blackfly, Simulium vittatum. Proc. Int'l. Colloq. Invert. Pathol., Kingston, Ontario, p. 341.

Fukada, T. and Clark, T. B. 1975. Transmission of the mosquito iridescent virus (RMIV) by adult mosquitoes of Aedes taeniorhynous to their progeny. J. Invertebr. Pathol., 25, 275-276.

Garzon, S. and Kurstak, E. 1976. Ultrastructural studies on the morphogenesis of the densonucleosis virus (parvovirus). Virology, 70, 517-531.

Granados, R. R., Nguyen, T. and Cato, B. 1978. An insect cell line persistently infected with a baculovirus-like particle. Intervirology, 10, 309-317.

Gouranton, J. 1972. Development of an intranuclear non-occluded rod-shaped virus in some midgut cells of an adult insect Tyrinus natator L. (Coleoptera). J. Ultrastruct. Res., 39, 281-294.

Haars, R., Zentgraf, H., Gateff, E. and Bautz, F. A. 1980. Evidence for endogenous reovirus like particles in a tissue culture cell line from Drosophila melanogaster. Virology, 101, 124-130.

Hall, D. W. and Anthony, D. W. 1971. Pathology of a mosquito iridescent virus infecting Aedes taeniorhyncus. J. Invertebr. Pathol., 18, 61-69.

Harrap, K. A. and Payne, C. C. 1979. The structural properties and identification of insect viruses. Adv. Virus Res., 25, 273-301.

Hasan, S., Vago, C. and Kuhl, G. 1971. Infection of Aedes detritus Hal. with a mosquito iridescent virus. Bull. World Health Org., 45, 268-269.

Heine, C. W., Kelly, D. C. and Avery, R. J. 1980. The detection of retrovirus particles in Drosophila melanogaster cell cultures. J. Gen. Virol., 49, 385-395.

Hibbin, J. A. and Kelly, D. B. 1980. Iridescent virus type 22 DNA. Arch Virol., submitted for publication.

Hall, B. J. 1976. Molluscan viruses: Their occurrence, culture and relationships. Proc. Int'l. Colloq. Invertebr. Pathol., Kingston, Ontario, pp. 25-30.

Huger, A. 1966. A virus disease of the Indian rhinoceros beetle Oryctes rhinoceros (L.) J. Invertebr. Pathol., 8, 35-51.

Jacobs, M. B., Gerstein, M. J. and Walter, W. G. 1957. Dictionary of Microbiology. D. Van Nostrand Co. Inc., London.

Kelly, D. C. 1976a. Iridescent virus type 22 replication in Aedes albopictus cells in culture. J. Invertebr. Pathol., 27, 415-418.

Kelly, D. C. 1976b. "Oryctes" virus replication: Electron microscopic observations on infected moth and mosquito cells. Virology, 69, 596-606.

Kelly, D. C. and Avery, R. J. 1974. The DNA content of four small iridescent viruses: Genome size, redundancy and homology determined by renaturation kinetics. Virology, 57, 425-435.

Kelly, D. C., Ayres, M. D., Lescott, T., Robertson, J. S. and Happ, G. M. 1979. A small iridescent virus (type 29) isolated from Tenebrio molitor. J. Gen. Virol., 42, 95-105.
Kelly, D. C. Ayres, M. D., Spencer, K. K. and Rivers, C. F. 1980a. Densonucleosis virus 3: A recent isolate from Agraulis vanillae. Microbiologica, in press.
Kelly, D. C., Barwise, A. H. and Walker, I. O. 1977. DNA contained by two densonucleosis viruses. J. Virol., 21, 396-407.
Kelly, D. C., Edwards, M-L. and Robertson, J. S. 1978. The use of enzyme linked immunosorbent assay to detect and discriminate between small iridescent viruses. Ann. Appl. Biol., 90, 369-374.
Kelly, D. C., Lescott, T., Ayres, M. D., Carey, D., Coutts, A. and Harrap, H. A. 1980b. Induction of a non-occluded baculovirus persistently infecting Heliothis zea cells by Heliothis armigera and Trichoplusia ni nuclear polyhedrosis viruses. Virology, (submitted for publication).
Kelly, D. C., Moore, N. F., Spilling, C. R., Barwise, A. H. and Walker, I. O. 1980c. Densonucleosis virus structural proteins. J. Virol., 36, 224-235.
Kelly, D. C. and Robertson, J. S. 1973. Icosahedral cytoplasmic deoxyriboviruses. J. Gen. Virol., 21, 17-41.
Kurstak, E. 1972. Small DNA densonucleosis virus (DNV). Adv. Virus Res., 17, 207-241.
Longworth, J. F. 1978. Small isometric viruses of invertebrates. Adv. Virus Res., 23, 103-157.
Longworth, J. F. and Carey, G. 1976. A small RNA virus with a divided genome from Heteronychus arator (F.) (Coleoptera: Scarabeidae). J. Gen. Virol., 33, 31-40.
Meynadier, G., Amargier, A. and Gentry, P. 1977a. Une virose de type densonucleose chez le lepidoptere Sibine fusca Stoll. Oleagineux 32, 357-361.
Meynadier, G., Matz, G., Veyrunes, J.-C. and Bres, N. 1977b. Virose de type densonucleose chez les orthopteres. Ann. Soc. Ent. Fr., 13, 487-493.
Meynadier, G., Galichet, P. F., Veyrunes, J. C. and Armagier, A. 1977c. Mise en evidence d'une densonuclease chez Diatraea saccharalis. Entomophaga, 22, 115-120.
Monsarrat, P., Duthoit, J.-L. and Vago, C. 1974. Mise en evidence de virions de type Baculovirus dans l'appareil genital du coleoptera Oryctes rhinoceros. C. R. Acad. Sci. Paris, (Ser. D), 278, 3259-3261.
Morris, O. N. 1970. Metabolic changes in diseased insects. III. Nucleic acid metabolism in Lepidoptera infected by densonucleosis and Tipula iridescent viruses. J. Invertebr. Pathol., 16, 180-186.

Nakagaki, M. and Kawase, S. 1980. DNA of a new parvovirus isolated from the silkworm Bombyx mori. J. Invertebr. Pathol., 35, 124-133.

Payne, C. C. 1974. The isolation and characterization of a virus from Oryctes rhinoceros. J. Gen. Virol., 25, 105-116.

Payne, C. C. and Kelly, D. C. 1980. Insect virus. In Virus Diseases of Insects and Mites, H. D. Burges, ed. Academic Press, London.

Printz, P. 1973. Relationship of sigma virus to vesicular stomatitis virus. Adv. Virus Res., 18, 143-157.

Quiot, J-M., Odier, F., Bergoin, M. and Vago, C. 1973. Etude de la pathogenese et titrage d'un virus de densonucleose en culture de cellules de Lepidopteres. C. R. Acad. Sci. Paris (Ser. D), 276, 2997-2999.

Rivers, C. F. and Longworth, J. F. 1972. A non-occluded virus of Junonia coenia (Nymphalidae: Lepidoptera). J. Invertebr. Pathol., 20, 369-270.

Reinganum, C., O'Loughlin, G. T. and Hogan, T. V. 1970. A non-occluded virus of the field crickets Teleogryllus oceanicus and T. commodus (Orthoptera: Gryllidae). J. Invertebr. Pathol., 20, 170-175.

Scali, V., Montanelli, A., Lafranchi, A. and Bedini, C. 1980. Nuclear alterations in a baculovirus-like infection of midgut epithelial cells in the stick insect, Bacillus rossius. J. Invertebr. Pathol., 35, 109-118.

Smith, K. M. 1967. Insect Virology. Academic Press, London and New York.

Suto, C. 1980. Characterization of a virus newly isolated from the smokey-brown cockroach Periplanata fuliginosa. Nagoya J. Med. Sci., 42, 13-26.

Stoltz, D. B. and Summers, M. D. 1971. Pathway of infection of mosquito iridescent virus. I. Preliminary observations on the fate of ingested virus. J. Virol., 8, 900-909.

Stoltz, D. B. and Vinson, S. B. 1979. Viruses and parasitism in insects. Adv. Virus Res., 24, 125-171.

Teninges, D., Ohanessian, A., Richard-Molard, C. and Contamine, D. 1979. Isolation and properties of Drosophila X virus. J. Gen. Virol., 42, 241-254.

Thomas, D. and Gouranton, J. 1975. Development of virus-like particles in the crystal containing nuclei of midgut cells of Tenebrio molitor. J. Invertebr. Pathol., 25, 159-169.

Tinsley, T. W. and Harrap, K. A. 1978. Viruses of invertebrates. Comprehensive Virology, 12, 1-101.

Tinsley, T. W., Robertson, J. S., Rivers, C. F. and Service, M. W. 1971. An iridescent virus of Aedes cantans in Great Britain. J. Invertebr. Pathol., 18, 427-428.

Vago, C. and Luciani, J. 1965. Développement du virus de la maladie des noyeux denses de Galleria mellonella en culture de tissus de Lepidopteres. Experientia, 21, 106-109.

Wagner, G. W. and Paschke, J. D. 1977. A comparison of the DNA of the "R" and "T" stains of mosquito iridescent virus. Virology, 81, 298-308.

Webb, S. R., Paschke, J. D., Wagner, G. W. and Campbell, W. R. 1974. Infection of Aedes aegypti cells with mosquito iridescent virus. J. Invertebr. Pathol., 23, 255-258.

Webb, S. R., Pashcke, J. D., Wagner, G. W. and Campbell, W. R. 1975. Bioassay of mosquito iridescent virus of Aedes taeniorhyncus in cell cultures of Aedes aegypti. J. Invertebr. Pathol., 26, 205-212.

Xeros, N. 1954. A second virus disease of the leatherjacket, Tipula palidosa. Nature, 174, 562.

Younghusband, H. B. and Lee, P. E. 1969. Virus-cell studies of Tipula iridescent virus in Galleria mellonella (L.) I. Electron microscopy of infection and synthesis of Tipula iridescent virus in hemocytes. Virology, 38, 247-254.

Younghusband, H. B. and Lee, P. E. 1970. Cytochemistry and autoradiography of Tipula iridescent virus in Galleria mellonella. Virology, 40, 757-760.

Yule, B. G. and Lee, P. E. 1973. A cytological and immunological study of Tipula iridescent virus infected Galleria mellonella larval hemocytes. Virology, 51, 409-423.

Chapter 3

CYTOPLASMIC POLYHEDROSIS VIRUSES

by C. C. Payne*

INTRODUCTION

The characteristic signs of infection by a cytoplasmic polyhedrosis virus (CPV) were first described by Ishimori (1934) when he observed inclusion bodies in the cytoplasm of midgut epithelial cells of diseased silkworms. However, it was not until the studies of Smith and Wyckoff (1950) that such inclusions were distinguished from the polyhedra produced following infection with nuclear polyhedrosis viruses. Since these early investigations, considerable progress has been made in studies of the characteristics of this virus group. A CPV was recognized as the causal agent of a debilitative disease of silkworm (Bombyx mori) cultures, and research interest culminated in the publication of a book largely devoted to studies of this virus (Aruga and Tanada, 1971). This serves as a useful review of CPV studies, particularly of the Japanese literature, up to 1970.

In the late 1960's and 1970's considerable advances were made towards an understanding of the biochemical properties and replication of these viruses (Miura et al., 1968; Kalmakoff et al., 1969; Lewandowski et al., 1969; Lewandowski and Traynor, 1972, Furuichi, 1974, 1978; Payne and Rivers, 1976; Mertens, 1979), as well as the numerous sublethal effects of virus infection on insect hosts (Ignoffo and Adams, 1966; Bullock et al., 1970; Magnoler, 1974; Bell and Kanavel, 1976; Bell, 1977; Sikorowski and Thompson, 1979). Recent reviews of biochemical properties and virus morphogenesis include those by Payne and Harrap (1977) and Harrap and Payne

*Glasshouse Crops Research Institute, Worthing Road, Littlehampton, Sussex, BN16 3PU, United Kingdom

(1979). Studies of general CPV pathogenesis have not been reviewed in detail since 1971 (Aruga and Tanada, 1971).

In the following review of pathogenesis, I have included information collected from studies of a large number of CPV isolates in an attempt to produce as complete a picture of the disease process as possible. In this way I hope that the review will define more clearly the common features of CPV pathogenesis. However, CPV's are a complex group of viruses including a range of "types" (Payne and Rivers, 1976). Thus, where differences have been observed in the infection process of some CPV's, it should be emphasized that they may reflect distinctive features of particular host-virus systems, as well as the different interpretations of individual research workers.

THE VIRUSES

CPV's are placed as a genus within the Reoviridae, a family of viruses whose members (including those from plant and vertebrate hosts) contain a segmented, double-stranded RNA genome composed of 10, 11 or 12 distinct segments (Matthews, 1979). CPV's, like members of the genera *Reovirus* and *Fijivirus*, contain 10 segments of RNA, each representing a single gene. Virus particles (virions) are icosahedral, approximately 60 nm in diameter with characteristic "spikes" or projections at the vertices of the particle. Although early studies suggested that the virions were composed of a core surrounded by two concentric protein shells, or capsids (Hosaka and Aizawa, 1964), it is now generally accepted that there is no distinct double shell to the particle (Stoltz, 1969; Lewandowski and Traynor, 1972; Payne and Tinsley, 1974; T. Hatta and R. I. B. Francki, pers. commun.). Virions contain a complex of enzymes including transcriptase, methylase, transferase and nucleotide phosphohydrolase activities which are associated with the synthesis and modification of messenger RNA produced during virus replication (Furuichi, 1974, 1978; Mertens and Payne, 1978; Storer et al., 1974a; Shimotohno and Miura, 1977). With rare exceptions (Stoltz, 1969) virus replication is restricted to the midgut epithelial cells of susceptible hosts producing a frequently chronic rather than lethal disease. In infected cells, many virions are occluded within proteinaceous inclusion bodies or "polyhedra" which are produced in the cytoplasm of infected cells. These polyhedra, and the cytoplasmic site of viral replication provide the basis for the name of this virus group.

Approximately 200 isolates of viruses with the morphological characteristics of CPV's have been observed in insect hosts (Martignoni and Iwai, 1977; C. C. Payne, unpublished observations). Twelve distinct virus "types" have been described among the 20-25% of these isolates which have been studied biochemically (Payne and Rivers, 1976; Payne et al., 1977; Matthews, 1979). The belief that some or all of these viruses may have a wide host range (Smith, 1963; Neilson, 1964) is supported by studies in which reliable methods of virus identification have been used (Payne and Rivers, 1976). The majority of isolates have been detected in Lepidoptera (Martignoni and Iwai, 1977). Other well-documented host orders are Hymenoptera (Longworth and Spilling, 1970) and Diptera (Clark et al., 1969; Stoltz and Hilsenhoff, 1969; Federici et al., 1973; Frederici, 1974; Bailey et al., 1975). Reports of CPV's in Neuroptera and Coleoptera (Smith et al., 1959; Sidor, 1970, 1971) need further investigation. The former may be attributable to an infection by a nuclear polyhedrosis virus; the latter appears unusual in that the reported virus replicated in the fat body of infected beetles rather than in the gut tissue. Without further details the description provided could also fit an entomopoxvirus infection (Sidor, 1971).

Outside the Insecta, distinct CPV-like particles and polyhedra have been observed in the freshwater crustacean Simocephalus expinosus (Federici and Hazard, 1975). Ibrahim et al. (1978) suggested that inclusion bodies and virus-like particles observed in swarming nematodes of Tylenchorhynchus martini might be attributed to a CPV-like infection. However, the dimensions of the particles associated with the nematodes are far smaller than values observed for typical CPV virions. Bird et al. (1972) detected CPV-like particles within vacuolated oocysts of the rodent malaria parasite, Plasmodium berghei yoelii. These intracellular oocysts were found in CPV-infected larval stages of the mosquito vector, Anopheles stephensi and the presence of virus within the parasite is insufficient evidence to believe that it replicated in the protozoan. Attempts to infect vertebrate cells with CPV's have produced no overt infection (Granados, 1979) and it would therefore appear from present evidence that this virus group is restricted to arthropod hosts.

CPV AND HOST: CELLULAR INTERACTIONS OF VIRUS MORPHOGENESIS

Accounts of the replication cycle of CPV's as observed with the electron microscope or by other histopathological techniques were summarized by Iwashita (1971) and Kobayashi (1971). Although little new information is available on morphogenesis since these reviews were written, an improved understanding of the biochemical properties of this virus group throws some doubts on the earlier interpretation of some stages in the replication process. Most investigations have been carried out with infected larval cells.

Virus entry

Under natural conditions it is most likely that virus is acquired by larvae following the ingestion of food contaminated with CPV polyhedra. Virus particles can be extracted *in vitro* from polyhedra by treatment with dilute alkali and, once in this form, will produce infection when artificially injected into the hemocoel (Hills and Smith, 1959; Faust and Cantwell, 1968; Miyajima and Kawase, 1968; Kawase, 1971b). The dissolution of polyhedra under these conditions presumably mimics polyhedral breakdown in the alkaline gut juice of susceptible hosts as observed by Vago (1959) and Vago et al. (1959). After feeding a concentrated suspension of CPV polyhedra to larvae of the pine processionary caterpillar, *Thaumetopoea pityocampa* and the cutworm, *Triphaena* (= *Agrotis*) *pronuba*, polyhedra were observed in the anterior midgut for 5-8 minutes. Polyhedra were rarely observed after this, consistent with their rapid breakdown during passage down the gut. When polyhedra were treated with gut juice *in vitro* (pH 8.5 - 9.5), they were observed to fragment and finally dissolve leaving particles of 40-70 nm diameter, consistent with the accepted size of CPV virions.

The mechanism of virus penetration into susceptible gut epithelial cells is less clear. In studies with primary cell cultures of *B. mori* midgut epithelial cells, Kobayashi (1971) suggested that CPV virus particles attach to the cell plasma membrane and release core material into the cell. Phagocytosis did not appear to play a role in virus penetration. However, from what is known of the biochemical properties of the virus, it is likely that retention of the structural integrity of the particle during entry is important for successful infection. Viral RNA is not infectious by itself (Lewandowski et al., 1969; Kawase, 1971a), presumably as host cells do not contain the enzymes capable of transcribing messenger RNA from the viral double-stranded RNA. However, intact virus particles contain a transcriptase enzyme capable of synthesizing messenger RNA *in vitro* (see below). This enzyme is functional only when the virions remain intact (Lewandowski et al., 1969; Storer et al., 1974b). For this reason, the mechanism of virus entry proposed by Kobayashi (1971) seems unlikely.

Studies with the related vertebrate virus group, *Reovirus*, have shown that virion uptake occurs into phagocytic vesicles which later fuse with lysosomes within the cell (Silverstein and Dales, 1968; Silverstein et al., 1972). Although CPV's and *Reovirus* are not completely analogous, this comparison suggests an alternative mechanism of CPV entry, for which there is some evidence in the literature. CPV virions, once released from polyhedra by the alkaline gut juices, may adsorb to host cells through interaction of the plasma membrane with the viral hemagglutinin (Miyajima and Kawase, 1969). Particles may then be absorbed by phagocytosis. Up to 15 hours after infection, Quiot and Belloncik (1977) observed virions only in cytoplasmic (possibly phagocytic) vacuoles in a gypsy moth, Lymantria dispar, cell line.

With Reovirus, the viral transcriptase is only activated by enzyme action within lysosomes (Silverstein et al., 1972). With CPV virions, however, activation of the transcriptase is completed by the release of virus from polyhedra (Lewandowski et al., 1969; Furuichi, 1974; Mertens, 1979) and messenger RNA synthesis could be initiated without further viral degradation in lysosomes. However, Bird (1965) and Kobayashi (1971) have observed an increase in the number of lysosome-like vacuoles in CPV-infected gut cells of the white-marked tussock moth, Orgyia leucostigma and of B. mori though this increase may not be associated with the early stages of infection.

Production and assembly of virion components

Miyajima and Kawase (1968) examined the titre of virus particles in midgut epithelial cells of B. mori larvae in a study of the time course of CPV infection. There was an observed decrease (eclipse phase) in infective particles up to 3 hours after virus infection presumably following virus entry and messenger RNA synthesis. From 3-24 hours after infection there was an almost linear increase in the number of infective particles. Polyhedra were first detected 15 hours post-infection.

One of the first visible signs of infection is the development of micronet structures within the cytoplasm of infected cells. These areas, commonly referred to as "virogenic stromata" (Xeros, 1956) or viroplasms (Sohi et al., 1971) appear as discrete structural entities, up to 10 μm x 2 μm in size (Fig. 3-1). They contain no cellular components such as mitochondria or endoplasmic reticulum (Xeros, 1956; Arnott et al., 1968; Longworth and Spilling, 1970; Sohi et al., 1971). Individual stromata form initially near the brush border (apical region) of infected columnar epithelial cells whether larvae are infected per os or by injection of virus particles into the hemolymph (Xeros, 1966; Kawase and Miyajima, 1969). The area of stroma extends to the base of the cell later in infection.

Mature stromata, produced by coalescence of several smaller stromata (Kobayashi, 1971) are heavily loaded with RNA and protein. They appear to represent active sites of protein synthesis as measured by autoradiography (Watanabe, 1968a) and by the detection of virus-specific proteins by immunofluorescence (Kawase and Miyajima, 1969). The observation by Xeros (1966) that the early stages of the stromata in CPV-infected Actias luna lacked cytochemically-

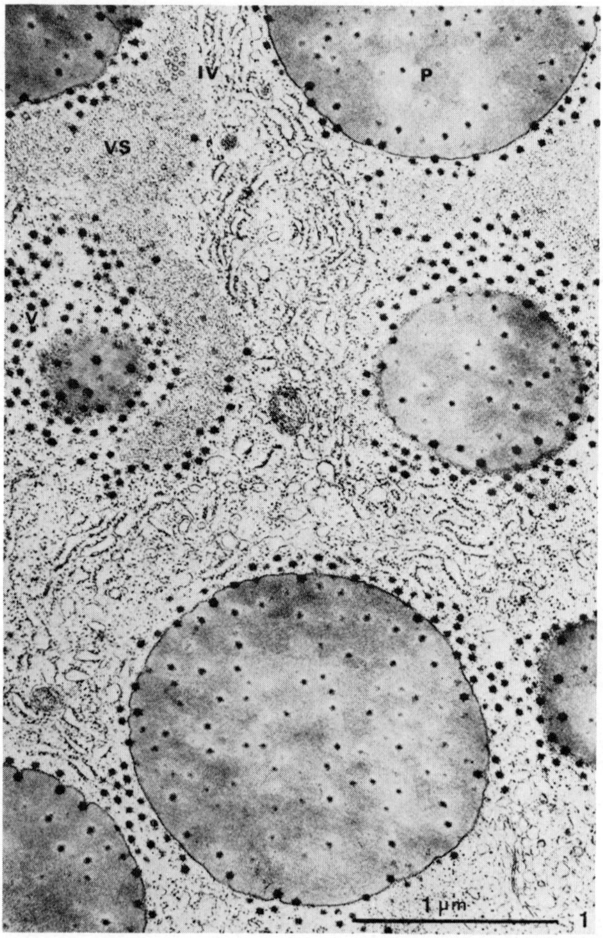

Figure 3-1. Cytoplasmic polyhedrosis in a larva of the monarch butterfly, Danaus plexippus. Larval midgut cell illustrating several polyhedra during a relatively early stage in the disease. Masses of rough endoplasmic reticulum are interposed between the polyhedra (P) and the virogenic stroma (VS). Dense bodies within the polyhedra, and associated with them in the cytoplasm, are complete virus particles (V). A number of incomplete (IV) particles are found associated with the virogenic stroma. At some points along the edge of the polyhedra, virus particles are in the process of being occluded.
From Arnott et al. (1968); reproduced by permission of Academic Press Inc., and Dr. K. M. Smith.

detectable RNA is consistent with several reports which suggest that the viral double-stranded RNA is synthesized in the nucleus (Hayashi and Kawase, 1965a; Watanabe, 1967b; Hayashi and Retnakaran, 1970). As all other aspects of replication appear to be confined to the cytoplasm, it is perhaps surprising that the nucleus may be involved. Most of the studies on the site of RNA synthesis have been indirect methods which do not directly characterize the RNA synthesized (Ralph, 1969) and it is therefore important that further studies should be conducted into the nature of the nuclear involvement in CPV replication.

Kobayashi (1971) provides a detailed account of CPV virus particle morphogenesis within the stromata of infected cells. He proposed that virus "cores" were derived from electron-dense membrane-bound structures found at the surfaces of stromata, or from other dense granular regions also observed in the cytoplasm. The cores are then enclosed within developing capsids. This interpretation is consistent with the studies of Bird (1965) with CPV infection in O. leucostigma. In contrast, Arnott et al. (1968) observed numerous empty capsids in infected cells and suggested that assembly occurred by the filling of these with core material. It is possible that the many empty particles observed in some infections may represent aberrant assembly and the formation of defective virus (Payne and Harrap, 1977). The characteristic "spikes" on CPV particles are not readily visible in thin section and their mode of development is uncertain. Kobayashi (1971) suggests that they are added subsequent to capside formation.

Polyhedra formation

During the later stages of virus replication, complete virus particles become occluded within large proteinaceous inclusion bodies or "polyhedra" (Fig. 3-1). The components of these polyhedra are virions and "polyhedron protein". This polyhedron protein is composed of aggregates of a single polypeptide of approximately 30,000 daltons (Payne and Rivers, 1976). Fluorescent antibody studies have shown that the protein in infected B. mori larvae is synthesized as early as 9 hours after infection at sites near the brush border of midgut epithelial cells (Kawase and Miyajima, 1969).

Polyhedron formation starts at several sites with the crystallization of protein around groups of mature virus particles often at the periphery of the virogenic stromata (Xeros, 1956, 1966; Granados et al., 1974; Quiot and Belloncik, 1977). Arnott et al. (1968) observed an area of increased electron density surrounding many polyhedra in a CPV infection of the monarch butterfly, Danaeus plexippus. This area, referred to as the "crystallogenic matrix" has been observed by others (Watanabe, 1973; Payne and Harrap, 1977) and may represent closely packed fibrils of uncrystallized polyhedron protein.

Polyhedra increase in size by a process of accretion and virions are occluded in a random fashion within the protein lattice (Wittig et al., 1960; Arnott et al., 1968; Kobayashi, 1971). In CPV-infected Lepidoptera and Hymenoptera many virus particles are occluded within a single crystal, which is sometimes as large as 10 µm in diameter and contains perhaps 10,000 virus particles (Arnott et al., 1968; Stairs et al., 1968; Longworth and Spilling, 1970; Kobayashi, 1971). In CPV infections of Diptera, two distinct types of inclusion bodies have been reported (Federici, 1974). Anthony et al. (1973) observed that virus particles were most frequently occluded singly within small cuboidal crystals in infected larvae of the mosquito, A. quadrimaculatus. There was some evidence that crystals later coalesced, the largest crystals containing 9 visible virus particles per thin section. Singly-occluded virus was also observed in Culex salinarius (Clark et al., 1969), and the apparent fusion or coalescence of singly-occluded particles was also reported by Stoltz and Hilsenhoff (1969) in Chironomus plumosus and Federici (1974) in Aedes taeniorhynchus.

The lattice of mature polyhedra appears as a simple cubic packing of protein molecules (Bergold and Suter, 1959) with a lattice spacing of 40-60Å (Arnott et al., 1968; Longworth and Spilling, 1970; Kobayashi, 1971). Occlusion of virus particles may cause slight alterations to the lattice spacing (Arnott et al., 1968) which may have led to the appearance of tail-like structures on virus particles reported by these authors, but not observed by other workers. Virus particles within polyhedra are composed of an electron-dense core surrounded by a capsid, and appear similar in basic structure to "non-occluded" particles. However, some authors have suggested that the diameter of the virus particle is increased by as much as 30% following occlusion (Arnott et al., 1968; Longworth and Spilling, 1970) while others observed that occluded particles are more compact (Anthony et al., 1973). These differences in size may be attributable to the clearer definition of the virus capsid layer in occluded virions (Longworth and Spilling, 1970).

The precise mechanism of polyhedron formation is unknown. As polyhedra do not contain cellular components it is likely that there is some relatively specific recognition (such as exists between enzymes and substrate) between virions and polyhedron protein (Arnott et al., 1968). Two serologically distinct viruses (classified as "types 2 and 3" (Payne and Rivers, 1976)) can replicate in a mixed infection of the tiger moth, Arctia caja, although polyhedron protein characteristic of only one of these viruses (type 3) is produced (Payne, 1976). In polyhedra derived from such an infection, proportionally fewer virions of type 2 CPV were occluded, implying some specificity in the process of polyhedra formation.

In many insects, a significant proportion of the virions may not be occluded and can be observed free in the cytoplasm at an advanced stage of infection. These can be readily isolated from the infected tissue. Hayashi (1970a) calculated that as few as 30% of the total virions were occluded in CPV infection of the forest tent caterpillar, Malacosoma disstria compared with 40-45% in B. mori CPV infection (Hayashi et al., 1970). Bird (1966) suggested that "non-occluded" particles were largely incomplete and coreless in the spruce budworm, Choristoneura fumiferana CPV. In contrast, Lipa (1977) observed that non-occluded virions of T. pronuba CPV appeared to be morphologically the same as occluded virus. A comparison of some of the biochemical properties of occluded and non-occluded virions of B. mori CPV did not reveal any fundamental differences between the two types (Payne and Kalmakoff, 1974). It is therefore unlikely that the failure of so many particles to be occluded is the result of major structural differences. Whereas polyhedra are the vehicle for the transmission of infection between insects, the occurrence of numbers of non-occluded virions would ensure that virus infection could spread within infected individuals.

The mechanisms controlling polyhedral size and shape are unknown. Polyhedra may range in size in different infections from less than 1 μm (Longworth and Spilling, 1970) to 10 μm (Stairs et al., 1968), more commonly 1-3 μm in diameter. They are not surrounded by a distinct membrane although their perimeter may appear somewhat more electron opaque (Arnott et al., 1968). The absence of any limiting membrane can be seen in scanning electron micrographs of purified polyhedra where the sockets remaining from the loss of virus particles at the surface are clearly visible (Fig. 3-2). Polyhedral shape (Fig. 3-2) also varies considerably although it may be largely characteristic of a particular virus strain. Hukuhara and Hashimoto (1966) reported three variants of B. mori CPV producing icosahedral, hexahedral or cubic, and triangular shaped polyhedra. Additional shapes have been identified by scanning electron microscope studies (Rao, 1973).

At the final stages of infection, many polyhedra may be found in a single cell. Arnott et al. (1968) observed up to 20 polyhedra in one section of a CPV-infected cell in D. plexippus larvae. In CPV-infected larch sawfly, Anoplonyx destructor, up to 100 smaller polyhedra may be present (Longworth and Spilling, 1970), such that the entire cytoplasm appears filled. By this stage the virogenic stromata have regressed, and cell lysis may occur as the plasma membrane disintegrates.

In most studies the distinct morphogenetic events in CPV replication have been restricted to the cytoplasm with little

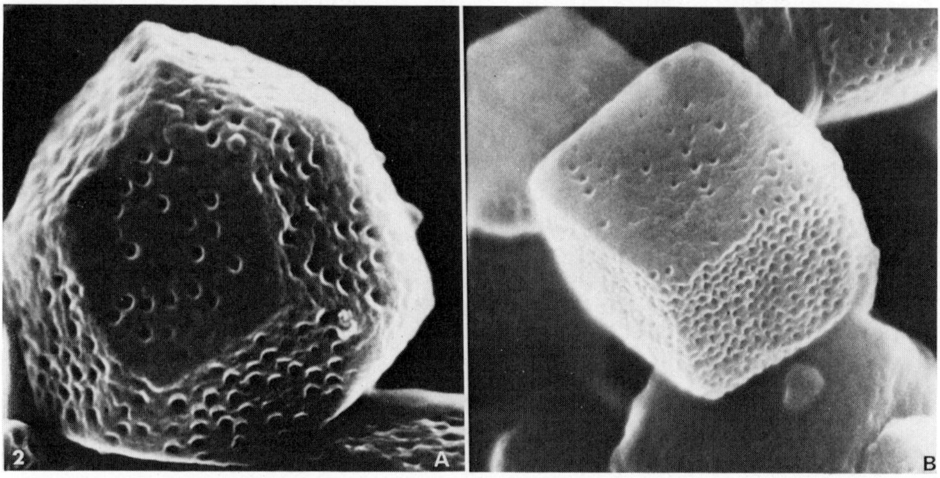

Figure 3-2. Scanning electron micrographs of purified CPV polyhedra. A. Polyhedron of icosahedral shape of the peacock butterfly, *Inachis io* ("type" 2 CPV; Mag. x 16,320. B. Polyhedron of cubic shape from the silkworm, *Bombyx mori* ("type" 1 CPV; Mag. x 12,240. The sockets at the surface of the polyhedra represent sites from which virus particles have been lost during purification. (Photographs supplied by Margaret Arnold).

apparent alteration in nuclear structure. However, Hukuhara and Yamaguchi (1973) and Kawase et al. (1973) observed CPV strains infecting B. mori which produced cubic inclusions within nuclei which superficially resembled polyhedra although they did not contain virus particles. Polyhedral formation in the cytoplasm was infrequent or absent.

Effect of infection on tissue and cell structure

The changes in cell and tissue structure induced by CPV infection have been reviewed extensively by Iwashita (1971) and Kobayashi (1971).

Virus infection is generally restricted in larvae to gut epithelial cells, the only recorded exceptions being reported from two CPV infections of Diptera. However Stoltz (1969) provided evidence for CPV replication in the fat body of a larva of C. plumosus. Kellen et al. (1966) also described a "CPV" in leg and wing buds of Cx. tarsalis. However, more recent investigations have shown that the inclusions described were paracrystalline arrays of another virus and not CPV polyhedra (Federici, 1974). Cell cultures, including both primary cultures and cell lines derived from ovarian tissues, are susceptible to infection (Grace, 1962; Vago and Bergoin, 1963; Sohi et al., 1971; Granados et al., 1974; Granados, 1976; Quiot and Belloncik, 1977). Although most descriptions of virus replication have been made from studies of larval stages, polyhedra have also been observed in the gut epithelia of pupae and adults (Tanada and Chang, 1960; Atger, 1962; Bird et al., 1972; Magnoler, 1974). Replication in embryonic gut cells has also been observed following virus particle injection (Iwashita and Seki, 1973).

Within the gut epithelium, the columnar epithelial cells of the midgut appear to be the primary sites of virus replication (Iwashita, 1971). Martignoni et al. (1969) have reported CPV replication in goblet cells of the Douglasfir tussock moth, Hemerocampa (= Orgyia) pseudotsugata, but there are no reports of virus replication in the regenerative cells situated at the basal region of the midgut epithelium (Iwashita, 1971). Virus infection may proceed from one discrete area of the midgut to another. Thus in CPV infected B. mori larvae, infection proceeds from the posterior midgut to the anterior region. In contrast, Tamashiro and Huang (1963) reported that a CPV infection in Cactoblastis cactorum occurred initially in the anterior portion of the midgut before proceeding to the posterior region. Both Bailey et al. (1975) and Weiser (1978) observed that the entire midgut of larval black fly species (Diptera: Simuliidae) is rarely infected, polyhedra being found most commonly in the anterior and posterior thirds of the midgut.

Apart from the formation of virogenic stromata and polyhedra, the most frequently reported intracellular changes include an increase in the number of lysosomes and/or pinocytotic vesicles (Arnott et al., 1968; Kobayashi, 1971). Whether the lysosomes serve in the resistance of the cell against virus infection or in the development of the disease, is not clear (Kobayashi, 1971). Another pronounced effect as disease progresses is the degradation and vesiculation of the rough endoplasmic reticulum (Xeros, 1956; Arnott et al., 1968; Kobayashi, 1971) together with a vacuolation of the cytoplasm as large pores develop in the stromata during polyhedra formation (Grace, 1962; Arnott et al., 1968). Mitochondria may also be enlarged at an advanced infection stage (Anthony et al., 1973) a feature in common with mitochondrial alterations observed in starved larvae (Kobayashi, 1971).

In general, few pathological changes have been observed in the nuclei of infected cells (Ishimori, 1934; Longworth and Spilling, 1970; Akutsu, 1971) other than in infections with B. mori CPV strains which produce inclusions in the nucleus (Hukuhara and Yamaguchi, 1973). Some changes in the distribution of chromatin have been seen (Steinhaus and Dineen, 1959; Stairs et al., 1968; Arnott et al., 1968) and signs of nuclear hypertrophy may be present late in infection (Vago and Bergoin, 1963; Harpaz et al., 1965).

At the final stages of infection, cellular hypertrophy is common both in larval cells (Xeros, 1966; Harpaz et al., 1965; Boucias and Nordin, 1978) and in infected cell cultures (Grace, 1962; Granados et al., 1974). In a study of the surface structure of CPV-infected midguts of the fall webworm, Hyphantria cunea, Boucias and Nordin (1978) observed that microvilli were completely or partially absent from infected cells while the cells were bulging with polyhedra at the microvillar surface. Undoubtedly these changes would reduce the adsorptive capacity of the infected midgut and explains why CPV-infected larvae show many of the symptoms of starved larvae (see below). Apical cell membranes eventually break down and polyhedra are released into the gut lumen (Kobayashi, 1971; Boucias and Nordin, 1978). The polyhedra are released in large numbers in the feces of diseased larvae (Ishimori, 1934; Smith, 1963; Watanabe, 1968b; Boucias and Nordin, 1978). In contrast to the cell lysis observed in infected larvae, no lysis of infected cells have been described in CPV infection of cell cultures (Granados et al., 1974; Quiot and Belloncik, 1977).

MOLECULAR INTERACTION OF CPV AND HOST: BIOCHEMISTRY OF REPLICATION

An understanding of the biochemical events associated with the replication of many viruses has come from studies of synchronously-infected cell cultures. However, the few studies of CPV-infected cell cultures have not included biochemical analyses. Thus the only direct information on the biochemistry of CPV replication comes from two sources: i) *in vivo* investigations on RNA and protein synthesis in the midguts of CPV-infected larvae (usually *B. mori*) and ii) *in vitro* studies of the enzyme activities, associated with CPV virus particles, which synthesize and modify single-stranded RNA. In addition, although CPV and *Reovirus* do not replicate in an identical manner, it is useful to compare what is known about CPV replication with the more extensive information available for *Reovirus*.

Messenger RNA synthesis

In a review of the replication of reoviruses, Joklik (1974) stated that the double-stranded (ds) genome RNA of Reoviridae cannot serve as messenger RNA and is, therefore, unable to initiate virus specific protein synthesis. Single stranded (ss) messenger RNA is provided by the action of a transcriptase enzyme which is a component of the virus capsid.

Lewandowski et al. (1969) were the first to report the presence of a transcriptase in *B. mori* CPV virions released from polyhedra. The enzyme activity was not inhibited by actinomycin D, an antibiotic which inhibits ssRNA transcription from DNA. Unlike *Reovirus* transcriptase, the expression of the enzyme activity of CPV virions did not require prior modification of the virus capsid by proteolytic enzyme degradation or heat shock (Joklik, 1974). Furuichi (1974) and Shimotohno and Miura (1974) confirmed these reports and also observed that the ssRNA products were complete transcripts of each of the 10 viral genome dsRNA segments. Thus the products re-annealed specifically to denatured genome RNA and the resulting double-stranded molecules had the same electrophoretic mobilities in polyacrylamide gels as the genome dsRNA (Furuichi, 1974). The ssRNA transcripts, however, did not self-anneal, indicating that only one strand of the genome dsRNA was transcribed (Lewandowski et al., 1969; Furuichi, 1974).

In vivo studies of viral RNA synthesis in CPV-infected insects have revealed evidence of similar ssRNAs produced during infection. Hayashi (1970b) and Hayashi and Donaghue (1971) observed that two size classes of ssRNA were synthesized in virus-infected, but not in healthy, larvae of *O. leucostima* and *M. disstria*, when host RNA synthesis was substantially reduced by the prior injection of actinomycin-D. These ssRNAs had sedimentation values (S) of 15 S and 22 S in sucrose density gradients. Under these conditions, the genome dsRNA segments of these viruses also separated into only 2 size classes and it seems likely that these ssRNAs represented messenger RNA transcripts of the viral genome. Furusawa and Kawase (1971, 1973) obtained similar results in CPV-infected *B. mori* larvae. Both Payne and Kalmakoff (1973) and Furusawa and Kawase (1973) observed other properties of the ssRNA produced *in vivo* which were comparable to those of the *in vitro* transcriptase products. Thus a large proportion of the ssRNA synthesized *in vivo* reannealed specifically to denatured viral dsRNA, and did not self-anneal. These similarities between ssRNAs synthesized *in vivo* and *in vitro* indicate that *in vitro* studies of RNA synthesis provide valuable information on the probable mechanisms of viral replication.

Although there are no more recent published studies on ssRNA synthesis *in vivo* than those described above, much more information has been obtained on the mechanisms of ssRNA synthesis from *in vitro* work. It is known that the transcriptase is only active in the intact virion (Lewandowski et al., 1969; Storer et al., 1974b) and it is therefore likely that little modification of the infecting virion occurs during entry into a susceptible cell. Virions are also capable of synthesizing several RNA molecules simultaneously (Furuichi, 1974; Shimotohno and Miura, 1977), although more of the smaller viral RNA segments are transcribed than the larger ones (Mertens, 1979). This difference in transcription could lead to some control in the rates of synthesis of different virus-specific polypeptides. Furuichi (1974) also observed that the synthesis of the transcriptase products of *B. mori* CPV was markedly stimulated by the addition of a methyl group donor-S-adenosylmethionine (SAM) - to the assay system. This led to the fundamental discovery (Furuichi and Miura, 1975) that the viral messenger RNA molecules contained a novel methylated nucleotide sequence, or "capping" group at their 5' terminus:

5' 3'

^{7m}Gppp Amp Gp ....

(G and A = Guanylic and Adenylic acid residues respectively; m = methyl group; p = phosphate residue)

in which a methylated guanylic acid residue is linked by a pyrophosphate bond to methylated adenylic acid. The formation of this capping group requires the activity of several other enzymes including nucleotide phosphohydrolase (NTPase), guanosyl transferase and methylases, all of whose activities have been found in CPV virus particles (Storer et al., 1974a; Shimotohno and Miura, 1977; Furuichi, 1974, 1978). It appears that "cap" formation is a pre-requisite of CPV messenger RNA synthesis as inhibition of the NTPase activity (which would prevent cap formation) halts ssRNA transcription (Furuichi, 1978).

Similar cap structures have been observed at the 5' termini of many virus-specified and cellular RNA molecules. Although the precise biological significance of this structure is still unclear, it appears to increase the resistance to degradation by ribonuclease of CPV messenger RNAs and may, therefore, prolong their life in infected cells (Furuichi and Miura, 1975; Shimotohno et al., 1977). In addition, CPV messenger RNA synthesized in the absence of a methyl group donor (Mertens, 1979) or after the chemical removal of the ^{7}mG residue (Shimotohno et al., 1977) is not as efficient in initiating protein synthesis in a cell-free wheat germ protein synthesizing system.

These studies are important in demonstrating that CPV virions contain the enzymatic machinery necessary for the initiation of infection, and are not dependent on cellular enzymes for messenger RNA production. Nonetheless, the nucleotide triphosphates and methyl group donors required for RNA synthesis must be provided by the cell. It could be envisaged that the cellular location and availability of these compounds could control the site and rate of viral RNA synthesis.

A comparison of ssRNA synthesis by two distinct CPV "types" ("type 1" from *B. mori* and "type 2" from the peacock butterfly, *Inachis io*) has shown differences in the response of the 2 viral transcriptases to certain stimuli, particularly SAM. This indicates that some CPV's may differ in the details of replication although the fundamental principles are unlikely to be different (Mertens, 1979).

Protein synthesis

Little information is available on virus-specific protein synthesis in CPV-infected cells in the insect host. Watanabe (1968a) used autoradiographic techniques to examine the incorporation of labelled amino acids in gut epithelial cells. Sites of incorporation, suggesting active protein synthesis, were in the apical regions of columinar epithelial cells, just below the brush border and coincident with the virogenic stromata. Using the fluorescent antibody technique, Kawase and Miyajima (1969) detected synthesis of virion structural proteins 6 hours after infection. In contrast, polyhedron protein was detected only weakly, 9 hours after infection, suggesting that its synthesis is a late function in replication. Synthesis of polyhedron protein 2-3 days after infection was also observed by Watanabe (1968c, 1970).

In some recent *in vitro* studies, M. McCrae and P. P. C. Mertens (unpublished observations) and Mertens (1979) have investigated the ability of *B. mori* CPV virus-specific ssRNA to synthesize virus proteins. By analogy with studies of *Reovirus* (Joklik, 1974; McCrae and Joklik, 1978), it was likely that each of the 10 CPV dsRNA genome segments represented a single gene and contained the genetic information for one primary protein product. One of the strands of each of the genome dsRNA segments has the same "cap" structure and nucleotide sequence as the messenger RNA produced by the viral transcriptase and should, therefore, be able to initiate protein synthesis (McCrae and Joklik, 1978). When individual segments of the genome RNA of *B. mori* CPV (denatured to a single-stranded form) were added to a reticulocyte cell-free protein synthesizing system, the proteins synthesized were examined by electrophoresis in polyacrylamide gels. The results showed, as expected, that some of the viral RNA segments coded for the structural virus polypeptides. Polyhedron protein was coded for by the smallest of the 10 RNA segments, confirming that this protein is virus specified rather than a host reponse to virus infection (Mertens, 1979). No information is available on the role of several non-structural proteins produced by the viral RNA *in vitro*. Presumably these are involved, perhaps together with cellular enzymes, in the processing and assembly of virus proteins and RNA. An extension of these investigations should make it possible to study the virus-coded proteins of each CPV RNA segment in detail. *In vivo* studies in infected cells are now essential to add to the information gained from *in vitro* observations.

Genome RNA synthesis

The precise mechanism of synthesis of genome (ds) RNA in CPV infections is unknown. However, it seems likely that the process may be similar to that described for *Reovirus*, where messenger RNA molecules serve as the template for the synthesis of a complementary RNA strand to produce a dsRNA (Joklik, 1974). The synthesis of these complementary strands in *Reovirus*-infected cells is associated with particles whose protein composition is similar to that of mature virus. The mechanism by which the 10 distinct genome segments are assembled into each virion is unclear in both *Reovirus* and CPV infections.

In CPV-infected larvae, several authors have observed a build up of dsRNA during infection, representing the synthesis of genome RNA. By 144 hours after infection, dsRNA (undetectable in healthy insects) amounted to 12% of the total RNA in midguts of CPV-infected B. mori (Hayashi and Kawase, 1965b). Active synthesis of dsRNA has been detected by labelled uridine incorporation in both the presence and absence of actinomycin D (Kawase and Kawamori, 1968; Hayashi, 1970b; Hayashi and Donaghue, 1971; Furasawa and Kawase, 1971, 1973; Payne, 1972), reaching a peak 72-96 hours after infection. Two size classes of dsRNA were synthesized (15 S and 12 S in sucrose density gradients) the same as the size classes of genome RNA extracted from purified virus particles (Hayashi, 1970b; Hayashi and Donaghue, 1971; Furusawa and Kawase, 1973).

As mentioned above, several authors have suggested that viral RNA synthesis occurs in the nuclei of infected cells. Hayashi and Kawase (1965a) demonstrated that viral dsRNA comprise 16%, 8.1%, and 5.2%, respectively, of the total RNA in nuclear, mitochondrial and microsomal fractions of CPV-infected silkworm midgut epithelial cells. Watanabe (1967b) and Hayashi and Retnakaran (1970) studied RNA synthesis in CPV-infected gut cells by autoradiographic techniques and observed high incorporation of labelled uridine in cell nuclei. In contrast, Morris (1968) obtained results supporting the view that the cytoplasm was the primary site of viral RNA synthesis.

The process of assembly of viral dsRNA and polypeptides in unknown.

Biochemical changes in infected cells and tissues

Several authors have observed an increase in the proportion of free nucleotides in CPV-infected midguts (Hayashi and Kawase, 1965b; Kawase and Kawamori, 1968; Furusawa and Kawase, 1971; Kawase and Furusawa, 1971; Payne, 1971). Such an increase may be partly responsible for the increased uptake of uridine observed in the nuclei of infected cells. Two to five days after infection, the total RNA content of infected midguts of silkworms was slightly increased, but not after that stage (Kawase and Hayashi, 1965). Ribosomal RNA content decreased from 70% of total RNA to 50-60%, late in infection (Kawase and Kawamori, 1968). DNA synthesis appears to be unaffected by infection (Watanabe, 1967b; Morris, 1968).

Watanabe (1968c) failed to observe any qualitative differences in the electrophoretic patterns on polyacrylamide gels of alkali-soluble proteins from healthy or CPV infected silkworms, at an early stage of infection. Polyhedron protein was detected 2-3 days post infection, but the resolution of the technique was insufficient to detect any other virus-specified proteins. Although there appeared to be enhanced synthesis of midgut proteins, some drop in the soluble protein content was observed in the posterior midgut 6-7 days after infection, probably accounted for by an increased protein turnover in the infected tissue (Watanabe, 1970). Five days after infection the amino acid content of the midgut was considerably reduced and more ammonia was detected, perhaps as a consequence of deamination (Kawase, 1965). This could be linked to the observed increases in uric acid content of midgut and feces. At a late stage of infection, uric acid content of infected silkworms midguts was up to five times higher than that of uninfected tissue (Watanabe, 1971a; 1972).

Vago et al. (1959) observed a reduction in the gut pH of CPV infected *T. pronuba* larvae of 1-1.5 pH units. This lowering of the gut pH at an advanced stage of infection may be responsible for the appearance of large numbers of polyhedra in feces. Without a reduction in gut pH, these would otherwise be dissolved in the gut juice once they were released from the infected cells.

THE RESPONSE OF INSECT HOSTS TO CPV INFECTION

Changes in the structure and biochemistry of gut tissue infected with CPV's have been described in previous sections. However, the effects of virus infection are not restricted to this tissue, as concurrent changes in host physiology may lead to altered behaviour, changes in the development of larval, pupal and adult stages, as well as mortality. In the following sections I have considered aspects of the susceptibility of insects to CPV infections and the responses of the different life stages to the disease.

Virulence of disease

Several authors have described that CPV infections produce a largely chronic disease without extensive mortality of larvae; many reaching the adult stage even though heavily diseased (Bird and Whalen, 1954; Tanada and Chang, 1968; Stairs et al., 1968; Bird, 1969; Magnoler, 1974; Bell and Kanavel, 1976; Bailey, 1977). The relationship between infection and mortality which confirmed the chronic nature of some CPV infections was clearly demonstrated by Bell and Kanavel (1976) with pink bollworm, *Pectinophora gossypiella*, CPV. Using neonate larvae they observed that the medium infective concentration (IC_{50}) of polyhedra was 1.9×10^2 polyhedra per ml diet. In contrast, the median lethal concentration (LC_{50}) was 1.7×10^5, approximately 1000 times greater than the IC_{50}.

Median lethal virus doses (LD_{50}s) or LC_{50}s have been estimated by bioassay for several CPV's. (Ignoffo and Adams, 1966; Vail et al., 1969; Martignoni et al., 1969). Considerable differences in host susceptibility exist between different species. Thus the LC_{50} for cabbage looper, Trichoplusia ni, CPV in first instar larvae of T. ni is 1.5 polyhedra mm^2 diet (Vail et al., 1969) compared with 2500 polyhedra mm^2 for P. gossypiella CPV in neonate larvae of its host (Ignoffo and Adams, 1966)). The former virus would appear to have the same order of virulence as H. pseudotsugata CPV (LD_{50} = 8 polyhedra; Martignoni et al., 1969). Magnoler (1970) observed that L. dispar CPV was also highly virulent, producing 97% mortality after feeding 500 polyhedra to individual third instar larvae.

Some CPV infections can be more virulent per polyhedron than some nuclear polyhedrosis virus (NPV) infections but larval mortality generally proceeds much more slowly with the CPV disease (Bird, 1969). Slizynski and Lipa (1973) calculated a median lethal time (LT_{50}) of as much as 31 days for fourth instar larvae of Dendrolimus pini fed 4 x 10^5 CPV polyhedra. Increasing the virus dose can reduce the duration of the LT_{50} (Martignoni et al., 1969).

Pupae and adults produced from infected individuals retain polyhedra in the gut epithelium, though whether the disease progresses during these stages is unclear. Although Bullock et al. (1970) observed significant levels of pupal mortality in CPV-infected P. gossypiella, as many moths of the tobacco budworm, Heliothis virescens were obtained from virus-infected pupae as from healthy pupae (Sikorowski and Thompson, 1979). The latter authors suggested that the low mortality during pupation may be explained by the fact that the midgut is not important in pupal physiology.

Reductions in the longevity of infected adults have also been reported. Severely-infected Nymphalis antiopa adults never survived for more than one month, compared with 7-10 months for healthy individuals (Neilson, 1965). The longevity of P. gossypiella adults was reduced from 27 to 14 days by infection during the larval stage (Bullock et al., 1970). However, Vail and Gough (1970) and Sikorowski and Thompson (1979) observed that diseased and healthy adults of T. ni and H. virescens survived for similar periods, although an earlier report had detected some difference in the survival of H. virescens adults (Simmons and Sikorowski, 1973).

Behavioral responses

The most frequently-reported behavioral response is the loss of appetite of CPV-infected larvae. Arnott et al. (1968) reported that D. plexippus larvae ceased feeding as early as 48 hours after infection and died after 10 days. As a consequence of this reduced feeding, larval body size is also reduced (Tamashiro and Huang, 1963; Ignoffo and Adams, 1966; Asayama and Osaki, 1970; Akutsu 1971; 1975). The larvae generally become inactive (Atger, 1962; Arnott et al., 1968; Longworth and Spilling, 1970; Watanabe, 1973), although CPV-infected larvae of Cactoblastis cactorum showed an increased tendency to wander (Tamashiro and Huang, 1963). In contrast, Weiser (1978) did not observe any visible change in feeding, mortality and body size in diseased Simulium spp.

At an advanced stage of infection, diarrhea is common (Asayama and Osaki, 1970; Akutsu, 1971, 1975), the feces being highly contaminated with polyhedra. In some insects the hindgut may be everted (Smith, 1963).

Development

The two most clearcut effects of CPV infection are a prolongation of larval development, and reduction in body size of all stages. Bullock et al. (1970) noted that 10 day old larvae of P. gossypiella, neonatally infected with cotton bollworm, Heliothis zea, CPV were less than one third the weight of healthy larvae. In a separate study, healthy dispausing larvae of the same species were found to be twice the weight of CPV infected individuals (Bell, 1977). Similar weight reductions were observed in CPV-infected larvae of H. virescens (Simmons and Sikorowski, 1973). When the infecting virus dose is increased, larval weights are progressively decreased. Thus, 10 day old larvae of P. gossypiella were 72% smaller when reared on diet containing 10^5 polyhedra/ml than on diet containing 10^3 polyhedra/ml (Bell and Kanavel, 1976).

As a consequence of reductions in larval size, pupae produced from infected individuals are also frequently smaller than healthy controls (Neilson, 1965; Ignoffo and Adams, 1966; Bird, 1969; Vail et al., 1969; Vail and Gough, 1970; Bullock et al., 1970; Simmons and Sikorowski, 1973; Magnoler, 1974; Bell and Kanavel, 1976) and like diseased larvae contain less lipid and protein (Bell and Kanavel, 1977). Magnoler (1974) observed that healthy pupae were 2-3 times heavier than those from virus-infected individuals of L. dispar. Average pupal weights of infected and healthy T. ni did not vary greatly although the weight range in infected populations was much greater, some being almost 3 times lighter than healthy pupae. These lightest pupae invariably produced deformed moths (Vail et al., 1969). Ignoffo and Adams (1966) suggested that the administration of increasing virus doses to larvae produced a greater decrease in pupal size although no clear relationships of this type were observed by Magnoler (1974) or by Bell and Kanavel (1976). Diseased adults are also generally smaller than uninfected individuals (Fig. 3-3).

Figure 3-3. The effect of CPV infection on adult stages of the gypsy moth, Lymantria dispar. A. Healthy (upper) and diseased (lower) males, showing size difference and normal and reduced pigmentation. B. Dissected CPV-infected female showing melanotic abdominal inclusions (From Magnoler (1974), reproduced by permission of Academic Press Inc. and Dr. A. Magnoler).

Among chronically infected larvae, the time taken to pupation may be extended considerably. The earlier the stage at which larvae are infected the greater is the extension of development. Bird (1969) found that larvae of C. fumiferana infected at the second instar took 8 days longer to pupate than healthy controls, while larvae infected as third instar took only 5 days longer. No increase in development time was noted when larvae were infected as fourth instar, although Vail et al. (1969) found that even when infection was delayed to this stage in the armyworm, Pseudaletia unipuncta, the time to pupation was significantly increased.

Both Ignoffo and Adams (1966) and Bell and Kanavel (1976) observed that increasing the infecting virus dose also increased the duration of the larval stage in P. gossypiella. In an extreme case, diseased newly-hatched larvae were observed to remain at first instar larvae for periods exceeding one generation, as moulting was delayed. Alternatively, moults occurred without body size increase (Ignoffo and Adams, 1966).

Kurata (1971) observed that the increased time to pupation in infected B. mori larvae was associated with an increased number of larval moults, particularly when larvae were infected as early instars. Magnoler (1974), who also noticed additional instars in CPV-infected Lymantria dispar, suggested that this was generated by hormonal imbalance. Once again, these signs of CPV infection are mirrored by the symptoms of starvation in healthy larvae. Thus starved uninfected larvae of L. dispar also show delays in larval development and additional instars (Magnoler, 1974).

In CPV-infected L. dispar pupae, the pupal period was reduced, a feature which was apparently characteristic of larvae which had developed via additional instars (Magnoler, 1974). No significant differences were observed in the lengths of the pupal period of CPV-infected H. virescens and P. gossypiella (Simmons and Sikorowski, 1973; Bell and Kanavel, 1976).

The profound sublethal effects of CPV infection are particularly noticeable in the adult stage. There is a high proportion of deformed adults among infected individuals, such that up to 90% of infected T. ni adults were malformed (Vail and Gough, 1970). Deformations commonly include a reduction in wing size and adults may be unable to emerge satisfactorily or are rendered flightless (Atgar, 1962; Neilson, 1965; Magnoler, 1974). This may have important consequences in affecting their ability to mate (Atgar, 1962; Neilson, 1965).

Several workers have observed reduced fecundity of CPV diseased females. An average of only 655 eggs were laid per diseased female H. virescens compared with 1693 by healthy moths (Simmons and Sikorowski, 1973). Such reductions are consistent with the decrease in oocyte development observed in some CPV-infected species (Neilson, 1965). Egg laying capacity appears to be linked to the size of individuals, as CPV treated females of T. ni emerging from small pupae laid 50% fewer eggs than did females of normal weight from either virus-treated or control groups (Vail and Gouch, 1970). The viability of eggs laid by CPV-infected females does not appear to be lowered (Bullock et al., 1970; Vail and Gouch, 1970). The likely significance of these effects of CPV infection are emphasized by Neilson (1965) who commented that, in four lepidopterous species examined, virus infection reduced the reproductive capacity of the hosts to almost nil.

Apart from direct effects on infected gut tissue (which are described in earlier sections), CPV infection may also influence the physiology of other tissues. Ignoffo and Adams (1966) and Longworth and Spilling (1970) noted that the fat body in infected larvae was reduced in size. In P. gossypiella and H. virescens, the lipid contents of larvae, pupae and adults are reduced (Bell, 1977; Bell and Kanavel, 1977; Sikorowski and Thompson, 1979). It is likely that this reduction may lead to decreased egg production in infected adults. In addition, the extensive breakdown of fat reserves during infection may also contribute to the profound increase in oxygen uptake observed in diseased larvae (47.1 μl O_2/mg body wt/h) compared with healthy individuals (7.6 μl O_2), (Wiygul and Sikorowski, 1978).

Although the hemolymph proteins of CPV-infected B. mori appeared unaffected by infection (Watanabe and Kobara, 1971), the amount of free amino acids in the hemolymph decreased markedly to between one tenth to one half of the quantity found in normal larvae. Several authors have also noted reduced larval and adult pigmentation (Tamashiro and Huang, 1963; Ignoffo and Adams, 1966; Magnoler 1974). Reduced pigmentation is due to the prevention of melanization, perhaps as a result of a tryosine deficiency (induced as a consequence of the reduced amino acid composition in diseased individuals) or a hormonal imbalance leading to reduced tyrosinase activity (Magnoler, 1974).

RESISTANCE TO INFECTION

Resistance to viral and other infections in insects is discussed in detail in Chapter 18. Resistance to CPV is briefly reviewed here.

Different populations or "strains" of one insect species may differ in their susceptibility to disease. The selection of silkworm strains for possible CPV resistance is an important commercial aspect of sericulture. Studies of B. mori larvae have demonstrated a hundred fold difference between the IC_{50}s of Okusa and Daizo strains and about a 1000 fold difference between the most resistant and most susceptible strains tested

(Watanabe, 1966a).

Older larvae are often less susceptible to infection than younger instars. Aruga and Watanabe (1964) found that the susceptibility of Chinese races of *B. mori* decreased with larval age from the first to third instar. In larvae of *C. fumiferana* fed on virus-contaminated diet, there was a progressive increase in the time taken to achieve 50% infection with second, third and fourth instar larvae, suggesting that resistance to infection increased with age (Bird, 1969). Susceptibility may also vary within an instar (Aruga, 1971).

The factors which influence the susceptibility of larvae, and control the insects' resistance to disease, are not clearly defined. Watanabe (1971b) and Aruga (1971) have summarized the available knowledge of the genetic basis of disease resistance in the silkworm. Resistance in one of the more resistant inbred-silkworm strains, Daizo, appears to be transmitted as a dominant major gene (Watanabe, 1965), although resistance in other strains may be generally polygenic. Aruga and Watanabe (1964) suggested that resistance of first instar larvae of F1 hybrids, derived from crosses of resistant and susceptible strains, was markedly similar to that of the maternal strain. However, this relationship is broken down in later instars and its significance is unclear.

Resistance to infection may be developed through a program of selection. When a selection pressure of greater than 60% virus mortality was imposed on strains of *B. mori*, the average median infective dose (ID_{50}) was increased by about one log_{10} over 5 generations (Watanabe, 1967a). The slopes of the dosage-infection responses in the selected resistant strains also increased, suggesting that the selected populations were more homogeneous in their response to virus than unselected strains.

Watanabe (1966b) concluded that host resistance to peroral infection by CPV depends much on the inhibitory mechanism in the gut lumen against invasion of the virus into the midgut cells. Whether there is any defensive mechanism to the virus in the gut lumen other than the possible antiviral activity of the gut juice is unknown, although Watanabe (1971b) speculated that different characteristics of the cell membrane receptors and peritrophic membrane may be responsible for interstrain differences in resistance. Recent studies of larval gut pH and virus inactivation with granulosis and nuclear polyhedrosis viruses (Ripa, 1978; Stiles and Paschke, 1980) have suggested that increase in gut pH which coincides with increasing larval age, may be significant in reducing the susceptibility of older larvae to infection. It is conceivable that such effects might also be important in modifying susceptibility to CPV infections.

Resistance to infection may also be induced by the application of a number of external treatments to larvae. In particular, heat effects, chemical treatment and interference by other viruses have been noted by several authors. Larvae fed with virus and reared at temperatures between 35° and 37° did not develop CPV infections (Tanada and Chang, 1968; Aruga, 1971; Bullock, 1972) although, up to these temperatures, disease incidence increased with increasing temperature (Aruga, 1971). When larvae of the alfalfa caterpillar, *Colias eurytheme* were reared (following exposure to virus) at 35° after 24 hours at 23°, most larvae developed uninfected, although larvae became infected when the 23° preincubation period was increased to 72 hours (Tanada and Chang, 1968). That early stages of virus development are restricted by high temperatures was confirmed by Kobayashi and Kawase (1980) who found that both polyhedra and viral dsRNA were undetectable in CPV-infected *B. mori* reared at 35°.

Although some chemical treatments, including $Ca(OH)_2$ reduce disease incidence in *B. mori* larvae (Aruga, 1971), other compounds including insecticides (Aratake et al., 1973) and dietary ingredients (Kunimi and Aruga 1974) may increase larval susceptibility. The mechanisms of such interactions are unknown.

INTERFERENCE

Interference by viruses, leading to depressed yields of a subsequently administered CPV, has also been studied as a means of reducing CPV infections. Particular emphasis has been placed on such interactions in the silkworm where the cross-protection induced may be advantageous in reducing disease incidence in silkworm stocks. Aruga et al. (1961) suggested that two strains of *B. mori* CPV producing icosahedral and hexahedral (cubic) polyhedra interfered with each other when fed to silkworm larvae. Similarly, a CPV from the pine caterpillar, *Dendrolimus spectabilis*, also interfered with *B. mori* CPV (Aruga et al., 1963b). Subsequent studies suggested that if heat-inactivated CPV polyhedra were fed to larvae, this reduced the infectivity of "live" virus administered later (Aruga et al., 1963a). However, the reduction in infectivity reported by the authors was only from 69.2% to 51.1% and it is therefore difficult to evaluate the significance of this observation without further details on the precision of the bioassay technique involved.

More recently, Aruga and Watanabe (1970) found that when virus inactivated by ultra violet irradiation was administered to *B. mori* larvae, protection from the pathological consequences of a challenge CPV was remarkable, suggesting that interference may be established even when the interval between applications is prolonged, as between second and fourth instars. A subsequent study gave more erratic results (Aruga and Watanabe, 1971). Further examples of interference in CPV infections are given in Tanada and Chang (1964) and Tanada (1971).

No clear explanations have been provided for interference phenomena in such mixed infections. It seems probable that, of four types of viral interference defined by Fenner at al (1974), interference mediated by interferon can be considered unlikely in insects as the induction of an interferon-like agent has not been detected in virus-infected insect cells (Kalmakoff et al., 1977).

It is important to be aware of the limitations of the CPV investigations described above. Measurement of interference in all these studies was dependent upon the recognition of polyhedra of distinct shape formed in infected cells. Rao (1973) observed considerable variation in polyhedral shape within previously-accepted virus strains, and comments that results obtained using strains of virus which have been identified according to the polyhedra they form should therefore be interpreted with some reservation.

TRANS-OVUM TRANSMISSION AND LATENCY

Several workers have suggested that the possible existence of "latent" or "occult" CPV infections in larvae arises in part as a result of the vertical transmission of virus from one generation to another (Aruga and Naganshima, 1962; Hukuhara, 1962; Aruga, 1971). Various stress treatments such as crowding (Steinhaus and Dineen, 1959), temperature shock (Hukuhara, 1962), chemicals (Aruga, 1971) or treatment with other pathogens (Smith and Rivers, 1956; Tanada et al., 1964) have been used to convert suspected "latent" CPV infections from a state where virus is not demonstrable, into an actively multiplying one.

The evidence for transovarian transmission of virus and the nature of latent infections must be critically examined. CPV's are among the most prevalent of microbial pathogens and are a major problem in the laboratory rearing of lepidopterous larvae (Mery and Dulmage, 1975). Transmission of virus with the egg could arise from external contamination (which would be prevented by surface disinfection of the egg) or by transfer within the egg.

Aruga and Nagashima (1962) observed that cold-stress treatment of larvae derived from eggs laid after mating of CPV-infected moths, produced 4-30% CPV infection. The highest disease incidence arose when the infected parent was female, and the author suggested that this might arise by transmission of virus through the egg cytoplasm to the next generation (trans-ovum transmission). However, no surface disinfection of eggs appears to have been used in this study. In experiments of similar type, Hukyhara (1962) treated eggs with 2% formalin for 5 minutes. Larvae from these eggs were allowed to develop to the fifth instar before temperature stress. Twenty-three per cent of larvae became CPV infected. However, no unstressed controls were included to check for mortality due to contamination.

The work of Bullock et al. (1969) emphasizes the need for particular care in the interpretation of such studies. They examined disinfection procedures for the inactivation of CPV painted onto the surface of eggs of P. gossypiella. Four per cent formaldehyde (15 minutes) proved ineffective and even 10% formaldehyde did not completely inactivate the virus in all the experiments. Where satisfactory egg disinfection is carried out, no trans-ovum transmission has been observed (Sikorowski et al., 1973). It must therefore be concluded that there is no good evidence that CPV may be transmitted from generation to generation within the egg, although surface contamination may be important as a method of vertical transmission of disease.

Data on the existence of "latent" infections must also be viewed with suspicion, bearing in mind that CPV infections in larvae are frequently chronic and may go undetected. Overt infections have been recognized only by the production of polyhedra and it is not inconceivable that frank infections, where many non-occluded virus particles are produced, would not be detected. The dangers of contamination should also be emphasized. These may not be appreciated without the use of adequate controls. It is particularly interesting that when B. mori larvae are reared under aseptic conditions on artificial diet, where risk of contamination should be eliminated, they do not respond to chemical or temperature stressors by producing an active CPV infection (Kurata, 1971).

However, the concept of latent CPV infections cannot be dismissed out of hand. Grace (1962) observed that B. mori NPV induced the appearance of CPV in early passage ovarian cell cultures of Antherea eucalpti. CPV also appeared spontaneously in some cultures of these cells. The conclusion was that the ovarian cells carried an "occult" CPV infection, implying that the tissues from which the cells were derived also carried this infection. As described above, CPV replication has, with few exceptions, been observed only in gut epithelial cells and never in ovarian tissue in vivo. Answer to such anomalies must await further investigation. With such considerations in mind, the reader is recommended to refer to Aruga and Tanada (1971) for further discussion on latency in CPV infections.

CONCLUSIONS

The pathological consequences of CPV infection of insect larvae are most clearly manifested by symptoms of starvation. These may arise as a result of reductions in the adsorptive capacity of the gut and the depletion of reserves of amino acids and other metabolites during virus replication and polyhedra formation. Cellular regeneration in infected gut epithelial tissue is probably most important in maintaining the infection as a frequently chronic disease. Although many larvae may survive infection and subsequently pupate, infected adults are frequently deformed and their reproductive potential is significantly reduced.

The overt signs of infection disguise a lack of information on the mechanisms of pathogenesis. Until additional information is available on the biochemical aspects of virus replication, the molecular interaction between virus and cell will remain unclear. Recent success in achieving productive CPV infections in insect cell culture (Quiot and Belloncik, 1977) should provide the basis for examining aspects of intracellular pathogenesis more simply than would be possible in the diseased organism. Together with complementary studies on the interactions of virus and insect host, these investigations should provide a more complete picture of CPV pathogenesis than is available at present.

ACKNOWLEDGEMENTS

I am particularly grateful to the following who responded to questions raised during the preparation of this chapter: Dr. J. P. Hollis, Dr. M. E. Martignoni, Dr. K. M. Smith, C. B. E., F. R. S. and Dr. D. B. Stoltz. Margaret Arnold and Dr. A. Magnoler kindly supplied the photographs for Figs. 3-2 and 3-3. Dr. N. E. Crook and Graham Allaway kindly read the draft manuscript, which was painstakingly typed by Mrs. S. Bewsey.

REFERENCES

Akutsu, K. 1971. Cytoplasmic polyhedrosis of the fall webworm, Hyphantria cunea Drury (Lepidoptera: Arctiidae). Appl. Entomol. Zool., 6, 198-205.

Akutsu, K. 1975. Cytoplasmic polyhedrosis of the cabbage armyworm, Mamestra brassicae Linne. Jap. J. Appl. Entomol. Zool., 19, 57-58.

Anthony, D. W., Hazard, E. I. and Crosby, S. W. 1973. A virus disease in Anopheles quadrimaculatus. J. Invertebr. Pathol., 22, 1-5.

Aratake, Y., Kayamura, T. and Ueno, H. 1973. Studies on factors affecting the susceptibility of the silkworm, Bombyx mori L. Bull. Seric. Exp. Stn. (Tokyo)., 25, 307-346.

Arnott, H. J., Smith, K. M. and Fullilove, S. L. 1968. Ultrastructure of a cytoplasmic polyhedrosis virus affecting the monarch butterfly, Danaus plexippus. J. Ultrastruct. Res., 24, 479-507.

Aruga, H. 1971. Cytoplasmic polyhedrosis of the silkworm - Historical, economical and epizootiological aspects. In The cytoplasmic-polyhedrosis virus of the silkworm, H. Aruga and Y. Tanada, eds. University of Tokyo Press, pp. 3-21.

Aruga, H. and Nagashima, E. 1962. Generation to generation transmission of the cytoplasmic polyhedrosis virus of Bombyx mori Linnaeus. J. Insect Pathol., 4, 313-320.

Aruga, H. and Tanada, Y. 1971. The Cytoplasmic-Polyhedrosis Virus of the Silkworm. University of Tokyo Press.

Argua, H. and Watanabe H. 1964. Resistance to per os infection with cytoplasmic-polyhedrosis virus in the silkworm Bombyx mori (Linnaeus). J. Insect Pathol., 6, 387-394.

Aruga, H. and Watanabe, H. 1970. Interference between UV - inactivated and active cytoplasmic-polyhedrosis virus in the silkworm, Bombyx mori L. II. Silkworm strain and time interval of inoculation. J. Sericult. Sci. Jap., 39, 382-386.

Aruga, H. and Watanabe, H. 1971. Interference between UV-inactivated and active cytoplasmic-polyhedrosis viruses in the silkworm, Bombyx mori L. J. Sericult. Sci. Jap., 40, 176-180.

Aruga, H., Yashotake, N. and Watanabe, H. 1963a. Interference between cytoplasmic polyhedrosis viruses in Bombyx mori (Linnaeus). J. Insect Pathol., 5, 1-10.

Aruga, H., Hukuhara, T., Fukuda, S. and Hashimoto, Y. 1963b. Interference between cytoplasmic-polyhedrosis viruses of the silkworm, Bombyx mori (Linnaeus), and the pine caterpillar, Dendrolimus spectabilis (Butler) J. Insect Pathol., 5, 415-421.

Aruga, H., Hukuhara, T., Yoshitake, N. and Israngkul, A. 1961. Interference and latent infection in the cytoplasmic polyhedrosis of the silkworm Bombyx mori (Linnaeus). J. Insect Pathol., 3, 81-92.

Asayama, T. and Osaki, N. 1970. A cytoplasmic-polyhedrosis of the cotton leafworm, Prodenia litura. J. Invertebr. Pathol., 16, 292-294.

Atger, P. 1962. Virose intestinalé chez la noctuelle du chou Mamestra brassicae L. (Lepidoptera). Ann. Epiphyt., 13, 263-265.

Bailey, C. H. 1977. Field and laboratory observations on a cytoplasmic polyhedrosis virus of blackflies (Diptera: Simuliidae). J. Invertebr. Pathol., 29, 69-73.

Bailey, C. H., Shapiro, M. and Granados, R. R. 1975. A cytoplasmic polyhedrosis virus from the larval blackflies Cnephia mutata and Prosimulium mixtum (Diptera: Simuliidae). J. Invertebr. Pathol., 25, 273-274.

Bell, M. R. 1977. Pink bollworm: Effect of infection by a cytoplasmic polyhedrosis virus in diapausing larvae. Annals Ent. Soc. Amer., 70, 675-677.

Bell, M. R. and Kanavel, R. F. 1976. Effect of dose of cytoplasmic polyhedrosis virus on infection, mortality, development rate and larval and pupal weights of the pink bollworm. J. Invertebr. Pathol., 28, 121-126.

Bell, M. R. and Kanavel, R. F. 1977. The effect of a cytoplasmic polyhedrosis virus on lipid and protein content of pupae of the pink bollworm (Lepidoptera: Gelechiidae). J. Kansas Entomol. Soc., 50, 359-362.

Bergold, G. H. and Suter, J. 1959. On the structure of cytoplasmic polyhedra of some Lepidoptera. J. Insect Pathol., 1, 1-14.

Bird, F. T. 1965. On the morphology and development of insect cytoplasmic polyhedrosis virus particles. Can. J. Microbiol., 11, 497-501.

Bird, F. T. 1966. The development of a cytoplasmic polyhedrosis virus in the spruce budworm, Choristoneura fumiferana (Clemends). Can. J. Microbiol., 12, 337-339.

Bird, F. T. 1969. Infection and mortality of spruce budworm Choristoneura fumiferana, and forest tent caterpillar, Malacosoma disstria, caused by nuclear and cytoplasmic polyhedrosis viruses. Can. Entomol., 101, 1269-1285.

Bird, F. T. and Whalen, M. M. 1954. A nuclear and a cytoplasmic polyhedral virus disease of the spruce budworm. Can. J. Zool., 32, 82-86.

Bird, R. G., Draper, C. C. and Ellis, D. S. 1972. A cytoplasmic polyhedrosis virus in midgut cells of Anopheles stephensi and in the sporogonic stages of Plasmodium berghei yoelii. Bull. Wld. Hlth. Org., 46, 337-343.

Boucias, D. G. and Nordin, G. L. 1978. A scanning electron microscope study of Hyphantria cunea CPV-infected midgut tissue. J. Invertebr. Pathol., 32, 229-233.

Bullock, H. R. 1972. Therapeutic effect of high temperature on tobacco budworms to a cytoplasmic-polyhedrosis virus. J. Invertebr. Pathol., 19, 148-153.

Bullock, H. R., Magum, C. L. and Guerra, A. A. 1969. Treatment of eggs of the pink bollworm, Pectinophora gossypiella, with formaldehyde to prevent infection with a cytoplasmic polyhedrosis virus. J. Invertebr. Pathol., 14, 271-273.

Bullock, H. R., Martinez, E. and Stuermer, C. W. Jr. 1970. Cytoplasmic-polyhedrosis virus and the development and fecundity of the pink bollworm. J. Invertebr. Pathol., 15, 109-112.

Clark, T. B., Chapman, H. C. and Fukuda, T. 1969. Nuclear-polyhedrosis and cytoplasmic-polyhedrosis virus infection in Louisiana mosquitoes. J. Invertebr. Pathol., 14, 284-286.

Faust, R. M. and Cantwell, G. E. 1968. Inducement of cytoplasmic polyhedrosis by intrahemocoelic injection of freed viral particles into the silkworm Bombyx mori.

J. Invertebr. Pathol., 11, 119-122.
Federici, B. A. 1974. Viral pathogens of mosquitoes and their potential use in mosquito control. In Le Contrôle des Moustiques/Mosquito Control, A. Aubin, J.-P. Bourassa, S. Belloncik, M. Pellissier and E. Lacoursiere eds. Les Presses de l'Universite du Quebec, pp. 93-135.
Federici, B. A. and Hazard, E. I. 1975. Iridovirus and cytoplasmic polyhedrosis virus in the freshwater daphnid, Simocephalus expinosus. Nature (London), 254, 327-328.
Federici, B. A., Hazard, E. I. and Anthony, D. W. 1973. A new cytoplasmic polyhedrosis virus from Chironomids collected in Florida. J. Invertebr. Pathol., 22, 136-138.
Fenner, F., McAuslan, B. R., Mims, C. A., Sambrook, J. and White, D. O. 1974. The Biology of Animal Viruses. Academic Press, N.Y.
Furuichi, Y. 1974. "Methylation - coupled" transcription by virus associated transcriptase of cytoplasmic polyhedrosis virus containing double stranded RNA. Nucleic Acid Res., 1, 809-822.
Furuichi, Y. 1978. 'Pretranscriptional capping' in the biosynthesis of cytoplasmic polyhedrosis virus mRNA. Proc. Natl. Acad., Sci. USA, 75, 1086-1090.
Furuichi, Y. and Miura, K. 1975. A blocked structure at the 5' terminus of mRNA from cytoplasmic polyhedrosis virus. Nature (London), 253, 374-375.
Furusawa, T. and Kawase, S. 1971. Synthesis of ribonucleic acid resistant to actinomycin D in silkworm midguts infected with the cytoplasmic-polyhedrosis virus. J. Invertebr. Pathol., 18, 156-158.
Furusawa, T. and Kawase, S. 1973. Virus-specific RNA synthesis in the midgut of silkworm, Bombyx mori, infected with cytoplasmic-polyhedrosis virus. J. Invertebr. Pathol., 22, 335-344.
Grace, T. D. C. 1962. The development of a cytoplasmic polyhedrosis in insect cells grown in vitro. Virology, 18, 33-42.
Granados, R. R. 1976. The infection and replication of insect pathogenic viruses in tissue culture. Adv. Virus. Res. 20, 189-235.
Granados, R. R. 1978. Biology of cytoplasmic polyhedrosis viruses and entomopoxviruses. In Viral Pesticides: Present Knowledge and Potential Effects on Public and Environmental Health, M. D. Summers and C. Y. Kawanishi, eds., Environmental Protection Agency Report No. 600/9-78-026. pp. 89-101.
Granados, R. R., McCarthy, W. J. and Naughton, M. 1974. Replication of a cytoplasmic polyhedrosis virus in an established cell line of Trichoplusia ni cells. Virology, 59, 584-586.
Harpaz, I., Zlotkin, E. and Ben Shaked, Y. 1965. On the pathology of cytoplasmic and nuclear polyhedrosis of the

Cyprus processionary caterpillar, Thaumetopoea wilkinsoni Tams. J. Invertebr. Pathol., 7, 15-21.

Harrap, K. A. and Payne, C. C. 1979. The structural properties and identification of insect viruses. Adv. Virus. Res., 25, 273-355.

Hayashi, Y. 1970a. Occluded and free virions in midgut cells of Malacosoma disstria infected with cytoplasmic polyhedrosis virus (CPV). J. Invertebr. Pathol., 16, 442-450.

Hayashi, Y. 1970b. RNA in midgut of tussock moth, Orgyia leucostigma infected with cytoplasmic-polyhedrosis virus. Can. J. Microbiol., 16, 1101-1107.

Hayashi, Y. and Donaghue, T. P. 1971. Cytoplasmic polyhedrosis virus: RNA synthesized in vivo and in vitro in infected midgut. Biochem. Biophys. Res. Commun., 42, 214-221.

Hayashi, Y. and Kawase, S. 1965a. Studies on the RNA in the cytoplasmic polyhedra of the silkworm, Bombyx mori L. IV. Subcellular distribution. J. Sericult. Sci. Japan. 34, 171-175.

Hayashi, Y. and Kawase, S. 1965b. Studies on the RNA in the cytoplasmic polyhedra of the silkworm, Bombyx mori L. V. Changes in fractionated RNA of the midgut. J. Sericult. Sci. Japan, 34, 244-250.

Hayashi, Y. and Retnakaran, A. 1970. The site of RNA synthesis of a cytoplasmic-polyhedrosis virus (CPV) in Malacosoma disstria. J. Invertebr. Pathol., 16, 150-151.

Hayashi, Y., Kawarabata, T. and Bird, F. T. 1970. Isolation of a cytoplasmic-polyhedrosis virus of the silkworm, Bombyx mori. J. Invertebr. Pathol., 16, 378-384.

Hills, G. J. and Smith, K. M. 1959. Further studies on the isolation and crystallization of insect cytoplasmic viruses. J. Insect Pathol., 1, 121-128.

Hosaka, Y. and Aizawa, K. 1964. The fine structure of the cytoplasmic-polyhedrosis virus of the silkworm Bombyx mori (Linnaeus). J. Insect Pathol., 6, 53-77.

Hukuhara, T. 1962. Generation-to-generation transmission of the cytoplasmic polyhedrosis virus of the silkworm, Bombyx mori (Linnaeus). J. Insect Pathol., 4, 132-135.

Hukuhara, T. and Hashimoto, Y. 1966. Studies of two strains of cytoplasmic-polyhedrosis virus. J. Invertebr. Pathol., 8, 184-192.

Hukuhara, T. and Yamaguchi, K. 1973. Ultrastructural investigation on a strain of a cytoplasmic-polyhedrosis virus with nuclear inclusions. J. Invertebr. Pathol., 22, 6-13.

Irahim, I. K. A., Joshi, M. M. and Hollis, J. P. 1978. Swarming disease of nematodes: Host range and evidence for a cytoplasmic polyhedral virus in Tylenchorhynchus martini. Proc. Helminth. Soc. Wash., 45, 233-238.

Ignoffo, C. M. and Adams, J. R. 1966. A cytoplasmic-polyhedrosis virus, Smithiavirus pectinophorae sp.n. of the pink bollworm, Pectinophora gossypiella (Saunders). J. Invertebr. Pathol., 8, 59-66.

Inoue, H. and Miyagawa, M. 1978. Regeneration of midgut epithelial cell in the silkworm, Bombyx mori, infected with viruses. J. Invertebr. Pathol., 32, 373-380.

Ishimori, N. 1934. Contribution of l'étude de la grasserie du ver à soie. C. R. Soc. Biol., 116, 1169-1170.

Iwashita, Y. 1971. Histopathology of cytoplasmic polyhedrosis. In The Cytoplasmic-Polyhedrosis Virus of the Silkworm, H. Aruga and Y. Tanada, eds. University of Tokyo Press, pp. 79-101.

Iwashita, Y. and Seki, H. 1973. The development of the cytoplasmic polyhedrosis virus in the midgut of the embryo and the young silkworm, Bombyx mori L. Bull. Coll. Agric. Utsonomiya Univ., 8, 27-42.

Joklik, W. K. 1974. Reproduction of Reoviridae. In Comprehensive Virology, H. Fraenkel-Conrat and R. R. Wagner, eds., Vol. 2. Plenum Press, N.Y. pp. 231-334.

Kalmakoff, J., Lewandowski, L. J., and Black, D. R. 1969. Comparison of the ribonucleic acid subunits of reovirus, cytoplasmic polyhedrosis virus and wound tumor virus. J. Virol., 4, 851-856.

Kalmakoff, J., Williams, B. R. G. and Austin, F. J. 1977. Antiviral response in insects? J. Invertebr. Pathol., 29, 44-49.

Kawase, S. 1965. Free amino acids in the hemolymph and midgut epithelium of the silkworm, Bombyx mori (Linnaeus), infected with cytoplasmic-polyhedrosis virus. J. Invertebr. Pathol., 7, 113-116.

Kawase, S. 1971a. Chemical nature of the cytoplasmic-polyhedrosis virus. In The Cytoplasmic Polyhedrosis Virus of the Silkworm, H. Aruga and Y. Tanada, eds., University of Tokyo Press, pp. 37-59.

Kawase, S. 1971b. Different infection response of silkworm larvae to the cytoplasmic-polyhedrosis virus with special reference to the locality of virus injection. II. Detection of the portion showing highest infection response to the virus. J. Sericult. Sci. Jap., 40, 459-462.

Kawase, S. and Furusawa, T. 1971. Effect of actinomycin D on RNA synthesis in the midgut of healthy and CPV-infected silkworms. J. Invertebr. Pathol., 18, 33-39.

Kawase, S. and Hayashi, Y. 1965. Nucleic acid and protein changes in blood and midgut of the silkworm, Bombyx mori (Linnaeus) during the course of cytoplasmic polyhedrosis. J. Invertebr. Pathol., 7, 49-54.

Kawase, S. and Kawamori, I. 1968. Chromatographic studies on RNA synthesis in the midgut of the silkworm, Bombyx mori infected with a cytoplasmic-polyhedrosis virus. J. Invertebr. Pathol., 12, 395-404.

Kawase, S. and Miyajima, S. 1969. Immunofluorescence studies on the multiplication of cytoplasmic polyhedrosis virus of the silkworm, Bombyx mori. J. Invertebr. Pathol., 13, 330-336.

Kawase, S., Kawamoto, F. and Yamaguchi, K. 1973. Studies on the polyhedrosis virus forming polyhedra in the midgut cell nucleus of the silkworm, Bombyx mori. I. Purification procedure and form of the virion. J. Invertebr. Pathol., 22, 266-272.

Kellen, W. R., Clark, T. B., Lindegren, J. E. and Sanders, R. D. 1966. A cytoplasmic-polyhedrosis virus of Culex tarsalis (Diptera: Culicidae). J. Invertebr. Pathol., 8, 390-394.

Kobayashi, M. 1971. Replication cycle of cytoplasmic-polyhedrosis virus as observed with the electron microscope. In The Cytoplasmic-Polyhedrosis Virus of the Silkworm , H. Aruga and Y. Tanada, eds. University of Tokyo Press, pp. 103-128.

Kobayashi, M. and Kawase, S. 1980. Absence of detectable accumulation of cytoplasmic-polyhedrosis viral RNA in the silkworm Bombyx mori, reared at a supraoptimal temperature. J. Invertebr. Pathol., 35, 96-98.

Kunimi, Y. and Aruga, H. 1974. Susceptibility to infection with nuclear and cytoplasmic-polyhedrosis viruses of the fall websorm, Hyphantria cunea Drury, reared on several artificial diets. Jap. J. Appl. Entomol. Zool., 18, 1-4.

Kurata, K. 1971. On symptoms of silkworms, Bombyx mori, infected with a cytoplasmic-polyhedrosis virus under aseptic conditions. J. Sericult. Sci. Jap., 40, 32-36.

Lewandowski, L. J. Kalmakoff, J. and Tanada, Y. 1969. Characterization of a ribonucleic acid polymerase activity associated with purified cytoplasmic-polyhedrosis virus of the silkworm, Bombyx mori. J. Virol., 4, 857-865.

Lewandowski, L. J. and Traynor, B. L. 1972. Comparison of the structure and polypeptide composition of three double-stranded ribonucleic acid containing viruses (diplorna viruses). Cytoplasmic polyhedrosis virus, wound tumor virus and reovirus. J. Virol., 10, 1053-1070.

Lipa, J. J. 1977. Electron microscope observations on the development of cytoplasmic polyhedrosis virus in Scotogramma trifolii Rott. (Lepidoptera: Noctuidae). Bull. Acad. Pol. Sci. Ser. Sci. Biol., 25, 155-158.

Longworth, J. F. and Spilling, C. R. 1970. A cytoplasmic polyhedrosis of the larch sawfly, Anoplonyx destructor. J. Invertebr. Pathol., 15, 276-280.

McCrae, M. A. and Joklik, W. K. 1978. The nature of polypeptides encoded by each of the double-stranded RNA segments of reovirus type 3. Virology, 89, 578-593.

Magnoler, A. 1970. Susceptibility of gypsy moth larvae to Lymantria spp. nuclear- and cytoplasmic-polyhedrosis viruses. Entomophaga, 15, 407-412.

Magnoler, A. 1974. Effects of a cytoplasmic-polyhedrosis on larval and post larval stages of the gypsy moth, Porthetria dispar. J. Invertebr. Pathol., 23, 263-274.

Martignoni, M. E. and Iwai, P. 1977. A catalog of viral diseases of insects and mites. USDA Forest Service General Technical Report. PNW-40.

Martignoni, M. E., Iwai, P. J., Hughes, K. M. and Addison, R. B. 1969. A cytoplasmic-polyhedrosis of *Hemerocampa pseudotsugata*. J. Invertebr. Pathol., 13, 15-18.

Matthews, R. E. F. 1979. Classification and nomenclature of viruses. Intervirology, 12, 131-296.

Mertens, P. P. C. 1979. A study of the transcription and translation (*in vitro*) of the genomes of cytoplasmic polyhedrosis viruses types 1 and 2. D. Phil. Thesis, University of Oxford, England.

Mertens, P. P. C. and Payne, C. C. 1978. S-adenosyl-L-homocysteine as a stimulator of viral RNA synthesis by two distinct cytoplasmic polyhedrosis viruses. J. Virol., 26, 832-835.

Mery, C. and Dulmage, H. T. 1975. Transmission, diagnosis and control of cytoplasmic polyhedrosis virus in colonies of *Heliothis virescens*. J. Invertebr. Pathol., 26, 75-79.

Miura, K., Fujii, I., Sakaki, T., Fuke, M. and Kawase, S. 1968. Double-stranded ribonucleic acid from cytoplasmic polyhedrosis virus of the silkworm. J. Virol., 2, 1211-1222.

Miyajima, S. and Kawase, S. 1968. Changes in virus infectivity titer in the hemolymph and midgut during the course of a cytoplasmic polyhedrosis in the silkworm. J. Invertebr. Pathol., 12, 329-334.

Miyajima, S. and Kawase, S. 1969. Hemaglutination with cytoplasmic polyhedrosis virus of the silkworm, *Bombyx mori*. Virology, 39, 347-348.

Morris, O. N. 1968. Metabolic changes in diseased insects. II. Radioautographic studies on DNA and RNA synthesis in nuclear-polyhedrosis and cytoplasmic-polyhedrosis virus infections. J. Invertebr. Pathol., 11, 476-486.

Neilson, M. M. 1964. A cytoplasmic polyhedrosis virus pathogenic for a number of lepidopterous hosts. J. Insect Pathol., 6, 41-52.

Neilson, M. M. 1965. Effects of a cytoplasmic polyhedrosis on adult lepidoptera. J. Invertebr. Pathol., 7, 306-314.

Payne, C. C. 1971. Properties and replication of some occluded insect viruses. D. Phil. Thesis. University of Oxford, England.

Payne, C. C. 1972. Actinomycin D and the replication of an invertebrate RNA virus. *In* Moving Frontiers in Invertebrate Virology, T. W. Tinsley and K. A. Harrap, eds. Monogr. Virol., 6, 11-15.

Payne, C. C. 1976. Biochemical and serological studies of a cytoplasmic polyhedrosis virus from *Arctia caja*: A naturally-occurring mixture of two virus types. J. Gen. Virol., 30, 357-369.

Payne, C. C. and Harrap, K. A. 1977. Cytoplasmic polyhedrosis viruses. In The Atlas of Insect and Plant Viruses, K. Maramorosch, ed. Academic Press, N.Y. pp. 105-129.

Payne, C. C. and Kalmakoff, J. 1973. The synthesis of virus-specific single-stranded RNA in larvae of Bombyx mori infected with a cytoplasmic polyhedrosis virus. Intervirology, 1, 34-40.

Payne, C. C. and Kalmakoff, J. 1974. Biochemical properties of polyhedra and virus particles of the cytoplasmic polyhedrosis virus of Bombyx mori. Intervirology, 4, 353-364.

Payne, C. C. and Rivers, C. F. 1976. A provisional classification of cytoplasmic polyhedrosis viruses based on the sizes of the RNA genomes segments. J. Gen. Virol., 33, 71-85.

Payne, C. C. and Tinsley, T. W. 1974. The structural proteins and RNA components of a cytoplasmic polyhedrosis virus from Nymphalis io (Lepidoptera: Nymphalidae). J. Gen. Virol., 25, 291-302.

Payne, C. C., Piasecka-Serafin, M. and Pilley, B. 1977. Properties of two recent isolates of cytoplasmic polyhedrosis viruses. Intervirology, 8, 155-163.

Quiot, J. M. and Belloncik, S. 1977. Caractérisation d'une polyedrose cytoplasmique chez lépidoptère. Euxoa scandens Riley (Noctuidae, Agrotinae). Etudes in vivo et in vitro. Arch. Virol., 55, 145-153.

Ralph, R. K. 1969. Double-stranded viral RNA. Adv. Virus Res., 15, 61-158.

Rao, C. B. J. 1973. Surface topography and shapes of polyhedral inclusion bodies of the cytoplasmic polyhedrosis virus. J. Ultrastruct. Res., 42, 582-593.

Ripa, R. 1978. Studies of the susceptibility of Pieris brassicae (L) to a granulosis virus. Ph.D. Thesis, University of London, England.

Shimotohno, K. and Miura, K. 1974. 5'-terminal structure of messenger RNA transcribed by the RNA polymerase of silkworm cytoplasmic polyhedrosis virus containing double-stranded RNA. J. Mol. Biol., 86, 21-30.

Shimotohno, K. and Miura, K. 1977. Nucleoside triphosphate phosphohydrolase associated with cytoplasmic polyhedrosis virus. J. Biochem. (Tokyo), 81, 371-379.

Shimotohno, K., Kodama, Y., Hashimoto, J. and Miura, K. 1977. Importance of 5'-terminal blocking structure to stabilize mRNA in eukaryotic protein synthesis. Proc. Nat'l. Acad. Sci. USA, 74, 2734-2738.

Sidor, C. 1970. Poliedarno virozno oboljenje topoling krasnika (Melanophila picta Pall, Coleoptera, Buprestidae). Topola, 77-78, 19-21.

Sidor, C. 1971. Virus diseases of some economically important harmful insects. Topola, 83-85, 65-69.

Sikorowski, P. P. and Thompson, A. C. 1979. Effects of cytoplasmic polyhedrosis virus on diapausing Heliothis virescens. J. Invertebr. Pathol., 33, 66-70.

Sikorowski, P. P., Andrews, G. L. and Broome, J. R. 1973. Trans-ovum transmission of a cytoplasmic polyhedrosis virus of Heliothis virescens (Lepidoptera: Noctuidae). J. Invertebr. Pathol., 21, 41-45.

Silverstein, S. C. and Dales, S. 1968. The penetration of reovirus RNA and initiation of its genetic function in L-strain fibroblasts. J. Cell. Biol., 36, 197-230.

Silverstein, S. C., Astell, C., Levin, D. H., Schonberg, M. and Acs, G. 1972. The mechanisms of reovirus uncoating and gene activation in vivo. Virology, 47, 797-806.

Simmons, C. L. and Sikorowski, P. P. 1973. A laboratory study of the effects of cytoplasmic polyhedrosis virus on Heliothis virescens (Lepidoptera: Noctuidae). J. Invertebr. Pathol., 22, 369-371.

Slizynski, K. and Lipa, J. J. 1973. Methodological instructions for calculation of median lethal time (LT_{50}) in insect pathology and toxicology. Prace Nauk., 15, 85-98.

Smith, K. 1963. Cytoplasmic polyhedroses. In Insect Pathology: An Advanced Treatise, E. A. Steinhaus, ed., Vol. 1. Academic Press, N.Y. pp. 457-497.

Smith, K. M. and Rivers, C. F. 1956. Some viruses affecting insects of economic importance. Parasitology, 46, 235-242.

Smith, K. M. and Wyckoff, R. W. G. 1950. Structure within polyhedra associated with insect virus diseases. Nature (London). 166, 861-862.

Smith, K. M., Hills, G. J. and Rivers, C. F. 1959. Polyhedroses in Neuropterous insects. J. Insect Pathol., 1, 431-437.

Sohi, S. S., Bird, F. T. and Hayashi, Y. 1971. Development of Malacosoma disstria cytoplasmic polyhedrosis virus in Bombyx mori ovarian and tracheal tissue cultures. Proc. Int'l. Colloq. Insect Pathol., College Park, Maryland, 1970. pp. 340-351.

Stairs, G. R., Parrish, W. B. and Allietta, M. 1968. An histopathological study involving a cytoplasmic polyhedrosis virus of the salt-marsh caterpillar, Estigmene acrea. J. Invertebr. Pathol., 12, 359-365.

Steinhaus, E. A. and Dineen, J. P. 1959. A cytoplasmic polyhedrosis of the alfalfa caterpillar. J. Insect Pathol., 1, 171-183.

Stiles, B. and Paschke, J. D. 1980. Midgut pH in different instars of three Aedes mosquito species and the relation between pH and susceptibility of larvae to a nuclear polyhedrosis virus. J. Invertebr. Pathol., 35, 58-64.

Stoltz, D. B. 1969. Observations on naturally occurring viruses in larvae of the midge Chironomus plumosus. Ph.D. Thesis. McMaster University, Canada.

Stoltz, D. B. and Hilsenhoff, W. L. 1969. Electron-microscopic observation on the maturation of a cytoplasmic-polyhedrosis virus. J. Invertebr. Pathol., 14, 39-48.

Storer, G. B., Shepherd, M. G. and Kalmakoff, J. 1974a. Enzyme activities associated with cytoplasmic polyhedrosis virus from Bombyx mori I. Nucleotide phosphohydrolase and nuclease activities. Intervirology, 2, 87-94.

Storer, G. B., Shepherd, M. G. and Kalmakoff, J. 1974b. Enzyme activities associated with cytoplasmic polyhedrosis virus from Bombyx mori. II. Comparative studies. Intervirology, 2, 193-199.

Tamashiro, M. and Huang, S. 1963. A cytoplasmic polyhedrosis of Cactoblastis cactorum (Berg). J. Insect Pathol., 5, 397-399.

Tanada, Y. 1971. Interactions of insect viruses, with special emphasis on interference. In The Cytoplasmic-Polyhedrosis Virus of the Silkworm, H. Aruga and Y. Tanada, eds. University of Tokyo Press, pp. 185-200.

Tanada, Y. and Chang, G. Y. 1960. A cytoplasmic polyhedrosis of the armyworm, Pseudaletia unipuncta (Haworth) (Lepidoptera, Noctuidae). J. Insect Pathol., 2, 201-208.

Tanada, Y. and Chang, G. Y. 1964. Interactions of two cytoplasmic-polyhedrosis viruses in three insect species. J. Insect Pathol., 6, 500-516.

Tanada, Y. and Chang, G. Y. 1968. Resistance of the alfalfa caterpillar Colias eurytheme, at high temperatures to a cytoplasmic-polyhedrosis virus and thermal inactivation point of the virus. J. Invertebr. Pathol., 10, 79-80.

Tanada, Y., Tanabe, A. M. and Reiner, C. E. 1964. Survey of the presence of a cytoplasmic-polyhedrosis virus in field populations of the alfalfa caterpillar, Colias eurytheme Boisduval, in California. J. Insect Pathol., 6, 439-447.

Vago, C. 1959. Sur le mode d'infection de la virose intestinale de Thaumetopoea pityocampa Schiff. Entomophaga, 4, 311-314.

Vago, C. and Bergoin, M. 1963. Development des virus a corps d'inclusion du lépidoptère Lymantria dispar en cultures cellulaires. Entomophaga, 8, 253-261.

Vago, C., Croissant, O. and Lepine, P. 1959. Processus de l'infection à virus a partir des corps d'inclusion de "Polyédrie Cytoplasmique" ingèrés par le lépidoptère sensible. Mikorskopie, 14, 36-40.

Vail, P. V. and Gough, D. 1970. Effects of cytoplasmic-polyhedrosis virus on adult cabbage loopers and their progeny. J. Invertebr. Pathol., 15, 397-400.

Vail, P. V., Hall, I. M. and Gough, D. 1969. Influence of a cytoplasmic polyhedrosis on various developmental stages of the cabbage looper. J. Invertebr. Pathol., 14, 237-244.

Watanabe, H. 1965. Resistance to peroral infection by the cytoplasmic-polyhedrosis virus in the silkworm, *Bombyx mori* (Linnaeus). J. Invertebr. Pathol., 7, 257-270.

Watanabe, H. 1966a. Relative virulence of polyhedrosis viruses and host-resistance in the silkworm, *Bombyx mori* L. (Lepidoptera: Bombycidae). Appl. Entomol. Zool., 1, 139-144.

Watanabe, H. 1966b. Some aspects on the mechanism of resistance to peroral infection by cytoplasmic-polyhedrosis virus in the silkworm, *Bombyx mori* L. J. Sericult. Sci. Japan, 35, 411-417.

Watanabe, H. 1967a. Development of resistance in the silkworm, *Bombyx mori*, to peroral infection of a cytoplasmic-polyhedrosis virus. J. Invertebr. Pathol., 9, 474-479.

Watanabe, H. 1967b. Site of viral RNA synthesis within midgut cells of silkworm, *Bombyx mori* infected with cytoplasmic polyhedrosis virus. J. Invertebr. Pathol., 9, 480-487.

Watanabe, H. 1968a. Light radio-autographic study of protein synthesis in the midgut epithelium of the silkworm, *Bombyx mori*, infected with a cytoplasmic-polyhedrosis virus. J. Invertebr. Pathol., 11, 310-315.

Watanabe, H. 1968b. Pathogenic changes in the feces from the silkworm, *Bombyx mori* L., after peroral inoculation with a cytoplasmic-polyhedrosis virus. J. Sericult. Sci. Japan, 37, 385-389.

Watanabe, H. 1968c. Disc-electrophoretic patterns of larval midgut proteins in normal and virus-infected silkworms, *Bombyx mori* L. (Lepidoptera: Bombycidae). Appl. Entomol. Zool., 3, 74-80.

Watanabe, H. 1970. Enhancement of protein metabolism in the midgut epithelium of the silkworm, *Bombyx mori*, by infection with a cytoplasmic-polyhedrosis virus. J. Invertebr. Pathol., 15, 247-252.

Watanabe, H. 1971a. Pathophysiology of cytoplasmic polyhedrosis in the silkworm. *In* The Cytoplasmic-Polyhedrosis Virus of the Silkworm, H. Aruga and Y. Tanada, eds. University of Tokyo Press, pp. 151-167.

Watanabe, H. 1971b. Resistance of the silkworm to cytoplasmic-polyhedrosis virus. *In* The Cytoplasmic-Polyhedrosis Virus of the Silkworm, H. Aruga and Y. Tanada, eds. University of Tokyo Press, pp. 169-184.

Watanabe, H. 1972. Pathophysiology of nitrogen catabolism in the midgut of silkworm. *Bombyx mori* L. (Lepidoptera: Bombycidae), infected with a cytoplasmic-polyhedrosis virus. Appl. Entomol. Zool., 6, 163-168.

Watanabe, H. 1973. On a cytoplasmic polyhedrosis of the smaller tea tortrix, *Adoxophyes fasciata* Walsingham. Jap. J. Appl. Entomol. Zool., 17, 1-4.

Watanabe, H. and Kobara, R. 1971. Synthesis of hemolymph proteins in the silkworm, Bombyx mori L., infected with a nuclear - and a cytoplasmic-polyhedrosis virus. Jap. J. Appl. Entomol. Zool., 15, 198-202.

Weiser, J. 1978. A new host, Simulium argyreatum Meig., for the cytoplasmic polyhedrosis virus of the blackflies in Czechoslovakia. Folia Parasitol., 25, 361-365.

Wittig, G., Steinhaus, E. A. and Dineen, J. P. 1960. Further studies of the cytoplasmic polyhedrosis virus of the alfalfa caterpillar. J. Insect Pathol., 2, 334-345.

Wiygul, G. and Sikorowski, P. P. 1978. Oxygen uptake in tobacco budworm larvae (Heliothis virescens) infected with cytoplasmic polyhedrosis virus. J. Invertebr. Pathol., 32, 191-195.

Xeros, N. 1956. The virogenic stroma in nuclear and cytoplasmic polyhedrosis. Nature (London), 178, 412-413.

Xeros, N. 1966. Light microscopy of the virogenic stromata of cytopolyhedroses. J. Invertebr. Pathol., 8, 79-87.

Yamaguchi, K. 1979. Natural recovery of the fall webworm, Hyphantria cunea to infection by a cytoplasmic-polyhedrosis virus of the silkworm, Bombyx mori. J. Invertebr. Pathol., 3, 126-128.

Chapter 4

ENTOMOPOXVIRUS INFECTIONS IN INSECTS

by Robert R. Granados*

INTRODUCTION

Entomopoxviruses are a complex and heterogenous group of insect viruses that are members of the genus *Entomopoxvirus* (EPV) in the family Poxviridae (Matthews, 1979). Since the discovery of the first EPV disease in the common cockchaffer, *Melolontha melolontha* (Hurpin and Vago, 1963; Vago, 1963) EPV's have been isolated from approximately 31 species of insects (Granados, 1973a; Kurstak and Garzon, 1977). In most cases cross infection, serological, or biochemical studies have not been made with these EPV's, consequently it is not known if they represent distinct virus species. Insects are the only known hosts of EPV's and four orders of insects, Lepidoptera, Coleoptera, Diptera, and Orthoptera, are affected, with most viruses isolated from Lepidoptera and Coleoptera. Some vertebrate poxviruses (e.g. myxomatosis) are known to be vectored by mosquitoes and fleas in a mechanical manner (i.e., no replication in vector). The EPV's have a widespread geographical distribution and have been found in North and South America, Europe, Asia, and Africa and Australia.

The *in vivo* developmental cycle of EPV's infecting larvae of *M. melolontha*, the saltmarsh caterpillar, *Amsacta moorei*, and the spruce budworm, *Choristoneura fumiferana* is relatively well known but only *A. moorei* EPV has been grown in cell culture. This review will therefore concentrate on the pathogenesis of these three viruses although other EPV's will be considered. For earlier general reviews on the morphology, developmental cycle, and chemistry of EPV's the readers are referred to the papers by Bergoin and Dales (1971), Granados

*Insect Pathology Resource Center, Boyce Thompson Institute, Cornell University, Ithaca, New York 14853

(1973a, 1978) and Kurstak and Garzon (1977).

GENERAL CHARACTERISTICS OF ENTOMOPOXVIRUSES

The EPV's are distinguished principally by their unusual morphology, large DNA genome, and cytoplasmic site of replication. EPV's are large, brick-shaped or ovoid particles 170-250 x 300-450 nm in dimension. The virions have a lipoprotein envelope with tubular or globular protein structures on the outer surface, enclosing one or two lateral bodies and a core that contains the genome. The genome consists of a single molecule of double-stranded DNA of molecular weight 130-240 x 10^6 daltons. EPV's may be occluded in an ordered protein matrix or occlusion body (OB) which is sphere-like or ellipsoidal in shape. Fusiform, proteinic bodies (spindles) devoid of virus particles are also produced during the developmental cycle of some of these viruses. Entomopoxviruses have morphological and developmental similarities to the vertebrate poxviruses but are biochemically distinct (Bergoin and Dales, 1971; Arif, 1976; Langridge et al., 1977; Langridge and Roberts, 1977; Granados, 1978).

EPV's are tentatively grouped into three genera (Fig. 4-1) based on insect host affected and virion morphology (Matthews, 1979). Genus A which includes the relatively large viruses affecting mostly Coleoptera and some Lepidoptera, has virions which are ovoid in shape, possess a characteristic unilateral concave core and one lateral body. Included in genus B are the intermediate-sized viruses affecting mostly Lepidoptera and some Orthoptera. The viruses are ovoid in shape and possess a cylindrical core surrounded by a sleeve-shaped lateral body. Genus C is comprised of viruses affecting Diptera and are the smallest of the EPV's. The brick-shaped virions possess a biconcave core and two lateral bodies. The assignment of EPV's to either of these three genera may prove difficult especially in instances where a particular virus possesses characteristics overlapping two genera. Milner and Beaton (1979) described an EPV from *Oncopera alboguttata* (Lepidoptera: Hepialidae) in Australia which possessed morphological and pathological properties peculiar to genera A and B. However, based on the ultrastructure of the virions, *O. alboguttata* EPV was provisionally placed in genus A. Until more is known about the biophysical and biochemical properties of EPV's we will have to rely on morphological and pathological characteristics to differentiate these viruses. Table 4-1 presents morphological and pathological characteristics of a few EPV's representative of the three genera.

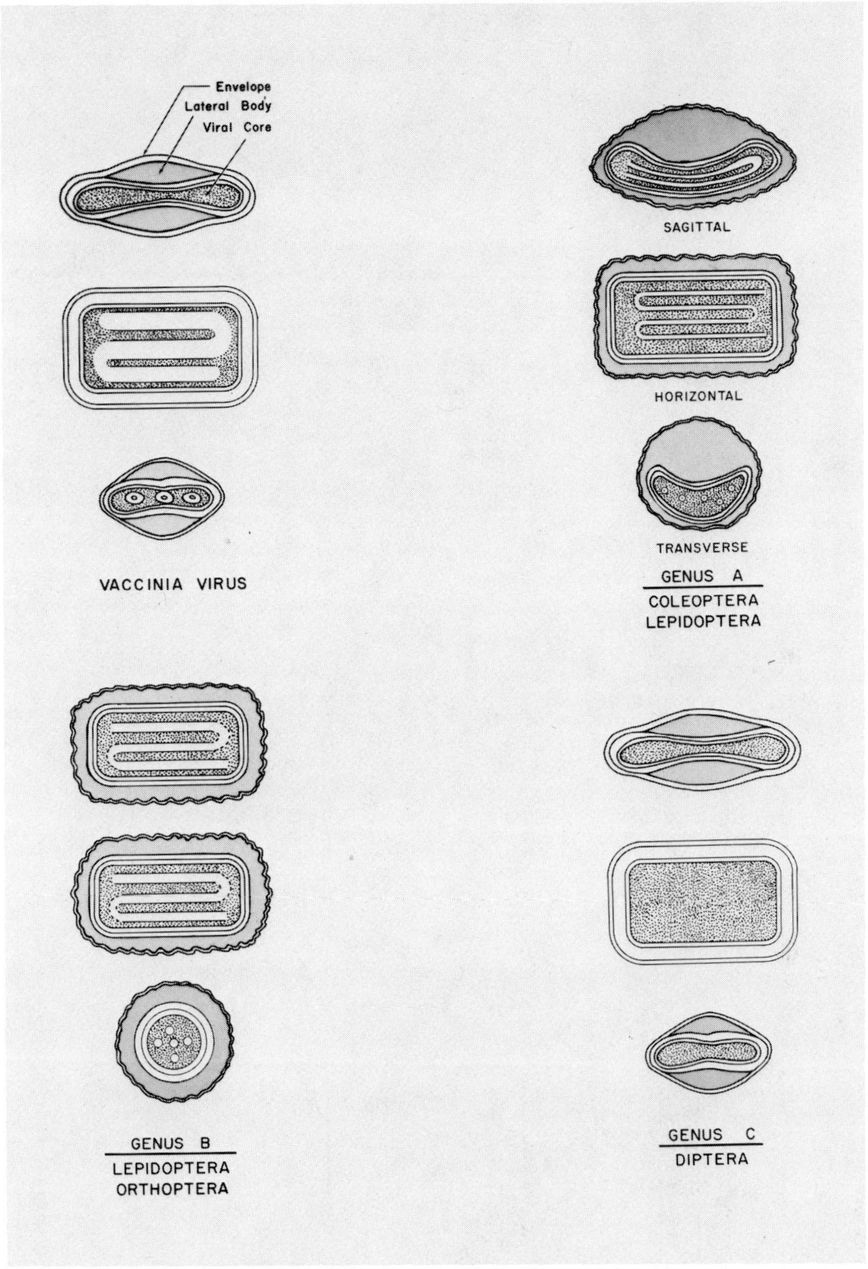

Figure 4-1. Diagrammatic representation of the morphology of 3 entomopoxvirus genera and the vertebrate vaccinia virus in 3 planes of symmetry. (Modified from Granados, 1973a).

Table 4-1. MORPHOLOGICAL AND PATHOLOGICAL CHARACTERISTICS OF SOME ENTOMOPOXVIRUS

EPV Genus and Insect Host	Virus size (nm)	Occlusion body size (μm)	Virus-free spindles	Tissues affected	Nuclear and/or Cytoplasmic fibrils	*In vitro* replication	References
GENUS A							
1. Coleoptera *Melolontha melolontha* (Scarabaeidae)	450 x 250	10-24	Present	Fat body Hemocytes	Absent	No	Vago, 1963 Bergoin et al., 1968a, b
Othnonius batesi (Scarabaeidae)	470 x 265	5-10	Present	Fat body Hypodermis	Absent	No	Goodwin and Filshie, 1969 Goodwin and Roberts, 1975
2. Lepidoptera *Oncopera alboguttata* (Hepialidae)	390 x 270 x 220	6- 8	Present	Fat body	Present	No	Milner and Beaton, 1979
GENUS B							
1. Lepidoptera *Amsacta moorei* (Arctiidae)	350 x 25	1- 4	Rare	Fat body Hemocytes Hypodermis Tracheoblasts Muscle Nerve cells Midgut	Present	Yes	Roberts and Granados, 1968 Granados, 1973a Granados and Naughton, 1975a, 1976a Quiot et al., 1975
Choristoneura beinnis (Tortricidae)	350 x 280	2- 3	Present	Fat body Midgut	Absent	No	Bird et al., 1971 Bird, 1974
2. Orthoptera *Melanoplus sanguinipes* (Acrididae)	320 x 250	2-11	Absent	Fat body	Absent	No	Henry et al., 1969
GENUS C							
1. Diptera *Chironomus luridus* (Chironomidae)	320 x 230 x 100	4- 7	Absent	Fat body Hemocytes Epidermis Oenocytes Imaginal discs of legs and reproductive organs, Connective tissue sheaths of muscle, nervous system, and intestine	Absent	No	Gotz et al., 1969; Huger et al., 1970
Chironomus attenatus	320 x 250	up to 6	Absent	Hemocytes	Absent	No	Stoltz and Summers, 1972

SYMPTOMATOLOGY AND PATHOLOGY

Coleoptera

Signs and Symptoms: In general, EPV infections of coleopteran species exhibit an exceptionally prolonged course of disease development and may be as long as 30 to 40 weeks depending upon the temperature (Hurpin, 1968; Milner and Lutton, 1975). The early macroscopic disease signs in infected *M. melolontha* larvae includes a distention or proliferation of diseased fat body, a chalky white appearance of the larvae, and a softening of the integument (Hurpin and Vago, 1963; Weiser, 1965). The lobes of the fat bodies have a foamy white appearance. Infected larvae exhibit a loss of turgidity and a decrease in mobility (Hurbin and Vago, 1963). Hemocytes are the only other cells infected and in the later stages of infection the hemolymph appears milky. All larval stages are susceptible to infection, the first two instars being the most susceptible (Robert, 1969).

In the case of black-soil scarab larvae (*Othnonius batesi*) signs of infections are not observed before 2 months post-infection (Goodwin and Roberts, 1975). Initially there is a proliferation of diseased fat body and hypodermis which leads to whitish appearance of the affected larva (Fig. 4-2). A characteristic change in color which takes place dorsoposteriorly over the dark rectal sac is a white median flecking which appears around the dorsal vessel in light or early infections and which becomes more pronounced in moderate or mid-stage infections. The color change results in a general chalky-white mottling in heavy or late infections. Goodwin and Roberts (1975) suggested that the presumed proliferation of tissue may be due to an abnormal whitening of diseased subdermal fat-body and hyperdermis. The whitening is a result of an increase of spindles and occlusion bodies (called spheroids or spherules by other authors) in these cells. Upon dissection, the fat body of diseased larvae appears a foamy white instead of the normal flat white of healthy insects. The foaminess is believed to be associated with the presence of many spindles in affected cells (Goodwin and Roberts, 1975). Milner and Lutton (1975) showed that *O. batesi* EPV infection had little effect on mortality, duration of instars, growth, or food uptake of *O. batesi* larvae. Infected larvae usually succumb shortly before or during pupation. Pupae and adults may also be infected. The disease signs in infected greyback cane beetle larvae (*Dermolepida albohirtum*) are similar to those described for *O. batesi* (Goodwin and Roberts, 1975).

Pathology: Fat body and hemocytes are the cell types most commonly affected by coleopteran EPV's. However hyperdermis and intestinal tracts may also be infected.

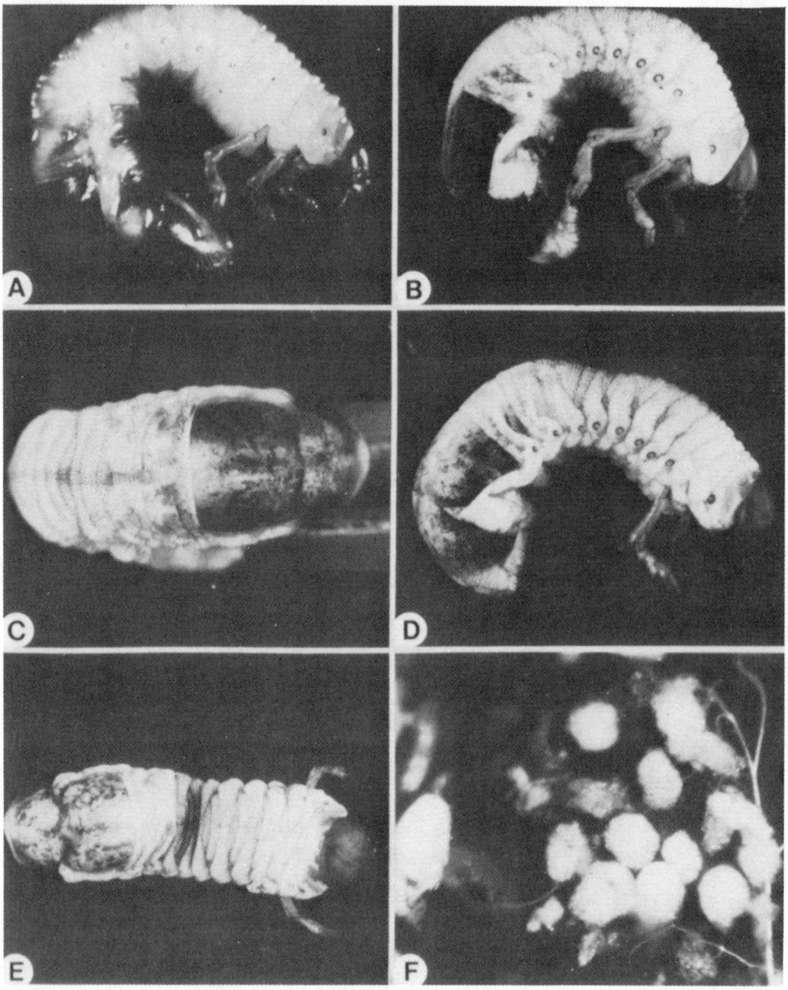

Figure 4-2. Healthy and entomopoxvirus - infected Othnonius batesi larvae. A. Healthy third-instar larva. B. Moderate virus infection of a third-instar larva showing a general whitening. C. Dorso-posterior view of larva in Fig. 2B, showing flecking due to presence of viral OB's and spindles in tissue adjacent to the cuticle over the rectal sac. D. and E. Lateral and dorsal views of a heavily infected third-instar larva. Note characteristic white mottling dorsoposteriorly over the rectal sac. F. Fat body lobes from diseased larvae showing a reflective foamy appearance typical of this infection. (from Goodwin and Roberts, 1975; reprinted by permission of Academic Press, Inc.)

Cytologically, the most striking feature of EPV infection is the presence of fusiform spindles and ovoid OB's in the cytoplasm of cells (Fig. 4-3). The length of spindles ranges from 5-10 μm, but maximum lengths of 15 to 23 μm have been reported (Goodwin and Roberts, 1975; Bergoin et al., 1976). The diameter of the OB's generally ranges between 5 to 20 μm. The most significant difference between these two inclusions, as revealed by electron microscopic investigations, is the presence of large ovoid virions occluded within the OB's whereas the spindles do not contain any particles. All but one described coleopteran EPV (i.e. D. albohirtum EPV) produce spindles in infected cells. In M. melolontha, O. batesi, and D. albohirtum the fat-body adjacent to the midgut and rectal sac is initially infected (Weiser, 1965; Goodwin and Roberts, 1975). In heavily infected fat-body of M. melolontha, fat droplets are entirely absent and in the late stages of infection, the fat bodies disaggregate and lyse (Hurpin and Vago, 1963; Weiser, 1965).

Lepidoptera

Signs and symptoms: In lepidopteran EPV infections the course of disease is relatively short, usually less than 3 weeks. Some species become whitish after infection (e.g. the army cutworm, Euxoa auxiliaris, and the winter moth, Operophtera brumata), but Estigmene acrea larvae remain dark in color (Weiser and Vago, 1966; Sutter, 1972; Granados, 1973a). Infected larvae may become lethargic and lose coordination and mobility during late stages of infection. Death of E. acrea larvae is often preceded by regurgitation or defecation of fluid containing OB's. Paralysis of the posterior two-thirds of the body shortly before death is typical in infected E. acrea larvae (Granados, 1970). The infection of pupae by some EPV's has been reported (Sutter, 1972; Retnakaran and Bird, 1972); however infections of the adult stage are not known. During late stages of infection the hemolymph of E. acrea (Granados, unpublished observations) and E. auxiliaris (Sutter, 1972) turns milky white in color due to the presence of many OB's from lysed cells. After death, O. brumata larvae may remain as hard and rigid cadavers (Weiser and Vago, 1966), while dead E. acrea larvae usually remain soft but intact (D. W. Roberts, unpublished data).

A peculiar hormonal imbalance syndrome was reported in C. fumiferana larvae infected with a 2 year-cycle spruce budworm, Choristoneura biennis EPV (Retnakaran and Bird, 1972). Wing buds were present in insects that died as sixth-instar larave and infected pupae retained prolegs. The last larval stadium was around 15 days as compared to 7 days for healthy larvae, and the diseased larvae were approximately twice the size of the control. This type of allometric development is

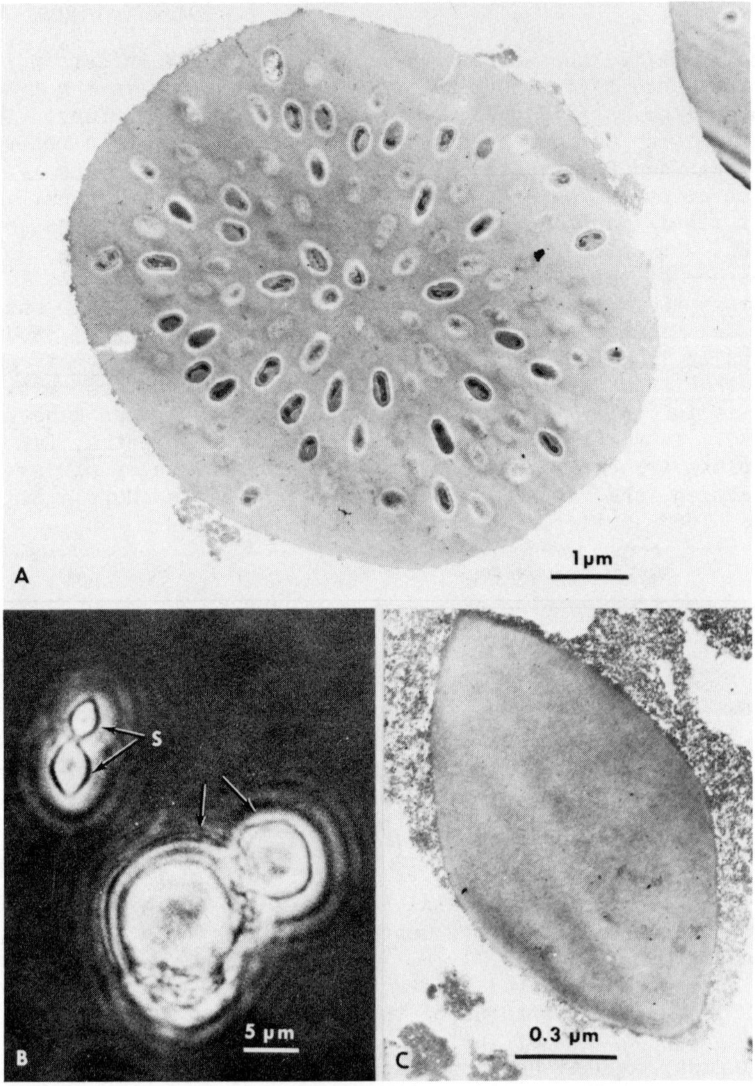

Figure 4-3. Occlusion bodies and virus-free spindles of Melolontha melolontha entomopoxvirus.

A. Thin section through a mature occlusion body. Note that the virions are embedded with their long axis oriented radially.

B. Phase contract photographs of spindle(s) and occlusion bodies (arrows).

C. Thin section through a virus-free spindle.

characteristic of insects having excess of juvenile hormone.

Pathology: The EPV affecting A. moorei is polytropic and may invade many cells and tissues including fat body, hemocytes, nerve cells, midgut cells, hypodermis, tracheoblasts, and muscle cells (Granados, 1973a). All other lepidopteran EPV's exhibit a more restricted cell and tissue specificity and in most instances only fat body cells are infected (Kurstak and Garzon, 1977). The most characteristic cytopathological effect is the presence of ellipsoidal OB's (usually 4-7 μm in the major axis) and virus-free spindles in infected cells. The spindle inclusions are not characteristic of all Lepidopteran EPV's, i.e., they are not present in E. auxiliaris EPV (Sutter, 1972) and rare in A. moorei EPV (Roberts, 1970). Spindles in coleopteran EPV infections are rather large and found free in the cell cytoplasm; however in C. biennis EPV, Choristoneura diversana EPV, and O. alboguttata EPV the spindles are small (microspindles) and are occluded together with the viral particles in the OB's (Bird et al., 1971; Bird, 1974; Cunningham et al., 1973; Katagiri, 1973; Milner and Beaton, 1979). The spindles produced by C. biennis EPV and C. diversana EPV measures about 100 x 500 nm whereas most coleopteran EPV spindles are approximately 5 x 10 μm in dimension. Cunningham et al. (1973) noted that three categories of spindles have been described in EPV infections - no spindles, microspindles, and macrospindles - and they suggested the possibility of using these categories as a taxonomic feature of EPV's.

The infection of midgut cells has been reported only for A. moorei EPV (Granados, 1973a, b) and C. biennis EPV (Bird et al., 1971); however, since the main portal of entry for EPV's is presumably via the midgut epithelia it is reasonable to conclude that there is probably limited replication in these cells by all EPV's. The degree of infection of E. acrea larval midgut cells by A. moorei EPV is limited, with the production of few OB's in most cells (R. Granados, unpublished observations). Although A. moorei EPV may infect various cells and tissues of E. acrea larvae, hemocytes and fat body appear to be most severely affected. In fat-body cells and hemocytes, numerous OB's occupy the cytoplasm within 7 to 9 days post-infection and in the late stages of infection, i.e. 12 to 14 days post-inoculation, these cells lyse and release their contents into the hemolymph. Phase contrast microscopy of wet mounts of either fat body or hemocytes (Fig. 4-4) shows the presence of coiled masses of long cytoplasmic and nuclear fibrils. These characteristic fibrils which have been reported only in lepidopteran EPV's were first described by Weiser and Vago (1966) and subsequently reported by others (Granados and Roberts, 1970; Sutter, 1972; Milner and Beaton, 1979; Weiser et al., 1970).

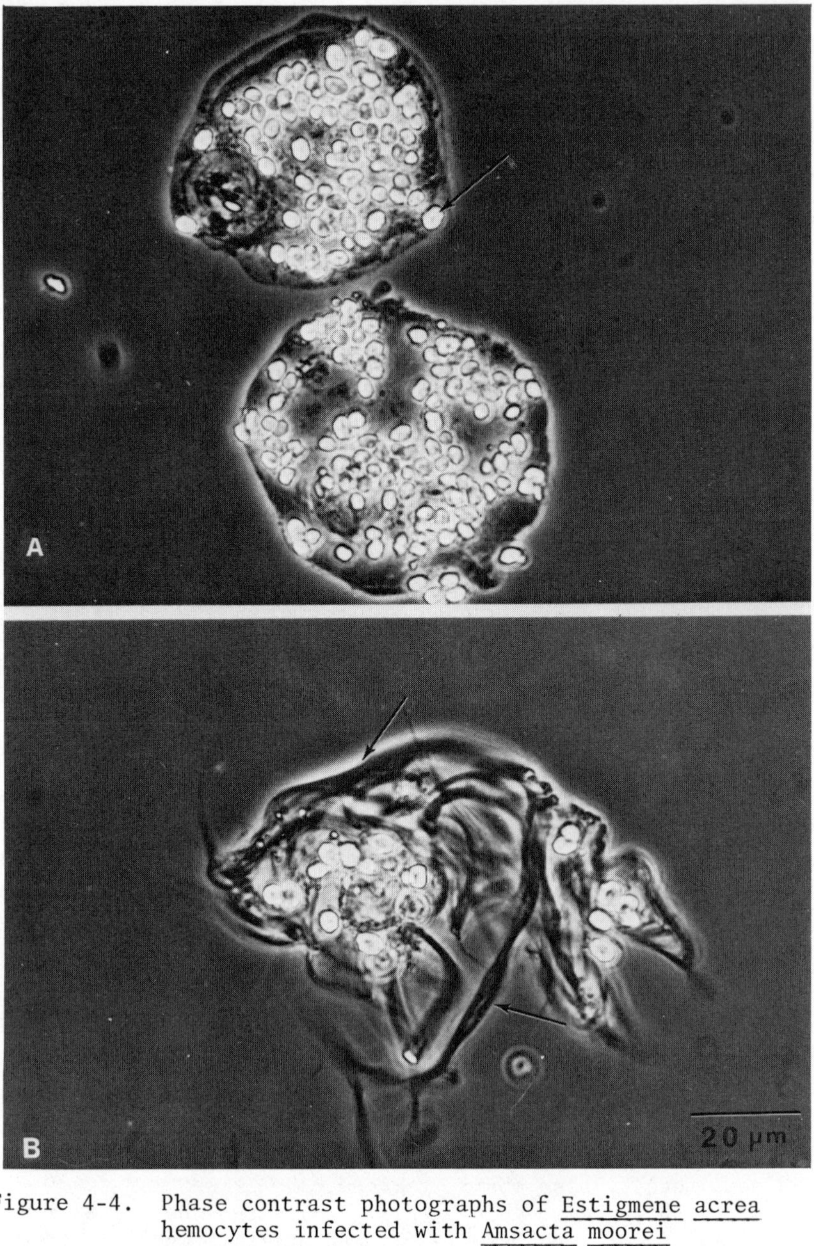

Figure 4-4. Phase contrast photographs of *Estigmene acrea* hemocytes infected with *Amsacta moorei* entomopoxvirus.

A. Two infected cells (72 hrs post-infection) showing numerous refractive occlusion bodies (arrows) in cytoplasm.

B. Lysed hemocyte showing numerous bundles of fibrils (arrows) being released from the cell. bar = 20 µm.

Diptera

Signs and symptoms: Diseased chironomid larvae are generally identified by the presence of irregularly distributed whitish areas beneath the integument (Huger et al., 1970; Stoltz and Summers, 1972; Federici et al., 1974; Harkrider and Hall, 1975). As the infection progresses the usual red color of the larvae gradually becomes whitish due to the accumulated masses of opaque white, infected hemocytes (e.g. *Goeldichironomus holoprasinus* and *Chironomus attenuatus*) or fat body (e.g. *Chironomus luridus*). Infected *Chironomus* sp. No. 51 larvae can be recognized in early stages of disease by the accumulation of infected hemocytes in the dorsal region of the 9th abdominal segment just posterior to the heart and infected 1st instar larvae may die 24 to 78 hrs post-inoculation (Harkrider and Hall, 1975, 1979). Diseased *G. holoprasinus* and *C. luridus* may live from 3 to 8 weeks although this has not been carefully documented (Federici et al., 1974; Huger et al., 1970). Infected *C. attenuatus* and *G. holoprasenus* larvae appear to be as active as uninfected ones until shortly before death (Stoltz and Summers, 1972; Federici et al., 1974).

Pathology: The *C. luridus* EPV appears to be polytropic, affecting various tissues including fat bodies, hemocytes, epidermis, oenocytes, imaginal discs of legs and genital organs, muscles, nervous system, and intestinal tract (Huger et al., 1970). Other chironomid EPV's show a high degree of tissue specificity and are known to infect only the hemoctyes of *C. attenuatus* (Stoltz and Summers, 1972), *G. holoprasinus* (Federici et al., 1974), and larval midges of the *Chironomus decorus* complex (Harkrider and Hall, 1975) or fat bodies of *Camptochironomus tentans* (Weiser, 1969). The most obvious sign of cytopathology is the presence of ellipsoidal cytoplasmic OB's ranging in size from 3 to 6 µm in diameter. In *G. holoprasinus*, the OB's may be ellipsoidal or spindle-shaped (Federici et al., 1974). Proliferation of affected cells has been reported for *C. luridus* EPV (Huger et al., 1970) and *C. tentans* EPV (Weiser, 1969), but Stoltz and Summers (1972) reported that mitosis could not be observed within infected hemocyte aggregates of *C. attenuatus*.

Orthoptera

Signs and symptoms: The migratory grasshopper, *Melanoplus sanguinipes*, is the only orthopteran reported to be infected by an EPV (Henry and Jutila, 1966; Henry et al., 1969). Infected nymphs may exhibit general torpor, a decreased rate of development, and a high rate of mortality. The abdomens of severely infected nymphs are often distended with protuding cervical membrane caused by large accumulations of OB's in the fat bodies. Diseased fat bodies become hypertrophied and eventually appear as irregular, grayish masses. Severely infected nymphs rarely develop beyond the last instar and remain there in a semimoribund condition. Adults that develop from less heavily infected nymphs appear "anemic" and are often sterile.

In general, third-instar nymphs that are inoculated *per os* with about 10^5 OB's die between 20 to 30 days post-inoculation. *Melanoplus sanguinipes* nymphs injected through the ventral thoracic membranes with virus inoculum may die as early as 7 to 8 days post treatment, with most insects dying 15 to 20 days after injection. Insects infected by injection usually have fewer and smaller OB's than insects infected *per os*. The OB's are usually discernible under light microscopy at 12 days after *per os* inoculation and at about 10 days after thoracic injection. The hemolymph of heavily infected nymphs often appears cloudy due to large amounts of OB's released from disrupted fat body cells.

Pathology: The lethal dose of infection is approximately 10^5 to 10^6 OB's per mg of grasshopper weight (Henry et al., 1969). The fat bodies of *M. sanguinipes* appear to be the only tissue affected by the EPV. Occlusion bodies have been observed in hemocytes but this is considered to result as a consequence of phagocytosis of OB's released from lysed fat body cells. Based on light and electron microscopic studies, Henry et al. (1969) stated that infections can occur in both the cytoplasm and nuclei of fat body cells; however, they did not present data showing virus and/or OB's within nuclei of infected cells.

THE REPLICATIVE CYCLE

Penetration

The attachment and entry of EPV's into susceptible cells may depend on the route of infection. If infection is initiated by peroral inoculation, the OB's are dissolved in the midintestine of inoculated larvae and penetration of midgut epithelial cells is initiated via fusion of envelopes of adsorbed virions with the membrane of microvilli (Granados, 1973b). This is followed by passage and migration of the virus core and lateral bodies into the cell cytoplasm proper. In intrahemocoelic infections due to injection of virus, however, viropexis was shown to be the process of entry into host cells (Devauchelle et al., 1971; Granados, 1973a; Kurstak and Garzon, 1977). Following viral attachment, the plasma membrane of the host cell invaginates and the virus enters the cell within a phagocytic vacuole (Fig. 4-5A and B). Within the phagocytic vacuole the viral envelope interacts with the vacuolar membrane in an as yet unspecified manner, resulting in dissolution of both membranes and release of the DNA-containing core into the cytoplasmic matrix. The second

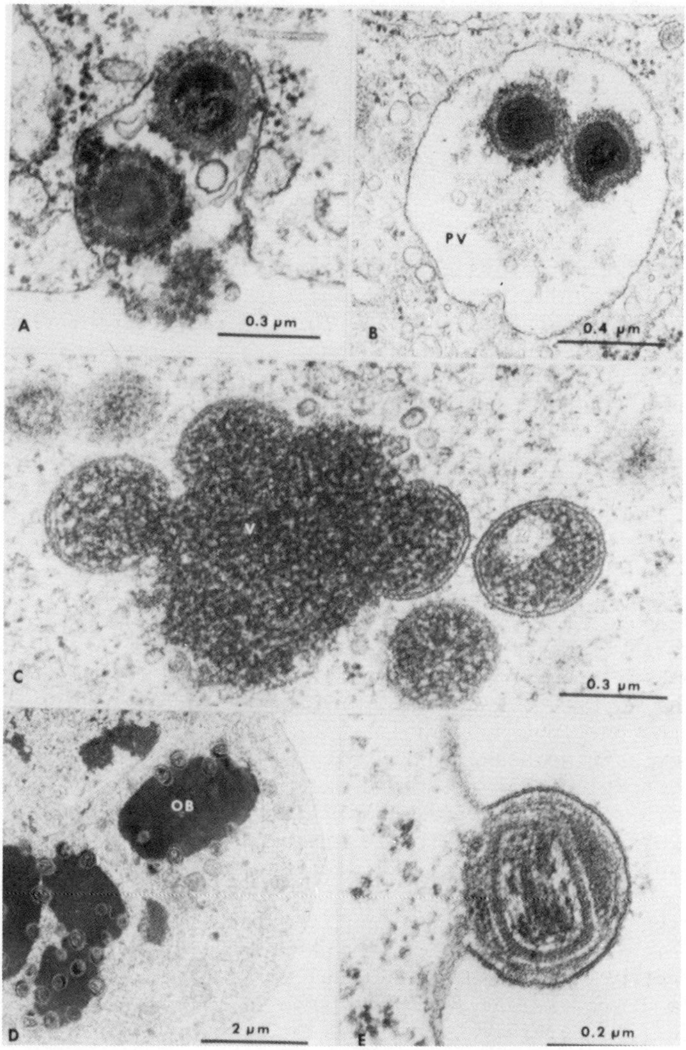

Figure 4-5. Entry and replication of _Amsacta moorei_ entomopoxvirus in _Estigmene acrea_ hemocytes.

A. Two virions being phagocytosed at the cell surface.

B. Virions in a phagocytic vacuole (PV) at ½ hr post-inoculation.

C. Viroplasm (V) and immature forms of the virus 24 hr post-infection.

D. Several developing occlusion bodies (OB) at 48 hr post-inoculation.

E. Mature virion budding at the plasma membrane.

stage of uncoating involves the release of the viral genome. Some electron microscopic images suggest that this might occur by rupture of the core (Devauchelle et al., 1971; Kurstak and Garzon, 1977).

Morphogenesis

EPV replication occurs within discrete cytoplasmic foci (viroplasms) in the vicinity of the nucleus (Bergoin et al., 1968a; Granados and Roberts, 1970). The initial stage of virus morphogenesis is the appearance of short arched segments which appear to consist of two unit membranes. The two membranes are approximately 75 Å in width and separated by a clear space of about 75-100 Å. During development of these viral membranes the surface area is increased while the curvature is maintained, resulting in the envelopment of immature virus particles in a spherical envelope (Fig. 4-5C). This envelope is a unique structure, discontinuous from any pre-existing host cell membranes, and it appears to arise from small membraneous spherical structures (600-800 Å in diameter), designated as micelles (Bergoin et al., 1969). The two membranes which comprise the envelope of immature particles have been observed in EPV's from Coleoptera (Bergoin et al., 1969) and Lepidoptera (Granados and Roberts, 1970). However, under certain conditions the envelope appears to consist of an inner trilaminar or unit membrane coated externally with a dense layer of short spicules (Stoltz and Summers, 1972; Kurstak and Garzon, 1977). Immature virus particles may be formed in association with 2 types of viroplasmic areas. Type I viroplasms appear as discrete, electron-dense amorphous masses (Fig. 4-5C), whereas type II viroplasms represent areas of loosely aggregated matrices of granular material. Immature particles which originate at the periphery of type I viroplasms give rise to dense spherical virus particles. In contrast, immature virions formed in type II viroplasms give rise to spherical particles, initially with a low density, but which change into dense particles by condensation of the enclosed material. Type I viroplasms have been reported for several EPV's (Goodwin and Filshie, 1969; Henry et al., 1969; Granados and Roberts, 1970; Kurstak and Garzon, 1977). Both viroplasm types have been reported for *M. melolontha* EPV, *A. moorei* EPV, *E. auxiliaris* EPV, *C. fumiferana* EPV and *O. alboguttata* EPV. Type II viroplasms are believed to be the principal site for replication of *M. melolontha* EPV (Bergoin et al., 1969) and *O. alboguttata* EPV (Milner and Beaton, 1979), whereas with *A. moorei* EPV, type I foci appear to predominate in infected cells. Only type II viroplasmic areas have been observed associated with *C. attenuatus* EPV (Stoltz and Summers, 1972).

All EPV's undergo differentiation from the spherical immature forms into mature particles by alterations occurring inside the virus envelopes. An early stage of differentiation is evident by the formation of nucleoids or viral cores inside the envelopes (Bergoin et al., 1969; Granados and Roberts, 1970; Stoltz and Summers, 1972). In later stages, the virus core develops into a rectangular-shaped core bounded by a core shell. This shell appears to be composed of an inner unit membrane and an outer layer of subunits (Stoltz and Summers, 1972; Granados, 1973a). Concomittant with viral core formation there is a lateral compression of the particle, the appearance of an electron-dense lateral body or bodies around the core, and the modification of the outer envelope into the beaded mulberry outline. Once the virions have matured, they move away from the viroplasmic areas and they may be occluded in OB's (Fig. 4-5D) or released from the infected host cell by exocytosis (Fig. 4-5E).

Devauchelle et al. (1971) compared the replication cycle of an EPV in fat bodies and hemocytes from M. melolontha larvae, and they determined that the infection cycle was different depending upon the cell type. In infected hemocytes the virions are rarely occluded within OB's and most often they are released at the cell surface by exocytosis. In contrast, the virions in diseased fat bodies are often occluded within OB's (Bergoin et al., 1969). Infection of the hemocytes was generally observed several weeks prior to infection of other insect tissues, and Devauchelle et al. (1971) suggested that an early replication cycle in hemocytes might assure a general contamination of other tissues. They also hypothesized that EPV's may occur as different virus types -- one type which multiplies and is occluded in OB's and another type which multiplies and undergoes release by budding at the cell membrane.

The development of M. sanquinipes EPV in both the cytoplasm and nucleus of insect cells (Henry et al., 1969) is contrary to the published data for almost all members of the family Poxviridae (Matthews, 1979) which show that replication of these viruses occurs within discrete cytoplasmic foci. However the possibility of intranuclear M. melolontha EPV morphogenesis was also suggested by Quiot et al. (1975) (see cell culture section). Other host or viral-induced cytopathological changes within nuclei of infected cells include virus-free crystalline or dense inclusions reported with EPV's from M. melolontha (Quiot et al., 1975), C. fumiferana (Kurstak and Garzon, 1977), A. moorei (Granados, 1973a), C. attenuatus (Stoltz and Summers, 1972) and C. luridus (Huger et al., 1970). At the present time, insufficient data is available to evaluate the role the nucleus plays in EPV replication.

Viral occlusion

The process of viral occlusion is very specific as only viral particles or microspindles (in the lepidopteran EPV's are occluded in paracrystalline OB's (Fig. 4-5D). Based on morphological and chemical properties these OB's may be compared to those of nuclear and cytoplasmic polyhedrosis viruses and the granulosis viruses (Bergoin et al., 1970). The process of nucleation and growth of OB's appears to commence only after formation of some mature virions and microspindles (when present). A characteristic peripheral structure was observed on OB's of C. attenuatus EPV (Stoltz and Summers, 1972) and the authors suggested that the crystallization or "nucleation" of OB protein might occur on the inner surface of this structure. In C. fumiferana it was observed that nucleation took place in various cytoplasmic sites and that virions could be the initial site of nucleation (Kurstak and Garzon, 1977). Growth of C. fumiferana EPV OB's occurred through continuous protein accumulation and crystallization. Fusion of small OB's may also lead to the formation of OB's (Kurstak and Garzon, 1977). Bergoin et al. (1968b, 1969) speculated that the accumulation of virions around developing OB's might arise from the "contraction of noncovalent links between free groups of the protein(s) of the viral envelope and those of the inclusion body protein(s) not yet crystallized."

In coleopteran EPV's the virus particles are commonly found occluded with their long axis oriented radially in relation to the center of the OB (Fig. 4-3A; Bergoin et al., 1971; Goodwin and Filshie, 1969), while with other EPV's the orientation of occluded particles appears to be random. The developing OB's have irregular outlines, with depressions in which the virions are inserted. The particles are normally confined to the central zones of fully formed OB's and are not embedded in the peripheral cortical area. In some instances, OB's may contain none or few occluded virions (Stoltz and Summers, 1972; Granados, 1973a). The shapes of fully formed OB's may vary from spheroid (coleopterous hosts) to ellipsoidal (lepidopterous hosts) and Goodwin (1968) has proposed the term spheroid as a useful general term for all virus-containing EPV OB's.

Further morphological changes of mature virions may occur after their occlusion within OB's (Goodwin and Filshie, 1969; Bergoin et al., 1968a; Stoltz and Summers, 1972; Sutter, 1972; Granados, 1973a). These structural transitions include: (1) formation of a "halo" of variable size around the virions, (2) the appearance of rope-like structures in the core, (3) a more electron dense core, (4) a wavy appearance of the outer virus envelope, (5) a reduction in size of the occluded virions; (6) a more biconcave and flattened core, and (7) disappearance of the electron-lucent layer surrounding the core.

Occlusion body and spindle formation induced by M. melolontha EPV was rare in infected hemocytes (Devauchelle et al., 1971) but is very common in infected fat cells. Stoltz and Summers (1972) observed that within given cells infected with C. attenuatus EPV OB formation and, to lesser extent, virus morphogenesis, appeared to be at a similar stage of development. They proposed that viral multiplication might not be a continuous process but may be synchronized in a manner as to maximize occlusion of virions.

Spindle formation

During the course of infection with some EPV's paracrystalline inclusion bodies devoid of virus particles are formed. Macrospindles (size range, 1-15 µm in their major axis) have been observed in the Coleoptera (Hurpin and Vago, 1963; Vago et al., 1968, 1969; Goodwin and Filshie, 1969) and in the Lepidoptera (Weiser and Vago, 1966; Meynadier et al., 1968). Microspindles (size range, 0.5-1.0 µm in their major axis) have only been reported from the Lepidoptera (Katagiri, 1973; Cunningham et al., 1973; Bird, 1974; Milner and Beaton, 1979). Unlike the macrospindles which are located only in the cytoplasm of infected cells, the microspindles are present in the cytoplasm and also frequently occluded within OB's .

The spindles are proteinaceous in nature and appear to be antigenically and chemically different than OB proteins (Bergoin et al., 1970; Croizier and Veyrunes, 1971). The structure and development of M. melolontha EPV spindles in fat body cells of M. melolontha larvae were described by Bergoin et al. (1976). Melolontha melolontha EPV spindles first appear in the cytoplasmic area next to the nuclear envelope approximately 1 month after inoculation. Spindles are present singly or in groups enveloped in ergastoplasmic cisternae. The origin of the spindle protein is believed to be the perinuclear space. This is suggested by the presence of masses of dense material accumulated between the inner and outer membranes of the nuclear envelope and the observation of spindles between these membranes. Spindle growth takes place within ergastoplasmic vesicles, and in highly infected cells the spindles are numerous and dispersed throughout the cytoplasm. According to Bergoin et al. (1976) the presence of small spindles near the nucleus suggests a continuous synthesis of these inclusions during the duration of the disease. On rare occasion, spindles were observed in the interior of the nucleus.

An entirely different mode of development of the spindles was described in the EPV of C. fumiferana (Bird, 1974). It was reported that spindles could arise from the fragmentation of proteinaceous cytoplasmic masses which were abundant in infected tissues. Subsequently, the spindles thus formed acquired a membrane and were incorporated into OB's at the same time as mature virions.

Spindle formation is not a unique expression of entomopoxvirus infections. Some cytoplasmic inclusions similar to the spindles associated with the entomopoxviruses have been described in insect cells infected with nuclear polyhedrosis viruses (Adams and Wilcox, 1968; Smirnoff, 1970; Cunningham, 1971).

Virus release

The release of enveloped virus progeny from infected tissues by exocytosis has been reported for several EPV's (Devauchelle et al., 1971; Stoltz and Summers, 1972; Granados, 1973a; Kurstak and Garzon, 1977). Nonoccluded, enveloped virus particles acquire a second envelope as they bud through the plasma membrane into the insect hemocoel (Fig. 4-5E). The virions liberated in this manner are infectious and probably contribute to the systemic infection of the host insect. Budding of virions also occurs under *in vitro* conditions (see cell culture section) and is probably a general mechanism for EPV release. The passage of *A. moorei* EPV by budding through the cell membrane of infected intestinal cells into adjacent cells has been observed (R. Granados, unpublished).

Pathophysiology

The physiological responses of insect hosts to EPV infections have been examined in only a few instances. The histochemical changes in DNA, RNA, protein, and glycogen in the fat body from *M. melolontha* larvae infected with *M. melolontha* EPV were observed from 5 to 195 days post-infection (Morris, 1966). Throughout the course of infection there was no apparent change in DNA content, although the average diameter of infected nuclei increased considerably. Cytoplasmic RNA increased progressively for approximately 75 days then gradually decreased, whereas the infection process was accompanied by an early increase and later reduction of nuclear RNA. Cytoplasmic protein content increased gradually throughout the disease process concomittant with an increase in formation of viral OB's in the cytoplasm. There was no significant change in glycogen reserves 23 days post-infection; subsequently, glycogen levels were greatly reduced. Acridine orange staining of *M. melolontha* EPV infected hemocytes showed an intense RNA synthesis in cells prior to the appearance of OB's and spindles (Kurstak and Garzon, 1977).

Huger et al. (1970) demonstrated by the periodic acid-Schiff's test that glycogen reserves were drastically reduced in C. luridus fat body cells during formation of EPV OB's.

Cell culture studies

Only A. moorei EPV has been successfully grown in vitro and the lack of other EPV-cell culture systems has been a major obstacle for studies on the molecular biology of these viruses. Amsacta moorei EPV has been grown in primary cultures from E. acrea larvae (Granados and Naughton, 1975a) and in established cell lines from several lepidopterous species (Table 4-2). Of the 2 hemocyte cell lines derived from E. acrea larvae, the one designated BTI-EAA is most commonly used since it allows a faster rate of virus development than BTI-EAH (Granados and Naughton, 1976a). It appears that the ultrastructural development of A. moorei EPV in cell cultures is similar to that observed in vivo. Entry of the virus into cells occurs by viropexis. Immature particles are formed at the periphery of viroplasms or within undifferentiated cytoplasmic areas. Following maturation of the spherical immature form to the mature form, the particles are either occluded in OB'sor released from the cell surface by exocytosis. As the nonoccluded virions bud through the plasma membrane they acquire a second viral envelope. Infected cells normally do not lyse. Fasicles of fibrils and crystalline nuclear inclusions which were observed in in vivo infections are also produced in infected cultured cells (Granados, 1973a; Quiot et al., 1975). An interesting observation reported by Quiot et al. (1975) was the apperance of structures resembling immature virus particles in the nuclei of infected L. dispar cell cultures. Thus, the possibility of intranuclear viral morphogenesis is suspected, but remains to be documented.

Various aspects of the growth characteristics of A. moorei EPV in BTI-EAA cell cultures were determined (Granados and Naughton, 1976b). The optimum temperature for infecting cultures was 28° C, while 25° C and 32° C were only slightly less suitable for infection. The infection rate at 20° C and 37° C was greatly reduced. Medium and cells from cultures infected for 5 to 7 days and centrifuged at 1000 xg for 10 min gave the highest yields, approximately 10^8 to 10^9 50% tissue culture infective dose ($TCID_{50}$) units/ml of cell-free supernatant virus. In addition to supernatant virus, tissue culture cell-associated virus, infectious hemolymph from E. acrea larvae, and virus purified from OB's were also infectious to cell cultures. A virus growth curve study showed that virus adsorption was followed by a latent period in which infectivity diminished and then by a period of exponential growth. An increase in infectious AMEPV was detectable at 24 hrs post-inoculation, and a maximum titer was obtained at 96 hrs for supernatant virus ($10^{9.25}$ $TCID_{50}$ units/ml)

TABLE 4-2

INSECT CELL LINES SUSCEPTIBLE AND NONSUSCEPTIBLE TO *AMSACTA MOOREI* ENTOMOPOXVIRUS

Species	Cell line designation	Virus replication[1/]	Reference
Invertebrate cells			
Estigmene acrea	BTI-EAA	Yes	Granados and Naughton, 1976a, b
E. acrea	BTI-EAH	Yes	Granados and Naughton, 1976a, b
Lymantria dispar	SCLD-135	Yes	Quiot et al., 1975
L. dispar	IPLB-65z	Yes	R. Granados, unpublished
Heliothis zea	IPLB-1075	Yes	R. Granados, unpublished
Spodoptera frugiperda	IPLB-21	Yes	R. Granados, unpublished
Bombyx mori	BM-N	Yes	R. Granados, unpublished
Trichoplusia ni	TN-368	No	R. Granados, unpublished
Aedes aegypti	-	No	R. Granados, unpublished
Vertebrate cells			
Human	HeLa	No	R. Granados, unpublished
Mouse	L-cells	No	R. Granados, unpublished

[1/] Virus replication based on light microscopy and ultrastructural observation of inoculated cultures.

and at 120 hrs for cell-associated virus ($10^{8.9}$ $TCID_{50}$ units/ml). In addition to the end-point dilution method for the assay of virus, a sensitive plaque assay technique for A. moorei EPV in BTI-EAA cells has been developed (Granados and Naughton, 1976; in preparation). Serial transfer of A. moorei EPV was carried out and after 32 to 35 serial passages, fewer OB's were produced in infected cells although other cytopathic effects including cell hypertrophy and granulation were evident. By the 47th passage only 1 to 5% of the infected cells appeared to support typical virus replication based on electron microscopical examination. In approximately 80 to 95% of the infected cells, viroplasmic foci and numerous aberrant, immature, and mature virus particles were observed. These cells did not appear to produce OB's.

The inhibition of A. moorei EPV replication in inoculated hemocyte cell cultures treated with rifampicin was demonstrated (Granados and Naughton, 1975b). Rifampicin has been shown to inhibit growth of vertebrate poxviruses by binding specifically to viral DNA-dependent RNA polymerase, with subsequent inhibition of transcription (Follet and Pennington, 1973). Treatment of inoculated cultures with 50 μg rifampicin per ml of medium did not affect A. moorei EPV morphogenesis; however, 100 μg rifampicin per ml blocked virus formation, and viral membranes with irregular contours accumulated around viroplasmic foci. Only a few small, dense inclusion bodies were detected. Cultures treated with 150 μg rifampicin contained only a few viroplasms and no virus particles. Inclusion bodies were not present.

CONCLUSION

Entomopoxviruses are the most recently discovered and least well investigated group of occluded-type insect viruses. They occur in insect populations worldwide, including agricultural and aquatic insects. They have been used in trials to control insect pests, but the interaction of EPV's with non-target hosts is poorly documented and requires greater attention. Studies on the biological properties of EPV's indicate that they are similar but distinct from poxviruses infecting vertebrate hosts.

The mode of action, specificity, and pathogenesis of only a few EPV's has been investigated. However, further stimulus for additional research on these viruses should result from the availability of susceptible insect cell lines (Table 4-2). These *in vitro* systems will permit controlled investigation on the cytopathological and biochemical changes induced by EPV infections and lead to a better understanding of this group of invertebrate viruses.

ACKNOWLEDGEMENTS

This work was supported in part by Grant No. PCM-7816121 from the National Science Foundation. The author thanks Ms. K. A. Lawler for her assistance in preparing this review.

REFERENCES

Adams, J. R. and Wilcox, T. A. 1968. Histopathology of the almond moth, Cadra cautella, infected with a nuclear-polyhedrosis virus. J. Invertebr. Pathol., 12, 269-274.

Arif, B. M. 1976. Isolation of an entomopoxvirus and characterization of its DNA. Virology, 69, 626-634.

Bergoin, M. and Dales, S. 1971. Comparative observations on poxviruses of invertebrates and vertebrates. In Comparative Virology, K. Maramorosch and E. Kurstak, eds., Academic Press, N.Y. pp. 169-205.

Bergoin, M., Devauchelle, G. and Vago, C. 1968a. Observations au microscope électronique sur le développement due virus de la "maladie à fuseaux" du coléoptère Melolontha melolontha L. C. R. Acad. Sci., Série D, 267, 382-385.

Bergoin, M., Devauchelle, G., Duthoit, J.-L. and Vago, C. 1968b. Etude au microscope électronique des inclusions de la virose à fuseaux des coléoptères. C. R. Acad. Sci., Série D, 266, 2126-2128.

Bergoin, M., Devauchelle, G. and Vago, C. 1969. Electron microscopy study of the pox-like virus of Melolontha melolontha L. (Coleoptera, Scarabaeidae). Virus morphogenesis. Arch. Gesamte Virusforsch. 28, 285-302.

Bergoin, M., Devauchelle, G. and Vago, C. 1971. Electron microscopy study of Melolontha poxvirus: The fine fine structure of occluded virions. Virology, 43, 453-467.

Bergoin, M., Veyrunes, J. C. and Scalla, R. 1970. Isolation and amino acid composition of the inclusions of Melolontha melolontha poxvirus. Virology, 40, 760-763.

Bergoin, M., Devauchelle, G. and Vago, C. 1976. Les inclusions fusiformes associeés à l'entomopoxvirus du coléoptère Melolontha melolontha. J. Ultrastruct. Res., 55, 17-30.

Bird, F. T. 1974. The development of spindle inclusions of Choristoneura fumiferana (Lepidoptera: Tortricidae) infected with entomopox virus. J. Invertebr. Pathol., 23, 325-332.

Bird, F. T., Sanders, C. J. and Burke, J. M. 1971. A newly discovered virus disease of the spruce budworm, Choristoneura biennis (Lepidoptera: Tortricidae). J. Invertebr. Pathol., 18, 159-161.

Croizier, G. and Veyrunes, J. C. 1971. Analyse immunochimique des inclusions de la "Maladie a Fuseaux" du coléoptère Melolontha melolontha L. Ann. Inst. Pasteur, 120, 709-715.

Cunningham, J. C. 1971. An ultrastructural study of the development of a nuclear polyhedrosis virus of the eastern hemlock looper, Lambdina fiscellaria fiscellaria. Can. J. Microbiol., 17, 69-72.

Cunningham, J. C., Burke, J. M. and Arif, B. M. 1973. An entomopoxvirus found in populations of the large aspen tortex. Choristoneura conflictana (Lepidoptera: Tortricidae) in Ontario. Can. Entomol., 105, 767-773.

Devauchelle, G., Bergoin, M. and Vago, C. 1971. Etude ultrastructurale du cycle de réplication d'un Entomopoxvirus dans les hémocytes de son hôte. J. Ultrastruct. Res., 37, 301-321.

Federici, B. A., Granados, R. R., Anthony, D. W. and Hazard, E. I. 1974. An entomopoxvirus and nonoccluded virus-like particles in larvae of the chironomid Goeldichironomus holoprasinum. J. Invertebr. Pathol., 23, 117-120.

Follett, E. A. C. and Pennington, T. H. 1973. The mode of action of rifamycins and related compounds on poxviruses. In Advances in Virus Research, M. A. Lauffer, F. B. Bang, K. Maramorosch, and K. Smith, eds., 18, 105-142.

Gotz, P., Huger, A. M. and Krieg, A. 1969. Uber ein insektenpathogens virus aus der gruppe der pickenviren. Naturwissenschaften, 56, 145.

Goodwin, R. H. 1968. Use of the term "virus inclusion body" and vernacular virus names. J. Invertebr. Pathol., 12, 479-480.

Goodwin, R. H. and Filshie, B. K. 1969. Morphology and development of an occluded virus from the blacksoil scarab, Othnonius batesi. J. Invertebr. Pathol., 13, 317-329.

Goodwin, R. H. and Roberts, R. J. 1975. Diagnosis and infectivity of entomopoxviruses from three Australian scarab beetle larvae (Coleoptera: Scarabaeidae). J. Invertebr. Pathol., 25, 47-57.

Granados, R. R. 1970. Development of an insect poxvirus in the nervous system of Estigmene acrea larvae. Proc. 4th Int'l Colloq. Insect Pathol., College Park, MD. pp. 110-114.

Granados, R. R. 1973a. Insect poxviruses: Pathology, morphology, and development. Misc. Publ. Entomol. Soc. Am., 9, 73-94.

Granados, R. R. 1973b. Entry of an insect poxvirus by fusion of the virus envelope with the host cell membrane. Virology, 52, 305-309.

Granados, R. R. 1978. The biology of cytoplasmic polyhedrosis viruses and entomopoxviruses. In Viral Pesticides: Present Knowledge and Potential Effects on Public and Environmental Health, M. D. Summers and C. Y. Kawanishi, eds., U. S. Environmental Protection Agency, Research Triangle Park, North Carolina, pp. 89-102.

Granados, R. R. and Naughton, M. 1975a. Development of Amsacta moorei entomopoxvirus in ovarian and hemocyte cultures from Estigmene acrea larvae. Intervirology, 5, 62-68.

Granados, R. R. and Naughton, M. 1975b. Effect of rifampicin on Amsacta entomopoxvirus morphogenesis. Abstract, 3rd Int. Virol. Congr., Madrid, Spain, p. 96.

Granados, R. R. and Naughton, M. 1976a. Replication of Amsacta moorei entomopoxvirus and Autographa californica nuclear polyhedrosis virus in hemocyte cell lines from Estigmene acrea. In Invertebrate Tissue Culture. Applications in Medicine, Biology, and Agriculture, K. Maramorosch and E. Kurstak, eds., Academic Press, N.Y. pp. 379-389.

Granados, R. R. and Naughton, M. 1976b. Replication of Amsacta moorei entomopoxvirus in a continuous cell culture from Estigmene acrea. Proc. First Int'l Colloq. Invertebr. Pathol., Kingston, Ont., Canada, pp. 113-117.

Granados, R. R. and Roberts, D. W. 1970. Electron microscopy of a poxlike virus infecting an invertebrate host. Virology, 40, 230-243.

Harkrider, J. R. and Hall, I. M. 1975. The occurrence of an entomopoxvirus in a field population of Chironomus sp. Proc. Calif. Mosq. Cont. Assoc., 43, 103-106.

Harkrider, J. R. and Hall, I. M. 1979. The effect of an entomopoxvirus on larval populations of an undescribed midge species in the Chironomus decorus complex under laboratory conditions. Environ. Entomol., 8, 631-635.

Henry, J. E. and Jutila, J. W. 1966. The isolation of a polyhedrosis virus from a grasshopper. J. Invertebr. Pathol., 8, 417-418.

Henry, J. E., Nelson, B. P. and Jutila, J. W. 1969. Pathology and development of the grasshopper inclusion body virus in Melanoplus sanguinipes. J. Virol., 3, 605-610.

Huger, A. M., Krieg, A., Emschermann, P. and Gotz, P. 1970. Further studies on Polypoxvirus chironomi, an insect virus of the pox group isolated from the midge Chironomus luridus. J. Invertebr. Pathol., 15, 253-261.

Hurpin, B. 1968. The influence of temperature and larval stage on certain diseases of Melolontha melolontha. J. Invertebr. Pathol., 10, 252-262.

Hurpin, B. and Vago, C. 1963. Une meladie à inclusions cytoplasmiques fusiformes chez le coléoptère Melolontha melolontha (L.). Rév. Pathol. Veg. Entomol. Agric. Fr., 42, 115-117.

Katagiri, K. 1973. A newly discovered entomopoxvirus of Choristoneura diversana (Lepidoptera: Tortricidae). J. Invertebr. Pathol., 22, 300-302.

Kurstak, E. and Garzon, S. 1977. Entomopoxviruses (Poxviruses of invertebrates). In The Atlas of Insects and Plant Viruses, Including Mycoplasma Viruses and Viroids, K. Maramorosch, ed., Academic Press, N.Y. pp. 29-66.

Langridge, W. H. R. and Roberts, D. W. 1977. Molecular weight of DNA from four entomopoxviruses determined by electron microscopy. J. Virol., 21, 301-308.
Langridge, W. H. R., Bozarth, R. F. and Roberts, D. W. 1977. The base composition of entomopoxvirus DNA. Virology, 76, 616-620.
Matthews, R. E. F. 1979. Classification and nomenclature of viruses. Third report of the International Committee on Taxonomy of Viruses. Intervirology, 12, 132-296.
Meynadier, G., Fosset, J., Vago, C., Duthoit, J.-L. and Bres, N. 1968. Une virose a inclusions ovides chez un lépidoptère. Ann. Epiphyt., 19, 703-706.
Milner, R. J. and Beaton, C. D. 1979. An entomopoxvirus from Oncopera alboguttata (Lepidoptera: Hepialidae) in Australia. Intervirology, 11, 341-350.
Milner, R. J. and Lutton, G. G. 1975. The pathogenicity of an entomopoxvirus from Othnonius batesi (Col.: Scarabaeidae) and its possible use as a control agent. Entomophaga, 20, 213-220.
Morris, O. H. 1966. Histochemical changes in Melolontha melolontha (Linnaeus) (Coleoptera) infected by a new virus. Can. J. Microbiol., 12, 965-971.
Quiot, J. M., Bergoin, M. and Vago, C. 1975. Développment et pathogenèse d'un entomopoxvirus en culture cellulaire de lépidoptère. C. R. Acad. Sci., Série D, 280, 2273-2275.
Retnakaran, A. and Bird, F. T. 1972. Apparent hormone imbalance syndrome caused by an insect virus. J. Invertebr. Pathol., 20, 358-360.
Robert, P. H. 1969. Étude en laboratorie de l'influence du stade larvaire de la quantité de virus et de la température sur la virose a vagoia-virus de Melolontha melolontha L. (Coléopt. Scarabaeidae). Ann. Zool. Ecol. Anim., 1, 289-307.
Roberts, D. W. 1970. Poxvirus of arctiid larvae: Infectivity for greater wax moth. Bacteriol. Proc. (Abstr.), p. 177.
Roberts, D. W. and Granados, R. R. 1968. A poxlike virus from Amsacta moorei (Lepidoptera: Arctiidae). J. Invertebr. Pathol., 12, 141-143.
Smirnoff, W. A. 1970. Rhombohedral twisted crystalloids found in the cytoplasm of nuclear-polyhedrosis-infected larvae of the ugly-nest caterpillar. Can. J. Microbiol., 16, 906-907.
Stoltz, D. B. and Summers, M. D. 1972. Observations on the morphogenesis and structure of a hemocytic poxvirus in the midge Chironomus attenuatus. J. Ultrastruct. Res., 40, 581-598.
Sutter, G. R. 1972. A pox virus of the army cutworm. J. Invertebr. Pathol., 19, 375-382.

Vago, C. 1963. A new type of insect virus. J. Insect Pathol., 5, 275-276.

Vago, C., Amargier, A., Hurpin, B., Meynadier, G. and Duthoit, J.-L. 1968. Virose a fuseaux d'un scarabaéide d'amérique de sud. Entomophaga, 13, 373-376.

Vago, C., Robert, P., Amargier, A. and Duthoit, J.-L. 1969. Nouvelle virose a spheroides et a fuseaux observeé chez le coléoptere *Phyllopertha horticola* L. Mikroskropie, 25, 378-386.

Weiser, J. 1965. *Vagoiavirus* gen. n., a virus causing disease in insects. J. Invertebr. Pathol., 7, 82-85.

Weiser, J. 1969. A pox-like virus in the midge *Camptochironomus tentans*. Acta Virol., 13, 549-553.

Weiser, J. and Vago, C. 1966. A newly described virus of the winter moth, *Operophtera brumata* Hübner (Lepidoptera: Geometridae). J. Invertebr. Pathol., 8, 314-319.

Weiser, J., Tchubianishvili, C. and Zizka, Z. 1970. Ultrastructure of the spindle-shaped virus, *Vagoiavirus operophterae*, in the winter moth, *Operophtera brumata* L. Acta Virol., 14, 314-317.

CHAPTER 5

VIRAL DISEASES OF INVERTEBRATES OTHER THAN INSECTS

by John A. Couch*

INTRODUCTION

Considerable recent interest has focused on the discovery of viruses that cause diseases in non-insect invertebrates. Since Vago (1966) reported the first virus disease of a marine crustacean, viruses and/or viral diseases have been reported for shrimps, crabs, daphnids, entoniscid isopods, oysters of several genera, octopods and squids, annelids, trematodes, nematodes, hydra, and a sponge. Most of the reports contained only brief descriptions of the virus with little information on pathogenesis in the hosts. A few reports have been more extensive and contain detailed descriptions of virus-related pathogenesis and virus-host cell relationships.

This chapter will describe some aspects of the pathogenesis of better known virus diseases in several exemplary non-insect invertebrate hosts. Also included will be brief descriptions of some lesser known viruses in other invertebrates to demonstrate the range of relationships among viral groups and non-insect invertebrates. Available information on the histopathogenesis, cellular pathogenesis, physiological, and behavioral responses to select host-virus interactions will be discussed under the headings of virus/host types (Table 5-1).

*United States Environmental Protection Agency, Environmental Research Laboratory, Sabine Island, Gulf Breeze, Florida 32561

PATHOGENESIS OF INVERTEBRATE MICROBIAL DISEASES. Davidson, ed. © Allanheld, Osmun (Totowa, N.J.)

TABLE 5-1

A PARTIAL LIST OF VIRUS/HOST PATHOLOGICAL RELATIONSHIPS IN NON-INSECT INVERTEBRATES (STARRED VIRUSES * ARE DISCUSSED IN GREATER DETAIL IN TEXT)

Virus[1]	Host	Tissue/cell type	Reference
*Baculovirus	Penaeid shrimps (*Penaeus* spp.)	hepatopancreas epithelial cells; midgut	Couch, 1974b
Baculovirus	Penaeid shrimp (*Penaeus monodon*)	hepatopancreas epithelial cells	Lightner & Redman, 1981 Lightner et al.,1981
*Baculovirus	Blue crab (*Callinectes sapidus*)	hepatopancreas	Johnson, 1976b
*Baculovirus	Crabs (*Carcinus mediterraneus*) (*Carcinus maenas*)	epithelial cells, hepatopancreas	Bazin et al., 1974
*Spherical virus	*Carcinus maenas*	hemocytes	Bang, 1971
Paramyovirus or Bunyavirus	Swimming Crab (*Macropipus depurator*)	hepatopancreas cells; hemocytes	Bonami & Vago, 1971
Paralysis virus (Reovirus)	Swimming Crab (*M. depurator*)	hemocytes	Bonami, J. R. 1973
*Picornavirus	Isopod (*Portunion conformis*) Shore crab (*Hemigrapsus oregonensis*)	hepatopancreas	Kuris et al, 1979
*Herpes like virus	Blue crab (*C. sapidus*)	hemocytes; gill epithelium	Johnson, 1976b
*Rhabdolike virus	Blue crab (*C. sapidus*)	cells of mandibular gland; hemocytes	Yudin & Clar, 1979
Paramyxovirus (enveloped helical virus)	Blue crab (*C. sapidus*)	hemocytes; hemopoietic tissue	Johnson & Farley, 1980
*Reolike virus	Blue crab (*C. sapidus*)	hemocytes; hemopoietic tissue	Johnson & Bodammer, 1975

Iridovirus	Daphnid (*Simocephalus expinosus*)	adipose and glandular tissue; nerve	Federici & Hazard, 1975
Cytoplasmic Polyhedrosis virus	Daphnid (*S. expinosus*)	adipose and glandular tissue; nerve	
*Herpes-like virus	American Oyster (*Crassostrea virginica*)	midgut epithelial cells; hemocytes (?)	Farley et al., 1972
*Hexagonal virus	Pacific oyster (larvae) (*Crassostrea gigas*)	velar epithelial cells	Elston, 1979, 1980
*Phage	Rickettsia parasitic in the marine bivalve, *Tellinia tenuis*	Rickettsia, host of phage in cytoplasm of digestive gland cells of bivalve	Buchanan, 1978
Reovirus	Squid (*Sepia officinalis*)	stomach epithelial cells	Devauchelle & Vago, 1971
*Hexagonal virus like particles	*Octopus vulgaris*	ventral mantle; arms; in tumors	Rungger et al., 1977
*Spherical virus like particles	Nematode (*Trichosomoides crassicauda*)	somatic cells (hypodermal cells)	
*Adenovirus	*Hydra vulgaris*	epithelial cells	Bonnefoy et al., 1972
Adenovirus	Sponge (*Verongia cavernicola*)	giant cells (possibly degenerating oocytes)	Vacelet and Gullissian, 1978

[1] Group or taxonomic designations for viruses are those used or suggested by original authors and may be subject to change as more is learned of these viruses.

BACULOVIRUS INFECTIONS IN PENAEID SHRIMPS

Until 1973 *Baculovirus* infections (including nuclear polyhedrosis viruses) were unknown in hosts other than insects and mites (see Chapter 1). During 1973 and 1974 two *Baculovirus* forms were reported, one in penaeid shrimps (Couch and Niumo, 1973; Couch 1974a) and one in crabs (Bazin et al., 1974). Work on the shrimp virus distribution (Couch et al., 1975; Couch, 1978), ecology (Couch et al., 1975), ultrastructure (Couch, 1974 a, b, 1978) cytopathic effects (Couch, 1974 a, b; Couch et al., 1975; Couch, 1978), and physical and biochemical characterization (Summers, 1977) has been considerable. Therefore, this virus-host system will be described in some detail.

Histopathogenesis

Hepatopancreatic epithelial cells are the major tissue type infected in the following penaeid shrimps: pink shrimp, *Penaeus duorarum*; brown shrimp, *Penaeus aztecus*; white shrimp, *Penaeus setiferus*; and possibly the Pacific penaeid, *Penaeus monodon*. Infections occur in the nucleus, and are evident with light microscopy (Couch, 1974 a, b; Couch, et al., 1975) because of the formation of large intranuclear, tetrahedral inclusion bodies (polyhedral inclusions - PIB's) in the presence of rod-shaped virions (Figs. 5-1, 5-2). The major results of histopathologic damage in infections are destruction of segments or areas of hepatopancreatic tubule epithelium by growth of PIB's in and eruptions from individual cells (Figs. 5-3, 5-4).

Cellular and ultrastructural pathogenesis

Table 5-2 presents the nuclear and cytoplasmic virus-associated changes that occur in the chronological order of their appearance in infected shrimp cells. Infected hepatopancreatic cells may show bizarre departures from normal in both the nucleus and cytoplasm (Figs. 5-5, 5-6).

The first indication of infection at the cellular level is hypertrophy of the hepatopancreatic epithelial cell nucleus to two to five times normal diameter (Fig. 5-7). Nuclear hypertrophy is not usually accompanied by cytoplasmic increase; therefore, the nucleus is obviously enlarged. Diminution and margination of heterochromatin next occurs (Figs. 5-8). Eventually all heterochromatin disappears from the nucleus (Fig. 5-9). Concurrent with loss of heterochromatin is the appearance of fibrillar and granular stromata in the nucleoplasm (Figs. 5-5, 5-8, 5-9). Soon after this, the virions appear and growth of the tetrahedral inclusion body begins (Fig. 5-6).

Cytoplasmic changes associated with virus replication and synthesis are extreme and unusual, even in comparison with known cytopathic effects of other viruses. The most striking cytoplasmic event in infection is the extreme proliferation of the outer nuclear membrane resulting in the formation of the membranous labyrinth within the cytoplasm (Couch, 1974a) (Figs. 5-5, 5-6, 5-8). This membrane proliferation is so distinctive that at the ultrastructural level it is pathognomic of Baculovirus infections in shrimp cells. Myelenoid body formation in the cytoplasm may result partially from the collapse of proliferated nuclear membranes into concentric masses (Fig. 5-6). Mitochondria become enlarged and distorted as nuclear replication of viruses occurs (Fig. 5-6); this probably results from loss of membrane permeability control and osmotic balance controls in the infected cell. The second most striking cytoplasmic change, after nuclear membrane proliferation, is the loss of larger organelles in advanced infections concomitant with all cytoplasmic space becoming filled with free ribosomes (Fig. 5-9). At this stage, the cell may lyse or become necrotic.

Physiological effects

Few physiological characteristics of shrimps have been investigated in relationship to Baculovirus infections. From studies to date, however, there appear to be two important functional effects.

The first is the direct, observable effect of viral infection, i.e. replication in, and destruction of vital hepatopancreatic epithelial cells. Extensive epithelial tissues are destroyed when infected cells are lysed (Figs. 5-1, 5-3). The loss of tubular embryonic cells, absorptive cells, and secretory cells in the hepatopancreas leads to the loss of the major functions of this important organ. The loss of tubular growth capacity, nutrient absorption, and secretory, digestive functions may constitute the major, direct dysfunctional results of infection. Further, infected larval shrimp may lose, at a critical period in their growth, the majority of their early, developing hepatopancreas.

The second functional effect is less directly observed and probably involves an interaction between the virus and stress moderated effects in the host. Couch and Courtney (1977) demonstrated that shrimp that were sublethally stressed with the toxic polychorinated biphenyl (PCB) Aroclor 1254 were more susceptible to infection and spread of the Baculovirus than were non-stressed shrimp. The process causing this loss of resistance to viral spread is not known, but may involve

Figure 5-1. Baculovirus polyhedral inclusion bodies (PIB's) in, and free of, epithelial cells from hepatopancreas of penaeid shrimp, *Penaeus duorarum* (Phase contrast).

Figure 5-2. Tetrahedral PIB's released from shrimp cells (Phase contrast).

Figure 5-3. Rupture and release of PIB's from heavily *Baculovirus* infected hepatopancreas epithelium of *P. duorarum*.

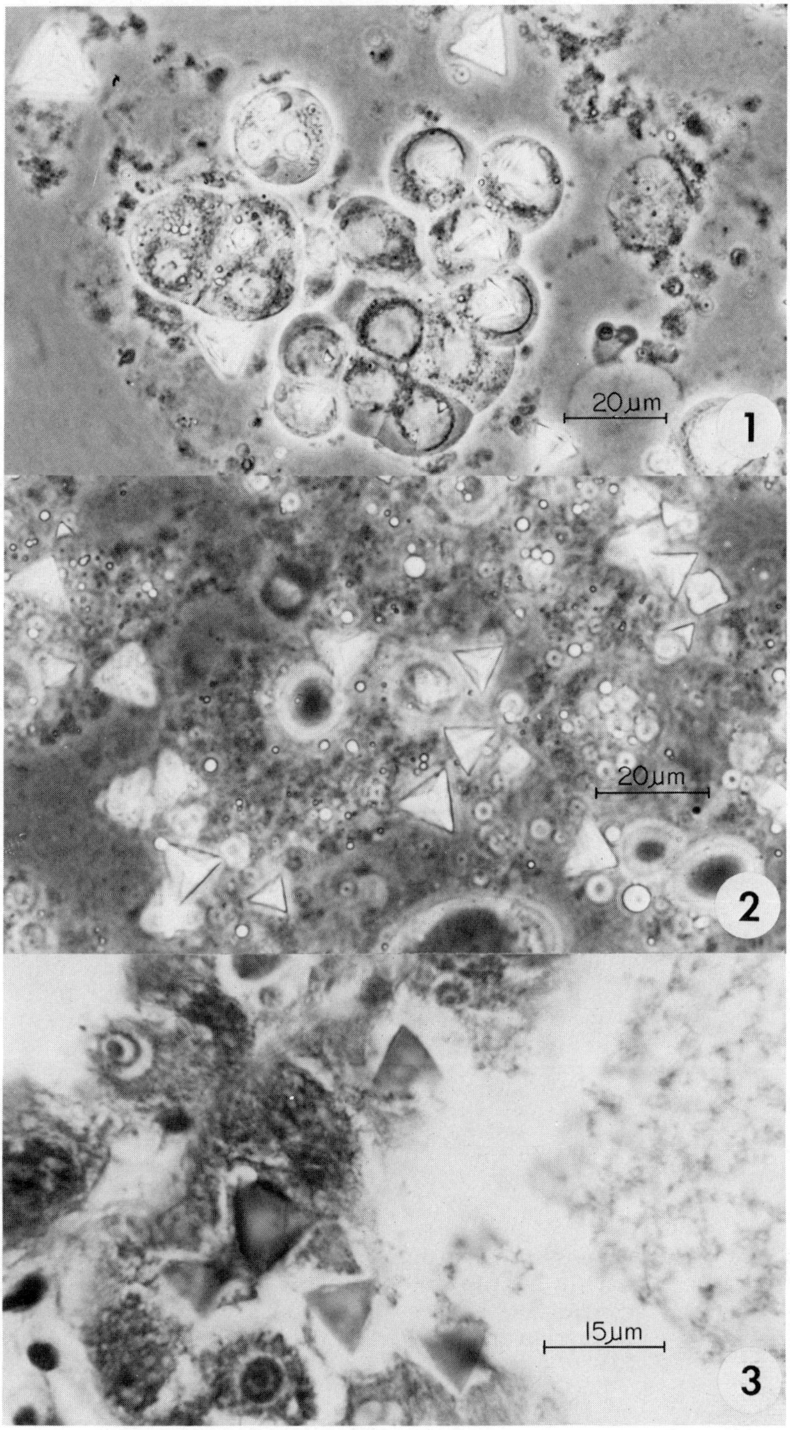
20 µm
1
20 µm
2
15 µm
3

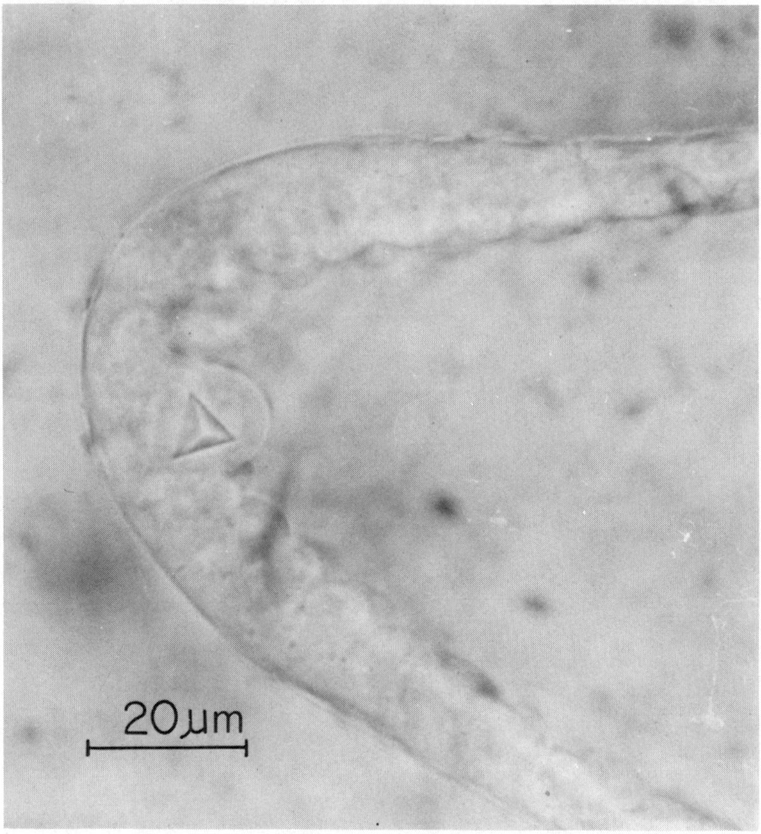

Figure 5-4. Acinus of hepatopancreas from *Penaeus duorarum* with a single embryonic cell nucleus (epithelium) containing a large PIB.Growth of the PIB will soon rupture the cell freeing the virion-laden PIB.

TABLE 5-2

CELLULAR PATHOGENESIS IN BACULOVIRUS-INFECTED SHRIMP CELLS

Nuclear	Cytoplasmic
1. Nuclear hypertrophy	1. Cytoplasmic diminution
2. Heterochromatin diminution	2. Formation of membranous labyrinth[1]
3. Heterochromatin margination	3. Myelenoid bodies formed
4. Disappearance of heterochromatin	4. Mitochondria enlarged, distorted
5. Nucleolar loss	5. ER vesicles enlarged
6. Nucleoplasm fibrillar stroma and granular stroma	6. Cytoplasm filled with free ribosomes
7. Nuclear membrane proliferation[1]	7. Disappearance of all organelles
8. Appearance and growth of polyhedral inclusion body (PIB)	
9. Virion-associated membranes and materials present in nucleus	

[1]These nuclear and cytoplasmic events are interrelated, i.e. the membranous labyrinth of the cytoplasm is generated by the proliferation of outer nuclear membrane (Fig. 5-5) (Couch, unpub.).

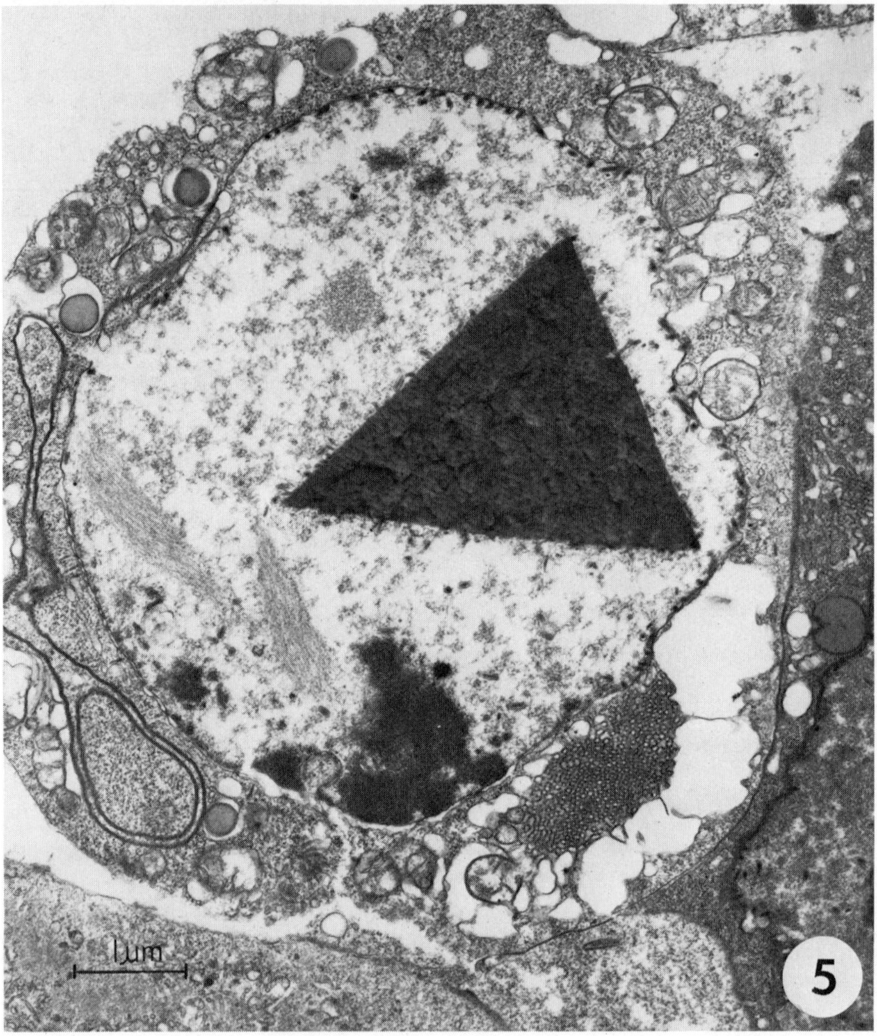

Figure 5-5. Electron micrograph of hepatopancreatic cell infected with Baculovirus penaei; note the nuclear hypertrophy, heterochromatin diminution, and growth of large PIB in nucleoplasm with fibrillar and granular stromata in nucleus; cytoplasmic changes are obvious and consist of membrane proliferation or formation of membranous labyrinth which is pathognomonic for the shrimp Baculovirus, mitochondrial swelling, and free ribosome increase. Note virions adjacent to inner layer of nuclear envelope.

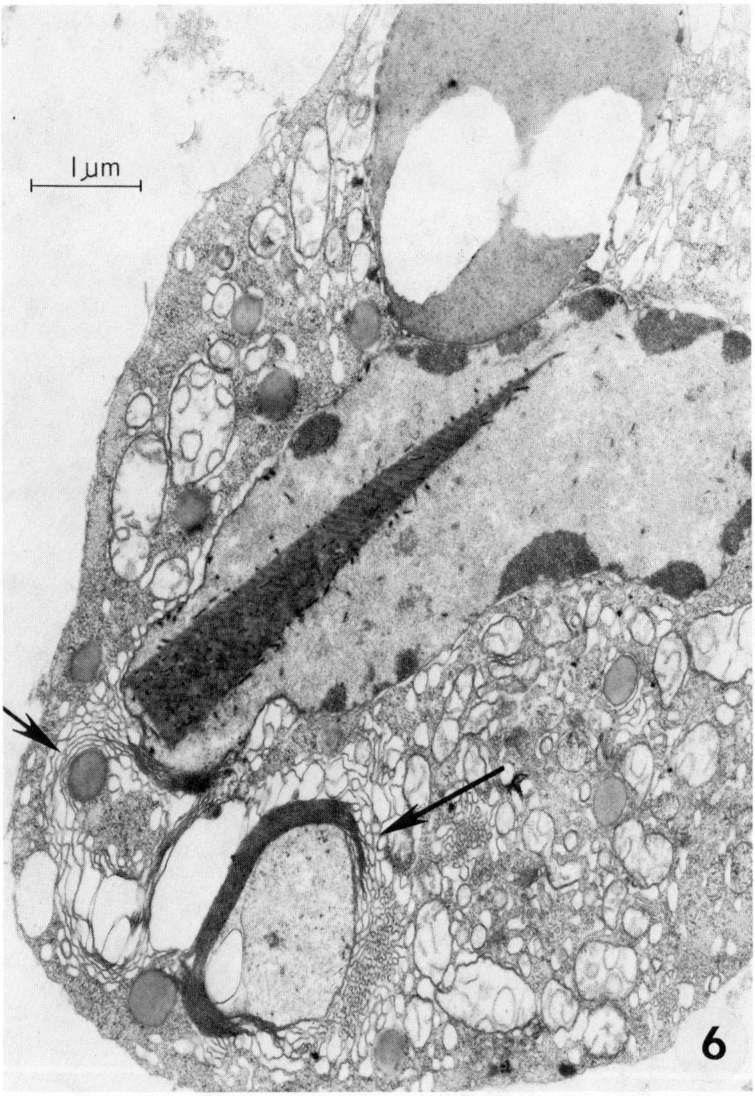

Figure 5-6. Electron micrograph of hepatopancreatic cell showing unusual place of section: note the peripheral section of PIB in nucleus; sharp edges and apices of PIB's serve to rupture nuclear and cytoplasmic membranes effecting the release of the virion-containing PIB. Cytoplasmic changes include bizarre membrane alterations and formation of myeloid bodies (arrow).

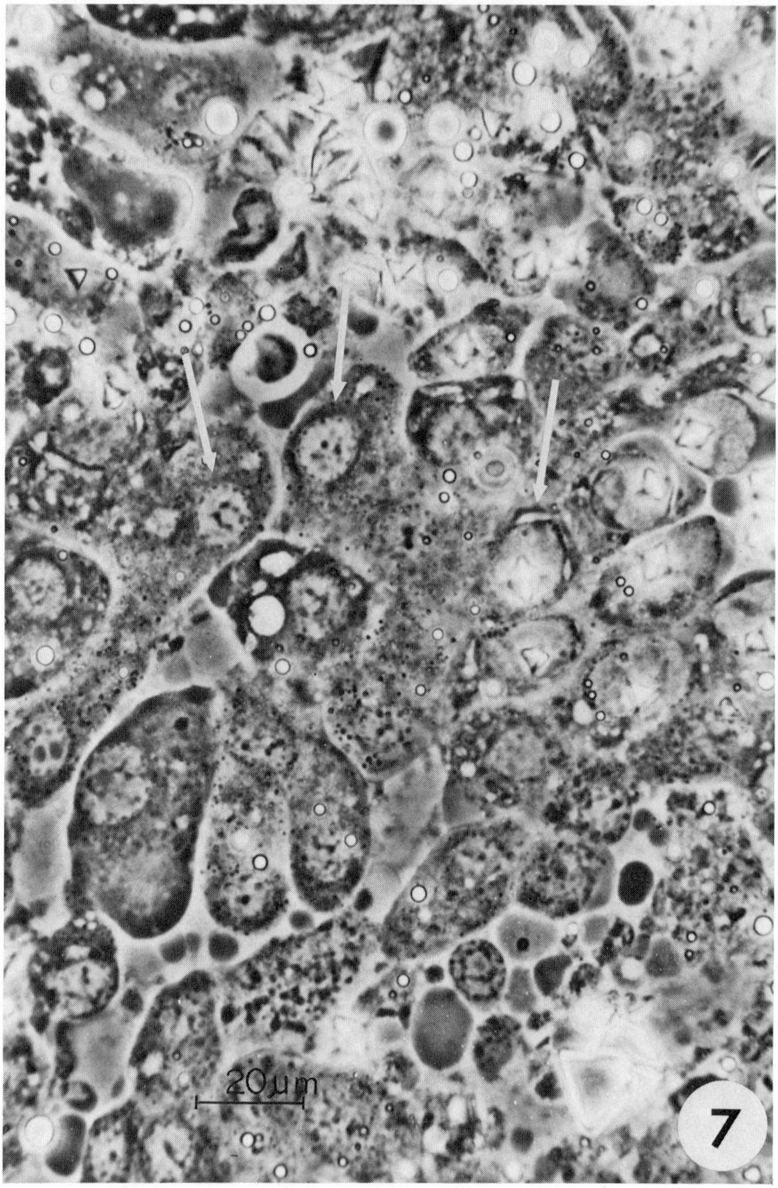

Figure 5-7. Nuclear hypertrophy of Baculovirus-infected hepatopancreas cells (arrows); note that this figure contains several stages of infected cells from non-patent to patent infections. (Phase contrast).

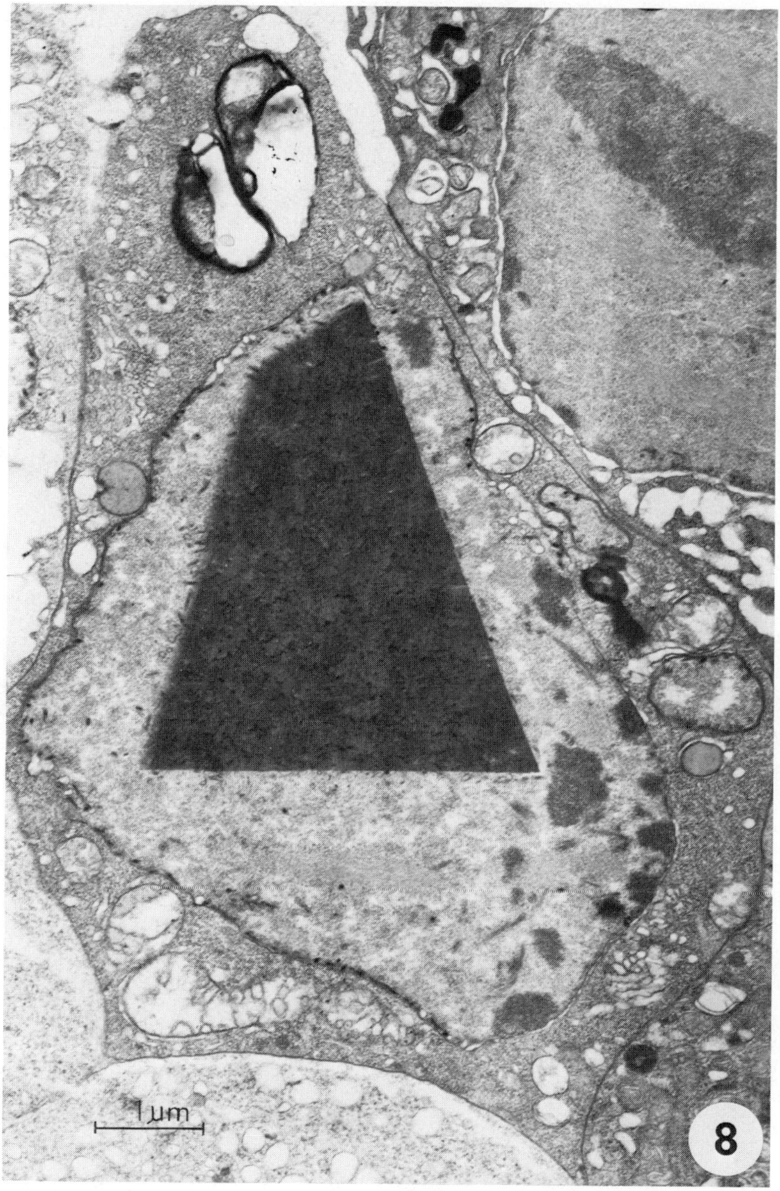

Figure 5-8. Margination and diminution of heterochromatin is apparent in this nucleus infected by Baculovirus penaei; note large, semitruncated PIB in section and virions distributed around PIB; enlarged or swollen mitochondria are apparent in cytoplasm.

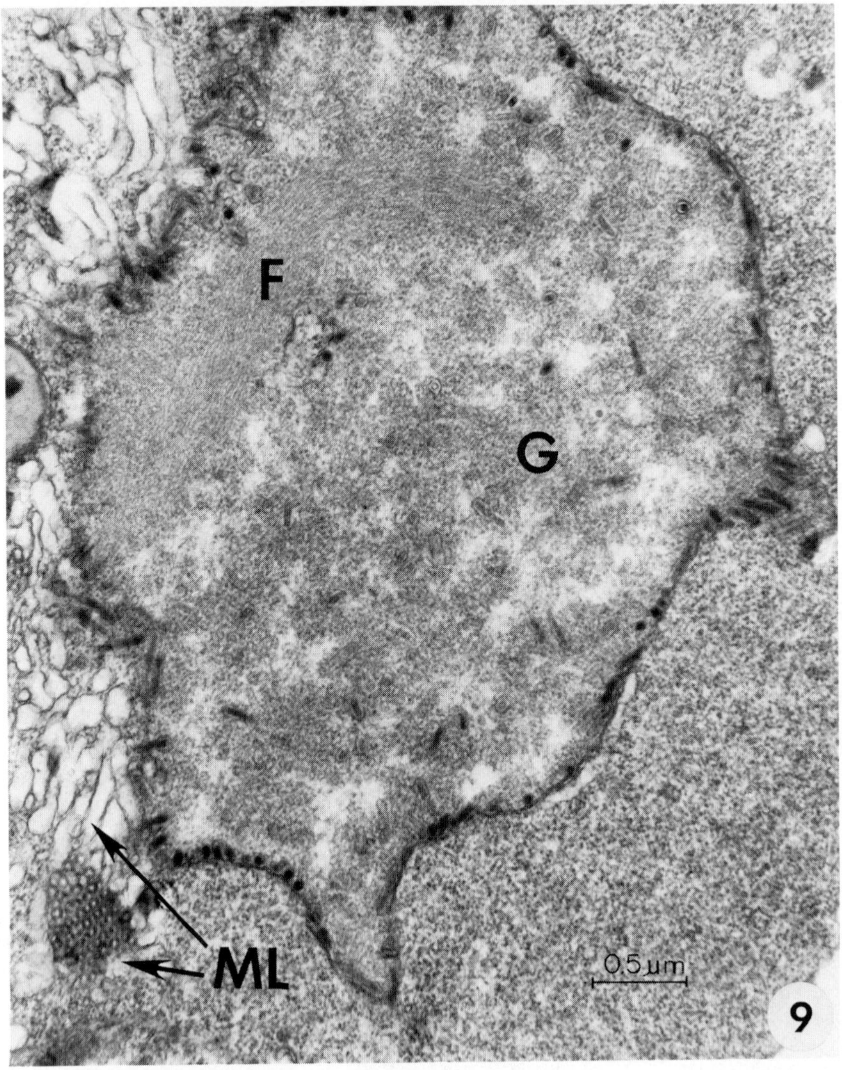

Figure 5-9. Oblique section through nucleus excluding the PIB in advanced, Baculovirus-infected cell from shrimp; note fibrillar (F) and granular (G) stromata in nucleus, with virions attached to inner nuclear envelope; cytoplasm is filled with free ribosomes and membranous labyrinth (ML:proliferated outer nuclear envelope) characteristic of this virus infection in shrimp.

chemical or toxic inhibition of innate cellular defense mechanisms, e.g. phagocytosis, interferon production, or secretory functions.

BACULOVIRUS INFECTIONS IN THE BLUE CRAB AND IN THE EUROPEAN SHORE CRAB

Johnson (1976a) reported a *Baculovirus* in nuclei of epithelial cells of the hepatopancreas in blue crabs (*Callinectes sapidus*) (Fig. 5-10). Earlier, Bazin et al. (1974) reported a *Baculovirus* in the nuclei of the hemocytes and connective tissue cells of the shore crab, *Carcinus maenas*. These two viruses appear to be more similar to one another than either is to the shrimp *Baculovirus*. A major difference between the shrimp *Baculovirus* and the crab baculoviruses is that the latter are not associated with inclusion bodies in the host cell nucleus (compare Fig. 5-5 with Fig. 5-10). There also appear to be differences as well as a few similarities in the host cell cytopathic responses between the crab and shrimp baculovirus infections. The major similarity between host cellular response to infection between the shrimp and crab baculoviruses is the chromatin diminution and margination concomitant with nuclear hypertrophy (Figs. 5-5, 5-10, and Fig. 1 in Bazin et al., 1974). Host cellular responses differ significantly, however, in lack of membrane proliferation, and other cytoplasmic changes in crab baculovirus infections. Perhaps not as much material has been examined in the crab infections compared to large numbers of infected shrimp studied, thus giving an incomplete picture of the crab cellular response.

Baculovirus infections in the above crabs have not yet been associated with fatal disease, whereas considerable evidence exists for lethal infections in shrimp, particularly larval shrimp (Couch, 1978). However, Pappalardo et al. (1979) have reported a *Baculovirus* in hepatopancreatic cells of the crab *Carcinus meditteraneus* that may cause lethal disease. Little detail of pathogenesis was given in the original report.

SPHERICAL VIRUS IN SHORE CRAB, *CARCINUS MAENAS*

Bang (1971) reported a transmissable infection associated with a spherical virus-like particle of varying size (55-125 nm) in the European shore crab, *Carcinus maenas*. It is considered here because Bang (1971) gave considerable information on pathologic changes in hemolymph, hemocyte, and tissue characteristics of the host, and on successful transmission of the agent to previously uninfected crabs in the laboratory. Hemolymph from an original infected crab was inoculated into many other crabs, producing disease similar to that in the

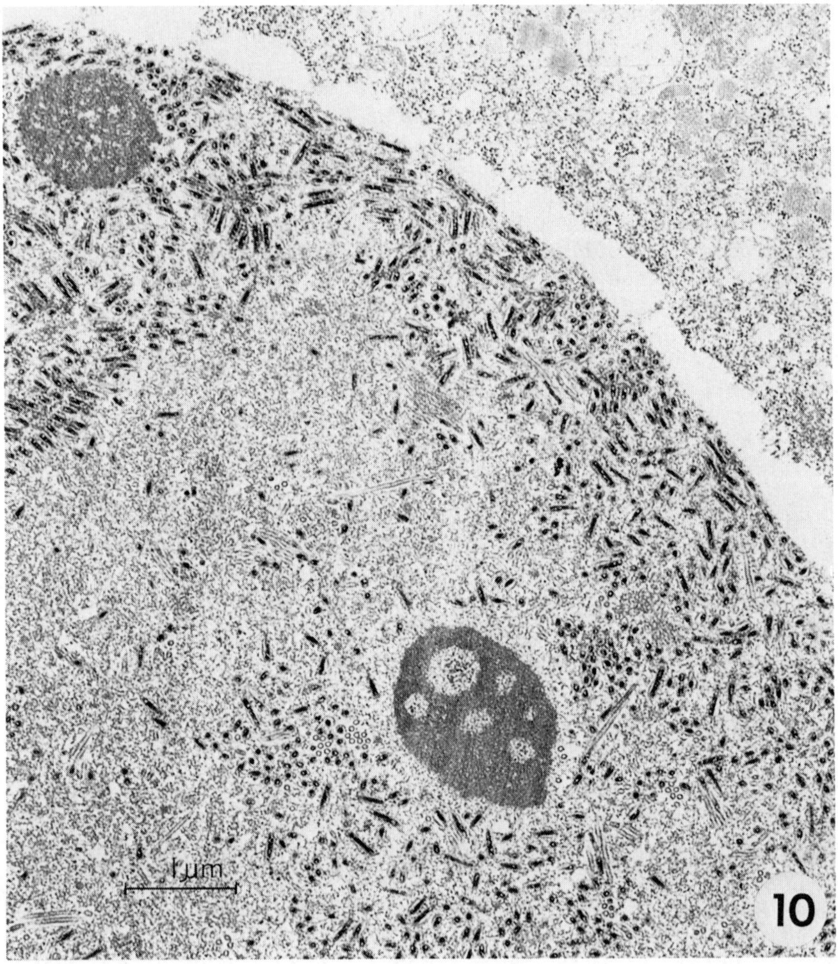

Figure 5-10. *Baculovirus* in nucleus of hepatopancreatic cell from blue crab (*Callinectes sapidus*); note lack of inclusion body - compare with Figs. 5-5, 5-6 and 5-8 of shrimp *Baculovirus*; note nucleolar remnants in nucleus (micrograph supplied by P. T. Johnson).

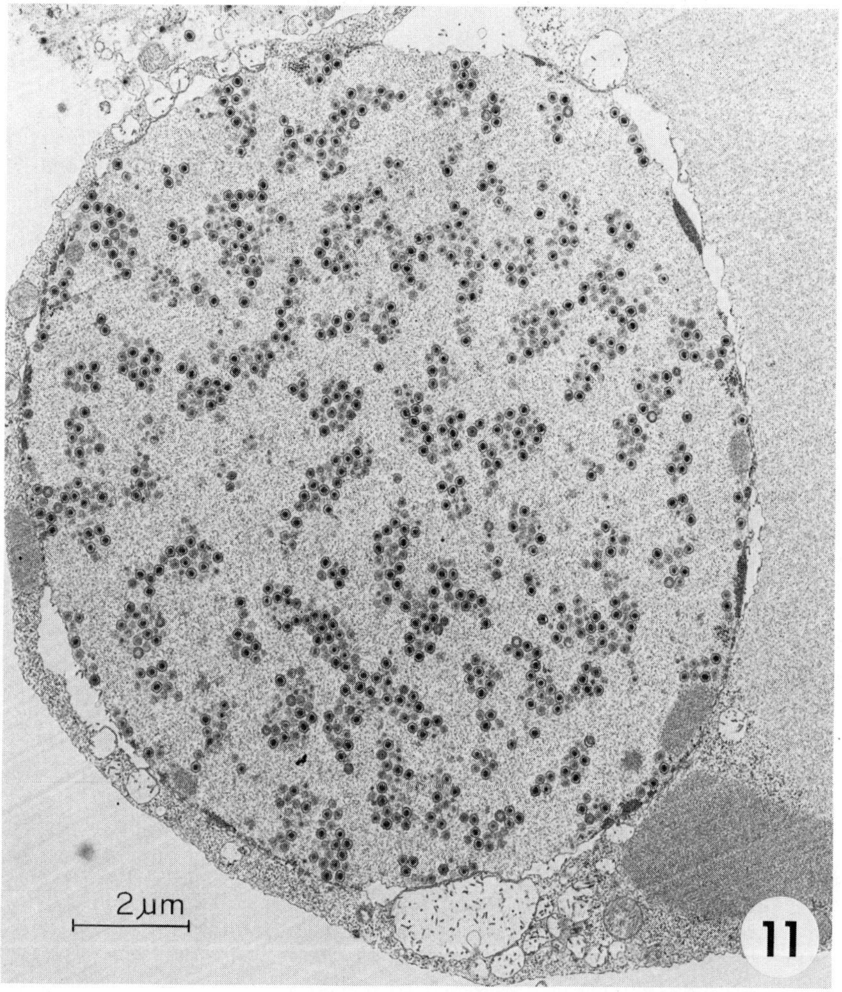

Figure 5-11. Herpes-like virus in hemocyte of the blue crab; note heterochromatin diminution, nuclear hypertrophy, and absence of nucleolus; virions occur in clusters as completely and partially formed particles; perinuclear vacuolization and cytoplasmic lysis are suggested in micrograph (Micrograph supplied by P. T. Johnson).

original crab.

General pathology

The major pathologic change in afflicted crabs was the loss of hemolymph clotting ability, which in crustacea is controlled or mediated by hemocytes or blood cells. Following inoculation, crabs that became infected showed loss of blood clotting in from 2 to 19 days. Seventy-four percent of 73 crabs inoculated with known infected blood developed a failure to form blood clots (Bang, 1971). The following differences were found between the blood cells of the normal and infected crabs: (i) decrease in blood cells numbers in affected crabs in comparison to blood cell numbers in normal crabs; (ii) morphological differences between infected crab blood cells and those of normal crabs consisting of fewer filaments, altered ectoplasmic form, fewer granules per cell and transition from a normal granular appearance to a hyaline appearance; and (iii) loss of ability to secrete fibrinogen. This latter abnormality probably provides the basis for loss of clotting capacity of the blood in infected animals.

Most evidence including some limited electron microscopy, indicates that a filterable, spherical virus is the agent responsible for the disease. However, only limited electron microscopy of this virus has been reported; therefore, little information is available on ultrastructural effects of the agent and or actual identification or affinity of the virus. With electron microscopy the virus was found in blood cells in the endoplasmic reticulum and associated with cell membranes (Bang, 1971).

HERPESLIKE VIRUS IN THE BLUE CRAB

Johnson (1976 a, b) reported herpeslike virus particles in nuclei of hemopoietic cells, hemocytes, and gill epithelial cells of the blue crab from Virginia, USA.

General pathology

Infected cells were recognized because of gross hypertrophy of nuclei visible with the light microscope. Cowdry Type-A inclusions were present in some of the nuclei. Nuclei lacking cytoplasm also were found in acellular spaces, and lysed or partially lysed cells were found. Electron microscopy revealed that enlarged, infected nuclei contained many typical herpeslike virions (Fig. 5-11). Infected nuclei were altered further in that most of the heterochromatin was gone in the hypertrophied nuclei, nucleoli were absent, and the nucleoplasm appeared fine and granular surrounding clusters of virions (Fig. 5-11). Cytoplasmic effects included vacuolization adjunct to nuclear membranes, swelling of perinuclear space (between nuclear envelope membranes), loss of cytoplasmic organelles or their reduction in number, and general lysis of cytoplasm to a granular homogeneity (Fig. 5-11).

No information was presented on the systemic effect of the virus in the crabs. Destruction of hemotopoietic cells might lead to anemia if lysis of blood cell progenitor cells were extensive enough. No evidence was reported for such a condition. Johnson (1976a) did report finding several juvenile crabs infected from Chincoteague Bay, Virginia.

RHABDOVIRUS-LIKE PARTICLES IN THE BLUE CRAB

Yudin and Clark (1979) have described a _Rhabdovirus_ that occurs in mandibular gland cells and hemocytes following ablation of eyestalks in the blue crab. The virus-like particles appeared within the nuclear envelope and tubular endoplasmic reticulum of gland cells during days 3, 4, and 5 following ablation.

General Pathology

At the light microscopic level infected gland cells and hemocytes could be distinguished from non-infected by the presence of large accumulations of virus particles in the cytoplasm. Though individual virions were not visible, their aggregates were conspicuous. No other histopathologic signs were reported for infected crabs.

With the electron microscope, Yudin and Clark (1979) found significant cellular changes associated with the presence and development of virus particles. They found that virus proliferation was associated with three organelle systems: the nuclear envelope, the tubular endoplasmic reticulum (TER) characteristic of the mandibular gland cells, and the cell membrane. Effects on these organelles or upon the cell involved the appearance of virus-filled vacuoles and enlargement of TER prior to or as it filled with particles. In hemocytes, two effects were the replacement of normal cytoplasmic organelles and necrosis of the nucleus. Fig. 5-12 shows the ultrastructural effects of virus proliferation in hemocytes of the blue crab.

The Rhabdovirus-like particles have been reported only in blue crabs that have undergone eyestalk ablation. Yudin and Clark (1979) alluded that a stress related condition or even an endocrine imbalance in these ablated animals may have enhanced or precipitated the virus expression and proliferation.

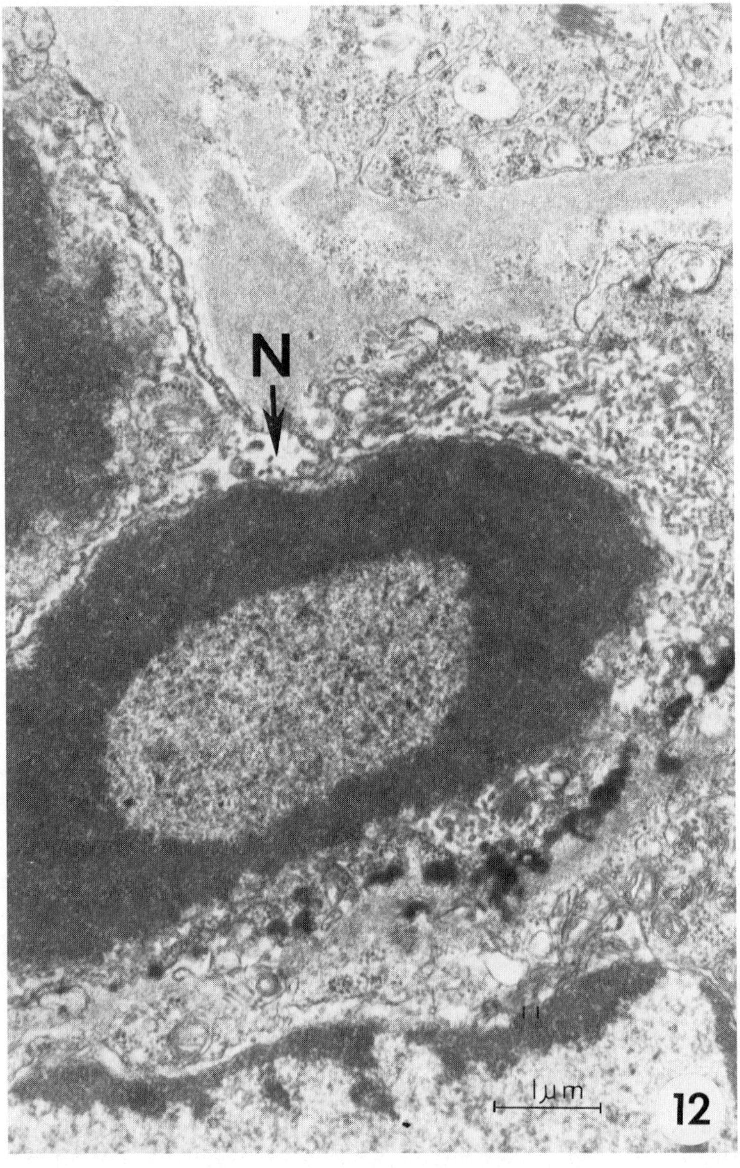

Figure 5-12. Electron micrograph of hemocyte infected with Rhabdo-like virus from the blue crab; note the necrotic nucleus (N) and the displacement of cytoplasmic organelles by the proliferation of virus in the nuclear envelope, and tubular endoplasmic reticulum (micrograph supplied by A. I. Yudin).

More reserach in the area of stress and enhancement of viral infections is needed. Such research may produce insight into virus-host relationships, particularly with regard to using virus-invertebrate systems which are more amenable to experimental manipulation (Couch and Courtney, 1977) than are many vertebrate-virus systems.

REOLIKE VIRUS IN THE BLUE CRAB

Johnson and Bodammer (1975) reported a disease in blue crabs from Chincoteague Bay, Virginia that caused death, and was associated with Reolike virus particles in hemocytes and hemopoietic tissues.

General pathology

Diseased crabs had discolored exoskeltons, and did not feed or molt. Blood from these crabs did not clot well, and paralysis of crabs occurred in a more advanced stage of the disease.

Histopathologic changes found in affected crabs were: necrosis and karyorhexis within hematopoietic tissues; presence of binucleate and multinucleate hemocytes; granular and hyaline hemocytes, and fibroblast like cells within the neural lamellae of the brain of the host; and mild to severe degeneration of nervous tissues (Johnson and Bodammer, 1975).

Ultrastructural changes were somewhat apparent in hematopoietic cells and hemocytes infected with the Reolike virus. Figure 5-13 demonstrates the crystalline arrays of virions in the cytoplasm of a hematopoietic cell. Loss or displacement of normal cytoplasmic organelles are effects of the virus. In some cells almost all cytoplasmic space is filled with either linear arrays of the virus or with randomly distributed virions (Fig. 5-13).

A complex of diseases in crabs is apparently associated with small (ca. 55-60 nm) icosahedral Reolike-viruses and includes the paralysis virus of *Macropipus depurator* (Bonami, 1973) from Europe, as well as the blue crab virus discovered by Johnson and Bodammer.

PICORNAVIRUS IN PARASITIC ISOPODS AND THEIR CRAB HOSTS

Kuris et al. (1979) described a presumed picornavirus in tissues of a parasitic isopod, *Protunion conformis*, and its crab host, *Hemigrapsus oregonenesis*. The hosts were collected from San Francisco Bay, California.

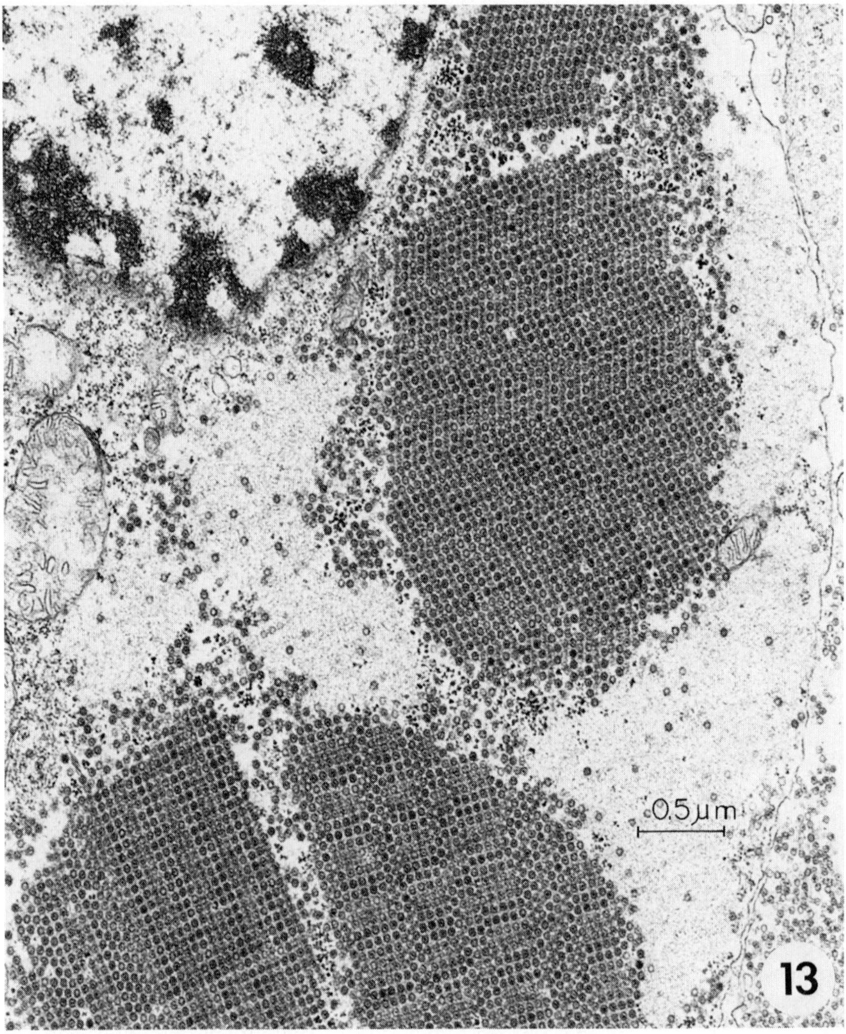

Figure 5-13. Reolike virus in cytoplasm of hemocyte from a blue crab; note loss or displacement of cytoplasmic organelles, and paracrystalline formation of virions within the cytoplasm; the nucleus appears unaffected (micrograph supplied by P.T. Johnson).

General Pathology

Cells of the epidermis and hepatopancreas of three specimens of P. conformis had cytoplasmic infection (Fig. 5-14). Two sizes of virus particles were observed in the cytoplasm in crystalline arrays: one was a small particle, 25 nm in diameter, and the other was larger, 58 nm in diameter. The major microscopic and ultrastructural effect of the virus was a complete replacement of normal cellular organelles by the extensive crystalline viral arrays. Nuclei were not affected in infected cells (Fig. 5-14).

Kuris et al. (1979) could not relate virus infection to death in the isopod or crab hosts. They did occasionally find dead isopods on several hosts from populations along the Pacific coast and suggested that studies should be conducted to correlate host mortalities with virus prevalence.

HEXAGONAL VIRUS IN LARVAE OF THE PACIFIC OYSTER

Elston (1979) reported large, hexagonal virus particles (ca. 222 nm) associated with cells in the velar epithelium of hatchery-reared larval Pacific oysters, Crassostrea gigas. The larvae were undergoing serious systemic disease-related changes. Major pathogenic changes in the virus infected cells consisted of: (i) large cytoplasmic vacuoles; (ii) poorly defined cell membranes; and (iii) loss of cytoplasmic homogeneity in the region of virus particles.

More recently, Elston (1980) has related the overall disease syndrome, demonstrated in the larval oysters, to a fungus-like organism that is associated with gastro-intestinal impaction. The virus particles, however, are associated with the above described pathologic changes in the cells in the velar epithelium. Thus, the virus cannot be excluded categorically as a disease agent at the cell level in oysters.

HERPESLIKE VIRUS IN THE AMERICAN OYSTER

Farley et al. (1972) briefly described a herpeslike virus (ca. 70-90 nm) with associated intranuclear inclusions in cells surrounding the blood sinuses in Crassostrea virginica. The oyster hosts were from estuarine waters with elevated temperatures near Wiscasset, Maine, USA. A correlation was found among elevated water temperature (due to effluent from a power plant cooling system), virus infection, and mortality of oysters. Oysters that died in the heated water and which possessed viral inclusion bodies, had the following histopathologic conditions: (i) dilated digestive diverticula; (ii) cellular infiltrates in the vescicular connective tissues about the blood sinuses; and (iii) in

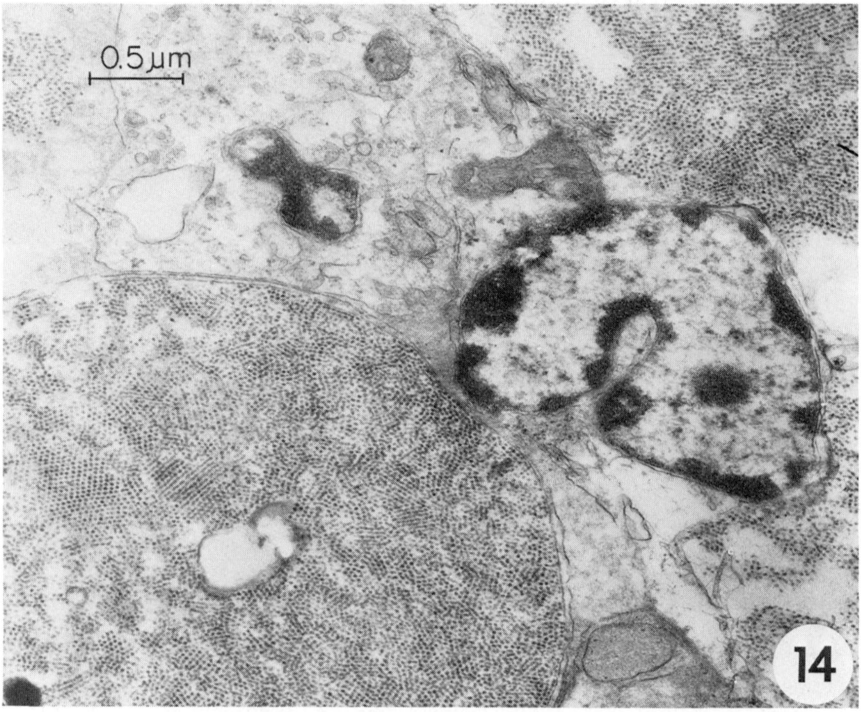

Figure 5-14. Picornavirus filling cytoplasm of epidermis cell from the parasitic isopod, Portunion conformis; note the extensive distribution and abundance of the particles; nucleus of host cell is unaffected. (MIcrograph supplied by A. M. Kuris).

advanced cases, massive cellular aggregates in connective tissues surrounding blood sinuses.

Ultrastructural changes found in the oyster cells consisted of development of virions in the nucleoplasm, presence of large intranuclear inclusions some containing tubules, loss of heterochromatin, and slight nuclear hypertrophy.

The authors suggested that the disease may have a proliferative component, and that elevated water temperature favored the spread or activation, or both, of the virus in oysters. To date, this virus disease has been reported in oysters only from Maine, USA.

PHAGE HYPERPARASITIC IN BIVALVE MOLLUSKS

Buchannan (1978) described a phage-like particle in the cytoplasm of a rickettsia (possibly *Coxiella* sp.) parasitic in digestive gland cells of the bivalve, *Tellina tenuis* from various coastal regions of Scotland. Harshbarger *et al.* (1977) had previously reported phage-like particles from a Chlamydia-like organism from clams and oysters from Chesapeake Bay. This section is concerned with the phage-like particles and related pathology in the rickettsia and its bivalve mollusk host from Scottland.

General pathology

The following descriptive pathologic features reflect both the action of the rickettsia on the bivalve host and the phage on the rickettsia. It is presented this way because future studies may reveal that the pathologic relationship of the rickettsia to the mollusk may be influenced either positively or negatively by the phage (i.e. toxin production mediated by lysogenic phage transfer of a *tox* gene from rickettsia to rickettsia), even though, at present, there is no evidence for such a role for the phage.

Sections of molluscan digestive gland revealed rickettsia in microcolonies in epithelial cells (Fig. 5-15). Electron microscopy of such microcolonies revealed many rickettsial organisms, most of which contained phage particles (67 nm in diameter) (Fig. 5-17). Some rickettsia were filled with phage (Fig. 5-17, arrow). The major effect of the phage on the rickettsia was to produce pleomorphic forms with low opacity, branching filamentous forms and unusual rod shaped forms of abnormally large size (2-4 µm).

Major pathologic effects of the rickettsia and its contained phage on the host appeared to be hypertrophy of the infected epithelial cell, karyorhexis and karyolysis in infected cells, and compression of the contents of the host cell cytoplasm (Fig. 5-16).

Figure 5-15. Light micrograph of _Tellinia tenuis_ digestive gland tubule from section of epoxy embedded tissue. Note the microcolonies of rickettsia in digestive gland cells (arrows) (micrograph supplied by J. S. Buchannan).

Figure 5-16. Microcolony of rickettsia in the digestive gland epithelium cell of _T. tenuis_. Note the severe compression of the internal membrane system of the host cell (arrow); note also the presence of phage, visible in this low magnification micrograph, within the rickettsia (micrograph supplied by J. S. Buchanan).

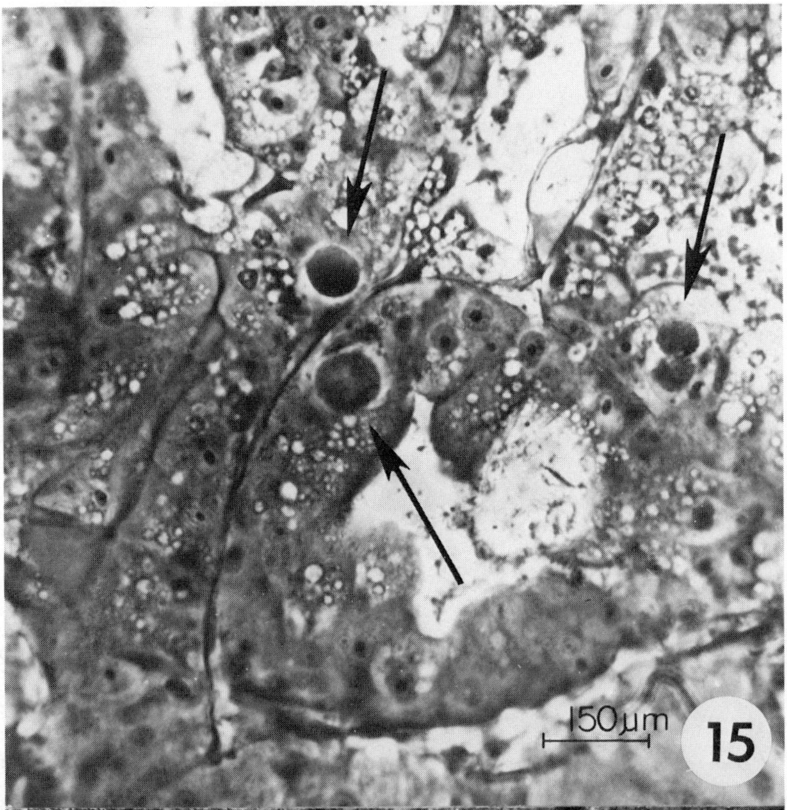
150 µm
15

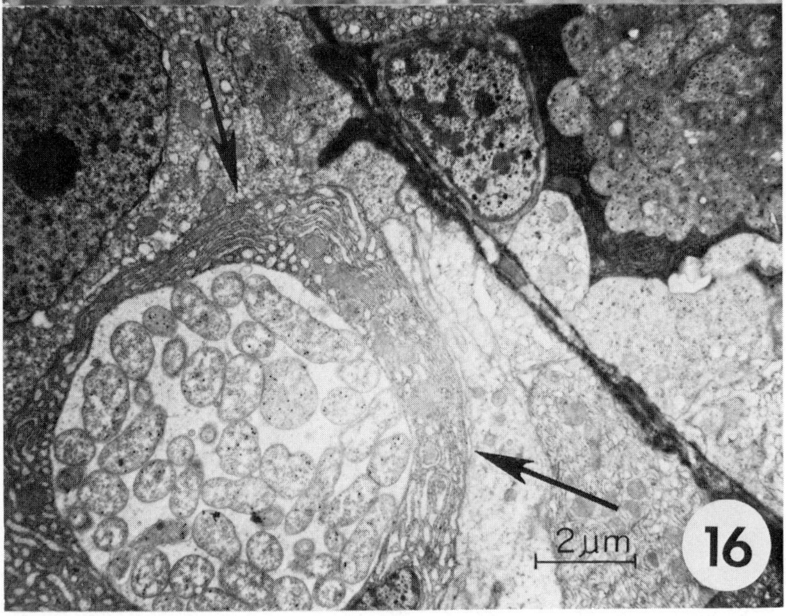
2 µm
16

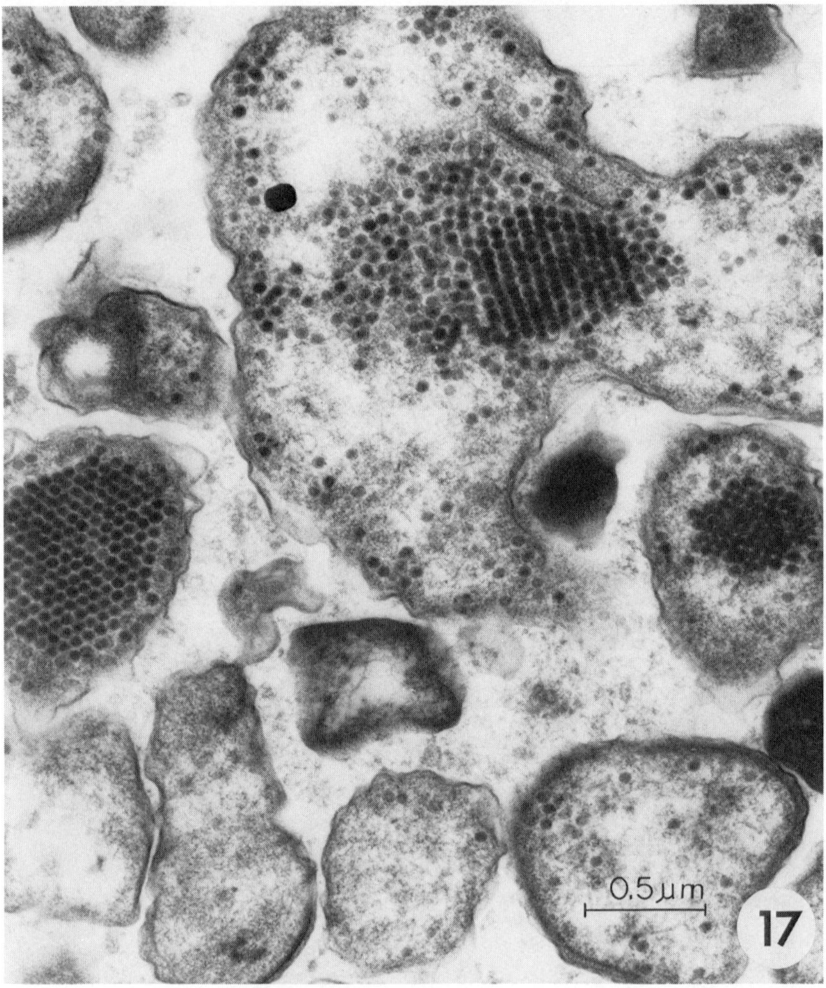

Figure 5-17. Electron micrograph showing rickettsia heavily to lightly infected by phage; note the large (giant) form of rickettsia in the microcolony is heavily infected by phage: also note the rod-like structures in the cytoplasm of the rickettsia associated with the phage (micrograph supplied by J. S. Buchanan).

Buchanan (1978) gave no evidence of lethality of the rickettsia-phage complex in the mollusk.

HEXAGONAL VIRUSLIKE PARTICLES IN TUMORS OF OCTOPUS VULGARIS

Runnger et al. (1971) described large, hexagonal, viruslike particles (ca. 100-400 nm) from edematous, nodular tumors on the arms and bodies of *Octopus vulgaris* from Naples, Italy. These authors gave a good detailed account of the histopathology associated with the virus and tumors.

General pathology

Edematous tumors from 1 to 10 mm in diameter appeared as nodules on the arms and body in estimated 8.4% of *Octopus vulgaris* at the Zoological Station in Naples. The tumors were imbedded in muscle tissue (subdermal) and the larger growths may have interfered with normal arm movement.

Histologically, tumors consisted of degenerate muscle fibers surrounded by other myofibers that were in varying stages of degeneration. Muscle cell nuclei in tumors were rounded, enlarged, and sometimes pycnotic. There was an overall picture of concentric necrosis with a fluid filled region at the center of the tumor. No mitotic figures were associated with the tumors. Proximal arteries, connective tissues, and interstitial septa were eventually caught up in the necrotic phenomenon which has its origin in the muscle. Hexagonal, viruslike particles were found between muscle fibers in the tumors. Cellular pathogenic changes associated with the large viruslike particles consisted of multilayered membranes forming around clusters of viruslike particles, and occasional formation of large extracellular inclusion bodies. The viruslike particles themselves resembled the profile of a section through a hexagonal icosahedron. Most particles contained an electron dense core, probably a nucleic acid.

SPHERICAL VIRUSLIKE PARTICLES IN SOMATIC CELLS OF A NEMATODE

Some recent reports have suggested nematodes as possible carriers or vectors of viruses of their hosts, particularly plants (Harrison et al., 1974). Very little has been published on viruses as pathogens of nematodes.

Foor (1972) reported the occurrence of viruslike particles in the somatic cells (hypodermal cells) of *Trichosomoides crassicauda*, a nematode from the urinary bladder of rats. The viruslike particles were spherical with an electron dense outer shell and were 15 nm in diameter. The particles were found in dense aggregates in the nucleus of somatic cells.

General pathology

Numerous, amorphous inclusions within the nucleus of hypodermal cells and oviduct cells contained viruslike particles. Cytoplasm of affected cells was vacuolated and contained dense granular elements. Overall appearances of affected cells was that of a degenerate, necrotic cell. The author concluded that there were similarities in degenerate appearance of the nematode cells to degenerate cells in certain insects infected with viruses. No conclusions were presented on the ultimate effects of the viruslike particles on the nematodes.

ADENOVIRUS-LIKE PARTICLES IN EPITHELIAL CELLS OF HYDRA

Viruslike particles that had some characteristics of adenovirus were found in the epithelial cell nuclei to the body of *Hydra vulgaris* by Bonnefoy et al. (1972). The particles were hexagonal in profile and were 70 to 75 nm in diameter.

General Pathology

Associated with the presence of virus was a general lack of growth and budding (asexual reproduction) in the Hydra. The cessation of growth has been associated with oxygen deficiency and elevated water temperature; therefore, the role of the virus, if any, is uncertain.

Pathogenesis in virus infected cells resulted in nuclear hypertrophy without concomitant changes in nuclear structure. The viruslike particles were found in irregular masses in the nucleoplasm forming plaques of particles that occupy considerable intranuclear space. Bonnefoy et al. (1972) did not speculate on the ultimate effect of the adenovirus-like particles in *Hydra*.

DISCUSSION

Only thirteen examples of pathologies related to virus or viruslike particles in non-insect invertebrates are described in this chapter. From consideration of these examples, it becomes obvious that detailed descriptions of pathogenesis of virus diseases in non-insect invertebrates has not kept pace with the frequency of new reports of viruses from these invertebrates (Table 6-1). When possible, it is important for authors reporting viral entities from invertebrates to include as much information as can be collected on the disease syndrome, cellular pathogenesis, physiological and behavioral effects and tissue distribution of the virus. In some cases, information has been so scant that readers are not even certain as to the prevalence of the virus in reported hosts. In some cases only a single report was made with no subsequent papers published. The virus/host examples selected here reflect the rather broad phylogenetic range of non-insect invertebrate hosts now reported to harbor one or more viruses.

Basic information on the nature of non-insect invertebrate virus diseases is needed for several reasons. Foremost is the need to know the range of hosts infected by a particular virus type or group. This not only aids in elucidation of disease etiologies for all animals but also provides clues for potential viral surrogate hosts, reservoirs, or vectors (c.f. papers on phage hyperparasitic in mollusks: Buchanan, (1978); Harshbarger et al. (1977)). Obviously the insights provided into mechanisms of pathogenesis and pathogenic effects in invertebrate viruses may give better understanding of vertebrate virus diseases. Finally, invertebrate virus diseases may be used to investigate fundamental aspects of cellular immunity and disease defense mechanisms in ways that vertebrate-virus systems do not permit.

The development of certain invertebrate virus groups such as the baculoviruses as biological pesticides dictates that the potential host limits of these viruses be determined particularly in non-target invertebrates of close phylogenetic affinity to the natural host. Concurrent with gaining this information will be the opportunity to examine the pathology of such virus groups in newly discovered hosts.

REFERENCES

Bang, F. B. 1971. Transmissible disease, probably viral in origin affecting the amoebocytes of the European shore crab. Infect. and Immun., 3, 617-623.

Bazin, F., Monsarrat, P., Bonami, J. R., Croizier, G., Mynadier, G., Quoit, J. M. and Vago, C. 1974. Particules de type Baculovirus observées chez le crabe Carcinus maenas. Rev. Trav. Inst. Pêches Marit., 38, 205-208.

Bonami, J. R. 1973. Recherche sur la paralysie virale du Crustacé décapode; Macropipus depurator L. Rev. Trav. Pêches Marit., 3, 387-389.

Bonami, J. R. and Vago, C. 1971. A virus of a new type pathogenic to Crustacea. Experientia, 27, 1363.

Bonnefoy, A. M, Kolenkine, X. and Vago, M. C. 1972. Pathologie des Invértebratés - Particules d'allure virale chez les Hydres. C. R. Acad. Sc. Paris, Ser. D, 275, 2163-2165.

Buchanan, J. S. 1978. Cytological studies on a new species of rickettsia found in association with a phage in the digestive gland of the marine bivalve mollusc, Tellina tenuis (da Costa). J. of Fish Dis., 1, 27-43.

Couch, J. A. 1974a. An enzootic nuclear polyhedrosis virus of pink shrimp: Ultrastructure, prevalence, and enhancement. J. Invertebr. Pathol., 24, 311-331.

Couch, J. A. 1974b. Free and occluded virus similar to Baculovirus in Hepatopancreas of pink shrimp. Nature (London), 247, 229-231.

Couch, J. A. 1978. Diseases, parasites, and toxic responses of commercial panaeid shrimps of the Gulf of Mexico and south Atlantic Coasts of North America. Fishery Bull., 76, 1-42.

Couch, J. A. and Courtney, L. 1977. Interaction of chemical pollutants and virus in a Crustacean. A novel bioassay. Ann. N.Y. Acad. Sci., 298, 497-504.

Couch, J. A. and Nimmo, D. R. 1973. Cytopathology, ultrastructure and virus infection in pink shrimp exposed to the PCB, Aroclor 1254. (Abst.) Proc. Int'l. Colloq. on Insect Pathol. Oxford, Engl. p. 105.

Couch, J. A., Summers, M. D. and Courtney, L. 1975. Environmental significance of Baculovirus infections in estuarine and marine shrimp. Ann. N.Y. Acad. Sci., 266, 528-536.

Devauchelle, M. and Vago, C. 1971. Virologie - Particules d'allure virale dans les cellules de l'estomac de la seiche, Sepia officinalis (Mollusques, Céphalopodes). C. R. Acad. Sc. Paris, Ser. D, 272, 894-896.

Elston, R. 1979. Virus-like particles associated with lesions in larval Pacific oysters (Crassostrea gigas). J. Invertebr. Pathol., 33, 71-74.

Elston, R. 1980. Ultrastructural aspects of a serious disease of hatchery reared larval oysters, Crassostrea gigas Thünberg. J. Fish Dis., 3, 1-10.

Farley, C. A., Banfield, W. G., Kasnic, G., Walter, J. and Foster, S. 1972. Oyster herpes-type virus. Science, 178, 759-760.

Federici, B. A. and Hazard, E. I. 1975. Iridovirus and cytoplasmic polyhedrosis virus in the freshwater daphnid Simocephalus expinosus. Nature (London), 254, 327-328.

Foor, W. E. 1972. Viruslike particles in a nematode. J. Parasitol., 58, 1065-1070.

Harshbarger, J. C., Chang, C. S. and Otto, S. V. 1977. Chlamydiae (with phages) mycoplasmas, and rickettsiae in Chesapeake Bay bivalves. Science, 196, 666-668.

Harrison, B. D., Robertson, W. M. and Taylor, C. E. 1974. Specificity of retention and transmission of viruses by nematodes. J. Nematol., 6, 155-164.

Johnson, P. T. 1976a. A baculovirus from the blue crab, Callinectes sapidus. Proc. Int'l. Colloq. Invertebr. Pathol., Kingston, Ontario, p. 24.

Johnson, P. T. 1976b. A herpes-like virus from the blue crab Callinectes sapidus. J. Invertebr. Pathol., 27, 419-420.

Johnson, P. T. and Bodammer, J. E. 1975. A disease of the blue crab Callinectes sapidus of possible viral etiology. J. Invertebr. Pathol., 26, 141-143.

Johnson, P. T. and Farley, C. A. 1980. A new enveloped helical virus from the blue crab Callinectes sapidus. J. Invertebr. Pathol., 35, 90-92.

Kuris, A. M., Poinar, G. O., Hess, J. R. and Morris, T. J. 1979. Virus particles in an internal parasite, Portunion conformes (Crustacea:Isopoda:Entoniscidae) and its marine crab host, Hemegrapsus oregonenesis. J. Invertebr. Pathol., 34, 26-31.

Lightner, D., and Redman, R. 1981. A Baculovirus caused disease of the Penaeid shrimp Penaus monodon. J. Invertebr. Pathol., in press.

Lightner, D., Redman, R., Danald, D., Williams, R. 1981. Major disease encountered in controlled environment culture of Penaeid shrimp in Puerto Penasco, Sonora, Mexico. Proc. U.S.-Japanese Meeting on Aquaculture, UJNR Coop. Prog. on Natural Resources, in press.

Pappalardo, R. and Bonami, J. R. 1979. Pathologie des Invertébrés-Infection des Crustacés marine due à un virus de type nouveau apparenté aux Baculovirus. C. R. Acad. Sc. Paris, Ser. D, 288, 535-537.

Rungger, D., Rastelli, M., Braendle, E. and Malsherger, R. G. 1971. A viruslike particle associated with lesions in the muscles of Octopus vulgaris. J. Invertebr. Pathol., 17, 72-80.

Summers, M. D. 1977. Characterization of shrimp *Baculovirus*. Eco. Res. Series EPA-600/3-77-130, Gulf Breeze Environmental Res. Lab., pp. 35.

Vacelet, J. and Gallissian, M. F. 1978. Virus-like particles in cells of the sponge *Verongia cavernicola* (Demospongiae, Dictyoceratida) and accompanying tissue changes. J. Invertebr. Pathol., *31*, 246-254.

Vago, C. 1966. A virus disease in Crustacea. Nature (London), *209*, 1290.

Yudin, A. I. and Clark, W. H., Jr. 1979. A description of Rhabdovirus-like particles in the mandibular gland of the blue crab, *Callinectes sapidus*. J. Invertebr. Pathol., *33*, 133-147.

SECTION II

BACTERIA

Chapter 6

PRINCIPLES OF PATHOGENESIS OF INSECT BACTERIAL DISEASES AS EXEMPLIFIED BY THE NONSPOREFORMING BACTERIA

by Oleg Lysenko*

INTRODUCTION

The ability to form spores, i.e. morphologically distinct dormant resting stages, is limited to about seven genera of bacteria and of these only two are found commonly in nature. Thus the nonsporeformers are the most frequently occurring bacteria in nature including insects. They are very heterogeneous in their morphological, metabolic and ecological characteristics. Their taxonomy is uneven; some groups are relatively well defined, while the others still lack a basic taxonomic arrangement. The taxonomic determination of bacteria occurring in insects depends to a great extent upon the knowledge of a particular group of bacteria, which may change from year to year. Literature such as Bergey's Manual (Buchanan and Gibbons, 1974) can be a good source of general information, but for some groups more up to date specialized articles are necessary. It is important to note that two strains of bacteria bearing the same name may differ in their pathogenicities as well as in other characteristics.

The main purpose of this chapter is to explain and discuss general principles. More complete literature references can be found in monographs by Steinhaus (1949, 1963), Krieg (1961), Weiser (1966), Burges and Hussey (1971), and Afrikian (1973).

Pathogenesis, the origin and development of a disease, is a product of a pathogen. Therefore it is useful to discuss first some of the theoretical aspects of pathogenicity, the features which make a bacterium a pathogen.

[1]Insect Pathology, Institute of Entomology, Czechoslovak Academy of Sciences, Flemingovo nam. 2, Prague, Czechoslovakia

PATHOGENICITY

Pathogenicity of a bacterium is usually defined as the ability to cause damage to a host; this damage is manifested as a disease. Conventionally speaking, pathogenicity is an attribute of a taxonomic species, while virulence is an attribute of a strain. This definition is rather onesided as it takes into consideration only the ability of the bacterium. In medicine, this usage has been quite practical; however, for general and theoretical study such a definition is inaccurate.

Classification of pathogenicity

Pathogenicity is an interrelationship between two organisms, in which the microorganism causes damage to the host. This interrelationship constitutes at least a two-variable system. By simplifying we can assume that each variable, the bacterium (B) and the host (H), can occur in only one of the two possible states. The bacterium can be virulent (V+) or avirulent (V-) for a host. The host can be resistant (R+) or sensitive (R-) to the bacterium. The result of these relationships is a disease (1) or an unaltered host (0).

Following this assumption, the outcome of interaction between the host and pathogen can be summarized in a conventional 2 x 2 frequency table:

	BV+	BV-
HR+	0	0
HR-	1	0

Under these circumstances the disease can occur only as a result of an interaction between a virulent bacterium and a sensitive host (the lower left quadrant). This requires the assumption, however, that the intensity of resistance is greater than the virulence (R > V). In this model where each variable may occur in only one state of a two-state system, pathogenicity would not be a common characteristic among bacteria and pathogens would be peculiarities. This agrees with the concept of the obligate pathogen.

If however one of the variables occurs in a multi-state system and the other one is a constant (e.g. different strains of bacteria with one host), there will be a series of interactions depending upon different levels of virulence from BV+ to BV-. These levels can be measured quantitatively and expressed in LD_{50} units or in other values. If another host or another developmental stage is used, the values may be quite different. Similarly, susceptibility of different hosts to one bacterial strain may vary, i.e. BV is a constant while HR is a variable. These considerations form the theoretical principles for biological standarization.

In nature both variables BV and HR exist in multi-state systems and their interactions produce an unlimited number of relationships. Thus, for example, a weakly virulent bacterium can produce a severe infection in a weakly resistant or susceptible host. From the above it can be seen that in determining the pathogenicity of a particular bacterium the susceptibility of the host must be determined as well. Only in this way can experimental data be compared.

Bacteria can be arranged in a continuous spectrum according to the intensity of damage caused to a particular host. Any boundary lines between the defined groups are a matter of agreement. When Koch's postulates were formulated, only two groups of bacteria were distinguished: a relatively small group of pathogens (now called obligate pathogens), and "nonpathogens". Later with the development of more accurate diagnostic criteria and improved bacterial taxonomy many of the "nonpathogens" were linked with certain pathological processes. For this large group of bacterial species new terms were proposed: facultative pathogens, opportunistic pathogens, conditional pathogens, etc. In medicine, the importance of facultative pathogens has become evident only since the 1940's.

In insect pathology some workers noticed that under certain conditions insect gut microflora can produce a disease (e.g. Pospelov, 1927) but no serious attempts to classify pathogens were made. In 1949 Steinhaus in Principles of Insect Pathology distinguished two groups: "opportunists" and "true" pathogens. The opportunists (sometimes called "secondary invaders") were harmful for the host when "under certain conditions the barrier or resistance that usually protects the host" was broken down.

Later Lysenko (1958a) characterized symbionts, nonpathogens, conditioned (or facultative) pathogens and obligatory pathogens, and defined some criteria for determination of the main types of pathogenicity. This concept of classification expressed pathogenicity as a dynamic "interrelation of a microorganism and a macroorganism" and stressed that there exist various intermediate categories. This classification must be understood in the relationship of a definite microorganism to a definite insect species in a definite stage of development. Conditioned pathogens were defined as requiring a provocation factor or stressor which substitutes for the missing phase of the pathogenic process.

Bucher (1960, 1963) classified bacteria as obligate pathogens, potential pathogens, facultative pathogens and pathogens of doubtful status. His classification was not exclusively based on interaction between parasite and host; e.g. "obligate pathogens require specialized conditions for growth and reproduction" and they "can be cultured *in vitro* only with difficulty". Potential pathogens were defined by the LD_{50} dose of bacteria which injected into the hemocoel produced a disease. Facultative pathogens differed from other types "in possessing some mechanism for invading" a host tissue. The last group contained pathogens of doubtful status which "have been isolated in association with insect disease but have not been experimentally demonstrated to be the causal agents". These examples demonstrate the differences between a theoretically based concept of pathogenicity and an empirical classification.

Interpretation of experimental data

Pathogenicity depends on a number of factors which are very difficult to recognize and separate from one another. Major methodological difficulties arise because virulence of a bacterium as well as resistance of a host can be measured only in living systems. Therefore many studies of model systems performed *in vitro* cannot be easily interpreted *in vivo*. Bacteria produce *in vitro* a number of compounds that are toxic for insects by injection. The yield of these toxins or toxic enzymes is dependent on the medium, conditions of cultivation, and other factors which can be determined by appropriate methods. These methods, however, cannot be used for determination of the same parameters in a living host. A few studies have attempted to correlate some results obtained *in vitro* with data obtained from the same system *in vivo*. Lysenko and Kučera (1968) tried to calculate the rate of digestion of specific proteins in hemolymph of *Galleria* larvae by a toxic proteinase of *Pseudomonas aeruginosa*. Even though their data roughly correlated with those data previously obtained *in vivo* (Lysenko, 1977), the authors were cautious in making any definite statement.

Chadwick (1973) examined *in vitro* growth patterns of three nonpathogenic and three pathogenic bacterial species in a nutrient broth and in tissue of *G. mellonella* larvae. Nonpathogenic bacteria were unable to grow in crushed tissue, while the pathogens grew equally well in both media. These differences were attributed to absence of growth factors and differences in protease production among the groups of bacteria.

In some respects these examples are in accordance with Garber's concept of pathogenicity (Garber, 1960), "the host as a nutrient medium for the parasite". From this point of view an infectious disease may be distinguished from a food poisoning. In an infectious disease, bacteria grow *inside* the host, multiply and perhaps form toxins; this complex produces a disease of the host. In contrast, in a food poisoning, bacteria grow *outside* the host, multiply and form toxins; a disease is caused by ingestion of this complex by the host. Under this definition, metabolites which are not formed by the bacterium during its growth in a host (even though they can be toxic for the host) should not be classified as bacterial toxins and cannot play a role in the pathogenesis of the particular infectious disease. Failure to understand this difference has led to misunderstanding of the role of some bacterial products in the pathogenesis of bacterial disease, and to problems in interpretation of *in vitro* experiments.

Conditions for disease development

To produce a true infectious disease a bacterium must be able to (i) enter the host, (ii) live and multiply in the host, (iii) interfere with host defense mechanisms and (iv) cause damage to the host tissues. To each of the virulence determinants corresponds one or more targets in the host: (i) epithelial cells, (ii) insect tissues and fluids as convenient bacterial substrates, (iii) humoral and cellular defense systems, (iv) susceptible tissues and their metabolic functions. Each of these determinants and targets take part in the pathogenesis of every true infectious disease. For instance, if a bacterium produces a disease in an insect but the bacterium cannot grow in insect tissues, either it has been administered in a very large dose, or it has multipled in another suitable substrate outside the host and the host only ingested a toxic metabolite. This is typical bacterial poisoning, not an infection.

The infection dose

The infection dose is defined as the number of bacteria which are required to produce a disease in a host. This definition includes neither the conditions of the bacterial cells, nor the method of administration into the host, nor the condition of the host (see above). The infection dose is usually expressed in LD_{50} or other convenient terms.

Most of the facultative nonsporeforming pathogens, e.g. *P. aeruginosa* or *Serratia marcescens*, are pathogenic for caterpillars in doses of LD_{50} = 5 to 100 bacteria by injection into the hemocoel and in doses of 10^4 to 10^6 cells by feeding (e.g. Lysenko and Slama, 1959; Lysenko, 1962; Bucher, 1963; Podgwaite and Cosenza, 1976a, b). Roughly similar values have been found for grasshoppers and leafhoppers (e.g. Stevenson, 1959; Bucher and Stephens, 1957; Whitcomb et al., 1966) and Coleoptera (e.g. Heimpel and West, 1959; Slatten and Larson, 1967). In general, the LD_{50} dose for feeding is two to six logs higher than for injection; this phenomenon will be discussed below. Bucher (1963) calculated the relationship between LD_{50} and LD_{99} for feeding P. aeruginosa to grasshoppers. One LD_{99} unit was equal to about 200 to 5000 LD_{50} units, or about 10^8 P. aeruginosa cells.

Lysenko and Weiser (1974) determined the following LD_{50} values for some bacteria by injection into G. mellonella larvae: P. flourescens - 66 cells; Achromobacter nematophilus - 1 cell; P. maltophilia - 5.0×10^4 cells; Alcaligenes odorans - 1.5×10^4 cells; and Achromobacter sp. N. 19 - 1.0×10^5 cells per larva. These data suggest that there might be bacterial species which even in doses of 10^4 to 10^5 cells cannot produce a disease when injected into the hemocoel.

Incubation period

Incubation period is usually defined as the interval between exposure of a host to a pathogen and appearance of a disease. Exposure to a pathogen can be easily determined experimentally, while it is more difficult to define appearance of a disease. In the majority of cases the most easily definable event is death of the host. In 1959 Lysenko and Slama tried to determine the incubation period of pronymphs of Cephaleia abietis infected with S. marcescens by measuring the rate of oxygen consumption of infected insects. The "clinical" beginning of the disease was characterized by increased oxygen consumption and the incubation period was inversely correlated with the infection dose. Twenty years later the same authors (Lysenko and Slama, 1978) monitored the course of infections in Tenebrio molitor caused by different bacteria and toxins by measuring hemolymph pressure pulsation of experimental animals. This method was developed by Slama (1976) and its principle and a schematic diagram of the instrument used are shown in Fig. 6-1.

The pattern of frequency and amplitudes of pulses in T. molitor pupae were recorded and typical hemolymph pressure curves (HPC) were obtained for infections caused by different doses of different bacteria. Changes of HPC related to different pathogens are discussed elsewhere (Slama and Lysenko, 1980); only the incubation period (t_i) is discussed here. This period is defined as the period between injecting the active agent and the first occurrence of depressed amplitudes of the pulses of the HPC. The death of the insect (t_e) was characterized by the total disappearance of these pulses. By injecting different doses (d) ranging from 10^1 to 10^6 bacterial cells they found: (i) t_i is inversely proportional to d, i.e. $t_i = K_d^{-1}$, where the constant K is equivalent to about 10^8 for $10^4 < d < 10^7$. (ii) At the time t_i animals infected with different d contained approximately the same number of bacteria ($\bar{N}$). (iii) Animals injected with some bacterial enzymes toxic for insects produced HPC resembling those

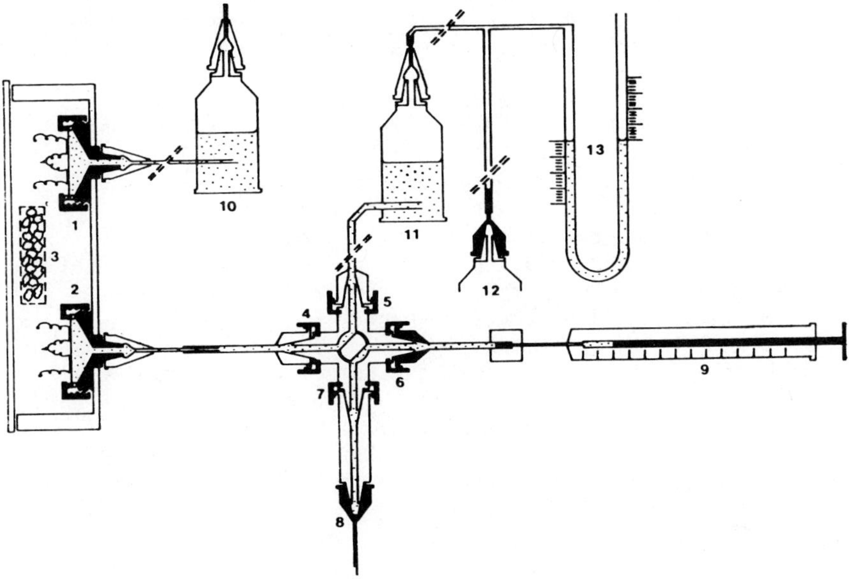

Figure 6-1. Schematic picture of the Slama apparatus for recording hemolymph pressure in insects. (1) (2) tensometric membranes, (3) moisture absorbant, (4) (5) (6) (7) 4-way inert valve, (8) injection needle connecting the insect hemocoel with the hydraulic system, (9) syringe for manipulating the hydraulic system and injecting through (8), (10) barometric pressure compensation reservoir, (11) (12) (13) manometric unit.

Tensometers (1) (2) form branches of a Wheatstone resistance tensometric bridge; electrical signal is amplified and recorded on a recorder. The insect specimen is inserted on needle (8) and the hemolymph pressure is measured by connecting 2-4-7-8; calibration of the unit is made via 2-4-5-11-13. Injection of a material deposited in (8) is performed by 8-7-6-9. For technical data and full description of this apparatus see Slama (1976). (Reprinted by permission of Acta Entomologica Bohemoslovacia, Prague).

obtained with bacterial infection. (v) Different bacterial species produced different HPC.

The authors found that numbers of bacteria N were about 10^9 cells per pupa for P. aeruginosa, S. marcescens and Flavobacterium sp.. The t_i values for P. aeruginosa ranged from 200 min ($\underline{d} = 10^7$) to 480 min (for $\bar{d} = 10^4$ bacteria per pupa). These results confirm that (i) the infection dose must be big enough to enable multiplication of the pathogen in the host, (ii) the pathogen must be able to multiply in the host, and (iii) the given symptoms of disease appear when the bacteria reach a certain concentration in the host.

Although no general agreement was found between the quantity of the bacterial toxin or enzyme and bacterial cells, the HPC records of bacterial infection and intoxication by a toxin produced by the same bacterium were highly similar. This suggests that in these cases a bacteremia is an intoxication. In a bacteremia the number of bacterial cells increases by multiplication which subsequently increases the concentration of toxic metabolites. There is also an incubation period between the administration of a dose of a toxic metabolite and the occurrence of pathological symptoms; this suggests that the toxin alone does not produce an immediate effect.

ENTRY OF BACTERIA INTO THE INSECT BODY

In order to produce a disease a pathogen must be able to enter into the host's body. Any organism must penetrate the barriers of cuticle, gut lining, or tracheae in order to penetrate the host insect. This may occur either by the pathogen's own activity (active penetration) or by the help of another organism (e.g. invasion of a nematode), or through a mechanical injury (passive penetration).

Each active penetration requires an active interaction between a covering membrane and the pathogen. Even when the parasite is microscopic and the host is macroscopic, the actual interaction of these systems is on a cellular level, where sizes of both partners are nearly the same. In this interaction surface adherance phenomena are of primary importance (reviewed by Smith, 1977).

Insect cuticle

In general insect cuticle is composed of layers of epicuticle, exocuticle, and endocuticle, covering epidermal cells. The epicuticle is formed of lipids, which prevent water loss. Exocuticle and endocuticle are complexes of chitin and proteins. Any active penetration of bacteria through the cuticular barrier requires first that the pathogen possess a mechanism for adhering to the epicuticle. It must then be able to produce extracellular enzymes (i.e. lipases, proteases and chitinase) capable of digesting the epi-, exo-, and endocuticle; these enzymes must be produced in adequate quantities and be active under *in vivo* conditions. Finally, the pathogen must have a mechanism by which it can actively penetrate the epithelial cells and into the hemolymph. In insects, active cuticular penetration has not yet been described for bacteria; it is known only for fungi. Bacteria apparently can penetrate the cuticular barrier only at the site of microscopic and macroscopic breaks in its continuity, as from wounds.

The intestinal tract

The insect intestinal tract is composed of pharynx, salivary glands, proventriculus, midgut, Malphighian tubules, hindgut, and rectum. The pharynx, proventriculus, hindgut and rectum are lined with cuticle. The midgut is lined with the peritrophic membrane, a thin tubular acellular membrane which surrounds the food bolus and protects the gut epithelial cells (reviewed by Richards and Richards, 1977). The peritrophic membrane is composed of chitin (ca. 10%), proteins (ca. 40%) and different mucopolysaccharide and hyaluronic acid-like compounds (ca. 15%); proportions vary in different species. There are several layers of particles and microfibrils formed of chitin-protein aggregations.

In this way the whole intestinal tract is lined with chitinous material, preventing direct contact between the epithelial cells and the gut contents. This phenonmenon plays an important role in the pathogenesis of insect intestinal diseases.

In 1956 Weiser and Lysenko found that larvae of *Bombyx mori* fed *S. marcescens* did not show any mortality while the same bacterium administered by injection caused a fatal septicemia. When larvae were fed the same bacterial suspension with triturated glass, mortality rapidly increased and reached values similar to that of injected larvae. On the basis of these experiments they postulated that the peritrophic membrane of insects is resistant to bacterial penetration in a manner similar to the cuticle. When mechanical injury to the peritrophic membrane occurs, microbes penetrate between the membrane and gut epithelial cells and operate as pathogens. Steinhaus (1958, 1959) repeated these experiments more extensively, and they formed part of the basis for his hypothesis of stress as a factor in insect disease.

The possibility of penetration of viruses and bacteria through the peritrophic membrane has been discussed by LeBerre (1967), Stohler (1961), and Maramorosh and Shope (1975). Pashke and Summers (1975) suggested that some microbial enzymes may damage the peritrophic membrane and assist penetration of pathogens. Brandt et al. (1978) showed by electron microscopy that there were no pores and other discontinuities in the peritrophic membrane of *Orgyia pseudotsuga* and that this membrane could function as a mechanical barrier. Structural changes in the membrane were found after it had been digested with chitinase and proteinase.

Though the membrane is irreparable, it is continuously synthesized, and interruptions in synthesis can provide the portal of entry for a pathogen persisting in the gut.

The likelihood of infection via the intestinal tract is affected by the digestive ability of enzymes, by pH of the gut contents and by interactions of the pathogen with normal microflora. Very little is known about these phenomena in insects.

The ability of a facultative pathogen to multiply and overgrow the gut must be the primary phase of these infections. This may result from a failure of the regulatory mechanisms of the gut and its microflora to maintain appropriate conditions in the gut.

Kodama and Nakasuji (1971) studied the mechanism of pathogenicity of D-streptococci and *Serratia piscatorum* in gnotobiotic silkworm larvae which lack their normal flora. Streptococci were associated with a chronic intestinal disease and also played an inductive role for the secondary invader, *Serratia*, which produced a hemocoelic septicemia. When D-streptococci predominated in the gut, the content of lactate and acetate was increased and the pH of the gut decreased. This resembles "*S. bombycis*" infection in silkworms (Lysenko, 1958); however, enterococci are likely normal microflora of many insect species (Lysenko, 1959; Martin and Mundt, 1972; Charpentier et al., 1978). Daoust and Gunner (1979) showed that enterococci isolated from an aquatic environment were ineffective in lowering the midgut pH of *Lymantria dispar* larvae, while fermentation bacteria selected from the gut flora decreased the midgut pH. The reduction of pH facilitated the action of chitinase on peritrophic membrane. Chitinase was produced by a microflora bacterium isolated from this insect. The normal gut microflora may have a regulatory function which is absent in gnotobiotic insects.

Mediated entry - Nematodes

If a microorganism is not able to actively penetrate through the membranous barriers of the insect body into the hemocoel, it can get there either as a mechanical contaminant or through interaction with another organism. The damage caused to the membranes can be either mechanical or chemical. Chemical injuries will be discussed later under toxins.

The entry of microorganisms into the host mediated by another organism may be accidental (essentially mechanical) for example, the case of some bacterial infections accompanying protozoal diseases (e.g. Hostounsky and Weiser, 1972). On the other hand, there may be a special persistant association between the bacterium and another organism, which is able to actively penetrate through the membranes. The association of neoplectanid nematodes with certain bacteria was perhaps first demonstrated by Dutky (in Welch, 1963) and later by Weiser (1955), who correlated a nematode invasion of Neoaplectana carpocapsae with the occurrence of septicemia in the insect host. More recently, different bacterial species have been described as associated with nematodes (e.g. Weiser, 1966; Poinar, 1973; Khan and Brooks, 1977). The nematode-bacterium association has been characterized as mutualism and pathogenesis of the disease has been explained as a septicemia caused by the symbiotic bacterium which was introduced into the insect hemocoel by a "living syringe" - the nematode. However, Lysenko and Weiser (1974) have shown that although nematodes are mediators enabling bacteria to penetrate through the peritrophic membrane, the nematode is not merely a "living syringe", and some other as yet undefined factors may be involved. The authors found that there was no correlation between the virulence of a particular bacterial strain and the level of mortality caused by the same strain introduced into the insect via the nematode. They did not find Achromobacter nematophilus which had been described as specific for this species of nematode, but instead they found different bacterial species in the nematodes with LD_{50} to Galleria larvae ranging from 1 to 10^5 cells/larva. Lysenko (1978) could not establish a correlation between the species of nematodes and their microflora.

Normal microflora

Normal microflora include all bacteria accompanying an insect under its normal living conditions, but not obligatorily bound to the insect. From the point of view of pathogenesis it is preferable not to include the obligatory symbionts (Brooks, 1963) in this category. The composition of normal microflora may depend on the insect's environment as well as on its food. For example, the guts of saw fly larvae collected and examined in their natural conditions (in mountains on conifers in summer) are mostly sterile. The same larvae when transferred to the laboratory develop a gut microflora within a few days (Kudler and Lysenko, unpub.).

Sometimes normal microflora are called commensals; the accuracy of this definition, however, is not so important as the correct determination of the species of the microflora.

Numerous articles have described the bacteria from different insect species. However only a few papers have investigated the role of these bacteria in the pathogenesis of bacterial disease.

The role of normal microflora is probably to exclude other microorganisms. As we discussed above, Kodama and Nakasuji (1971) indirectly suggested a "buffering power" of normal microflora to keep the gut and its contents in an appropriate physiological state. Experimental evidence of the role of normal microflora requires experiments with germ-free animals. This type of study has rarely been made with insects.

Thompson et al. (1977) calculated that bacteria as contaminants in food of the boll weevil, *Anthonomus grandis*, can decrease the amount of fatty acids in the food and be a stress on oviposition.

In many cases there seems to be no essential difference between the species comparing the normal microflora and facultative pathogens as discussed above. Under certain conditions, nearly every species of gut bacteria can become pathogenic for the insect host. It may be concluded that normal microflora can play both roles in insect life: under normal conditions they may be useful; under altered conditions they may be harmful for the host.

BACTERIAL TOXINS

Bacterial toxins in insects have been reviewed by Lysenko (1977), Lysenko and Kučera (1971), and Somerville (1973). Bacterial toxins have been divided by different criteria. The most commonly used criterion has been the place where the toxin occurs. Those liberated into the medium are known as exotoxins and those associated with the cell, usually with the cell wall, are endotoxins. Toxins can also be divided into categories according to their role in the pathogenesis, e.g. toxins of overriding importance in infectious disease, or toxins which are not the only factors responsible for a particular disease (Smith, 1978). Any such classification is a simplification, and is dependent on the information available at the time.

Since the first classical work of MacFarlane and Knight (1941) who showed that α-toxin of *Clostridium welshii* was an enzyme (phospholipase C), knowledge of bacterial toxins has rapidly increased as a result of improvements in biochemistry and physiology. It is now generally accepted that most microbial toxins are in reality enzymes or inhibitors. Because it is difficult to correctly appreciate the role of a toxin *in vivo*, the term "toxic enzymes" is introduced, referring to those purified bacterial enzymes which are toxic by injection or feeding.

For the microorganism, toxins seem to be normal metabolic products. Their significance as pathogenic tools seems to be of secondary importance (c.f. Lysenko and Kučera, 1971). A single bacterial strain can produce several different enzymes all of which can exhibit different pathogenicities in different hosts. This multivariable system explains various kinds of pathogenicity as well as host specificity. A particular disease is a result of an interaction between these enzymes and metabolic pathways or cell structures of the host. The "mosaic of toxic enzymes" can differ from strain to strain and taxonomic definition of a bacterial species cannot be based on toxins alone.

Cytolytic bacterial toxins

In vertebrate pathology these toxins are defined as "bacterial products capable of causing physical dissolution of mammalian or other cells *in vitro*" (Bernheimer, 1974). About 37 cytolytic toxins of bacterial origin have been described; all are lytic by virtue of interaction with plasma membranes.

Toumanoff (1953) showed that lecithinase C was produced by strains of *Bacillus cereus*. Its toxicity to insects was studied by Heimpel (1955), Kushner (1962), and Lysenko (1972 a, b) (c.f. Davidson, Chapter 10). The phospholipase C (PLC) of *B. cereus* was isolated and biochemically defined. This enzyme had the same toxicity to *Galleria* larvae as that of *Clostridium perfringens*. Both produced marked changes in the esterase pattern of larval hemolymph (Lysenko, 1972b). Both damaged hemocytes, which lost their shape and were changed into "ghost cells", presumably caused by membrane damage (Madziara-Borusiewicz and Lysenko, 1971). Lysenko (1973) showed that the PLC of *Pseudomonas fluorescens* was nearly nontoxic for *Galleria* larvae. The only difference between the biochemical characteristics of the three enzymes was the molecular weight: *B. cereus* PLC - 24,000 daltons; *C. perfringens* PLC - 33,000 daltons; and *P. fluroescens* PLC - 55,000 daltons. Smaller molecules were apparently able to attack the sensitive substrate in the molecular structure while the larger *P. fluorescens* PLC was unable to contact the substrate. A phospholipase D of *Corynebacteriium bovis* (16,000 daltons) was also toxic for *Galleria* larvae. Toxicity of an enzyme or toxin is therefore not only an attribute of a type of enzymatic activity, but is also limited by the availability of the substrate to the enzyme.

From studies with vertebrates it is known that phospholipases (i.e. lecithinases, sphyngomyelinases, etc.) primarily attack membranes of epithelial cells, erythrocytes, microsomes, fibroblasts, etc. (e.g. Sugahara et al., 1977;

Yamakawa and Ohsaka, 1977; Stahl, 1973; reviewed by Jeljaszewicz and Wadstrom, 1978). Phospholipases of various bacterial species may differ in substrate specificity causing selective cytological activity. The mechanism of toxicity of phospholipases in insects remains unexplained. The primary effect, damage to the sensitive structure of the membrane, is followed by a subsequent release of cell material which can cause secondary damage. However, the secondary damage can be more lethal for the host than the primary one.

Enterotoxins

As shown by Turnbull (1979), there is a difference between "food poisoning" and "food-bourne infection". By this definition the crystals of *B. thuringiensis* cause a "food poisoning". The extent to which the other insect diseases of the alimentary tract can be included in this category is unknown.

Typical enterotoxins are cholera toxin and *Escherichia coli* thermolabile enterotoxin. Both cause an enhanced intestinal secretion mediated by cyclic AMP (e.g. Kantor, 1975). Cyclic AMP is a regulator of cellular expression of the effect of hormones and of the rate of cell division. Yousten and Guerrant (1976) found that cholera toxin produces dramatic ballooning of cells and distention of the midgut in caterpillars of the cabbage looper, *Trichoplusia ni*. Denlinger (1976) demonstrated stimulation of the adenylate cyclase system by cholera toxin in larvae of *Sarcophaga crassipalpis*; injection of the toxin into larvae prior to pupariation prevented diapause.

Although *Vibrio cholerae* does not normally occur in insects, *E. coli* infections may occur in flies living in contaminated substrate (Lysenko and Povolny, 1961). Many enterobacteria are insect facultative pathogens or microflora. Since the ability to form the *E. coli* enterotoxin can be transmitted by a plasmid, it is possible that toxin-producing plasmid transmission may occur in the insect gut and that some epizootics described in the literature (e.g. "*Coccobacillus acridiorum*") originated in this way.

Proteinases

In 1962 Lysenko isolated a proteinase of *P. aeruginosa* toxic by injection to *G. mellonella* larvae (Lysenko, 1963 a, b, c, d). Proteases were shown to be toxic to insects; different proteinases exhibited different toxicities in the same insect (Lysenko, 1964). When several proteinases were isolated, biochemically characterized (Kučera and Lysenko, 1968), and tested for toxicity it was found that the most toxic proteinases had a specific activity to some proteins of Galleria hemolymph (Lysenko and Kučera, 1968). The enzyme attacked hemocytes selectively, and the insect had no defensive mechanism (Madziera-Borusiewicz and Lysenko, 1971). When Galleria hemolymph protein fractions were used as the substrates for the enzymes, four protein fractions showed specific susceptibility and clotting effect (Kučera and Lysenko, 1971; 1977).

The mechanism of proteinase toxicity was explained as follows: (i) Proteinase makes the phagocytes unable to digest bacterial cells; ingested cells can even multiply inside the phagocytes. (ii) Proteinases exhibit a specific affinity to some proteins of the hemolymph. The toxic effect is due neither to changing pH of the hemolymph, nor to exhausting hemolymph proteins, nor to toxic products released from these proteins. (iv) The activity of each proteinase varies by insect species it is tested against; conversely each insect species shows different sensitivity to different proteinases. There is not a direct correlation between proteolytic activity and toxicity. (v) Proteinases showing high milk-clotting activity are most toxic for larvae; hemolymph proteins possessing highest sensitivity to a particular proteinase are also intensively clotted (Lysenko, 1974).

Similar mechanisms of toxicity were shown by proteinases of S. marcescens (Kaska, 1976; Kaska et al., 1976) and of B. thuringiensis (Krieg and Lysenko, 1979; Lysenko et al., 1979). An electron microscopic study by Dorn (1977) described the enzymatic cellular breakdown caused by P. aeruginosa in Oncopeltus fasciatus. Later proteinases were shown to play an important role in pathogenesis of some human infectious diseases (e.g. Botzenhart and Ebel, 1973; Takeuchi and Suto, 1976; Haschke et al., 1979; Knosel and Nimitan, 1976; Wretlind et al., 1977).

Chitinase

Chitinase hydrolyses free chitin, as is found in the peritrophic membrane (Jeuniaux, 1963), while digestion of bound chitin requires previous treatment. Chitinase preparations have been used to enhance the effect of some B. thuringiensis preparations (e.g. Smirnoff et al., 1973).

Chitinase from S. marcescens was purified, biochemically characterized, and the mechanism of its production in vitro was determined. The toxicity of this chitinase for Galleria larvae by injection was comparatively higher than that of simultaneously formed proteinase. The pH optimum of this chitinase was near 9.0 (Lysenko, 1976). Daoust and Gunner (1979) described a bacterium which produced chitinase with optimum pH of 5.5. In order for this chitinase to be toxic in the gut of *Lymantria dispar*, the alkaline gut pH must be lowered by growth of fermentative bacteria.

"Not yet discovered" toxins

There are many exoenzymes of bacteria whose activities have not yet been tested on insects. Some of these play an important role in the pathogenesis of human and vertebrate diseases.

Hemolysins are either phospholipases enzymatically damaging sensitive cell membranes, or other proteins or even small peptides acting as biological detergents. Some of them are normally synthesized by bacteria, while others may be induced by bacteriophages, plasmids, etc.

Toxins inhibiting protein synthesis are represented by the diphtheria toxin and *P. aeruginosa* A toxin. Both attack the elongation factor EF-2, which is required for the translocation step in protein synthesis (Vasil et al., 1976, 1977; Vasil and Iglewski, 1978; Collier, 1979).

Hyaluronidase is an invasive factor of streptococci, but it is formed also by some other bacteria. The insect body contains hyaluronic acid.

Desoxyribonucleases, lipases, phenylalanine oxidases and other enzymes which may drastically interfere with some insect metabolic pathways are produced by many bacteria. Other compounds such as pigments can also be toxic; e.g. Kalasperis et al. (1975) showed that prodigiosin, a pigment of *S. marcescens*, was toxic for chick embryo.

HOST RESPONSE TO INFECTION

Host response is an integral part of any infectious disease. These responses have been intensively studied in vertebrate pathology, and some general principles have been formulated. In invertebrate pathology this field has been inadequately studied, and thus no generalizations can be drawn. A brief outline of this topic will be given in the following paragraphs; much information can be found in reviews edited by Maramorosch and Shope (1975) and Bulla and Cheng (1977). In vertebrates the immune response is usually divided into antibody- and cell-mediated components.

Humoral immunity

The humoral mechanisms in insect immunity have been recently reviewed by Chadwick (1975). In contrast to vertebrates, insects lack immunoglobulins, but instead have other inducible immunity systems. One of the most intensively studied models is that of Boman and his coworkers (Boman et al., 1979), who demonstrated a potent, cell-free antibacterial activity in the hemolymph of the giant silk moth. Immune proteins were synthesized *de novo* when induced by injection of bacteria. This defense system was composed of three separate bacteria-killing mechanisms with different specificities. *Bacillus thuringiensis* produced two different inhibitors interacting with these immune mechanisms; zymosan had a similar effect. This immunity system seems to be related to that described by Stephens and Marshall (1962).

Several other antibacterial compounds were described from insect hemolymph, but to what extent they are related to the Boman system is still unknown (e.g. Bakula, 1970; Hink and Briggs, 1968).

Lysozyme is produced in insect hemolymph and intestine (Mohring and Mesner, 1968 a, b), and in snails it is released from phagocytes into the hemolymph (Cheng et al., 1977). Jarosz (1975) found that lysozymelike lytic enzyme was produced by D-enterococci in *Galleria* larval gut and suggested that it may act as a gut defense mechanism.

Cellular immunity

Phagocytosis, one of the most commonly occuring immunological mechanisms in animals, was discovered by Mechnikoff in invertebrates. It is one of the most effective defense mechanisms in invertebrates and has been intensively studied in molluscs (cf. Cheng, 1975). Phagocytosis is associated with changes in hemocyte count (e.g. Werner and Jones, 1969). Examining electron micrographs of bacterial cells encapsulated by hemocytes, Rizki (1968) suggested that bacterial lysis within capsules was the mode of the hemocytes' defense reaction rather than intracellular phagocytic destruction.

The hemolymph is effectively cleared of injected bacteria by formation of clumps, composed of granular hemocytes, plasmatocytes and bacteria. Clumps are attached to the internal surface of insects and within a few hours are melanized to nodules. Granular hemocytes show high polyphenoloxidase activity and are probably associated with rapid discharge of granular cells (Ratcliffe and Gagen, 1976; Schmit et al., 1977). The role of the polyphenol-phenoloxidase system was discussed by Taylor (1969). Although the basic function of insect hemocytes resembles that of mammalian leukocytes and both have a glycolytic pathway providing energy, there are several fundamental differences in their mechanism of phagocytosis: hemocytes have no respiratory increment associated with phagocytosis;

the hexosamonophosphate pathway activity is not stimulated by phagocytosis; and antimicrobial systems other than the myeloperoxidase-H_2O_2-halide systems of mammals is utilized (Anderson et al., 1973 a, b).

As has been shown above, insects possess defense mechanisms which are in many respects analogous to those in vertebrates. To what extent these mechanisms are genetically controlled is unknown. It is also not known whether insects can exhibit immunopathological responses such as those known in mammals (i.e. inflammation, cell infiltration and tissue destruction) which can sometimes cause major pathological damage.

REFERENCES

Afrikian, E. K. 1973. Entomopatogenous bacteria and their importance. 420 pp. Editinghouse of the Armenian Acad. Sci. Erevan, (in Russian).

Anderson, R. S., Holmes, B. and Good, R. A. 1973a. Comparative biochemistry of phagocytizing insect hemocytes. Comp. Biochem. Physiol., 46B, 595-602.

Anderson, R. S., Holmes, B. and Good, R. A. 1973b. In vitro bactericidal capacity of Blaberus cranifer hemocytes. J. Invertebr. Pathol., 22, 127-135.

Bakula, M. 1970. Antibacterial compounds in the cell-free haemolymph of Drosophila melanogaster. J. Insect Physiol., 16, 185-197.

Bernheimer, A. W. 1974. Interactions between membranes and cytolytic bacterial toxins. Biochim. et Biophys. Acta, 344, 27-50.

Boman, H. T., Faye, I., Pye, A. and Rasmuson, T. 1978. The inducible immunity of giant silk moths. Compar. Pathobiol., 4, 145-163.

Botzenhart, K. and Ebel, E. A. 1973. Cytotoxic activity of Pseudomonas aeruginosa with regard to lecithinase, lipase, proteinase, hemolysin, and surface activity. Abl. Bakt. Hyg., A225, 387-397.

Brandt, C. R., Adang, M. J. and Spence, K. D. 1978. The peritrophic membrane: Ultrastructural analysis and function as a mechanical barrier to microbial infection in Orgyia pseudotsugata. J. Invertebr. Pathol., 32, 12-24.

Brooks, M. A. 1963. The microorganism of healthy insects. In Insect Pathology, An Advanced Treatise, E. A. Steinhaus, ed., Vol. 1. Academic Press, N.Y. pp. 215-250.

Buchanan, R. E. and Gibbons, N. E. 1974. Bergey's Manual of Determinative Bacteriology, 8th ed. 1246 pp. Williams & Wilkins Co., Baltimore.

Bucher, G. E. 1960. Potential pathogens of insects and their characteristics. J. Insect Pathol., 2, 172-195.

Bucher, G. E. 1963. Nonsporulating bacterial pathogens. In Insect Pathology, An Advanced Treatise, E. A. Steinhaus, ed., Vol. 2. Academic Press, N.Y. pp. 117-147.

Bucher, G. E. and Stephens, J. S. 1957. A disease of grasshoppers caused by the bacterium Pseudomonas aeruginosa (Schroeter) Migula. Can. J. Microbiol., 3, 611-625.

Bulla, L. A. and Cheng, T. C., eds. 1977. Invertebrate immune response. In Comparative Pathobiology, Vol. 3, Invertebrate Immune Responses. Plenum Press, N.Y. and London, 192 pp.

Burges, H. D. and Hussey, N. W., eds. 1971. Microbial Control of Insects and Mites. Academic Press, London and N.Y. 861 pp.

Chadwick, J. S. 1973. In vitro growth of bacteria in tissue of Galleria mellonella. J. Invertebr. Pathol., 22, 238-241.

Chadwick, J. S. 1975. Hemolymph changes with infection or induced immunity in insects and ticks. In Invertebrate Immunity, K. Maramorosch and R. E. Shope, eds. Academic Press, N.Y. pp. 241-271.

Charpentier, R., Charpentier, B. and Zether, O. 1978. The bacterial flora of the midgut of two Danish populations of healthy fifth instar larvae of the turnip moth, Scotia segetum. J. Invertebr. Pathol., 32, 59-63.

Cheng, T. C. 1975. Functional morphology and biochemistry of molluscan phagocytes. Ann. N.Y. Acad. Sci., 266, 343-379.

Cheng, T. C., Chorney, M. J., and Yoshino, T. P. 1977. Lysozymelike activity in the hemolymph of Biomphalaria glabrata challenged with bacteria. J. Invertebr. Pathol., 29, 170-174.

Collier, R. J. 1979. Toxins that inhibit protein synthesis. Soc. Gen. Microbiol. Quarterly, 6, 144.

Daoust, R. A. and Gunner, H. B. 1979. Microbial synergists pathogenic to Lymatria dispar: Chitinolytic and fermentative bacterial interactions. J. Invertebr. Pathol., 33, 368-377.

Denlinger, D. L. 1976. Preventing insect diapause with hormones and cholera toxins. Life Sciences, 19, 1485-1490.

Dorn, A. 1977. Studies of the fat body on Oncopeltus fasciatus invaded by Pseudomonas aeruginosa. J. Invertebr. Pathol., 29, 347-353.

Garber, E. D. 1960. The host as a growth medium. Ann. N.Y. Acad. Sci., 88, 1187-1194.

Haschke, K. D., Schaeg, W., Blobel, H. and Burckler, J. 1979. Purification of protease from Staphylococcus aureus. Zbl. Bakt. Hyg., A238, 300-309.

Heimpel, A. M. 1955. Investigation of the mode of action of strains of Bacillus cereus Fr. and Fr. pathogenic for the larch sawfly, Pristiphora erichsonii. Can. J. Zool., 33, 311-326.

Heimpel, A. M. and West, A. S. 1959. Notes on the pathogenicity of Serratia marcescens Bizio for the cockroach Blatella germanica L. Can. J. Zool., 37, 169-172.

Hink, W. F. and Briggs, J. D. 1968. Bactericidal factors in haemolymph from normal and immune wax moth larvae, Galleria mellonella. J. Insect Physiol., 14, 1025-1034.

Hostounsky, Z. and Weiser, J. 1972. Production of spores of Nosema plodiae Kellen and Lindegren in Mamestra brassicae after different infective dosage. I. Vest. Csl. Spol. Zool., 36, 97-100.

Jaroz, J. 1975. Lysozymelike lytic enzyme of Streptococcus faecalis and its role in the larval development of wax moth, Galleria mellonella. J. Invertebr. Pathol., 26, 275-281.

Jeljaszewicz, J. and Wadstrom, T., eds. 1978. Bacterial Toxins and Cell Membranes. Academic Press, London.

Jeuniaus, C. 1963. Chitin et chitinelyse. Masson & Cie., Paris. 181 pp.

Kalesperis, G. S., Prahlad, K. V. and Lynch, D. L. 1975. Toxigenic studies with the antibiotic pigments from Serratia marcescens. Can. J. Microbiol., 21, 213-220.

Kantor, H. S. 1975. Enterotoxins of Escherichia coli and Vibrio cholerae: Tools for the molecular biologist. J. Infect. Dis., 131, S22-S32.

Kaska, M. 1976. The toxicity of extracellular proteases of the bacterium Serratia marcescens for larvae of greater wax moth, Galleria mellonella. J. Invertebr. Pathol., 27, 271.

Kaska, M., Lysenko, O. and Chaloupka, J. 1976. Exocellular proteases of Serratia marcescens and their toxicity to larvae of Galleria mellonella. Folia Microbiol., 21, 465-473.

Khan, A. and Brooks, W. M. 1977. A chromogenic bioluminescent bacterium associated with the entomophilic nematode Chromonema heliothidis. J. Invertebr. Pathol., 29, 253-261.

Knosel, D. and Nimitan, E. 1976. In vitro - Versuche uber extrazellulare proteolytische Aktivitat bei Pseudomonas aeruginosa in Zusammenhang mit phytopathologischen Untersuchungen. Z. F. Allg. Mikrobiol., 16, 609-614.

Kodama, R. and Nakasuji, Y. 1971. Further studies on the pathogenic mechanism of bacterial diseases in gnotobiotic silkworm larvae. IFO Res. Comm., 5, 1-9.

Krieg, A. 1961. Grundlagen der Insektenpathologie. Steinkopff Verlag, Darmstadt. 304 pp.

Krieg, A. and Lysenko, O. 1979. Toxine und Enzyme bei einigen Bacillus-Arten unter besonderer Berucksichtigung der B. cereus-thuringiensis-Gruppe. Zbl. Bakt. Abt. II. 134, 70-88.

Kučera, M. and Lysenko, O. 1968. The mechanism of pathogenicity of Pseudomonas aeruginosa. V. Isolation and properties of the proteinases toxic for larvae of the greater wax moth Galleria mellonella L. Folia Microbiol., 13, 288-294.

Kučera, M. and Lysenko, O. 1971. The mechanism of pathogenicity of Pseudomonas aeruginosa. V. Isolation and hemolymph proteins from Galleria mellonella larvae and their digestibility by the toxic protease. J. Invertebr. Pathol., 17, 203-210.

Kuĉera, M. and Lysenko, O. 1977. The mechanism of pathogenicity of Pseudomonas aeruginosa: Milk-clotting activity of the proteolytic enzymes toxic for Galleria mellonella larvae. J. Invertebr. Pathol., 29, 388-389.

Kushner, D. J. 1962. Formation and release of lecithinase activity by growing cultures of Bacillus cereus. Can. J. Microbiol., 8. 674-688.

LeBerre, R. 1967. Les membranes peritrophique cher les arthropodes. Cah. ORSTOM Ser. Entomol. Med. Parasitol., 5, 147-204.

Lysenko, O. 1958a. Ecology of microorganisms and the microbiological control of insects. Proc. 1st Int'l Conf. Insect Pathol. and Biolog. Control, Prague, 109-114.

Lysenko, O. 1958b. "Streptococcus bombycis", its taxonomy and pathogenicity for silkworm caterpillars. J. Gen. Microbiol., 18, 775-781.

Lysenko, O. 1959. Report on diagnosis of bacteria isolated from insects (1954-1958). Entomophaga, 4, 15-22.

Lysenko, O. 1962. The nature of pathogenicity of Pseudomonas aeruginosa for Galleria mellonella. Col. Int. Pathol. Insects, Paris; Entomophaga, Mem. Hors. No. 2, p. 35 (1964).

Lysenko, O. 1963a. The mechanism of pathogenicity of Pseudomonas aeruginosa (Schroeter) Migula. I. The pathogenicity of the strain N 06 to wax moth larvae. J. Insect Pathol., 5, 78-82.

Lysenko, I. 1963b. The mechanism of pathogenicity of Pseudomonas aeruginosa (Schroeter) Migula. II. The toxic substance produced into the filtrate of the culture. J. Insect Pathol., 5, 83-88.

Lysenko, O. 1963c. The mechanism of pathogenicity of Pseudomonas aeruginosa (Schroeter) Migula. III. The effect of the N 06 toxin on the oxygen consumption of the wax moth prepupae. J. Insect Pathol., 5, 89-93.

Lysenko, O. 1963d. The mechanism of pathogenicity of Pseudomonas aeruginosa (Schroeter) Migula. IV. The antigenic character of the toxin produced by the strain N 06. J. Insect Pathol., 5, 94-97.

Lysenko, O. 1964. Some aspects of the mechanism of pathogenicity in bacteria pathogenic on insects. Proc. XII. Int. Congr. Ent., London, p. 717.

Lysenko, O. 1967. Bacterial toxins. In Insect Pathology and Microbial Control, P. A. Van der Laan, ed., North Holland Publ. Co., Amsterdam. pp. 219-237.

Lysenko, O. 1972a. Pathogenicity of Bacillus cereus for insects. I. The production of phospholipase C. Folia Microbiol., 17, 221-227.

Lysenko, O. 1972b. The pathogenicity of Bacillus cereus for insects. II. Toxicity of phospholipase C for Galleria mellonella. Folia Microbiol., 17, 228-231.

Lysenko, O. 1973. Phospholipase C of Pseudomonas chlororaphis and its toxicity for insects. Folia Microbiol., 18, 125-132.

Lysenko, O. 1974. Bacterial exoenzymes toxic for insects; proteinase and licithinase. J. Hyg. Epidemiol. Microbiol. Immunol., 188, 347-352.

Lysenko, O. 1976. Chitinase of Serratia marcescens and its toxicity for insects. J. Invertebr. Pathol., 27, 385-386.

Lysenko, O. 1978. Taxonomic relationship of yellow pigmenting bacteria isolated from entomophilic nematodes. Proc. Int'l. Colloq. Invert. Pathol., Prague, Czech.

Lysenko, O., Kandybin, N. V. and Stus, A. A. 1979. Toksigneost i entomopatogenost raznych stammov bakterij gruppy Bacillus cereus-thuringiensis. Doklady Vaschnil, 2, 17-18 (in Russian).

Lysenko, O. and Kucera, M. 1968. The mechanism of pathogenicity of Pseudomonas aeruginosa. VI. The toxicity of proteinase for larvae of the greater wax moth, Galleria mellonella L. Folia Microbiol., 13, 295-299.

Lysenko, O. and Kucera, M. 1971. Microorganisms as sources of new insecticidal chemicals: Toxins. In Microbial Control of Insects and Mites, H. D. Burges and N. W. Hussey, eds., Academic Press, London and N.Y. pp. 205-227.

Lysenko, O. and Povolny, D. 1961. The microflora of synantropic flies in Czechoslovakia. Folia Microbiol., 6, 27-32.

Lysenko, O. and Slama, K. 1959. The relation between oxygen consumption and bacterial infection in sawflies. J. Insect Pathol., 1, 184-187.

Lysenko, O. and Slama, K. 1978. Monitoring the course of bacterial infections using hemolymph pressure pulse records. Proc. Int'l. Colloq. Invert. Pathol., Prague, Czech.

Lysenko, O. and Weiser, J. 1974. Bacteria associated with the nematode Neoaplectana carpocapsae and their pathogenicity of this complex for Galleria mellonella larvae. J. Invertebr. Pathol., 24, 332-336.

Macfarlane, M. G. and Knight, B. C. J. G. 1941. The biochemistry of bacterial toxins. 1. The lecithinase activity of Cl. welchii toxins. Biochem. J., 35, 884-902.

Madziara-Borusiewicz, K. and Lysenko, O. 1971. The mechanism of pathogenicity of Pseudomonas aeruginosa VII. The influence of toxic proteinase on haemocytes of Galleria mellonella. J. Invertebr. Pathol., 17, 138-140.

Maramorosch, K. and Shope, R. E., eds. 1975. Invertebrate Immunity: Mechanisms of Invertebrate Vector-Parasite Relations. Academic Press, N.Y. 365 pp.

Martin, J. D. and Mundt. J. O. 1972. Enterococci in insects. Appl. Microbiol., 24, 575-580.

Mohrig, W. and Messner, B. 1968a. Immunreaktionen bei Insekten. I. Lysozym als grundlegender antibakterieller Faktor im hummoralen Abwehrmechanismus der Insekten. Biolog. Zbl., 87, 339-470.

Mohrig, W. and Messner, B. 1968b. Immunreaktionen bei Insekten. Biolog. Zbl., 87, 705-718.

Pashke, J. D. and Summers, M. D. 1975. Early events in the infection of the arthropod gut by pathogenic insect viruses. In Invertebrate Immunity, K. Maramorosch and R. E. Shope, eds. Academic Press, N.Y. pp. 75-112.

Podgwaite, J. D. and Cosenza, B. J. 1976a. A strain of Serratia marcescens pathogenic for larvae of Lymantria dispar: Characterization. J. Invertebr. Pathol., 27, 185-190.

Podgwaite, J. D. and Cosenza, B. J. 1976b. A strain of Serratia marcescens pathogenic for larvae of Lymantria dispar: Infectivity and mechanisms of pathogenicity. J. Invertebr. Pathol., 27, 199-208.

Poinar, G. O. 1973. Use of nematodes for microbial control of insects. In Microbial Control of Insects and Mites, H. D. Burges and N. W. Hussey, eds. Academic Press, N.Y. pp. 181-203.

Pospelov, W. 1927. Septicemia of the cutworm (Agrotis segetum Schiff). Izvest. Otd. Prikl. Entomol. GIOA (Leningrad), 3, 1-23.

Ratcliffe, N. A. and Gagen, S. J. 1976. Cellular defense reactions of insect hemocytes in vivo: Nodule formation and development in Galleria mellonella and Pieris brassicae larvae. J. Invertebr. Pathol., 28, 373-382.

Richards, A. G. and Richards, P. A. 1977. The peritrophic membranes of insects. Ann. Rev. Entomol., 22, 219-240.

Rizki, T. M. 1968. Hemocyte encapsulation of Streptococci in Drosophila. J. Invertebr. Pathol., 12, 339-343.

Schmit, A. R., Rowley, A. F. and Ratcliffe, N. A. 1977. The role of Galleria mellonella hemocytes in melanin formation. J. Invertebr. Pathol., 29, 232-234.

Slama, K. 1976. Insect haemolymph pressure and its determination. Acta. Ent. Bohemoslov., 73, 65-75.

Slama, K. and Lysenko, O. 1980. Monitoring of the course of bacterial infections by hemolymph pressure pulses in insects. J. Invertebr. Pathol., in press.

Slatten, B. H. and Larson, A. D. 1967. Mechanism of pathogenicity of Serratia marcescens. I. Virulence for the adult boll weevil. J. Invertebr. Pathol., 9, 78-81.

Smirnoff, W. A., Randall, A. P., Martineau, R., Haliburton, W. and Juneau, A. 1973. Field tests of the effectiveness of chitinase additive to Bacillus thuringiensis Berliner against Choristoneura fumiferana (Clem.). Can. J. Forest. Res., 3, 228-236.

Smith, H. 1977. Microbial surfaces in relation to pathogenicity. Bacteriol. Rev., 41, 475-500.

Smith, H. 1978. The determinants of microbial pathogenicity. In Essays in Microbiology, J. R. Norris and M. H. Richmond, eds. J. Wiley & Sons, Chichester, N.Y. 520 pp.
Somerville, H. J. 1973. Microbial toxins. Ann. N.Y. Acad. Sci., 217, 93-108.
Stahl, W. L. 1973. Phospholipase C purification and specificity with respect to individual phospholipids and brain microsomal membrane phospholipids. Archiv. Biochem. Biophys., 154, 47-55.
Steinhaus, E. A. 1949. Principles of Insect Pathology. McGraw-Hill, N.Y. 757 pp.
Steinhaus, E. A. 1958. Stress as a factor in insect disease. Proc. 10th Inter. Cong. Ent., Montreal, 4, 725-730.
Steinhaus, E. A. 1959. Serratia marcescens Bizio as an insect pathogen. Hilgardia, 28, 351-380.
Steinhaus, E. A., ed. 1963. Insect Pathology, An Advanced Treatise, 2 Vol. Academic Press, N.Y. 661 pp. + 689 pp.
Stephens, J. S. and Marshall, J. H. 1962. Some properties of an immune factor isolated from blood of actively immunized wax moth larvae. Can. J. Microbiol., 8, 719-725.
Stevenson, J. P. 1959. An infection of the desert locust, Schistocerca gregaria Forskal with a nonchromogenic strain of Serratia marcescens Bizio. J. Insect Pathol., 1, 129-141.
Stohler, H. 1961. The peritrophic membrane of blood-sucking Diptera in relation to their role as vectors of blood parasites. Acta Trop., 18, 263-266.
Sugahara, T., Takahashi, T., Yamaya, S. and Ohsaka, A. 1977. Vascular permeability increase by α-toxin (phospholipase C) of Clostridium perfringens. Toxicon, 15, 81-87.
Takeuchi, S. and Suto, T. 1976. Behaviors of a vigorous protease-producing strain of Staphylococcus aureus in the skin tissue of mice. Japan. J. Microbiol., 20, 155-162.
Taylor, R. L. 1969. A suggested role for the polyphenol-phenoloxidase system in invertebrate immunity. J. Invertebr. Pathol., 30, 274-275.
Thompson, A. C., Sikorowski, P. P. and Wyatt, J. M. 1977. The effect of bacterial load on fatty acids in the boll weevil, Anthonomus grandis (Coleoptera: Curculionidae). J. Invertebr. Pathol., 30, 274-275.
Toumonoff, C. 1953. Description de quelques souches entomophytes de Bacillus cereus Frank. et Frank. avec remarques sur leur action et celle dautres baciles sur le jaunce d'oeuf. Ann. Inst. Pasteur, 85, 90-99.
Turnbull, C. P. B. 1979. Enterotoxins in food poisoning. Soc. Gen. Microbiol. Quarterly, 6, 144.
Vasil, M. L. and Iglewski, B. H. 1978. Comparative toxicities of diphtherial toxin and Pseudomonas aeruginosa exotoxin A: Evidence for different cell receptors. J. Gen.

Microbiol., 108, 333-337.

Vasil, M. L., Kabat, D. and Igelwski, B. H. 1977. Structure-activity relationship of an exotoxin of Pseudomonas aeruginosa. Infect. Immun., 16, 353-361.

Vasil, M. L., Liu, P. V. and Iglewski, B. H. 1976. Temperature-dependent inactivating factor of Pseudomonas aeruginosa exotoxin A. Infect. Immun., 13, 1467-1472.

Weiser, J. 1955. Neoaplectana carpocapsae n. sp. (Anguilluleta, Steinernematinae), novy cizopasnik housenek obalece jablecneho, Carpocapsa pomonella L. Vestnik Ceskoslov. Zool. Spol., 19, 44-52.

Weiser, J. 1966. Nemoci hmyzu. 554 pp. Academia, Praha (In Czech).

Weiser, J. and Lysenko, O. 1956. Septikemie bource morusoveho. Ceskoslov. Mikrobiol., 1, 216-222.

Welch, H. E. 1963. Nematode infections. In Insect Pathology, An Advanced Treatise, E. A. Steinhaus, ed., Vol. 2. Academic Press, N.Y., pp. 363-392.

Werner, R. A. and Jones, J. C. 1969. Phagocytic haemocytes in unfixed Galleria mellonella larvae. J. Insect Physiol., 15, 425-437.

Whitcomb, R. F., Shapiro, M. and Richardson, J. 1966. An Erwinialike bacterium pathogenic to leafhoppers. J. Invertebr. Pathol., 8, 299-307.

Wretlind, B., Sjoberg, L. and Wadstrom, T. 1977. Protease-deficient mutants of Pseudomonas aeruginosa: Pleiotropic changes in activity of other extracellular enzymes. J. Gen. Microbiol., 103, 329-336.

Yamakawa, Y. and Ohsaka, A. 1977. Purification and some properties of phospholipase C (α- toxin) of Clostridium perfringens. J. Biochem., 81, 115-126.

Yousten, A. A. and Guerrant, R. L. 1976. Comparative activities of Bacillus thuringiensis δ-endotoxin, cholera toxin, and Escherichia coli enterotoxin. J. Invertebr. Pathol., 28, 395-397.

Chapter 7

INSECT DISEASES CAUSED BY BACILLI WITHOUT TOXIC MEDIATED PATHOLOGIES

by Clara M. Splittstoesser* and C. Y. Kawanishi†

INTRODUCTION

The organisms to be discussed in this chapter are entomopathogens, mainly of the genus *Bacillus*, from which no toxic compounds have been identified as important factors in the infective process.[1] They are *Bacillus larvae*, the causative agent of American foulbrood (AFB) of the honey bee, *Apis mellifera*, and members of the *Bacillus popilliae-Bacillus lentimorbus* group, causative agents of milky diseases in scarabaeid beetles. In addition, European foulbrood of the honey bee will be discussed briefly since there has been some debate as to whether a bacillus, *Bacillus alvei*, is involved in the pathology as well as the major pathogen, *Steptococcus pluton*. These organisms are classified as obligate pathogens according to the characteristics proposed for this group by Bucher (1960); i.e., they are found in nature associated with a specific insect disease, they have a narrow host range, they are readily transmitted *per os*, require specialized conditions for growth, and probably multiply only within the bodies of specific insects in nature. Both foulbrood and milky disease are initiated by larval ingestion of the bacterial spores which germinate within the alimentary canal, followed by vegetative rod penetration of mesenteric epithelium. Since

---

[1]The suggestion by some authors that toxins are formed by the milky or foulbrood disease organisms will be discussed in the appropriate section.

*Geneva, New York, 14456
†United States Environmental Protection Agency, Research Triangle Park, North Carolina 47711

several aspects of the ensuing disease processes differ for those pathogens, they will be discussed separately.

THE MILKY DISEASES

Taxonomy

Although taxonomy is not directly pertinent to a discussion of the pathogenesis of milky disease bacteria, a brief statement is in order. Several strains have been isolated from different scarabaeid hosts since the original work of Dutky (1940), who described and named two species, *B. popilliae* and *B. lentimorbus*, from the Japanese beetle, *Popillia japonica*. They have been reported from Europe, Australia, New Zealand, and the United States. New specific or varietal designations were assigned on the basis of host of isolation, *per os* host range, morphology of the spore and parasporal body if present, and for a few, serological and cultural characteristics. Unfortunately, the difficulty in culturing many of these isolates in artificial media and the lack of enough strains for adequate comparison permits only a provisional classification. Varietal considerations for the isolates available at the time were discussed by Gordon et al. in 1973, who recognized *B. popilliae* and *B. lentimorbus* as separate species. Scanning electron microscopy was used by Bulla et al. (1969) to show differences in surface topography of free spores of the latter two species. On the basis of detailed physiological and morphological characters, Wyss (1971) proposed that *B. popilliae*, *B. lentimorbus*, and the European isolates be combined into one species, *B. popilliae*, with three varieties, *popilliae*, *lentimorbus*, and *melolonthae*. This is supported by the serological studies of Krywienczyk and Lüthy (1974) but disputed by those of Hrubant and Rhodes (1968) using a different serological system. *Bacillus euloomarahae*, isolated by Beard (1956) in Australia, was retained as a separate species in the Wyss classification scheme because of its distinctive morphology. Until *B. eluoomarahae* and the morphologically similar *B. lentimorbus* var. *maryland* (Dutky, 1963) and *Ataenius* isolate (Kawanishi et al., 1974) are successfully cultured *in vitro*, the designation of these organisms remains uncertain. Other taxonomic criteria may prove to be of value in elucidating the relationships of the milky disease pathogens. The spore and sporangium of the *Ataenius* isolate were found by Kawanishi et al. (1974) to have different ultrastructural characteristics when compared to those of *B. popilliae* and *B. lentimorbus*. Kawanishi (Fig. 7-1) has found differences in the fine structure of the spores of certain isolates examined, suggesting two distinct subgroups. In one group,

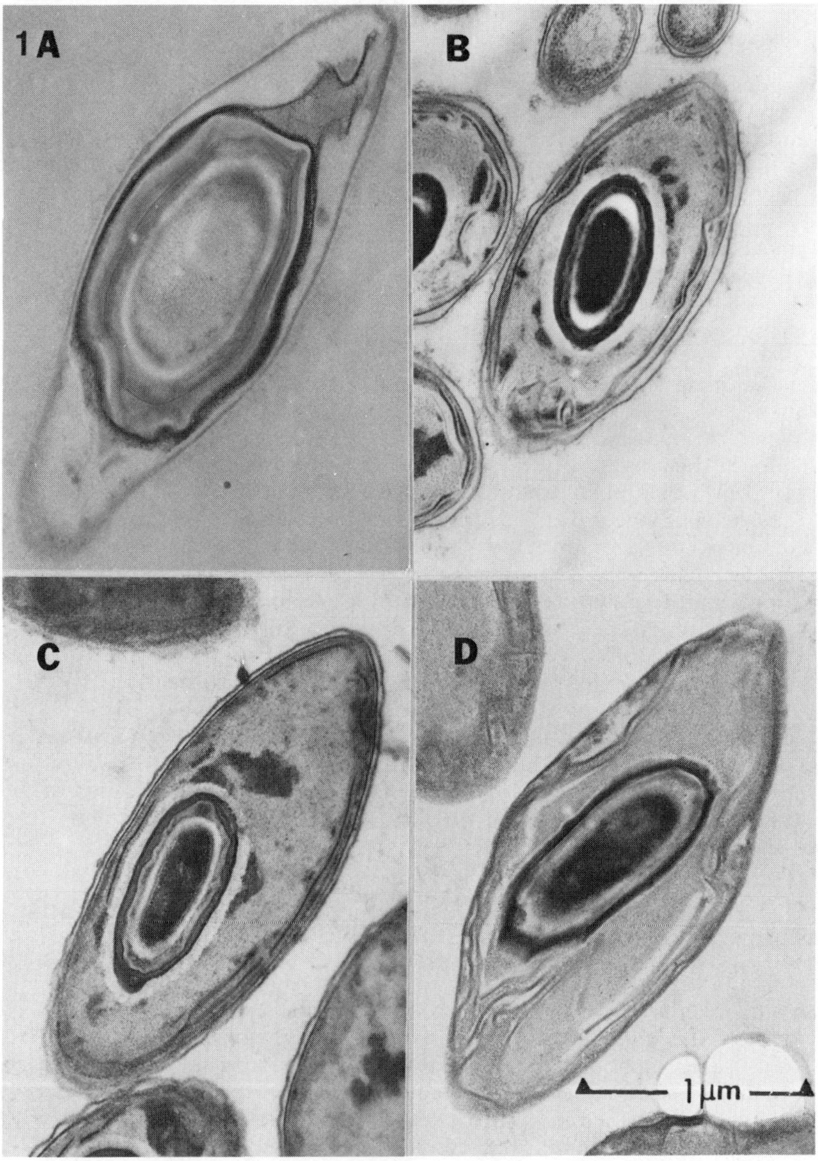

Figure 7-1. Spores of milky disease pathogens. Differences in structure suggest at least two different subgroups of bacteria may cause milky diseases. A. *Bacillus lentimorbus* B. *Bacillus lentimorbus* var. *maryland* C. *Bacillus euloomarahae* D. *Ataenius* isolate

the sporangium contains a small endospore with a few layers of exosporium and no parasporal body (*Ataenius* isolate, *B. euloomarahae*, *B. lentimorbus* var. *maryland*). The other group has a sporangium with a larger endospore with complex integument, and may or may not possess a parasporal body (*B. popilliae*, *B. lentimorbus*). This tends to support the classification proposal of Wyss.

Gross symptoms

The name "milky disease" is descriptive of the group as a whole, regardless of scarabaeid host species or strain of pathogen. Development of the disease within the host has been described by Dutky (1963) and other workers. Bacterial multiplication and sporulation in the hemolymph produces an obvious turbidity which obscures the normally visible dorsal blood vessel. This opacity is the basis for the name. Typically, diseased larvae remain alive for a long period of time and continue to feed until they become moribund. During this time, an initial period of vegetative multiplication followed by a period of concomitant vegetative multiplication and sporulation result in the ultimate accumulation of $6 \times 10^7 - 5 \times 10^8$ spores per larva before the insect dies. Selected strains of milky disease bacteria may be so virulent that death of the host occurs prior to sporulation. However, as pointed out by Dutky (1963), strain virulence under natural conditions probably adjusts to the level that produces the maximum number of spores in the host.

Cellular interactions of pathogens and host

Since the milky diseases were first described, it has been accepted that ingestion of large numbers of spores is necessary for the induction of the disease process and that by some means spores, or more probably vegetative rods, appear in the hemolymph to initiate vegetative multiplication and subsequent sporulation. The process by which the organisms invaded or were transferred to the hemolymph was unknown. In 1945, Beard showed that disease in the Japanese beetle could be induced by oral or anal injection of spores of *B. popilliae* and suggested on the basis of ligature experiments that the Malpighian tubules were the most probable site of bacterial penetration from the alimentary canal to the hemolymph. This remained the only proposal on the mode of infection until 1973, when Splittstoesser et al. reported the first histological investigation using the European chafer, *Amphimallon majalis*, as host and *B. popilliae* as pathogen. They showed that in perorally challenged grubs the spores germinated in the lumen of the alimentary canal and vegetative rods initiated a primary cycle of multiplication on the luminal side of the basement membrane in regenerative nidi areas prior to penetration of the hemocoel and further multiplication in the hemolymph. The process by which the bacilli entered the mesenteric epithelium and penetrated to regenerative nidi sites was not observed, and further investigations were made by cytological examination of alternate thick and thin plastic sections using both light and electron microscopy. Details of midgut alterations caused by epithelial infection, wound repair, and hemocytic encapsulation were elucidated and certain host defense reactions were defined (Splittstoesser et al., 1978; Kawanishi et al., 1978).

In plastic sections, vegetative rods were located within membrane-limited vacuoles of midgut columnar cells prior to establishing primary infection foci in regenerative nidi areas. This, plus the fact that intercellular bacilli were never present, suggests that penetration of columnar cells occurs by a phagocytosis-like process. The presence of bacteria in cells of the mesenteric epithelium resulted in focal necrosis and columnar cell desquamation which exceeded the normal turnover rate (Fig. 7-2). These events led to enhanced mitotic activity of regenerative nidi cells within and adjacent to the affected areas, and increased numbers of noncolumnar immature epithelial cells appeared. Vegetative rods were observed in such immature replacement cells and in regenerative cells with the subsequent accumulation of large masses of bacilli in nidi areas. The presence of such infection foci in mesenteric cells evoked a defensive or inflammatory reaction by certain circulating hemocytes, which adhered to the hemocoelic surface of the basement membrane immediately adjacent to the sites of bacterial proliferation. Extensive infiltration by such granule-bearing hemocytes resulted ultimately in the formation of a compact mass or capsule attached to the mesenteron-investing basal lamina. In one case, granular hemocytes with phagocytized bacilli were observed within a mesenteronic lesion.

Degenerating vegetative cells of B. popilliae were found in vacuoles of columnar cells of the midgut epithelium, immature replacement cells of the midgut, and in hemocytes composing the capsule. All of these cell types encountered by the bacilli during invasion were capable of intracellular digestion of the bacteria. Epithelial cells which overcame the infection were distinguished by residual bodies containing bacterial cell wall fragments. In hemocytes, granules identical to those in the cytoplasm were frequently present together with bacilli within vacuoles and endocytic invaginations of the plasma membrane (Fig. 7-3). Material closely resembling the matrix of intact granules was often observed together with deteriorating bacilli, suggesting degranulation and indicating a possible intracellular digestive function for these granules.

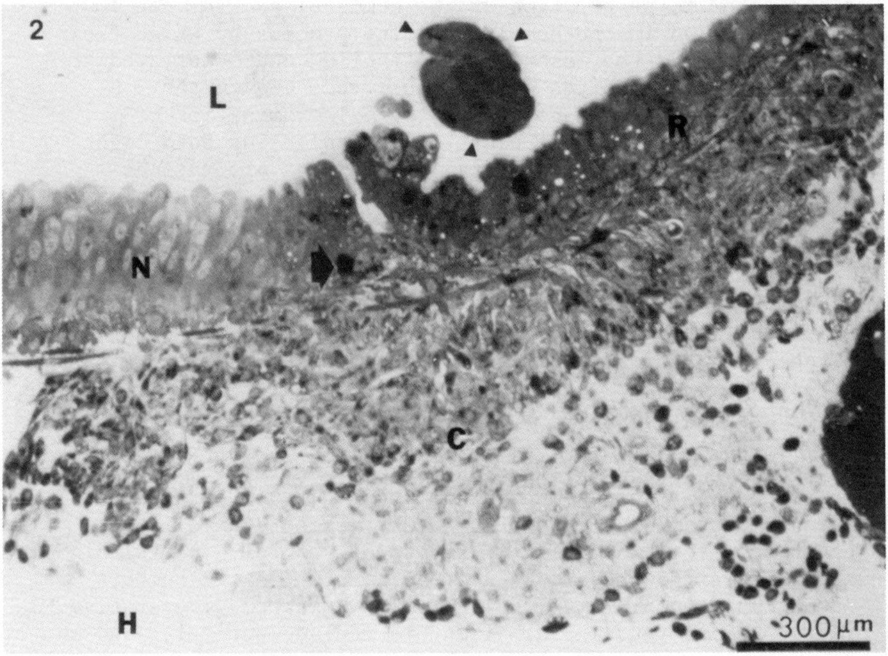

Figure 7-2. Region of a larval European chafer midgut infected by Bacillus popilliae and a subjacent inflammatory hemocyte capsule (C) on the hemocoelic surface (H). Masses of bacilli within midgut cells (arrow), exfoliated necrotic columnar epithelial cells (arrowheads) in the gut lumen (L), low replacement (R) and normal (N) columnar cells are shown. Light micrograph.

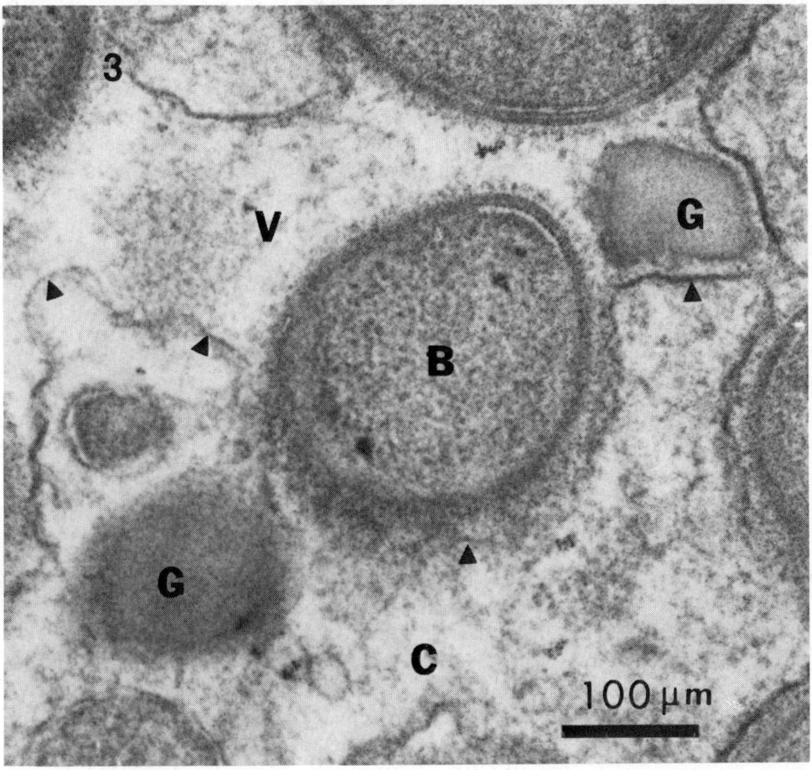

Figure 7-3. Degranulation of bacilli-containing capsule hemocyte. Hemocyte-granules (G) are present within the cytoplasm (C) and vacuole (V) containing Bacillus popilliae (B). Arrowheads indicate vacuolar membrane. Electron micrograph.

Vacuoles containing normal-appearing bacilli were also noted in all the above-mentioned host cell types, and apparently B. popilliae was capable of overcoming the bactericidal action of some cells and penetrating the basement membrane. The actual penetration process was not defined, but masses of bacilli were frequently seen in areas of greatly diminished or discontinuous basal lamina. Bacilli on the hemocoelic side of the basal lamina within the capsule were observed both within hemocytes and in intercellular spaces.

Once in the hemolymph, the vegetative rods were not phagocytized, even though hemocytes morphologically identical to those that formed the capsule were present and circulating. The reason for this phenomenon is not clear. Phagocytic ability was not inhibited as evidenced by the fact that the hemocytes were still capable of taking up carmine particles. Lüthy et al. (1970) reported that there was no interaction between vegetative rods of B. fribourgensis and hemocytes of Melolontha melolontha in vivo as well as in a tissue culture system. Schwartz and Townshend (1968) also stated they did not observe phagocytosis of B. popilliae by hemocytes of the Japanese beetle. In contrast, mature spores are subject to phagocytosis.

The effect of milky disease on total or differential hemocyte counts in perorally challenged scarabaeids has not been reported, but the numbers of phagocyte hemocytes involved in capsule formation and the degradation of some by internal bacilli would appear to indicate alterations. Schwartz and Townshend (1968) described the effect of infection of the Japanese beetle by injection on these blood parameters. A temporary increase in total hemocyte count occurred when there was a heavy concentration of vegetative rods in hemolymph prior to sporulation. They postulated that the increase may function either to absorb toxins formed by the bacilli or to offset the drain of hemolymph nutrients being utilized by the bacteria. Beard (1945) also reported an increase in total hemocyte count, but at a later stage of infection, i.e., after spores were present. Both investigations recorded a return of the hemocyte count to values observed in normal larvae, and the average total hemocyte counts were not significantly different.

Molecular interactions

No molecular interactions of milky disease bacteria and their insect hosts have been defined. However, the involvement of toxin(s) in the lethality of the disease has been suggested. On the basis of antibiotic interception of growth and development of B. popilliae in the hemocoel of Japanese beetle larvae, Dutky (1963) concluded that toxic products have an effect on the disease process. When dihydrostreptomycin was injected 48 hours after an infective dose of spores of *B. popilliae*, the larvae survived as healthy insects. Administration of the antibiotic 72 hours or more after the spores inhibited further development of the organisms, but the insects died at the same rate as those not receiving the antibiotic. He concluded from these tests that once the disease is established, the lethal process cannot be reversed by destroying the organism or inhibiting its further development. Also, cell free culture filtrates of *B. popilliae* in small amounts caused death of the host. These lethal products were produced by cultivation of the bacteria *in vitro* but are of unknown importance in the progression of the disease *in vivo*, and it is not known whether they are produced *in vivo*.

Ebersold (1976) reported that the parasporal protein of *B. popilliae* when dissolved with NaOH and added to hemocyte cultures derived from *M. melolontha* at concentrations of $\geq$10 μg/ml caused irreversible damage to the insect cells. The study of Weiner (1978) using similar preparations suggested toxicity to injected third instar Japanese beetle larvae. In contrast to that of *Bacillus thuringiensis*, the function of the parasporal body of *B. popilliae* is unknown.

The prolonged course of milky disease would appear to be inconsistent with the production of toxins, although it does not rule them out as auxiliary factors.

Physiological responses of host to pathogen

Knowledge of the biochemical changes which occur in a host in response to infection by *B. popilliae* has resulted primarily from the investigations of Bennett and Shotwell with the Japanese beetle. Their results have been reviewed by Bennett and Shotwell (1973) and Bulla et al. (1975). Proteins, organic acids, individual amino acids, lipids and carbohydrates were determined in hemolymph from healthy and diseased larvae. Larvae were challenged by injection and assays were performed after sporulation of the bacillus had occurred. The primary effect of infection on those hemolymph constituents was an overall reduction in the concentration of lipids, proteins, and carbohydrates.

Lipids decreased 45%, but there was no selective utilization of any specific fatty acid by the pathogen. The total amount of hydrocarbons in extractable lipids was reduced as a result of infection, but the relative concentrations of individual hydrocarbons was not significantly altered (Bennett et al., 1972). Also, only minor quantitative differences were observed in the major hydrocarbons from the cuticle of healthy and diseased larvae (Bennett and Shotwell, 1976), indicating that the biosynthesis and transport of larval hydrocarbons are unaffected by infection. The authors concluded that since the total quantity of hydrocarbons is reduced as a result of milky disease, as is the total lipid content, the lipid synthesis mechanisms of the larvae are impaired or blocked.

Hemolymph from infected larvae contained higher concentrations of malic, glycolic, tartaric, pyruvic, and glyoxylic acids compared to hemolymph from healthy larvae. No decrease in any organic acid was observed (Stubblefield et al., 1966).

The total free amino acids were not significantly different between healthy and diseased insects, but individual amino acids varied (Shotwell et al, 1965). Concentrations of glutamic acid and β-alanine were markedly higher in hemolymph from infected larvae and there was some increase in aspartic acid, phenylalanine, threonine, serine, and lysine. Glycine, tyrosine, and histidine levels decreased. The significance of these changes, whether brought about directly by the metabolism of the pathogen or as modifications in host physiology as a response to infection, is uncertain, as they do not correspond to known amino acid requirements of the bacteria *in vitro*. Also, as pointed out by Schwartz and Townshend (1968), the observed decrease in tyrosine (and possibly other components as well) may have resulted from different handling of samples. Hemolymph from infected insects was centrifuged, thus removing hemocytes as well as spores, while the hemolymph from healthy insects was not so treated.

Alterations in amino acid composition during milky disease may depend on the infective organism. With *B. lentimorbus*, the only amino acid to increase during infection was glutamic acid, while the amounts of histidine, proline, glycine, alanine, valine, isoleucine, tyrosine, and arginine decreased (Shotwell et al., 1965). Also, in contrast to *B. popilliae*, the total amino acid content was considerably lower in larvae infected with *B. lentimorbus*. Benz (1963) found that hemolymph of *M. melolontha* infected with *B. fribourgensis* had higher concentrations of β-alanine, histidine, valine, and leucine, and lower concentrations of serine, threonine, and alanine than hemolymph of healthy larvae.

During the course of the milky disease of the Japanese beetle infected with *B. popilliae*, the major protein in the hemolymph, a lipoglycoprotein, decreased markedly. The amino acid complex of the lipoglycoprotein in hemolymph from healthy and diseased larvae did not differ significantly (Bennett and Shotwell, 1970).

The concentration of trehalose, the major sugar in hemolymph, was found to decrease from 7.1 $\pm$ 1.3 mg/ml to 4.7 $\pm$ 0.2 mg/ml in injected larvae after 7 days. Naturally infected larvae were found to contain 1.0 mg/ml.

In 1966, Weiner et al. found that the hemolymph of Japanese beetle larvae with only vegetative cells of B. popilliae present had significantly lower dissolved oxygen content than in normal larvae. As sporulation occurred, and oxygen demand of rapidly growing vegetative cells decreased, dissolved oxygen returned to levels found in healthy larvae. The skin permeability and hemolymph pH were not changed by milky disease. Steinkraus (1957) also reported no change in the pH of hemolymph of European chafers infected by B. popilliae. In 1969, Weiner et al. found that the metabolic requirement for oxygen of milky larvae was similar to that of healthy larvae and that there was no correlation between the concentration of microorganisms in diseased larvae and the larval requirement for oxygen, possibly because of the narrow range of cell populations in the insects examined.

Steinkraus (1957) observed a downward trend in the oxidation reduction potential of hemolymph as the disease progressed in European chafers, but no abrupt changes occurred. Beard (1945) reported the oxidation reduction potential of hemolymph from diseased Japanese beetle larvae to be significantly lower than that from healthy larvae. He also found that hemolymph from diseased larvae did not cause a solution of tyrosine to melanize and described the effects of infection on ecdysis. Larvae containing mature spores did not pupate, regardless of how long they lived. Molting of first and second instar insects was also affected. With few exceptions, once the hemolymph of an infected larvae became milky, further metamorphosis of the host was prevented. From these observations, he suggested that the disturbance of tyrosine oxidase or some similar enzyme system could account for the inhibition of molting and metamorphosis.

Sharp and Detroy (1979) proposed that depletion of fat body reserves during milky disease is the reason diseased third instar larvae cannot undergo metamorphosis to the pupal stage.

Steinkraus et al. (1973) studied the effect of infection of the European chafer with B. popilliae on the concentration of ions in the hemolymph and excised tissues of the alimentary tract. Hemolymph of diseased larvae contained higher levels of Na^{+}, K^{+}, Mg^{++}, Fe^{+++}, and Zn^{+} than that of healthy larvae. This increase was accompanied by a corresponding decrease of ions in midgut and anterior intestine tissues. Although the decrease in tissue concentrations of the ions did not correlate quantitatively with the increase in hemolymph concentrations, they suggested that B. popilliae may affect the permeability of the anterior intestine and midgut.

BACTERIAL DISEASES OF BEES

The present review mainly concerns studies of host-pathogen interaction in bacterial diseases of the honey bee, about which little specific data exist. The more general discussions of Heimpel and Angus (1963) on sporeforming bacterial diseases of insects, including those of bees, are, for the most part, still pertinent. Bailey's review (1968) can be consulted for additional information on honey bee pathology. Bee disease diagnosis, treatment, culturing of pathogens, and biological control of infections were discussed by Gochnauer et al. (1975). Bacterial diseases of bees were reviewed by Shimanuki (1978).

The American Foulbrood disease (AFB) and, to a limited extent, European Foulbrood (EFB) are of primary concern. A detailed study of AFB host-pathogen relationships was reported by Davidson (1973). The only study of this aspect of EFB appears to be the histopathological investigation of Tarr (1938). The powdery scale disease and its putative etiological agent, *Bacillus pulvifaciens*, recently reported on by Hitchcock et al. (1979) is mentioned here in passing and will be discussed where data permit.

Bacillus larvae is well established as the etiological agent of AFB. There is some question about the "curly whip" structures characteristic of *B. larvae* cultures grown in larvae or on certain culture media. Gochnauer et al. (1975) state that these structures are diagnostic of AFB and consist of coiled masses of bacterial flagella. Krizanova et al. (1978) however, report the formation of feeble but separate colonies of these forms on solid media and suggest the possibility of their independent existence.

EFB is clearly a disease of much more complex etiology. Studies by Bailey (1963) indicated that *Streptococcus pluton* is the major pathogen responsible for this disease, although some isolates of *S. faecalis* may also be pathogenic. *Bacterium eurydice* was suggested to have a "supplementary effect" in EFB. Pure cultures of *Bacillus alvei*, originally thought to be the causative agent of EFB, were demonstrated to be non-pathogenic to honey bee larvae by Bailey and this organism therefore is likely a secondary invader. The concensus appears to be that *S. pluton* is the primary cause of the disease. However, because of the variable nature of the signs and symptoms of the EFB conditions and its occurrence frequently in weakened colonies, the possibility of multiple or mixed etiology cannot be excluded.

Cellular interactions of pathogens and host

Tarr's report (1938) appears to be the only histopathological study of EFB pathogenesis. The author concludes that the pathogen, *S. pluton*, and midgut cells of the larvae did not physically interact until the host was either dead or moribund. Detailed studies utilizing modern techniques such as electron microscopy or immunocytochemistry appear to be in order.

In contrast, cellular interactions between the AFB pathogen, *B. larvae*, and host have been studied histopathologically by Bamrick (1964) and by electron microscopy by Davidson (1973). Spore germination and rapid vegetative growth was observed to occur in the lumen of the midgut in both studies. Bamrick (1964) found spore germination usually occurred within one day in younger larvae of both susceptible and resistant strains of bees. Spores seemed to germinate a little slower in older larvae which are no longer susceptible to the pathogen.

The study of Davidson (1973) indicated the following sequence of pathogen-host interaction as observed with the electron microscope. The electron dense luminal edge of the peritrophic membrane (PM) was breached easily by *B. larvae* bacilli, however the matrix of the PM, which undergoes an age-dependent thickening, appeared to be a greater barrier to penetration (Fig. 7-4). This thickening appeared to parallel the increasing resistance to AFB disease with the age of the larvae.

A small number of bacilli was observed to penetrate the PM matrix in regions of high bacterial concentration in the lumen and contacted the midgut cell membrane. No evidence of enzymatic digestion of PM was observed.

The major areas of midgut penetration were in the anterior and posterior regions of the migut where large concentrations of bacteria occurred. In the anterior region, bacilli entered the space between the cuticular intima of the foregut neck and the midgut cells forming the valve ring. Bacilli penetrated the valve ring cells. In the posterior regions, at the imperforate juncture of the mid- and hindguts, bacilli entered the midgut cells but were limited from further penetration by the underlying cuticular intima of the hindgut plug.

Microvilli of the midgut epithelial brush border were displaced by bacilli during invasion and *B. larvae* entered midgut cells in areas where the cell surface apparently invaginated to accomodate the penetrating pathogen by a phagocytosis-like process (Fig. 7-5). Endocytic closure of the midgut cell membranes apparently followed, as evidenced by the observation of bacilli in vacuoles within the host cell. Occasionally, sections were obtained where closure was incomplete and cell membrane-limited furrows behind (luminal) the penetrating bacilli were observed. Flagella appeared to be copiously shed during the entry process.

Both degraded and intact bacilli were observed within some vacuoles. Degradation may have been due to lysosomal activity, as small vesicles were observed fusing with the

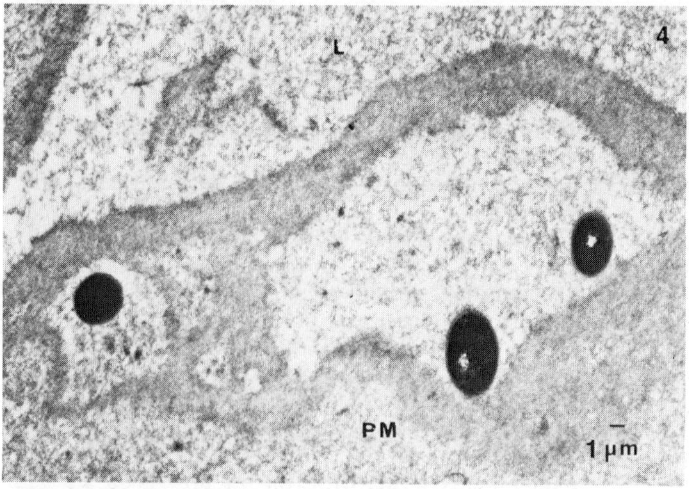

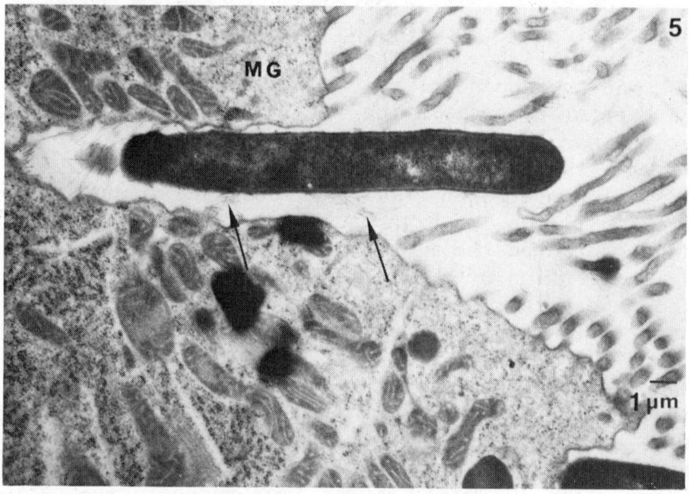

Figure 7-4. Bacillus larvae within the lumen (L)-bounding edge of the peritrophic membrane (PM) of a 75-hr larva of the honey bee, Apis mellifera. Bacilli are surrounded by material resembling the food mass of the lumen.

Figure 7-5. Entry of Bacillus larvae into a midgut cell (MG) of 111-hr honey bee larva. Microvilli are not present on that portion of the cell membrane within the invagination. Flagella (arrows) are visible surrounding the bacterium. Figures 7-4 and 7-5 from Davidson (1973); reprinted by permission of Academic Press, Inc.

vacuole membrane. Intact bacteria enclosed within vacuoles came to lie within the basal infoldings of midgut cells or on the basal cell membrane above regenerative nidi. This indicates that not all bacteria are degraded. Eventually infected cells degenerated and finally lysed. Bacteria without enclosing vacuole membranes were seen in such affected regions.

Penetration of B. larvae into the hemocoel was not observed by electron microscopy, but light microscopic studies revealed egress of large clusters of bacilli into the hemocoel. Location of hemocoelic bacteria was difficult during the prepupal stage. A systemic bacteremia eventually developed. Certain hemocoelic tissues such as fat bodies, tracheal cells, and hypodermis were centers of rapid bacterial proliferation and release, while muscle cells and oenocytes disrupted more slowly. Hemocyte cytoplasmic membranes were observed to be ruptured in the vicinity of the enclosed bacteria.

Davidson (1973) suggested that tissue tropism of hemocoelic bacilli may have been one of the causes of the scarcity of observable hemocoelic bacteria during the prepupal stage. Multiplication and release of bacilli from these sites may be one cause of the subsequent rapidly-developing systemic bacteremia. Another possibility is the abrogation of the midgut epithelial barrier by metamorphic reorganization, as was suggested by Bailey and Lee (1962).

Molecular interactions

Definitive characterization of specific molecular interactions have not been carried out with any of the bacterial diseases of bees. Patel and Gochnauer (1958) isolated proteolytic enzymes from sporulating B. larvae. The role of these enzymes in the pathogenesis of AFB was discounted by Bamrick (1964) on the basis that most of the histopathological destruction that occurred took place before sporulation. He believed that such enzymes were more important in the postmortem decomposition of the brood. Bamrick, however, suggested a toxic or enzymatic effect of B. larvae vegetative cells to explain the midgut epithelial penetration. Davidson (1973) saw no ultrastructural indication of such toxicity or enzymatic effect during the early interaction between midgut cells and B. larvae. The only fine structural observation that could be ascribed to toxicity was the eventual lysis of cells harboring bacilli. The gradual buildup of toxic materials was suggested as one of the possible causes.

Gochnauer (1976) studied the glycogen content of dried scales of bee larvae killed by AFB. From these results, he concluded that *B. larvae* infection neither strongly inhibited synthesis nor markedly depleted glycogen reserves. Because AFB-diseased larvae normally died at about the time glycogen reached maximal levels (white-eyed pupal stage), it was hypothesized that glycogen, like starch in culture, inactivates some substances (e.g., fatty acids, amino acids, etc.) inhibitory to growth and sporulation of *B. larvae*. If glycogen does function in this manner, this could be another contributing factor to the rapidly developing systemic bacteremia after a lag in observable hemocoelic *B. larvae* in the prepupal stages.

Behavioral response of host to pathogen

In EFB, abnormal movements of infected larvae were reported by White (1912) and Phillips (1918). Whether such movement is the cause of diseased larval ejection from the hive remains unresolved. Bailey (1960) believed the abnormal demand for food by larvae harboring *S. pluton* results in their ejection by nurse bees.

Extensive ejection was also reported by Hitchcock et al. (1979) of bee larvae treated with the putative etiological agent of powdery scale disease, *Bacillus pulvifaciens*. However, the results left unanswered the question of whether the ejection was disease induced.

CONCLUSION

It is evident that the pathogenesis of milky disease has been the subject of much more intensive investigation than that of the foulbrood diseases, and therefore, only limited comparisons can be drawn. From the available information, it appears the most striking similarity is the phagocytosis-like method of entry of *B. popilliae* and *B. larvae* into the mesenteric epithelium during *per os* infection of their respective hosts. The multiplication of *B. popilliae* in regenerative nidi areas of *A. majalis* midgut with subsequent cell proliferation and host response of encapsulation at these areas were not observed in the *B. larvae/A. mellifera* system. Following entry into the hemocoel, the tissue tropism and delay in cell proliferation of *B. larvae* appears unique. Massive sporulation occurs in the hemolymph in both milky and foulbrood diseases. The factor(s) which allow or promote sporulation have not been elucidated despite the intensive efforts of many researchers. This will undoubtedly continue to be investigated, since only limited sporulation of *B. popilliae* has been obtained *in vitro*, and it has a very real potential as a control agent.

It is also unknown whether the cellular interactions described for B. popilliae and A. majalis hold true for other milky disease organisms and their scarabaeid hosts. As discussed earlier, it has been reported that some strains elicit different physiological responses, i.e., in the amino acid composition of hemolymph. It would be of interest to know whether the various interactions described are host dependent or pathogen dependent.

REFERENCES

Bailey, L. 1960. The epizootiology of European foulbrood of the larval honey bee, Apis mellifera L. J. Insect Pathol., 2, 67-83.

Bailey, L. 1963. The pathogenicity to honey-bee larvae of microorganisms associated with European foulbrood. J. Insect Pathol., 5, 198-205.

Bailey,L. 1968. Honey bee pathology. Ann. Rev. Microbiol., 13, 191-212.

Bailey, L. and Lee, D. C. 1962. Bacillus larvae: Its cultivation in vitro and its growth in vivo. J. Gen. Microbiol., 29, 711-717.

Bamrick, J. F. 1964. Resistance to American foulbrood in honey bees. V. Comparative pathogenesis in resistant and susceptible larvae. J. Insect Pathol., 6, 284-304.

Beard, R. L. 1945. Studies on the milky disease of Japanese beetle larvae. Conn. Agr. Exp. Sta. Bull., 491, 505-583.

Beard, R. L. 1956. Two milky diseases of Australian Scarabaeidae. Can. Entomol., 88, 640-647.

Bennett, G. A. and Shotwell, O. L. 1970. A lipoglycoprotein from hemolymph of healthy and diseased larvae of the Japanese beetle, Popillia japonica. J. Invertebr. Pathol., 15, 157-164.

Bennett, G. A. and Shotwell, O. L. 1973. Isolation and characterization of constituents of in hemolymph from healthy and diseased Japanese beetle larvae. Biotech. Bioeng. 15, 1023-1037.

Bennett, G. A. and Shotwell, O. L. 1976. Cuticular hydrocarbons of healthy and diseased Japanese beetle larvae. Insect Biochem., 6, 345-346.

Bennett, G. A., Kleiman, R. and Shotwell, O. L. 1972. Hydrocarbons in haemolymph from healthy and diseased Japanese beetle larvae. J. Insect Physiol., 18, 1343-1350.

Benz, G. 1963. Physiopathology and histochemistry. In Insect Pathology, An Advanced Treatise, E. A. Steinhaus, ed., Vol. 1. Academic Press, N.Y., pp. 299-338.

Bucher, G. E. 1960. Potential bacterial pathogens of insects and their characteristics. J.Insect Pathol., 2, 172-195.

Bulla, L. A., St. Julian, G., Rhodes, R. A. and Hesseltine, C. W. 1969. Scanning electron and phase-contrast microscopy of bacterial spores. Appl. Microbiol., 18, 490-495.

Bulla, L. A., Rhodes, R. A. and St. Julian, G. 1975. Bacteria as insect pathogens. Ann. Rev. Microbiol., 29, 163-190.

Davidson, E. W. 1973. Ultrastructure of American foulbrood disease pathogenesis in larvae of the worker honey bee, Apis mellifera. J. Invertebr. Pathol., 21, 53-61.

Dutky, S. R. 1940. Two new spore-forming bacteria causing milky diseases of Japanese beetle larvae. J. Agric. Res., 61, 57-68.

Dutky, S. R. 1963. The milky diseases. In Insect Pathology, An Advanced Treatise, E. A. Steinhaus, ed., Vol. 2. Academic Press, N.Y., pp. 75-115.

Ebersold, H. R. 1976. Vegetatives waschstum und sporenbildung von Bacillus popilliae in gewebekulturen von invertebraten. Ph.D. Dissertation, ETH 5699. Eidgenossischen Technischen Hochschule Zurich.

Gochnauer, T. A., Furgala, B. and Shimanuki, H. 1975. Chapter XXI. Diseases and enemies of the honey bees. In The Hive and the Honey Bee, Dadant and Sons, eds., Rev. Ed. Dadant and Sons, pp. 615-662.

Gochnauer, T. A. 1976. Glycogen content of honey bee larvae killed by American foulbrood disease. J. Apicult. Res., 15, 53-58.

Gordon, R. E., Haynes, W. C. and Pang, C. H-N. 1973. The Genus Bacillus. Agric. Handbook No. 427. Washington, D.C. 283 pp.

Heimpel, A. M. and Angus, T. A. 1963. Diseases caused by certain sporeforming bacteria. In Insect Pathology, An Advanced Treatise, E. A. Steinhaus, ed., Vol. 2. Academic Press, N.Y. pp. 21-73.

Hitchcock, J. D., Stoner, A., Wilson, W. T. and Menapace, D. M. 1979. Pathogenicity of Bacillus pulvifaciens to honey bee larvae of various ages (Hymenoptera: Apidae). J. Kansas Entomol. Soc., 52, 238-246.

Hrubant, G. R. and Rhodes, R. A. 1968. Agglutinability of sporeforming insect pathogens with antiglobulins to milky disease bacteria. J. Invertebr. Pathol., 11, 371-376.

Kawanishi, C. Y., Splittstoesser, C. M. Tashiro, H. and Steinkraus, K. H. 1974. Ataenius spretulus, a potentially important turf pest, and its associated milky disease bacterium. Environ. Entomol., 3, 177-181.

Kawanishi, C. Y., Splittstoesser, C. M. and Tashiro, H. 1978. Infection of the European chafer, Amphimallon majalis, by Bacillus popilliae: Ultrastructure. J. Invertebr. Pathol., 31, 91-102.

Krizanova, H., Koppel, Z., Horvath, J., Beseda, M. and Halasa, M. 1978. Study of the spiral and spindle-shaped formations in cultures of Bacillus larvae (White, 1906) causing foulbrood in bees. Vet. Med. (Praha), 23, 305-315.

Krywienczyk, J. and Lüthy, P. 1974. Serological relationship between three varieties of Bacillus popilliae. J. Invertebr. Pathol., 23, 275-279.

Lüthy, P., Wyss, C. and Ettlinger, L. 1970. Behavior of milky disease organisms in a tissue culture system. J. Invertebr. Pathol., 16, 325-330.

Patel, N. G. and Gochnauer, T. A. 1958. Separation of two proteolytic enzymes from honeybee larvae infected by Bacillus larvae. Bacteriol. Proc. Soc. Am. Bacteriologists, pp. 118-119.

Phillips, E. F. 1918. The control of European foulbrood. U.S. Dept. Agr. Farmer's Bull., 975, 3-16.

Schwartz, P. H. and Townshend, B. G. 1968. Effects of milky disease on hemolymph coagulation and on the number of hemocytes in infected larvae of the Japanese beetle. J. Invertebr. Pathol., 12, 288-293.

Sharpe, E. S. and Detroy, R. W. 1979. Fat body depletion, a debilitating result of milky disease in Japanese beetle larvae. J. Invertebr. Pathol., 34, 92-94.

Shimanuki, H. 1978. Bacteria. In Honey Bee Pests, Predators, and Diseases, R. A. Morse, ed. Cornell University Press, Ithaca, N.Y., pp. 43-61.

Shotwell, O. L., Bennett, G. A., Hall, H. H., Stubblefield, R. D., Peters, J. E., Van Etten, C. H. and Jackson, R. W. 1965. Amino acids in the haemolymph of diseased Popilliae japonica (Newman) larvae. J. Insect Physiol., 11, 671-682.

Splittstoesser, C. M., Tashiro, H., Lin, S. L., Steinkraus, K. H. and Fiori, B. J. 1973. Histopathology of the European chafer, Amphimallon majalis, infected with Bacillus popilliae. J. Invertebr. Pathol., 22, 161-167.

Splittstoesser, C. M., Kawanishi, C. Y. and Tashiro, H. 1978. Infection of the European chafer, Amphimallon majalis by Bacillus popilliae: Light and electron microscope observations. J. Invertebr. Pathol., 31, 84-90.

Steinkraus, K. H. 1957. Studies on the milky disease organism. I. Parasitic growth and sporulation of Bacillus popilliae. J. Bacteriol., 74, 621-624.

Steinkraus, K. H., Field, C. C., Kochansky, M. C., Kaegebein, M. E. and Tashiro, H. 1973. Cations in hemolymph and alimentary tract tissues of healthy and milky diseased European chafer (Amphimallon majalis) larvae. Appl. Microbiol., 26, 72-77.

Stubblefield, R. D., Bennett, G. A., Shotwell, O. L., Hall, H. H. and Jackson, R. W. 1966. Organic acids in the haemolymph of healthy and diseased Popillia japonica (Newman) larvae. J. Insect Physiol., 12, 949-956.

Tarr, H. L. A. 1938. Studies on European foulbrood of bees. IV. On the attempted cultivation of Bacillus pluton, the susceptibility of individual larvae to inoculation with this organism and its localization in its host. Ann. Applied Biol., 25, 815-821.

Weiner, B. A. 1978. Isolation and partial characterization of the parasporal body of Bacillus popilliae. Can. J. Microbiol., 24, 1557-1561.

Weiner, B. A., Kowlek, W. F., St. Julian, G., Hall, H. H. and Jackson, R. W. 1966. Oxygen concentration in larval hemolymph of the Japanese beetle, Popillia japonica, infected with Bacillus popilliae. J. Invertebr. Pathol., 8, 308-313.

Weiner, B. A., St. Julian, G. and Kwolek, W. F. 1969. Rate of oxygen uptake of healthy and diseased larvae of the Japanese beetle. J. Invertebr. Pathol., 13, 250-255.

White, G. F. 1912. The cause of European foulbrood. U.S. Dept. Agr. Bur. Entomol. Circ., 157, 1-15.

Wyss, C. 1971. Sporulationsversuche mit drei Verietaten von Bacillus popilliae Dutky. Zentralbl. Bakteriol. Parasitenk. Infektionskr. Hyg. II., 126, 461-492.

Chapter 8

BACILLUS THURINGIENSIS DELTA-ENDOTOXIN: COMPOSITION AND ACTIVATION

by Hans E. Huber* and Peter Lüthy*

INTRODUCTION

Bacillus thuringiensis belongs to the aerobic sporeformers of the genus *Bacillus* which, together with the anaerobic *Clostridium* constitute the only genera of the family of the Bacillaceae. Following the vegetative growth phase, the members of the Bacillaceae form endospores which are resistant against extreme environmental conditions and which are able to preserve viability over many years.

Numerous species of the Bacilli and of the Clostridia produce secondary metabolites exhibiting biological activities. Such compounds include antibiotics and toxins as well as enzymes. The production of secondary metabolites occur during vegetative growth or during the sporulation phase.

Bacillus thuringiensis had previously been assigned to the *Bacillus cereus* group because of far-reaching similarities in the biochemical properties which are used for taxonomic classification. The increasing importance of this organism due to its unique insecticidal activity, however, justified the establishment of a separate species. The crystalline inclusion of *B. thuringiensis*, which is formed during sporulation, is a gut poison for larvae of more than one hundred insect species belonging to the orders Lepidoptera and Diptera. The species *B. thuringiensis* is divided into serotypes based on antigenic differences of the flagella. The biochemical differences of strains with the same serotype are used for a further division into varieties or subspecies. At the end of 1979, 19 varieties had been classified belonging to 14 serotypes. The activity of 13

*Swiss Federal Institute of Microbiology and Technology, Zurich, Switzerland

 © Allanheld, Osmun (Totowa, N.J.)

serotypes is primarily toward larvae of lepidopterous insects while serotype 14, variety israelensis, has its spectrum within the Diptera and is highly pathogenic for mosquito and black fly larvae. This new variety, discovered in 1977 by Goldberg and Margalit, is essentially inactive against Lepidoptera.

In addition to the crystalline inclusions (also known as the parasporal body, crystal, or as δ-endotoxin in its activated state), some varieties release a nucleotide during the vegetative growth phase which is called β-exotoxin. This toxin has a much broader spectrum of activity covering not only invertebrates but also vertebrates (see Lecadet and de Barjac, Chapter 11). Other insecticidal metabolites of B. thuringiensis strains, the α-exotoxin and the γ-exotoxin as well as enzymes are of only secondary importance for insect control (reviewed by Faust (1974)).

GENERAL FEATURES OF THE DELTA-ENDOTOXIN

Biogenesis

Angus (1964) succeeded in proving that the crystalline inclusion formed during sporulation was responsible for the insecticidal action. The unusual method of biosynthesis and deposition in a crystalline form must have been the reason that more than 50 years elapsed between the isolation of the microorganism and the discovery of the toxin.

Today, 25 years later, we still have little information about the biosynthesis and regulation of crystal production. At the end of logarithmic vegetative growth or under unfavorable environmental conditions, B. thuringiensis forms a heat-resistant endospore. The biosynthesis of the parasporal crystal takes place concomitant with sporulation, starting simultaneously with septum formation in Stage II and finishing after sporulation Stage IV. About 80% of the crystal protein is synthesized "de novo" within the protein turnover that takes place during spore formation. Radiolabelling of parasporal crystals is easily achieved by addition of ^{3}H or ^{14}C labelled amino acids to sporulating cultures. For comprehensive descriptions of sporulation and crystal formation the reader is referred to Ribier and Lecadet (1973) and Bechtel and Bulla (1976). It is an interesting and so far unexplained fact that no crystal antigens in a dissolved state are detectable in the cytoplasm at any of the sporulation stages (Monro, 1961). Synthesis probably takes place at a site which is very close to the surface of the growing parasporal body. Somerville and James (1970) found an association of the parasporal inclusion with the exosporium and Somerville (1978) suggested that the exosporium might serve as template for crystal formation. Other authors did not observe an involvement of the exosporium or of the forespore membrane with the synthesis of the crystal.

At the last phase of sporulation the cell wall is lysed, and the mature spore and the crystals are released separately into the medium. The variety *finitimus* is an exception; there, the parasporal body is formed within the exosporium. Since the exosporium is not lysed, spore and crystal remain together as an entity.

Crystal protein synthesis and sporulation are tightly linked events which can be desynchronized by the addition of antibiotics such as erythromycin (Arescaldino, 1969) and chloramphenicol or by a short term temperature shift-up (Mennakshi and Jayaraman, 1979). Clones of *B. thuringiensis* which have lost their capability for crystal formation can no longer be distinguished from *B. cereus*. Acrystalliferous mutants appear at a much higher rate than would be expected from the natural mutation probability. As yet, however, final proof has not been presented that the genetic code for the crystal protein is located on extrachromosomal DNA which would explain the increased prevalence of acrystalliferous mutants. Plasmids have been isolated from all the varieties. A possible involvement of plasmids in the formation of the crystal protein was shown by Stahly et al. (1978) who found that acrystalliferous mutants from heat-treated spores were lacking at least six plasmids present in the wildtype. Debabov et al. (1977) and Galushka and Azizbekyan (1977) obtained acrystalliferous strains following treatment with ethidium bromide, a plasmid-eliminating agent. On the other hand Lecadet et al. (1981) failed to correlate the presence of plasmids with any physiological functions. The high number of plasmids found in many *B. thuringiensis* serotypes is astonishing. The number of plasmids seems to be constant for a given variety or even serotype. Fischer et al. (1981) developed a two-dimensional separation method for the identification of *B. thuringiensis* plasmids since they can be present in at least two different configurations. At present the interest is focused on the high molecular weight plasmids between 40 and 80 megadaltons which seem to occur with some regularity in all *B. thuringiensis* serotypes.

Genetic studies with *B. thuringiensis* are in progress in several laboratories since the genes for a protein that is produced in such quantities (about 1/3 of the sporulating cell) must be of general interest. Efficient transfer systems are a basic requirement for genetic studies. Systems for transduction have been developed and described by Thorne (1978) and by Lecadet et al. (1980).

Strains with the opposite characteristics, asporogenous but able to form crystals, are difficult to obtain. They do not occur spontaneously, but must be induced by mutagenic agents. Unfortunately, these asporogenic strains do not seem to fulfill the requirements for industrial application though they would be of major importance. Most of these mutants are not stable and return to sporulation upon repeated transfer. In our laboratory, such asporogenic strains yielded less crystal protein with low specific activity. Often they were exotoxin producers, even if the wildtype was free of exotoxin. The isolation of sporeless mutants lacking these disadvantages were reported by Nishijtsutsuji-Uwo et al. (1975).

TOXICITY AND CLASSIFICATION

The most interesting feature of the δ-endotoxin is its spectrum of activity, which is centered on the insect order Lepidoptera, with the exception of the variety *israelensis* which kills mosquito and black fly larvae (Diptera). However, the activity spectra of the individual varieties or even strains differ widely. Thus, each variety or strain has its own specific host spectrum within Lepidoptera.

The generally accepted classification system which was introduced by de Barjac and Bonnefoi (1962, 1973), is based on the above mentioned flagellar or H-antigen serotypes. *Bacillus thuringiensis* strains belonging to the same variety or serotype show a general correspondence in activity spectrum. This shows the usefulness of the serotype classification for *B. thuringiensis* strains isolated in nature, mostly from diseased insects. It is very unlikely that laboratory produced mutants with altered δ-endotoxin still retain a relationship between specific activity, insect spectrum and flagellar antigens.

A classification system for mutant strains with an altered δ-endotoxin could be based on the crystal antigens. Krywienczyk (1977) investigated the crystal antigens of about 300 *B. thuringiensis* strains. Among these, she found for about 15% of the strains an antigenic pattern which lay outside of the assigned variety. This study shows the need for a strain-specific classification system within the varieties or serotypes since the crystal antigens do not necessarily correspond with the flagellar serotype, and they can mutate independently.

CRYSTAL STRUCTURE

Parasporal crystals are as a rule bipyramidal in shape. There are however strains with irregular parasporal bodies. The shape of the crystals is strain specific but it does not correlate with the toxicity (Pendleton, 1969; Grigorova and Kalucheva, 1966). The fine structure of crystals has been thoroughly investigated by Norris (1971). The subunits proved to be dumbbell- or rod-shaped with a molecular weight of several hundred thousand daltons.

The parasporal bodies of variety israelensis differ in several respects. An average of three parasporal bodies are formed per sporulating cell. Each of the parasporal bodies consists of two to three proteinaceous particles with different solubilization and staining properties. These particles are surrounded and held together by a coat-like envelope. Figure 8-1 shows a freeze-fractured crystal of the variety thuringiensis (H-1) in comparison to Figure 8-2 which represents an israelensis crystal with a quite different fine structure. In Figure 8-3 the typical parasporal bodies of the variety israelensis are shown surrounded by an envelope. After treatment with alkali (0.1 N NaOH), the parasporal material has dissolved while the envelopes resist the alkali treatment (Figure 8-4).

Ovoid inclusions, in addition to the bipyramidic parasporal bodies have been described by Bechtel and Bulla (1976) and by Aronson and Fitz-James (1976). The chemistry and function of these particles are unknown.

CHEMISTRY OF THE DELTA-ENDOTOXIN

Purification

A profound knowledge of the composition of the parasporal crystal and of the toxic moiety is a prerequisite for the understanding of the mode of action. Since the solvents used for the solubilization of the crystals may extract at the same time some spore coat protein, separation of spores and crystals is necessary. Despite the similarity in size, density and surface characteristics, several effective techniques for the purification of crystals have been developed. A convenient method is based for example on a two-phase system proposed by Delafield et al. (1968). Other authors have described centrifugation techniques such as density gradient centrifugation in CsCl (Fast, 1972), isopycnic density gradient centrifugation in Renografin (Sharpe et al., 1975), and finally zonal gradient centrifugation with NaBr (Ang and Nickerson, 1978). All these techniques yield crystal suspensions of more than 99% purity.

Composition

The first analyses of the parasporal bodies in the early 1950s revealed their proteinaceous nature. To date, the crystal protein of many B. thuringiensis strains has been

Figure 8-1. Parasporal crystal of *Bacillus thuringiensis* var. *thuringiensis*. Vertical distance of repetitive subunits approximately 30 nm. Freeze fracture.

Figure 8-2. Parasporal bodies of *B. thuringiensis* var. *israelensis*. Freeze fracture. P = parasporal bodies, C = crystalline area.

Figure 8-3. Parasporal bodies of *B. thuringiensis* var. *israelensis*. Different staining proteinaceous particles are surrounded by a coat-like envelope. Thin section of freeze substitution. P = parasporal bodies, EN = envelope.

Figure 8-4. Envelopes of parasporal bodies of the variety *israelensis*. Protein has been extracted with 0.1 N NaOH. Thin section of freeze substitution. EN = envelope.

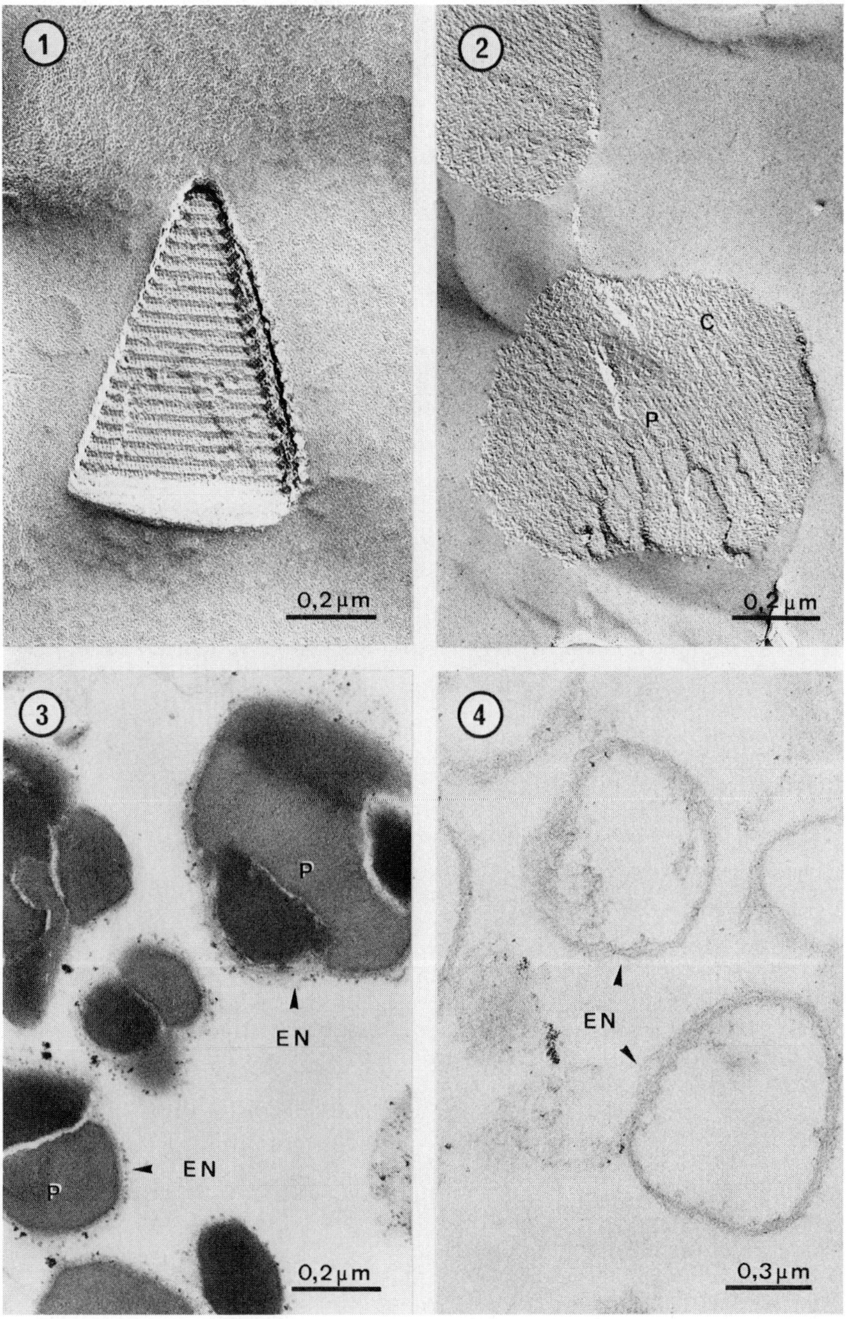
1
0,2 μm
2
C
P
0,2 μm
3
P
EN
EN
P
0,2 μm
4
EN
0,3 μm

subjected to amino acid analysis. No significant strain- or variety-specific differences in the amino acid content of the crystal protein could however be detected. The data obtained with the same strain by various research groups often differ more than the results reported for different serotypes by the same laboratory. The amino acid analyses of crystal protein have been discussed and summarized by Cooksey (1971). Noteworthy are the high content of glutamic and aspartic acid which account for more than 20% of the total amino acids. These two amino acids are responsible for the low isoelectric point of 4.4. The half-cystine content of only 1.3% is comparatively low. An unidentified basic amino acid, probably ornithine, has been reported by Aronson and Tillinghast (1977). According to Guillet and de Barjac (1979) the amino acid composition of the *israelensis* crystal protein differs quantitatively from all other varieties.

Several authors have found carbohydrates associated with the crystals. The results are however contradictory. Bateson and Stainsby (1970) reported a carbohydrate content of 12% for the variety *thuringiensis* while Bulla et al. (1977) determined 5.6% for the variety *kurstaki*. Nickerson (1980) examined 16 strains and measured carbohydrate contents between 1 and 5%. It was demonstrated that the carbohydrates were covalently bound to the protein fraction. Lecadet (1965) did not detect carbohydrate associated with the crystals. Similarly, in our own laboratory carbohydrates could not be discovered in crystals of variety *thuringiensis*. The varying proportions of carbohydrates and the lack of correlation with toxic activity (Nickerson, 1980) do not suggest an involvement in the mode of action. Other molecules like DNA, RNA or lipids are, according to Bulla et al. (1977), not present in the parasporal crystals.

Stability and the role of disulfide groups

It is rare in nature a protein is produced in an already crystalline form. The high purity and the easy production of the crystal protein would have been an ideal starting point of analytical investigations. Unfortunately, study of the crystal protein was greatly impaired by complete insolubility in standard buffer solutions.

The first solvent which was found to solubilize crystals quantitatively was NaOH. Concentrations of up to 0.1 N were used, corresponding to a pH of 13. Later, solvents combined with reducing agents such as thioglycolate, cysteine, mercaptoethanol or diethiothreitol proved to be very effective under much milder conditions, namely in a pH range between 9 and 11. Solubilization even at a neutral pH was achieved when denaturating agents, for example urea, guanidine hydrochloride or sodium dodecyl sulfate (SDS) were added. These agents alone induce reversible swelling of the crystals with an increase in volume by a facter of up to 100.

The choice of the solvent greatly affects the toxic activity of the crystal protein. In particular, NaOH and denaturing agents (e.g. SDS) lead to partial or even complete inactivation and to the formation of artifacts, for example by cleavage of peptide bonds. The best solvents have proved to be buffer solutions of pH 7 to 10 containing 0.1 M dithiothreitol (Nagamatsu et al., 1978), or 0.2 M thioglycolate (Nishiitsutsuji-Uwo et al., 1977). Toxicity is completely preserved in comparison to the intact crystal.

The surprisingly high stability of the parasporal crystals can be attributed to disulfide bonds and to noncovalent interactions. The latter can be overcome by denaturing agents or alkaline pH and the disulfide bonds can be broken by reducing agents. The half-cystine content of the crystals is however very low (1.3% to 1.99%) and can be compared to that of the readily soluble bovine serum albumin (1.5%). On the other hand, the solubility properties of the crystals are very similar to keratin where the half-cystine content reaches 10%. Several other suggestions have been presented to explain the stability of the crystals. Bateson and Stainsby (1970) made the carbohydrate fraction responsible for the insolubility while Faust and Estes (1966) proposed that the crystal protein was embedded within a silicone framework. Neither theory can explain the central role of reducing agents in the solubilization process.

Nickerson (1980) reviewing the protein crystal of B. thuringiensis presented evidence that disulfide crosslinks are the main factors for the stability. In recent experiments conducted in our laboratory we tried to calculate the number of sulfhydryl and disulfide groups present in the subunit (MW 230,000, see below) of the B. thuringiensis var. thuringiensis crystal. Determination was carried out by titration with 5,5'-dithiobis (2-nitrobenzoate) (DTNB) and p-chloromercuribenzoate (PCMB) (Boyer, 1954). 20 cysteine residues per subunit reacted with DTNB and also with PCMB; 15 rapidly formed intramolecular disulfide bonds when titrated with DTNB. We concluded that the fast and stochiometric reaction with DTNB indicates that the 15 S-H groups must be exposed on the protein surface and therefore available for intermolecular disulfide formation. Taking the highest reported amount of half-cystine in the crystal protein we obtain 45 residues per molecule of MW 230,000. Twenty of these or at least 45% are accessible to both DTNB and PCMB in the native protein. That all the sulfide groups buried within the polypeptide are derived from cystine has been shown by Nickerson (1980).

We also estimated the number of intermolecular S-S bonds per subunit in the intact crystal. The crystals were dissolved under defined conditions with ^{35}S-labelled sulfite. At pH 9.5 the noncovalent forces between the crystal subunits are broken and they become separated from each other by several Å which is microscopically visible as swelling. In this partially disaggregated state intermolecular disulfide bonds must be under considerable strain and thus more active than intramolecular S-S bonds which would probably be also accessible to sulfite. In addition the sulfite concentration used was the lowest possible which still yielded complete dissolution of the crystals. An average of 6.1 disulfides per molecule was cleaved under these conditions. This number still exceeds that required for complete crosslinking of all the subunits. Results published earlier (Huber et al., 1981) on the crosslinking deviate somewhat from our latest more quantitative findings.

The formation of intermolecular disulfide bonds by oxidation of sulfides or by a rearrangement of intramolecular S-S bonds could take place during the biosynthesis of the crystals. Concomitantly, the S-S bond translocation may be accompanied by changes in the tertiary structure of the polypeptides. This could explain the fact that no free crystal-specific antigens are present in the cytoplasm during crystal formation.

The 6 disulfides per molecule involved in linking the subunits are not necessarily a constant since an interchange of S-S bonds from an originally intramolecular to an intermolecular position may continue after the crystallization process, adding to the stability of the crystalline structure. Young and Fitz-James (1959) observed an increasing insolubility of the crystals during storage, and suggested that the formation of disulfide bonds was responsible for the so-called "ripening" of the crystals. Similar processes are known to occur in spore coats. Aronson and Fitz-James (1976) found that the inner part of the spore coat of *B. cereus* is stabilized by disulfide bridges. The outer coat consisting of identical protein develops from the inner coat by disulfide rearrangement.

The free reactive sulfide groups which are exposed following dissolution are responsible for the tendency to aggregate, especially in the presence of oxygen. The aggregation can be prevented either by maintaining a high concentration of thio reagents in the buffer solution, by blocking the sulfide groups with p-mercuribenzoate, or by alkylation, for example with iodoacetic acid.

The subunit of the crystal

It has to be emphasized that the parasporal crystal as well as its subunits represent a protoxin bare of biological activity. For example, host larvae do not react upon injection of protoxin into the hemocoel. The protoxin needs proteolytic activation which occurs under natural conditions in the gut juice of susceptible insect larvae.

Electron micrographs of the parasporal crystal usually reveal a regular crystalline lattice. Based on X-ray diffraction data, Holmes and Monro (1965) calculated a molecular weight (MW) of 230,000 daltons for the crystal subunit of the variety *thuringiensis*. Numerous MW determinations have been carried out with the components of solubilized crystals. The following factors seem to be responsible for the divergent results which are found in the literature:

i. The MW of native and denatured crystal proteins differ significantly;
ii. The MW of the same protein varies according to the determination method;
iii. The size of the subunit might be specific for each *B. thuringiensis* variety; and
iv. The presence of contaminating proteases may influence the MW by producing smaller molecular artifacts.

Only the recent literature is reviewed here. For earlier publications the reader is referred to Cooksey (1971). Experiments conducted in our laboratory and a study of the newer literature indicated that the crystal subunit consists of a dimer with a MW of 230,000, composed of two identical polypeptide chains. The dimer is obtained if solvents are used which preserve the native structure of the protein. Huber et al. (1981) solubilized crystals of the variety *thuringiensis* with 10 mM DTT and pH 9.5. The dissolved protein was alkylated with iodoacetic acid in order to prevent reaggregation. Polyacrylamide gel electrophoresis (PAGE) with acrylamide concentrations ranging between 3 and 8% yielded a MW of 231,000 for the subunit. Nagamatsu et al. (1978) investigated the native crystal protein of the variety *dendrolimus*, and obtained a MW of 232,000. When the same subunit was subjected to ultracentrifugation, a molecular weight of 177,000 was calculated (Huber et al., 1981), while gel filtration yielded MW above 500,000. It can be concluded that the same crystal protein fraction differs greatly in MW, depending on the determination methods. We feel however that ultracentrifugation is more accurate for MW determination since the results obtained with molecular sieving are influenced by the structure of the protein. Some results of crystal protein analysis published by different authors are summarized in Table 8-1.

Table 8-1

Molecular Weight of the Subunits of the Parasporal Crystal of

Bacillus thuringiensis

Method used	MW of the subunit native	MW of the subunit denatured	References
PAGE[1]	232,000		Nagamatsu et al. (1978)
	231,000		Huber et al. (1981)
in SDS[2]		142-157,000[3]	Nickerson (1980)
		130-145,000	Zalunin et al. (1979)
		120-140,000[3]	Chestukhina et al. (1977a)
		135,000	Bulla et al. (1977)
		130,000	Huber et al. (1981)
		120,000[3]	Herbert et al. (1971)
Gel filtration	>500,000		several authors
	235,000		Faust et al. (1974)
in GuHCl[4]		130,000	Bulla et al. (1977)
Ultracentrifugation	177,000		Huber et al. (1981)
in GuHCl[4]		101,000	Bulla et al. (1977)
		84,000	Huber et al. (1981)
		80,000	Glatron et al. (1972)
X-ray diffraction	230,000		Holmes and Monro (1965)

1) polyacrylamide gel electrophoresis

2) sodium dodecyl sulphate

3) in addition to the listed MW smaller components in the 60,000 daltons range have been reported

4) guanidine hydrochloride

The presence of SDS causes denaturation of the crystal protein which means a separation of the subunit into the two polypeptides. Thus, the denatured protein yields approximately half of the MW with the corresponding separation methods. Denaturation is also achieved when crystals are solubilized, for example with guanidine hydrochloride or urea in combination with a reducing agent. SDS-PAGE of the denatured protein yields a component with a MW of 130,000; at least 80% of the total protein is in this fraction. Similar results were obtained by several authors (Fast, pers. commun.; Bulla et al., 1977; Huber et al., 1981; Chestukhina, 1977a). The MW determination of the denatured protein by analytical ultracentrifugation gave MW's which were again 30% to 60% lower than the corresponding results of PAGE. Bulla et al. (1977) reported 101,000 daltons, Huber et al. (1981) 84,000 daltons, and Galtron et al. (1972) calculated a MW of 80,000 for their repetitive unit.

The above results were obtained either with strains of the variety thuringiensis, kurstaki or galleriae. It is difficult to estimate the strain or variety specific differences in the MW of the subunits. A paper by Zalunin et al. (1979) and our own investigations suggest however a great similarity between the subunits of different varieties. Zalunin and co-workers analyzed the crystal proteins of the serotypes H-2 and H-12 and found the main polypeptide fraction to be in the MW range of 130,000 to 145,000. Our results, obtained with strains of the serotypes H-1 to H-9 yielded in each case a main fraction of MW 130,000. These results indicate that the majority of the crystals derived from the different varieties show a structural uniformity which is based on the subunit. It must be added, however, that not all the published results fit into the proposed model. Polypeptides of other MW were reported by Herbert et al. (1971) and to some extent by Nickerson (1980). Also, low molecular components obtained by chemical solubilization have been reported by several authors. Sayles et al. (1980) as well as Aronson and Tillinghast (1977) claimed that the crystal of the variety thuringiensis consists of a heterogenous mixture of small fragments with an average MW of 1,400. Unfortunately, the biological activity of these compounds was not measured. In a recent study by Fast and Martin (1980) it was found that DTT solubilized crystal protein incubated with KSCN yielded low molecular weight peptides (below 5,000 daltons). The activity was retained in assays with B. mori. The role of KSCN in the generation of small toxic peptides is not clear, especially since no covalent bonds are broken. A uniform fraction of MW 130,000 was described by Prasad and Shetna (1974) after dissolution of parasporal bodies of the variety thuringiensis in 1 N NaOH. This component could be crystallized and showed insecticidal activity as well as an antitumor reaction.

Some authors have observed proteolytic activity associated with the crystal of *B. thuringiensis* (Bulla et al., 1977; Chestukhina et al., 1977b). The latter found two types of protease, namely a serine- and a metallo-protease in crystals of two *B. thuringiensis* strains. Chestukhina and co-workers regarded the proteases as contaminants of the crystal and they described a method for the removal of the proteolytic activity. The presence of proteases caused degradation of the 130,000 dalton unit to lower MW peptides. It seems that the proteases associated with the crystals are remnants of the bacterial cell. They are very likely adsorbed to the surface of the crystals. The action of such proteases as impurities may be one of the reasons for the occurrence of low molecular weight compounds following dissolution of crystals.

Bacillus thuringiensis var. *israelensis*

As demonstrated above, *B. thuringiensis* var. *israelensis* possesses not only a different host spectrum but also a distinct morphological structure of the parasporal inclusions. Experiments conducted in our laboratory on var. *israelensis* indicated that we were working with a basically new type of δ-endotoxin produced by the species of *B. thuringiensis*. Only partial solubilization was achieved with 8 M urea containing 0.1 M DTT. The electron dense particles of the parasporal inclusions still remained intact. The solubilized protein yielded a main component with MW 25,000. Complete solubilization was only achieved with 0.1 N NaOH, a solvent which should be avoided whenever possible because of the risk of artifact formation. The envelope which surrounds the parasporal inclusion was the only remnant after the NaOH-treatment. Besides components with higher MW, the main fraction was again 25,000 daltons. De Barjac (1978) was able to extract toxicity with NaOH in order to show that the parasporal inclusions were reponsible for the biological activity against larvae of *Aedes aegypti*.

Serological differences in the H-14 protein were mentioned briefly by Guillett and de Barjac (1979) and were fully demonstrated by Krywienczyk and Fast (1980). The latter authors showed that none of the *B. thuringiensis* varieties shared common crystal antigens with the var. *israelensis* protein.

THE ACTIVATION OF THE DELTA-ENDOTOXIN

In vivo activation

As stated earlier, the parasporal crystals and their subunits are protoxins, and they are not toxic upon injection into the hemocoel of host insects or when tested in tissue cultures (see Chapter 9). The subunit polypeptides require activation of proteolytic enzymes. Under natural conditions, the activation step takes place in the intestine of susceptible insect larvae, following uptake of the crystals along with the food. In most cases, the crystals are dissolved by the combined action of the alkalinity and the proteases of the gut juice. The role of the proteases has been investigated by Lecadet and Dedonder (1966). Two serine proteases isolated from the gut juice of *Pieris brassicae* were able to cleave disulfide bonds. Although the *P. brassicae* proteases were found to be similar to the mammalian enzymes, trypsin and chymotrypsin, they are less specific and split a wider range of peptide bonds. It can be assumed that the gut juice proteases of other susceptible insect species share related properties.

Among the host species of *B. thuringiensis*, larvae of the silkworm (*Bombyx mori*) and the Mediterranean flour moth (*Anagasta kuehniella*) represent exceptions. The gut juice of *B. mori* has a pH higher than 10, and dissolution of the crystals may occur in the absence of proteolytic enzymes. On the other hand, crystals are not readily dissolved and activated in the intestine of *A. kiehniella* despite of the presence of proteases. The low pH of the gut juice which is just above neutrality must be responsible for this phenomenon. *Anagasta kuehniella* is susceptible if the crystals are either applied together with the spores, or if they are dissolved and activated prior to ingestion (Yamvrias, 1962).

Although the proteolytic modification of the protoxin to the actual δ-endotoxin is essential, enzymes do not determine the specific activity. The different levels of activity of a given *B. thuringiensis* variety against various host species must be based on the site of action, namely the gut epithelium. In our laboratory, protoxin of var. *kurstaki* was digested with gut juice of *P. brassicae* and *Spodoptera littoralis* (cotton leafworm) as well as with trypsin. All three digests exhibited the same high specific activity against *P. brassicae* whereas *S. littoralis*, an insect which is not affected by *kurstaki* crystals, did not show any reaction (Huber, 1977).

In vitro degradation studies

The activation of the δ-endotoxin has been intensively studied in our laboratory with crystals of variety *thuringiensis*, larvae of *P. brassicae*, and an insect cell culture derived from *Choristoneura fumiferana* (spruce budworm) (for details see Chapter 9). Starting material was the protoxin in the form of the crystal subunit with MW 230,000. Proteolysis of the subunit with diluted gut juice of *P. brassicae* or with trypsin leads to a fraction with MW 80,000 which represents the first biologically active intermediate. Activity is not only obtained by feeding but also by injection into the hemocoel as well as upon incubation with the *C. fumiferana* cell culture (Geiser, 1979). Under the well defined conditions of the cell culture assay, any further nontarget site modification of the toxin is excluded.

The reaction rate of the activation step is extremely high. In undiluted gut juice of *P. brassicae* it takes only seconds, as was shown with radio-labelled protoxin. This explains the rapid appearance of the first symptoms which are manifested as early as 2 min following uptake of crystal protein. The larvae respond by cessation of feeding.

In further steps, but at a much slower speed, the MW 80,000 fraction is degraded by gut juice proteases but not by trypsin to fragments with a MW range of 60,000 and 40,000. The biological activity is almost fully retained by both products, and the smaller is very resistant to further proteolytic degradation. Low molecular weight polypeptides (MW $\leq$ 10,000) appear as by-products during proteolysis of the protoxin. These compounds never showed significant toxicity in our assay systems; on a molar basis they were at least 500 times less active than the protoxin. We calculated the specific activities of the different degradation stages using as experimental systems larvae of *P. brassicae* and insect tissue culture. The results are summarized in Table 8-2 (for definitions see Chapter 9).

When protoxin was incubated in diluted or undiluted gut juice, the entire toxicity was retained for several hours. On a molar basis however, the activity of the active fragments decreased by approximately 50% (Table 8-2). This is in support for our proposed dimeric subunit model. Thus, one mole of protoxin yields two moles of activated toxin, corresponding to 100% toxicity.

Interpretation of the published results concerned with the biological activation of the protoxin presents some difficulties because methods, *B. thuringiensis* strains, and test insects differ from one author to the other.

A variety of toxic fractions ranging from MW 80,000 down to MW 500 have been reported (Lecadet and Martouret, 1967; Cooksey, 1968; Fast and Angus, 1970; Pendleton, 1973).

Table 8-2. Toxicities of crystal protein at different stages of degradation, tested against larvae of *Pieris brassicae* and against a cell culture of *Choristoneura fumiferana*. *Bacillus thuringiensis* strain: LBG B 4412, var. *thuringiensis*.

MW of protein	*Pieris brassicae* larvae			Cell culture Cf-1 (*Choristoneura fumiferana*)	
	FID_{50}[1] ng/larvae	relative toxicity[2] per weight	relative toxicity[2] per mole	ED_{50}[1] µg/ml	toxicity[2] per mole
crystals	2.8	46%	-	-	-
230,000[3]	2.9	45%	100%	>125,000	<0.1%
80,000[4]	2.1	62%	48%	28	100 %
60,000[5]	1.5	87%	50%	-	-
40,000[5]	1.3	100%	39%	-	-
≤10,000[4]	>150	<1	<0.2%	>2,500	<0.2%

[1] FID_{50} (feeding inhibition dose 50%), ED_{50} (effective dose 50%). Data from a dose-response line (probit-transformed, standard deviation 7.5%) (Cordier 1978; Geiser 1979).

[2] The toxicity per weight or mole of protein is expressed as percentage of the most active fraction. Toxicity per weight or mole is calculated as $1/FID_{50}$.

[3] Alkali dissolved crystal protein.

[4] Degradation products obtained by incubating protoxin with gut juice of *P. brassicae* or with trypsin. Separation of fractions on Sephadex G-200 columns.

[5] Degradation products obtained by incubating protoxin with gut juice of *P. brassicae*.

According to Faust et al. (1974) the MW and the quantitative distribution of the degradation products depend largely on the type of protease used. The formation of significant amounts of toxic low molecular weight moieties is still doubtful and needs additional proof. Where a quantitative estimate of published results is possible, the activity values of fractions with MW below 30,000 are reduced by a factor of at least 10 to 100. Furthermore, only approximate MW have been attributed to small toxic fragments, although the smaller a compound is, the easier it should be to determine the MW. The amino acid composition and sequence analysis of an oligopeptide containing 5 to 20 amino acids should not have imposed major difficulties.

Recent publications favor as toxic moiety a compound in the range of MW 30,000 to 80,000 (Travers et al., 1976; Lilley and Somerville, 1976; Somerville, 1978; Geiser, 1979). Our own investigations have led us to the opinion that we are not confronted with a single, well defined toxic compound. As in the case of chymotrypsin, the activation process includes several steps. The precursor chymotrypsinogen A is hydrolysed sequentially to pi-chymotrypsin, δ-chymotrypsin and finally to α-chymotrypsin. The first step is catalysed by trypsin and the following two by chymotrypsin. All three chymotrypsin species have a different MW and are enzymatically active. A comparable process seems to take place during protelysis of the crystal protein. Our results demonstrated the presence of at least three distinct stages of degradation (Cordier, 1978). These compounds can be designated as true toxins since they do not need further protelytic modification for activation.

The molecular size of the active δ-endotoxin moiety may thus depend on experimental conditions, especially the type of protease, and on the incubation time. With trypsin, for example, the degradation comes to a stop at a MW of 80,000, while with gut juice proteases it proceeds further to MW 60,000 and 40,000. Each of these toxins consist of an active center which is surrounded by polypeptide chains of varying lengths which may determine the degree of resistance against further proteolytic attack. Prolonged exposure to proteases must finally lead to the loss of toxicity. The critical MW range seems to be 30,000. Below this point activity is rapidly lost.

While the mode of activation can be generalized for the serotypes 1 to 13, variety specific differences in the size of the activated polypeptides may occur. The variety specific degress of activity against a given insect must be based on the polypeptide structure.

The above presented model, which is supported by experimental data, explains most of the results reported in the literature, with the exception of low molecular weight fragments bearing toxic activity. Nevertheless, a successful isolation of a low molecular weight compound which is toxic and stable, is possible. It would of course facilitate studies of the molecular mode of action.

It has been demonstrated that a limited proteolysis is a necessary condition for the activation of the protoxin. The question arises however, if this is the only requirement, or if additional modifications are necessary. In this context, the following three possibilities are discussed:

i) Alkaline modification: The solubilization of the crystals brings about a cleavage of the disulfide bonds. Under alkaline conditions, the S-S bonds can be split without reducing agents, and several reactive intermediates can be formed. One of them, dehydroalanine, readily reacts with lysine or cysteine residues, leading to the formation of intra- and intermolecular crosslinks such as lysinoalanine and lanthionine. Dastidar and Nickerson (1978) have detected large amounts of lysinoalanine in alkali-solubilized crystal protein. Since lysinoalanine was absent from intact crystals, the authors suggested that such crosslinks were involved in the formation of the toxic moiety. More recently, Nickerson (1980) claimed that such an alkaline modification was essential for the ultimate toxin structure. His suggestion was mainly supported by the fact that solubilization of crystals at neutral pH-values always results in a great loss of toxic activity. We have shown however, that under carefully controlled conditions, using 6 M guanidine hydrochloride and 0.01 M DTT at pH 7.5, crystals can be dissolved with full retention of toxicity (Huber et al., 1981). The dissolved crystal protein can then be activated with proteases at neutral pH. The resulting degradation product, which is still heat labile, exhibits normal toxicity against *C. fumiferana* cells, in a test conducted at pH just below 7. Dissolution, activation and bioassay are thus all carried out with avoidance of alkaline conditions.

ii) Enzymatic modification: The bioassay system with *C. fumiferana* excludes the presence of free enzymes. Whether the action of a membrane-bound enzyme is essential for a final activation step cannot be decided on the basis of the present knowledge.

iii) Modification based on the carbohydrate moiety: This is unlikely, since the sugar content, as already mentioned, cannot be correlated with toxicity (Nickerson, 1980).

To summarize, it may be assumed that solubilization of the crystals and limited proteolysis of the polypeptides are the only steps required for activation.

Little is known about the active center and possible binding sites of the δ-endotoxin. Some information can be obtained with inactivation studies. Alkylation of the protoxin with iodoacetate leads to a complete loss of activity (Schesser et al., 1977), indicating a participation of S-H groups. An interpretation of such results is difficult since alkylation is often accompanied by changes in the tertiary structure.

The inactivation of crystals and toxin by heat treatment, or exposure to extreme pH-values and denaturing agents, demonstrates the need of a native, well defined tertiary structure (Watanabe et al., 1967; Cooksey, 1968; Nishiitsutsuji-Uwo et al., 1977). The same is true for the low molecular weight toxic compounds reported by Lecadet and Martouret (1967). Finally, inactivation of the toxin can be obtained by incubation with crystal-specific antibodies. Information is at present not sufficient to allow detailed conclusions about the action of the δ-endotoxin. A more precise knowledge about the active site will however be one of the keys for the elucidation of the molecular mode of action.

ACKNOWLEDGEMENTS

Our Bacillus thuringiensis project is supported by the Swiss National Science Foundation, Project Nr. 3.357-0.78. We thank Dr. H. R. Ebersold for the electron microscopy and for the preparation of the micrographs.

REFERENCES

Ang, B. J. and Nickerson, K. W. 1978. Purification of the protein crystal from Bacillus thuringiensis by zonal gradient centrifugation. Env. Microbiol., 36, 625-626.

Angus, T. A. 1954. A bacterial toxin paralyzing silkworm larvae. Nature, 173, 545.

Arescaldino, J. 1969. Sensibilité differentielle a l'erythromycine de la sporogenèse et de la cristallogenèse de Bacillus thuringiensis. C. R. Acad. Sc. Paris, Série D, 269, 1132-1135.

Aronson, J. N. and Fitz-James, P. 1976. Structure and morphogenesis of bacterial spore coat. Bact. Rev., 40, 360-402.

Aronson, J. N. and Tillinghast, J. 1977. A chemical study of the parasporal crystal of Bacillus thuringiensis. In Spore Research, Vol. I, A. N. Barker, J. Wolf, D. J. Ellar, G. J. Dring and G. W. Gould, eds., Academic Press, London, pp. 351-357.

de Barjac, H. 1978. Un nouveau candidat à la lutte biologique contre les moustiques: Bacillus thuringiensis var. israelensis. Entomophaga, 23, 309-319.

de Barjac, H. and Bonnefoi, A. 1962. Essai de classification biochimique et sérologique de 24 souches de Bacillus du type Bacillus thuringiensis. Entomophaga, 7, 5-31.

de Barjac, H. and Bonnefoi, A. 1973. Mise au point sur la classification des Bacillus thuringiensis. Entomophaga, 18, 5-17.

Bateson, J. B. and Stainsby, G. 1970. Analysis of the active principle in the biological insecticide Bacillus thuringiensis Berliner. J. Food Technol. 5, 403-415.

Bechtel, D. B. and Bulla, L. A. 1976. Electron microscopic study of sporulation and parasporal crystal formation in Bacillus thuringiensis. J. Bact., 127, 1472-1481.

Boyer, P. D. 1954. Spectrophotometric study of the reaction of protein sulfhydryl groups with organic mercurials. J. Am. Chem. Soc., 76, 4331-4337.

Bulla, L. A., Kramer, K. J. and Davidson, L. I. 1977. Characterization of the entomocidal parasporal crystal of Bacillus thuringiensis. J. Bact., 130, 375-383.

Chestukhina, G. G., Kostina, L. I., Zalunin, I. A., Kotova, T. S., Katrukha, S. P., Kutnetsov, Y. S. and Stepanov, V. M. 1977a. Proteins of Bacillus thuringiensis δ-endotoxin crystals. Biokhimiya, 44, 693-698.

Chestukhina, G. G., Kostina, L. I., Zalunin, I. A., Kotova, T. S., Katrukha, S. P., Kutnetsov, Y. S. and Stepanov, V. M. 1977b. Proteinase bound to crystals of Bacillus thuringiensis. Biokhimiya, 43, 857-964.

Galushka, F. P. and Azizebekyan, R. R. 1977. Investigation of plasmids of strains of different variants of Bacillus thuringiensis. Doklady Akademii Nauk SSSR, 236, 1233-1235.

Geiser, P. 1979. Versuche in vitro zum Nachweis einer Wirkung des δ-endotoxins von Bacillus thuringiensis. Diss. ETH Zurich, Switzerland, Nr. 6411.

Glatron, M. F., Lecadet, M. M. and Dedonder, R. 1972. Structure of the parasporal crystal of Bacillus thuringiensis Berliner: Characterization of a repetitive unit. Eur. J. Biochem., 30, 330-338.

Goldberg, L. J. and Margalit, J. 1977. A bacterial spore demonstrating rapid larvicidal activity against Anopheles sergentii, Uranotaenia unguiculata, Culex univittatus, Aedes aegypti and Culex pipiens. Mosquito News, 37, 355-358.

Grigorova, R. and Kalucheva, I. 1966. Electron microscope investigations on the shape of the crystals of six strains of Bacillus thuringiensis Berliner. C. R. Acad. Sc. Bulg. 19, 1075-1078.

Guillet, P. and de Barjac, H. 1979. Toxicité de Bacillus thuringiensis var. israelensis pour les larves de Simulies vectrices de l'Onchocercose. C. R. Acad. Sci. Paris, Série D, 289, 549-551.

Herbert, B. N., Gould, H. J. and Chain, E. B. 1971. Crystal protein of Bacillus thuringiensis var. tolworthi. Subunit structure and toxicity to Pieris brassicae. Eur. J. Biochem., 24, 366-375.

Holmes, K. C. and Monro, R. E. 1965. Studies on the structure of parasporal inclusions from Bacillus thuringiensis. J. Molec. Biol., 14, 572-581.

Huber, H. 1977. Zur Spezifität des δ-endotoxins von Bacillus thuringiensis, Diploma thesis, Faculty Nat. Sci., ETH, Zurich, Switzerland.

Huber, H. E., Cordier, J. L. and Luethy, P. 1981. The crystal of Bacillus thuringiensis: Structure and size of subunit. Arch. Microbiol., in press.

Krywienczyk, J. 1977. Antigenic composition of δ-endotoxin as an aid in identification of Bacillus thuringiensis varieties. Technical Report, Canadian Forestry Service, IP-X-16.

Krywienczyk, J. and Fast, P. 1980. Serological relationships of the crystals of Bacillus thuringiensis var. israelensis. J. Invertebr. Pathol., 36, 139-140.

Lecadet, M. M. 1965. La toxine figureé de Bacillus thuringiensis. Technique de séparation et composition en acides aminés. C. R. Acad. Sc. Paris, Série D, 261, 5693-5696.

Lecadet, M. M. and Dedonder, R. 1966. Les protéases de Pieris brassicae. II. Spécificité. Bull. Soc. Chim. Biol., 48, 661-691.

Lecadet, M. M., Lereclus, D., Blondel, M. O. and Ribier, J. 1981. Bacillus thuringiensis: studies on chromosomal and extrachromosomal DNA. In Spores VIII, American Society for Microbiology, Washington, D. C., in press.

Lecadet, M. M., Blondel, M. O. and Ribier, J. 1980. Generalized transduction in Bacillus thuringiensis var. berliner 1715 using bacteriophage CP-54Ber. J. Gen. Microbiol., 120, in press.

Lecadet, M. M. and Martouret, D. M. 1967. Hydrolyse enzymatic de l'inclusion parasporal de Bacillus thuringiensis par les protéases de Bombyx mori. C. R. Acad. Sc. Paris, Série D, 265, 1543-1546.

Lilley, M. and Somerville, H. J. 1976. The action of proteolytic enzymes on the crystalline insect toxin of Bacillus thuringiensis. Proc. Soc. Gen. Microbiol. Vol. III, p. 62.

Meenakshi, K. and Jayaraman, K. 1979. On the formation of crystal protein during sporulation in Bacillus thuringiensis var. thuringiensis. Arch. Microbiol., 120, 9-14.

Monro, R. E. 1961. Serological studies on the formation of protein parasporal inclusions in Bacillus thuringiensis. J. Biophys. Biochem. Cytol., 11, 321-331.

Nagamatsu, Y., Tsutsui, R., Ichimaru, T., Nagamatsu, M., Koga, K. and Hayashi, K. 1978. Subunit structure and toxic component of δ-endotoxin from Bacillus thuringiensis. J. Invertebr. Pathol., 32, 103-109.

Nickerson, K. W. 1980. Structure and function of the Bacillus thuringiensis protein crystal. Biotechnol. and Bioeng. 22, 1305-1333.

Nishiitsutsuji-Uwo, J., Wakisaka, Y. and Eda, M. 1975. Sporeless mutants of Bacillus thuringiensis. J. Invertebr. Pathol., 25, 355-361.

Nishiitsutsuji-Uwo, J., Aiako, O. and Nishimura, M. S. 1977. Factors affecting the insecticidal activity of δ-endotoxin of Bacillus thuringiensis. J. Invertebr. Pathol., 29, 162-169.

Norris, J. R. 1971. The protein crystal toxin of Bacillus thuringiensis: Biosynthesis and physical structure. In Microbial Control of Insects and Mites, H. D. Burges and N. W. Hussey, eds. Academic Press, London, pp. 229-246.

Pendleton, I. R. 1973. Characterization of crystal protoxin and an activated toxin from Bacillus thuringiensis var. entomocidus. J. Invertebr. Pathol., 21, 46-52.

Pendleton, I. R. 1969. Comparative toxicities of crystal serotypes of Bacillus thuringiensis for larvae of Philosamia cynthia var. ricini. J. Invertebr. Pathol., 13, 423-428.

Prasad, S. S. V. and Shetna, Y. I. 1974. Purification, crystallization and partial characterization of the antitumor and insecticidal protein subunit from the δ-endotoxin of Bacillus thuringiensis var. thuringiensis. Biochim. Biophys. Acta, 363, 558-566.

Ribier, J. and Lecadet, M. M. 1973. Etude ultrastructural et cinetique de al sporulation de Bacillus thuringiensis var. Berliner 1715. Remarques sur la sporulation et l'inclusion parasporale. Ann. Microbiol., 124A, 311-344.

Sayles, V. B., Aronson, J. N. and Rosenthal, A. 1970. Small polypeptide components of the Bacillus thuringiensis parasporal crystalline inclusion. Biochem. Biophys. Res. Commun, 41, 1126-1133.

Schesser, J. H., Kramer, K. J. and Bulla, L. A. 1977. Bioassay for homogenous parasporal crystal of Bacillus thuringiensis using the tobacco hornworm, Manduca sexta. Appl. Environ. Microbiol., 33, 878-880.

Sharpe, E. S., Kenneth, N. W., Bulla, L. A. and Aronson, J. N. 1975. Separation of spores and parasporal crystals of Bacillus thuringiensis in gradients of certain X-ray contrasting agents. Appl. Microbiol., 30, 1052-1053.

Somerville, H. J. 1978. Insect toxin in spores and protein crystals of Bacillus thuringiensis. TIBS, May 1978.

Somerville, H. J. and James, C. R. 1970. Association of the crystalline inclusion of Bacillus thuringiensis with the exosporium. J. Bacteriol., 102, 580-583.

Stahly, D. P., Dingman, D. W., Irgens, R. L., Field, C. C., Feiss, M. G. and Smith, G. L. 1978. Multiple extrachromosmal deoxyribonucleic acid molecules in Bacillus thuringiensis. FEMS Microbiol. Letters, 3, 139-141.

Thorn, C. B. 1978. Transduction in Bacillus thuringiensis. Appl. Env. Microbiol. 35, 1109-1115.

Travers, R. S., Faust, R. M. and Reichelderfer, C. F. 1976. Effects of Bacillus thuringiensis var. kurstaki δ-endotoxin on isolated lepidopteran mitochondria. J. Invertebr. Pathol., 28, 351-356.

Watanabe, T., Tsutsui, R., Iwahana, H. and Abe, K. 1967. Chemistry of the toxic crystals produced by Bacillus thuringiensis. Proc. Joint US-Japan Seminar on Microbial Control of Insect Pests, Fukuoka, Japan, pp. 71-78.

Yamvrias, C. 1962. Contribution of l'étude du mode d'action de Bacillus thuringiensis vis-a-vis de la teigne de la farine, Anagasta kuehniella Zell. (Lepidoptera). Entomophaga, 7, 101-159.

Young, E. I. and Fitz-James, P. 1959. Chemical and morphological studies of bacterial spore formation. II. Spore and parasporal protein formation in Bacillus cereus var. alesti. J. Biophysic. and Biochem. Cytol., 6, 483-489.

Cooksey, K. E. 1968. Purification of a protein from Bacillus thuringiensis toxic to larvae of Lepidoptera. Biochem. J., 106, 445-454.

Cooksey, K. E. 1971. The protein crystal toxin of Bacillus thuringiensis: Biochemistry and mode of action. In Microbial Control of Insects and Mites, H. D. Burges and N. W. Hussey, eds., Acacemic Press, London, pp. 247-274.

Cordier, J. L. 1978. Das Parasporalkorperprotein von Bacillus thuringeinsis: Versuche zur Erfassung der toxischen Komponente nach Darmasaftinkubation. Diploma Thesis, ETH Zurich.

Dastidar, P. G. and Nickerson, K. W. 1978. Lysinolalanine in alkali-solubilized protein crystal toxin from Bacillus thuringiensis. FEMS Microbiol. Letters, 4, 331-333.

Debabov, V. G., Azizbekyan, R. R., Khlebalina, O. I., D'Yachenko, V. V., Galushka, F. P. and Belykh, R. A. 1977. Isolation and preliminary characterization of extrachromosomal elements of Bacillus thuringiensis DNA. Genetika, 13, 496-501.

Delafield, F. P., Somerville, H. J. and Rittenberg, S. C. 1968. Immunological homology between crystal and spore protein of Bacillus thuringiensis. J. Bact., 96, 713-720.

Fast, P. G. 1972. Delta-endotoxin of Bacillus thuringiensis III: A rapid method for separating parasporal bodies from spores. J. Invertebr. Pathol., 20, 139-140.

Fast, P. G. and Angus, T. A. 1970. The δ-endotoxin of Bacillus thuringiensis var. sotto: A toxic low molecular weight fragment. J. Invertebr. Pathol., 16, 465.

Fast, P. G. and Martin, W. C. 1980. Bacillus thuringiensis crystal toxin: Dissociation into toxic low molecular weight peptides. Biochem. Biophys. Res. Commun., 95, 1314-1320.

Faust, R. M. and Estes, Z. E. 1966. Silicon content of the parasporal crystal of several crystalliferous bacteria. J. Invertebr. Pathol., 8, 141-144.

Faust, R. M. 1974. Bacterial diseases. In Insect Diseases, Vol. I, G. E. Cantwell, ed., Marcel Dekker, Inc., New York, pp. 87-183.

Faust, R. M., Hallam, G. M. and Travers, R. S. 1974. Degradation of the parasporal crystal produced by Bacillus thuringiensis var. kurstaki. J. Invertebr. Pathol., 24, 365-373.

Fischer, H. M., Watson, J. and Lüthy, P. 1981. Discrimination of covalently closed and open circular plasmid conformations by two-dimensional gel electrophoresis, demonstrated with Bacillus thuringiensis. Experientia, in press.

Zalunin, I. A., Chestukhina, G. G. and Stepanov, V. M. 1979. Protein composition of δ-endotoxin crystals in various serotypes of *Bacillus thuringiensis*. Biokhimiya, 44, 693-698.

CHAPTER 9

BACILLUS THURINGIENSIS DELTA-ENDOTOXIN: HISTOPATHOLOGY AND MOLECULAR MODE OF ACTION

by Peter Lüthy* and Hans Rudolf Ebersold*

INTRODUCTION

Bacillus thuringiensis can be placed among those pathogens which produce a potent toxin which disrupts host metabolism. Thus, it compares with the medically important pathogens such as Vibrio cholerae, Corynebacterium diphtheriae, Clostridium botulinum, Clostridium tetani and Staphylococcus aureus. The proteinaceous toxins of these microorganisms are the principle factors in the pathogenesis. They are either preformed outside the host or they are excreted by the bacteria during colonization of certain host tissues. Investigations of the mode of action of medically important pathogens are very advanced in comparison to the study of δ-endotoxin of B. thuringiensis.

It has been shown that all these toxins interact at one point with the cellular membranes although the mode of action by which the physiology of the host is impaired differs greatly. Specific receptors for the toxins located in the cell membranes play a key role in many cases. For example, the toxins of V. cholerae and C. diphtheriae only gain access to the site of action inside the cell if one portion of the toxic protein (B-fraction) is able to bind to specific membrane receptors.

Bacillus thuringiensis is not frequently encountered in nature. A study conducted by Delucca et al. (1979) revealed that the prevalence of B. thuringiensis among spore formers isolated from soil and lepidopterous insects was only 0.05%. Although most of the B. thuringiensis strains have been isolated originally from insects, decaying organic material

*Institute of Microbiology, Swiss Federal Institute of Technology, Zurich, Switzerland

represents the primary habitat as it is for the closely related *Bacillus cereus*. The production of δ-endotoxin with the potency to kill insects provides an additional and specific habitat. The δ-endotoxin is supported by other metabolites such as the β-exotoxin and exoenzymes. The colonization of a host starts with the ingestion of a sporulated culture of *B. thuringiensis* which contains the crystalline δ-endotoxin in the form of an inactive protoxin. Proliferation of *B. thuringiensis* occurs after destruction of the gut epithelium by the activated δ-endotoxin.

It is very important for commercial production that *B. thuringiensis* grows saprophytically with full preservation of the insect pathogenic properties. In the laboratory, after many transfers on suitable media, the insecticidal activity is still intact. Although we do not find obligate ties, a specific relationship between pathogen and host seems to exist; the pathogenicity of a certain *B. thuringiensis* isolate is usually higher against the original host than against other target insects. The high activity of the variety *sotto* against the silkworm (*Bombyx mori*) and the variety *thuringiensis* against the flour moth (*Anagasta kuehniella*) are only two examples.

It is surprising that epidemic proliferation of *B. thuringiensis* within insect populations occurs rarely or not at all under natural conditions. Epizootics caused by *B. thuringiensis* may be called man-made since they have been reported in sericulture and in stored-product insects. Even after field application *B. thuringiensis* is not able to proliferate within the population of target insects. Thus a carry-over effect to the next generation or to the following growing season cannot be expected. The initial high concentration of spores, reaching about 10^{13} per acre, is rapidly decreased by environmental conditions such as sunlight and weathering. Although the insect body represents an excellent medium for the proliferation of vegetative cells, the *in vivo* formation of spores seems to occur very slowly, probably due to a deficiency of oxygen.

The role of the bacterial cell during pathogenesis is difficult to evaluate. The δ-endotoxin is of primary importance since its absence leads to the loss of the pathogenic properties of *B. thuringiensis*. After ingestion by lepidopterous larvae, the spores find themselves in the alkaline gut juice which is an unfavorable environment for germination and vegetative growth. Following the destruction of the gut epithelium by the δ-endotoxin, the pH of the gut juice is lowered by leakage of hemolymph components, and spores are then able to germinate. The proliferating vegetative cells penetrate into the hemocoel. *Bacillus thuringiensis* cells seem to resist all host defense mechanisms since upon intra-hemocoelic injection a single cell is able to initiate a lethal septicemia. An exception to the above are the so-called Type III insects (see below), represented by A. kuehniella, the Mediterranean flour moth. Insects belonging to this group have a gut pH just above neutral where B. thuringiensis spores are able to germinate. The vegetative cells multiply in the intestine and start a lethal septicemia supporting the action of the δ-endotoxin, probably by the production of phospholipase C and other enzymes with insecticidal activity (Burgerjon and Martouret, 1971). Crystals without spores are apparently not activated in the intestine of A. kuehniella.

The systematic classification by serological and biochemical means (de Barjac and Bonnefoi, 1973) is presently the best method for keeping order among the many B. thuringiensis isolates which have been described and which will be detected or produced in the years to come. Up to 1980 sixteen serotypes of B. thuringiensis have been identified (for the characterization of serotype 15 and 16 see Delucca et al., 1979). With one exception (serotype 14) the activity spectrum of the δ-endotoxin is limited to insect larvae of the order Lepidoptera. Serotype 14 (Goldberg and Margalit, 1977; de Barjac, 1978a) of the variety israelensis is active against Diptera. From the many B. thuringiensis strains which have been tested and from the more than one hundred lepidopterous species which were reported to be susceptible, we know that the δ-endotoxin exhibits considerable variety-specific activity within the host range. The degree of specificity, however, is not only determined by the type of δ-endotoxin but also by the insect host. For example, larvae of B. mori are very susceptible to the δ-endotoxin of the variety sotto whereas the response of Pieris brassicae (large white butterfly) to the same endotoxin is moderate.

The activity spectrum of serotype 14 is surprisingly narrow and limited to a few families within the suborder Nematocera (order: Diptera). The highest susceptibility is shown by larvae in the genera Culicidae and Simulidae. Activity at a reduced level against aquatic Chironomidae and Tipulidae has been reported by Weiser and Vanková (1978), Sinègre et al. (1979) and Schnetter et al. (1981). These authors consider the side effects of B. thuringiensis var. israelensis to the non-target aquatic fauna as negligible.

Bacillus thuringiensis activity outside the host order has been found by several authors. No definite evidence however has been presented that the δ-endotoxin plays a role in the pathogenicity. For example, Hall et al. (1977) reported that some B. thuringiensis strains, classified among the above mentioned 13 serotypes, exhibited activity against mosquito larvae. In comparison with serotype 14, the activities seem to be low and the required periods of exposure seem to be very long for δ-endotoxin activity. It is possible that at least some of the mosquito mortality was caused by secondary effects e.g. the development of a vegetative cell population producing insecticidal metabolites such as exotoxin or phospholipases.

The development of insect cell lines derived from Choristoneura fumiferana (spruce budworm) and Malacosoma disstria (forest tent caterpillar) with susceptibility to the δ-endotoxin provided an in vitro assay (Sohi, 1971; Fast et al., 1976). An in vitro system allows the assay of δ-endotoxin samples without further uncontrolled modification by gut juice components.

Prasad and Shetna (1976) reported in a series of publications that the δ-endotoxin of B. thuringiensis possesses antitumor activity against Yoshida ascites sarcoma cells. Antitumor activity of the δ-endotoxin was also found by Seki et al. (1978) using sarcoma 180 ascites cells. On the other hand, Geiser (1979) showed that a cell line of chicken fibroblasts did not respond to the presence of δ-endotoxin.

A number of representative microorganisms tested by Geiser (1979) did not react upon incubation with δ-endotoxin. Vesicles derived from the plasmalemma of yeasts (Saccharomyces cerevisiae) did not alter their structure or permeability properties when they were exposed to the δ-endotoxin. A microbial assay would of course be the most convenient method to determine δ-endotoxin activity.

SYMPTOMS INDUCED BY THE DELTA-ENDOTOXIN

According to the manifestation of symptoms, susceptible insect species have been grouped into three categories. Type I insects, represented by B. mori, and Type II species (P. brassicae), are susceptible to the crystal protein alone whereas Type III insects (A. kuehniella) are killed only by the spore-crystal complex. The mode of action in Type III insects is not yet well understood; nevertheless the crystal protein must play a key role since spores alone are ineffective. In Type III insects crystal proteolysis may proceed slowly and may need the support of the proliferating vegetative cells in the intestine. Details of the different types are presented in a comprehensive review on the proteinaceous crystal of B. thuringiensis by Lecadet (1970). In this chapter emphasis will be on Type I and II insects since they are the classical target for the δ-endotoxin.

The first symptom of intoxication, cessation of feeding, appears within minutes following the uptake of crystals of B. thuringiensis. The uptake of food by P. brassicae is blocked as early as two minutes after the administration of crystals, while B. mori larvae cease feeding within seven to ten minutes after ingestion of crystal protein. In comparison to other microbial toxins the response to the δ-endotoxin is extremely rapid, especially when we consider that the crystals have to be dissolved and activated in the gut juice, and then the active moiety has to reach the site of action. It is probable that the δ-endotoxin acts without a lag phase. In contrast, a lag phase which lasts several hours is known for the diphtheria toxin (Murphy, 1976). Although symptoms are absent during the lag-phase there are no means to reverse the action of diphtheria toxin.

The symptoms of δ-endotoxin intoxication which follow the cessation of feeding depend on the insect species. The whole range of symptoms has been described by Heimpel and Angus (1960) in a review covering the early pioneering work with *B. thuringiensis*. Total paralysis in the case of the silkworm or at least a slowing of the movements of the larvae of other species are characteristic of the progressive poisoning process. The paralysis can be explained by the loss of permeability control of the damaged gut epithelium, leading to a reduction of the pH-gradient between the hemolymph and the intestine. A rise in the hemolymph pH causes paralysis, and the reduction of the alkalinity of the gut juice favors the germination of the *B. thuringiensis* spores, leading to the development of a vegetative cell population which finally breaks through the basal membrane into the hemocoel. The destruction of the gut epithelium often causes a drain of body fluid, which in the case of *P. brassicae* is manifested by a reduction of the larval size.

Larvae having ingested lethal doses of δ-endotoxin can remain alive for several days. When a population of *P. brassicae* has received an LD_{50}, mortality does not begin before 24 h and reaches its final level after 72 h. The primary site of action of the δ-endotoxin is apparently limited to the gut epithelium as other vital organs and the general metabolism are not directly affected.

Intrahemocoelic injection is only effective if activated δ-endotoxin is used (Lecadet and Martouret, 1967; Geiser, 1979), whereas whole crystals and NaOH dissolved protoxin do not induce symptoms of poisoning. The site of action of δ-endotoxin applied intrahemocoelically is not known. The minimum lethal dose is increased by a factor of one hundred compared to peroral application. According to results obtained by Benz (1966), the permeability of the gut epithelium remains unchanged after injection of δ-endotoxin into the hemocoel, indicating that other tissues are the primary targets.

The rapid response to the δ-endotoxin, blocking the uptake of food, is a disadvantage for field application, as sublethal doses induce cessation of feeding. Since the turnover of epithelial cells in the gut is high, damaged cells are sloughed and replaced by healthy ones originating from the regenerative nidi. According to our observations, the recovery of larvae treated with sublethal doses of δ-endotoxin may start after only three days. In the meantime, the *B. thuringiensis* concentrations on the plants may have been reduced by weathering or the insects may find fresh untreated foilage.

ASSAYS FOR THE DETERMINATION OF DELTA-ENDOTOXIN ACTIVITY

Highly sensitive and reproducible assays are needed for the investigation of the δ-endotoxin. Insect larvae are still the best test systems as long as only the amount of δ-endotoxin which is present in a sample has to be determined. An *in vitro* assay is required if we want to know whether a component of the crystal protein is present in an activated form.

For basic research *B. mori* and *P. brassicae* are the most frequently used lepidopterous species. Both insects are relatively easy to rear and are very susceptible to the δ-endotoxin. The test samples can be applied with a microsyringe to larvae in the last instar. Several methods are available to measure the effect of the toxin. In the case of *P. brassicae*, activity is determined by reduction in feeding, expressed by the amount of consumed leaf material (Burgerjon, 1962) or by the weight of frass pellets (Lüthy, 1975) following *ad libitum* consumption. If *B. mori* larvae are used as test insects, general paralysis within a fixed period of time or mortality can be used for the quantitative measurement of the δ-endotoxin. An example of a dose-response assay with *P. brassicae* larvae and crystals of *B. thuringiensis* var. *thuringiensis* is given in Fig. 9-1a (solid squares).

For serotype 14, bioassays are conducted with readily available mosquito larvae. Assays with 50 larvae in the same stage of development per toxin concentration give reproducible results. The dose is expressed as amount of crystal protein/ml of water. A representative assay with *Aedes aegypti* larvae is shown in Fig. 9-1a (open squares). Since it is known that the δ-endotoxin of serotype 14 also acts very fast, the time of the assay should not exceed 24 h in order to avoid increased mortality due to bacterial contamination.

Quantitative *in vitro* determination of the δ-endotoxin is now possible with the cell lines developed by Sohi (1971). Fast et al. (1978) found that the interaction between the δ-endotoxin and tissue culture cells leads to a rapid loss of cellular ATP. ATP concentrations can be determined by the method of Strehler (1968) using the luciferin-luciferase system. Geiser (1979) established a quantitative assay and showed that a normal dose-response relationship is also valid for the *in vitro* test. As can be seen from Fig. 9-1b, the sensitivity of this assay is lower by a factor of 10^3 compared to the *in vivo* test. It should therefore only be used if the state of activation of the δ-endotoxin toxin is relevant.

HISTOPATHOLOGY *IN VIVO*

Between 1954 and 1960 a series of papers was published by Angus and Heimpel, demonstrating step by step that the crystal protein was responsible for the insecticidal activity, and that the gut epithelium of target insects underwent drastic histopathological changes in the presence of δ-endotoxin. These investigations have been reviewed by Heimpel (1960). In the past, the histopathological events have been studied predominantly with larvae of *B. mori* and *P. brassicae*. The results of the early investigations can be summarized as follows: Within minutes of ingestion δ-endotoxin induces an increase in volume of the cells of the gut epithelium. The continuous enlargement of the cells is paralleled by the formation of bubble-like protrusions on the surface of the epithelium (Angus, 1970). The bubbles originate from the microvilli which lose their lamellated structure. The balloon-like protrusions are still an integrated part of the whole cell. Later, the epithelial cells separate from each other and leave the tissue. Cells often burst and release their cytoplasmic content into the lumen. All these observations led to the assumption that the gut epithelium is the primary target of the δ-endotoxin.

In our laboratory histopathological studies were continued using electron microscopy to evaluate the histopathological changes induced by the δ-endotoxin. The investigations were conducted with *P. brassicae* larvae and sublethal doses of δ-endotoxin (10^4 crystals per larva = 5-8 ng) derived from *B. thuringiensis* var. *thuringiensis*. The assay methods have been described by Ebersold et al. (1977). Investigation of histopathological events is only possible if sublethal doses are applied which prevent a rapid and total destruction of the cells.

The results of the experiments refer only to columnar cells because their organelles are more clearly defined than in goblet cells. A general view of a cross section of the mid-gut epithelium is presented in Figure 9-2. An early stage of the histopathological changes is shown in Figure 9-3 where the increasing vacuolization within the cell is the most distinct symptom. The following pictures show the events which occur under the influence of the δ-endotoxin at a higher magnification. Figure 9-4 is a control section showing the most important organelles of the cytoplasm. The first recognizable changes are changes in the rough endoplasmic reticulum which is the site where the majority of proteins are synthesized. Within five to ten minutes the intermembraneous spaces of this reticular network are significantly enlarged. Subsequently, the connections between the membranes are disrupted, leading to a vesicular arrangement of the endoplasmic reticulum. The vesicles

Figure 9-1a. Dose-response lines of two bioassays with the δ-endotoxin of *Bacillus thuringiensis*. Solid squares - feeding inhibition of L5 *Pieris brassicae*, tested against the δ-endotoxin of *B. thuringiensis* var. *thuringiensis*. The dose is expressed as ng of crystal protein force-fed per larva. Open squares - mortality of *Aedes aegypti* (L2), tested against the δ-endotoxin of *B. thuringiensis* var. *israelensis*. The dose is expressed as ng of crystal protein per ml of water.

Figure 9-1b. Example of an *in vitro* assay with activated δ-endotoxin of *B. thuringiensis* var. *kurstaki* tested against the Cf-1 cell line. The dose is expressed as μg of δ-endotoxin per ml of culture medium, the effect as reduction of cellular ATP.

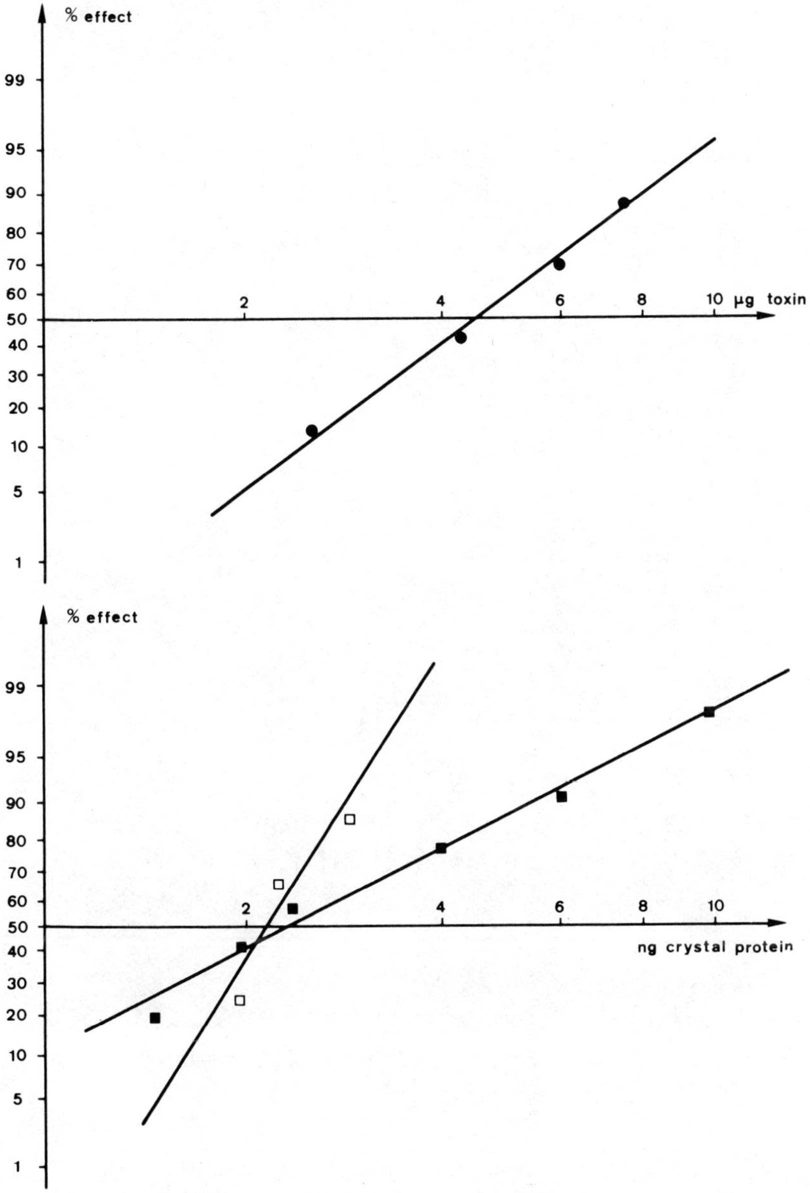
% effect
99
95
90
80
70
60
50
40
30
20
10
5
1
2
4
6
8
10 µg toxin
% effect
99
95
90
80
70
60
50
40
30
20
10
5
1
2
4
6
8
10
ng crystal protein

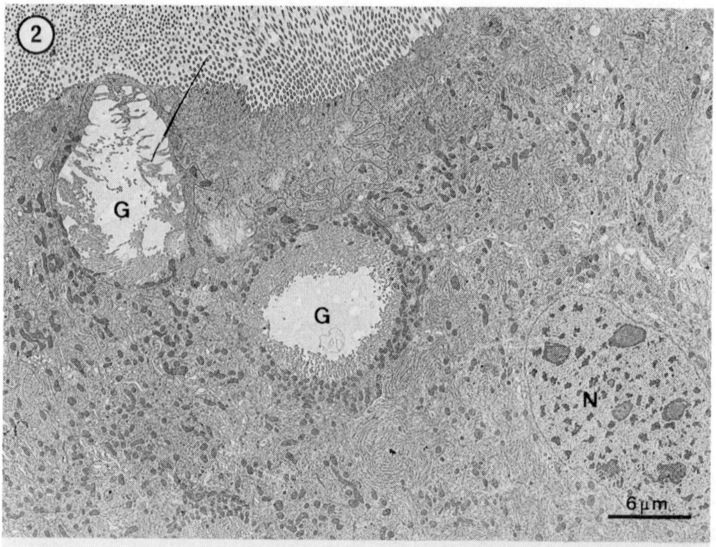

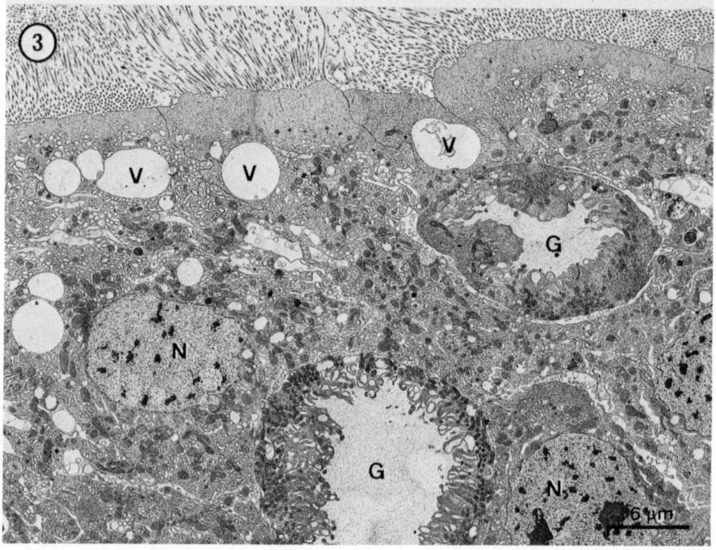

Figure 9-2. Section from a healthy midgut epithelium of a *Pieris brassicae* larva. G = goblet cells, N = nucleus.

Figure 9-3. Cross section from midgut epithelium of a δ-endotoxin treated larva. The appearance of vacuoles (V) are the early symptoms of the δ-endotoxin action. 10 minutes after feeding.

continue to increase in size but they remain intact as long as the cells do not burst. The vacuolization is particularly intense near Golgi complexes. The reaction of the mitochondria to the δ-endotoxin is similar to that of the endoplasmic reticulum. The mitochondria swell continuously while the cristae gradually disintegrate. At the final stage, the content of the mitochondria is dissolved, leaving only membranous fragments. In Figure 9-5 an early stage of the δ-endotoxin action is presented while in Figure 9-6 the disruption of the cell organelles have developed further. In addition, a significant increase in the number of lytic vacuoles can be observed (Figure 9-7).

Concomitant with the intracellular damage of the organelles, deformation of the microvilli takes place. They lose their internal fibrillar fine structure, followed by an extrusion, leading to the formation of the above mentioned bubbles. Figure 9-8 shows microvilli of an untreated midgut cell while in Figure 9-9 considerable damage by δ-endotoxin has occurred.

No histopathological changes can be observed in the nuclei of gut epithelial cells. Only the outer nuclear membrane which is a constituent of the rough endoplasmic reticulum exhibits symptoms of vacuolization (Figure 9-5). That the nuclei are not directly involved should not be generalized, since they were affected to a certain degree by the δ-endotoxin when the experiments were conducted with a cell line derived from the spruce budworm (see below).

The rapid and general vacuolization of the cell organelles indicates that the brush border membrane, which is the first structure exposed to the δ-endotoxin, plays a very important role. An intact membrane is a highly selective permeability barrier, responsible for the active and passive transport of ions and metabolites. As a rule, only low molecular weight compounds pass the membranes. The regulatory functions must be of special importance between the alkaline, enzyme-rich gut juice and the cytoplasm of the epithelial cells. It was shown using ruthenium red, an inorganic stain of molecular weight 852, that the permeability of the brush border membrane increased within a few minutes under the influence of the δ-endotoxin (Ebersold et al., 1978). Ruthenium red is not able to penetrate intact membranes. Figure 9-10 shows a section of two columnar cells in an early stage of exposure to δ-endotoxin where the cell on the left has already become permeable to the stain. Increased permeability of the gut epithelium was earlier observed by Benz (1966) using red kale juice.

A histopathological study on the mode of action of the δ-endotoxin was also made by Endo and Nishiitsuitsuji-Uwo (1980) using the midgut epithelium of _B. mori_. The goblet cells were found to react somewhat differently than columnar

Figure 9-4. A control section from a cell of the midgut epithelium at higher magnification ER = endoplasmtic reticulum, GA = Golgi apparatus, M = mitochondria.

Figure 9-5. Details of early histopathological changes induced by the δ-endotoxin: Vacuolization of the endoplasmic reticulum, vacuolization of the nuclear envelope (NE). 15 minutes after feeding.

Figure 9-6. A further step in the destruction and disintegration of cell organelles with a pronounced enlargement of the vacuoles. 10 minutes after feeding.

Figure 9-7. Lytic vacuoles appear in a more advanced stage of the δ-endotoxin action: A = autolytic vacuoles, MB = multivesicular body, P = peroxisomes, IS = intercellular separation. Note that the magnifications of Figures 9-4, 9-5, 9-6 and 9-7 are identical. 30 minutes after feeding.

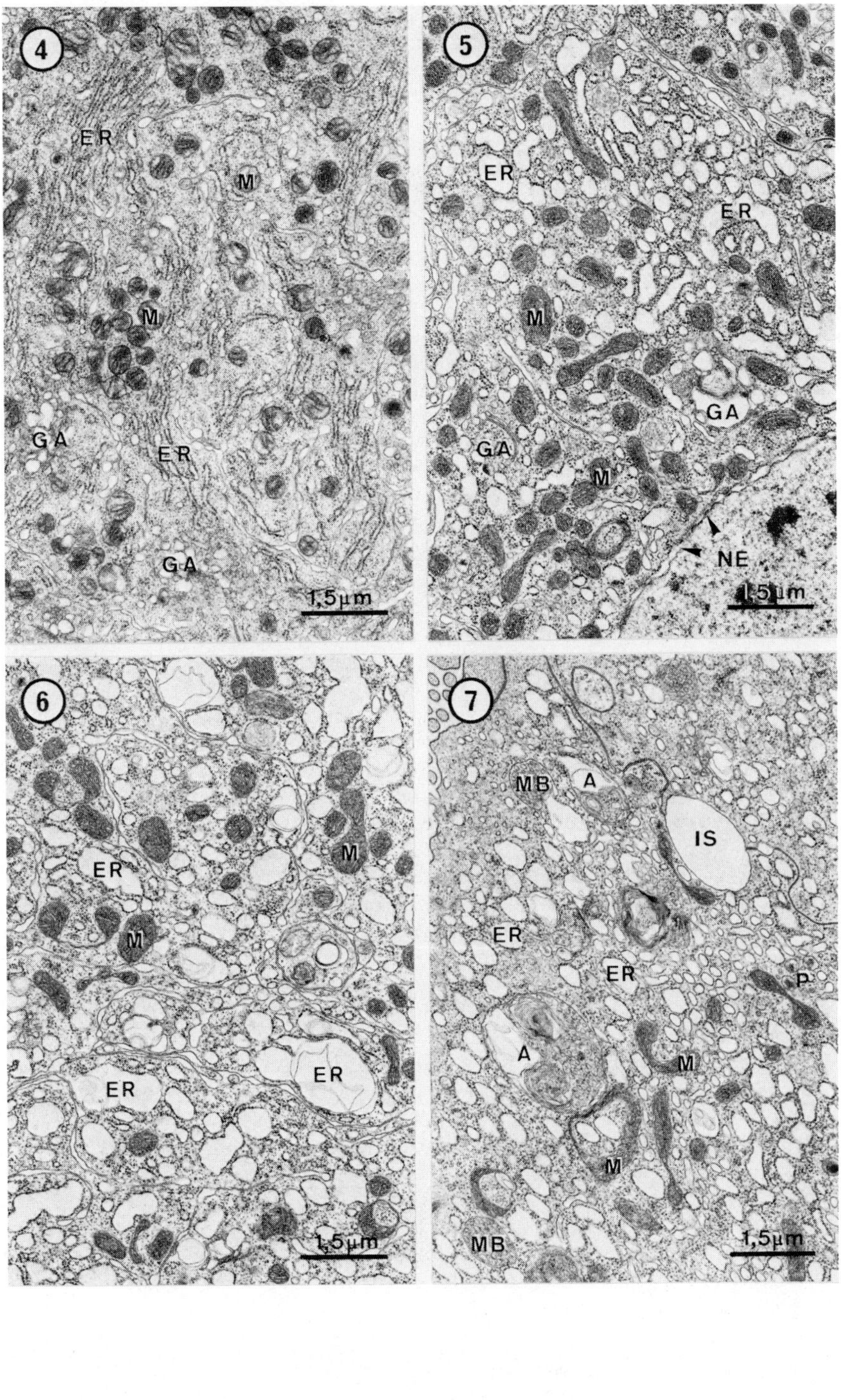
4
ER
M
M
GA
ER
GA
1,5μm
5
ER
ER
M
GA
GA
M
NE
1,5μm
6
ER
M
M
ER
ER
1,5μm
7
MB
A
IS
ER
ER
P
A
M
M
MB
1,5μm

cells. The transformation of mitochondria into condensed forms was described. Such condensed forms (which were also found in our investigation) are caused by the detachment of the inner from the outer membrane. We did not realize that they were toxin-induced since we assumed that the condensed forms were due to the preparation of the specimens.

As mentioned above, single cells or whole agglomerates separate from the epithelial tissue. Heimpel and Angus (1959), as well as Estes and Faust (1964) suggested that the δ-endotoxin dissolves the cell-cementing substances. Our micrographs however revealed that the intercellular connections remain intact for a long time. This can be demonstrated with lanthanum which is a specific stain for contrasting the links between cells (Figure 10-11). The intercellular connections consist of the so-called tight junctions, the desmosomes and the gap junctions (Staehelin and Hull, 1978). The tight junctions seal off the lateral cell membrane from the brush border, thus preventing the infiltration of gut juice between the membranes. The desmosomes are responsible for the structural stability, and the gap junctions regulate the intercellular exchange of metabolites with a molecular weight not exceeding 1000.

The location of the δ-endotoxin during the early phase of the poisoning process has not yet been identified. The δ-endotoxin either interacts with the brush border membrane or the activated toxic moiety, (perhaps only a fragment of it), penetrates through the membrane and interferes with a constituant of the cytoplasm. Fast et al. (1978) using an *in vitro* system, presented evidence that the δ-endotoxin interacts with the cell membrane. They bound δ-endotoxin to Sephadex G-25 beads in order to prevent penetration into the cells. Autoradiography conducted in our laboratory with ^{3}H-labeled δ-endotoxin indicates that no major part of the toxin is present within the cells for at least half an hour following ingestion of δ-endotoxin by the larvae. A significant labeling of the cells could only be observed after one hour. It must be added that the sensitivity of the radioactive labeling of the crystal protein might have been too low to trace only a few molecules which could be lethal for a single cell. The same could be true for other standard tracing methods.

To date no reports have been published which show that the δ-endotoxin has a primary target within the cytoplasm of the cells, though Travers et al. (1976) proposed that the δ-endotoxin induces uncoupling of oxidative phosphorylation. Their results were obtained in an *in vitro* study with mitochondria isolated from the gut epithelium of *B. mori*. Since a primary target within the cells cannot be excluded it would be of interest to investigate, for example, a possible interaction between the δ-endotoxin and lytic vacuoles. Lytic vacuoles are responsible for the lysis of old cells which are on the point of being replaced by young ones. The histopathological changes during the lysis of old cells very much resemble those induced by the δ-endotoxin.

HISTOPATHOLOGY *IN VITRO*

In 1976 Murphy et al. succeeded for the first time in showing activity of enzymatically digested δ-endotoxin against three cell lines derived from the spruce budworm (*Choristoneura fumiferana*), from the forest tent caterpillar (*Malacosoma disstria*) and from the cabbage looper (*Trichoplusia ni*). The cells responded with swelling, lysis and vesicle formation. However, despite heavy doses of up to 680 μg/ml of activated crystal protein, only an average of 45% of the cells were affected. Apparently the tested cell lines were heterogeneous and only certain cell types responded to the δ-endotoxin. This phenomenon is a further indication of the high specificity of the δ-endotoxin. The histopathological changes induced by the δ-endotoxin in the Cf-1 cell line of *C. fumiferana* were investigated in our laboratory by Geiser (1979). Figure 9-12 shows a representative part of a control culture, whereas Figure 9-13 shows a thin section of a Cf-1 cell which had been incubated with activated δ-endotoxin. The typical histopathological alterations can be found. Vacuoles have been formed in the cytoplasm; the mitochondria, the Golgi complexes and the endoplasmic reticulum are swollen and have lost their characteristic structure. In contrast to gut epithelial cells of *P. brassicae*, the nuclei were affected, especially by a dislocation of the chromatin towards the peripheral regions. Very pronounced vacuolization occurred between the outer and the inner membrane of the nuclei (arrow, Figure 9-13). Intact connections are only maintained at sites where pores link the nucleus with the cytoplasm.

Other insect cells were tested by Graf (personal communication) and by Geiser (1979). Graf demonstrated that epithelial cells isolated from the midgut of *P. brassicae* developed the characteristic protrusions when δ-endotoxin was added to the culture medium. On the other hand, Geiser did not find a histopathological response of primary cell cultures of hemocytes which were also derived from *P. brassicae*. Nishiitsutsuji-Uwo et al. (1980) performed an *in vitro* study with insect as well as with mammalian cell lines. They observed that insect cells generally reacted by swelling and bursting caused by the breakdown of ion regulation whereas no cytotoxic effect could be demonstrated in mammalian cells. The authors were thus not able to confirm the results obtained by Seki et al. (1978) and by Prasad and Shetna (1976), and

Figure 9-8. Intact microvilli of a columnar cell from the gut epithelium of *Pieris brassicae*; control.

Figure 9-9. Microvilli under the influence of δ-endotoxin. 10 minutes after feeding.

Figure 9-10. A few minutes after contact with δ-endotoxin the cells of the midgut epithelium become permeable to ruthenium red. The cell on the right has not yet lost permeability control. 10 minutes after feeding.

Figure 9-11. Lanthanum staining of a gap junction 30 min following application of δ-endotoxin to the larva. The intercellular connections visible as fine white lines are still intact.

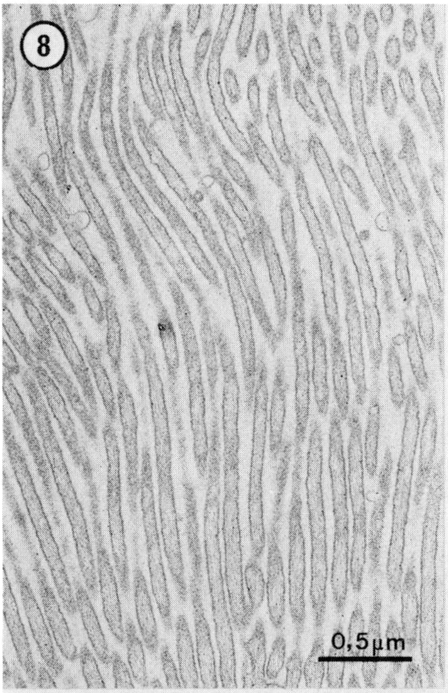
8
0,5μm

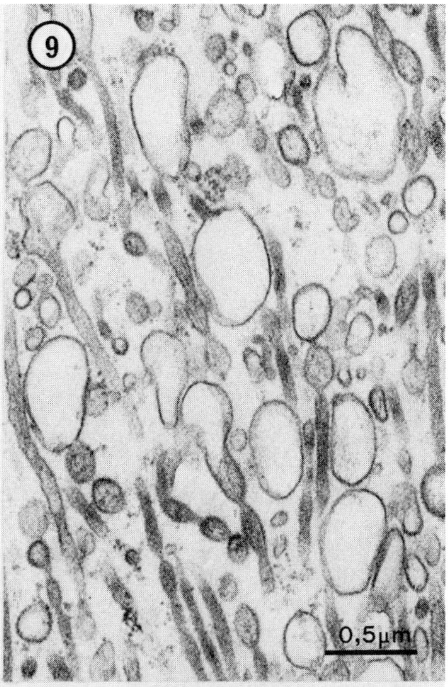
9
0,5μm

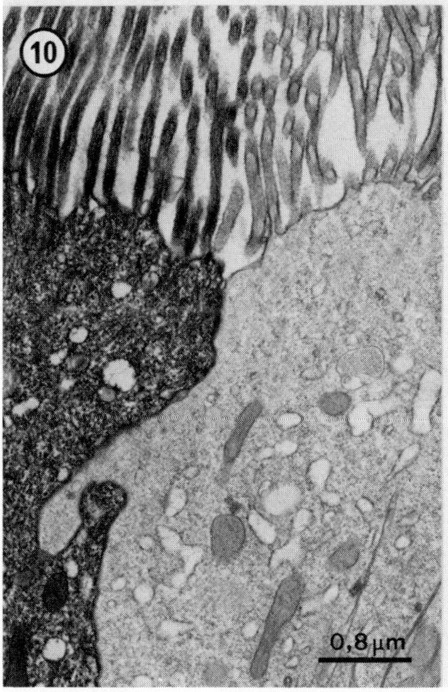
10
0,8μm

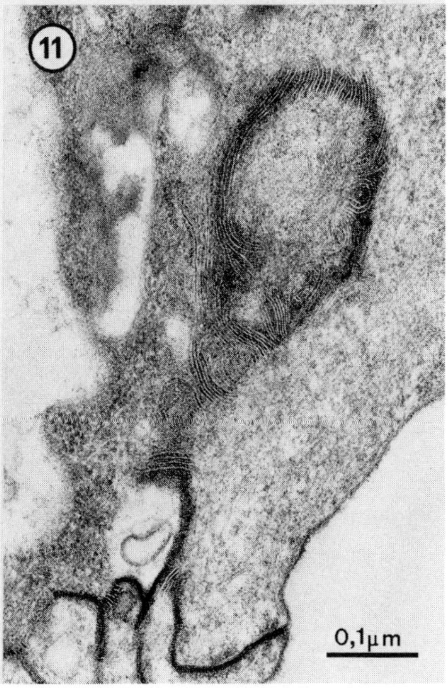
11
0,1μm

concluded that δ-endotoxin is not able to interfere with mammalian cells.

MOLECULAR MODE OF ACTION *IN VIVO*

Within the complex system of an insect larva it is extremely difficult to obtain valid results on the molecular mode of action of a toxin. The rapid sequence of events caused by the δ-endotoxin adds to the complexity since a clear-cut differentiation between primary and secondary pathological effects is difficult. Publications on this subject are therefore primarily dealing with changes in the transport activities for certain compounds, and with alterations in the permeability of the brush border membrane or the whole gut epithelium. Fast and Donaghue (1971) measured the transport of glucose from the intestine to the hemolymph in the presence of the δ-endotoxin. The uptake of glucose by the gut epithelium was markedly enhanced with 1 min reaching a peak after 5 min. In a later stage of the intoxication process, the transport of glucose was inhibited. Fast and Morrison (1972) reported that the levels of potassium, sodium, calcium and magnesium remained unchanged during the first 10 min following application of δ-endotoxin. Thereafter, an increase in the potassium concentration of the hemolymph could be measured. Based on these results, the authors concluded that the δ-endotoxin does not act as a ionophore. This study was initiated because the symptoms caused by the δ-endotoxin can be mimiced with valinomycin, which is a classical ionophore (Angus, 1968). Thus, the reason for potassium accumulation in the hemolymph lays in the general breakdown of permeability control of the gut epithelium. Fast and Videnova (1974) working with ^{3}H-labeled crystal material found low molecular weight fragments in the hemolymph of larvae of *C. fumiferana* and *M. disstria* within 1 to 5 min. These results are to some extent in contrast with the hypothesis that the δ-endotoxin acts on the surface of the cell membrane. The presence of crystal protein fragments within the cells could be explained by a rapid proteolytic degradation in the gut juice followed by an immediate uptake of small oligopeptides by the still functioning transport system. Fast and Videnova (1974) do not state that the derivatives of the crystal protein isolated from the hemolymph are biologically active.

MOLECULAR MODE OF ACTION IN VITRO

The development of insect cell lines susceptible to the δ-endotoxin opened the possibility of investigating the molecular mode of action *in vitro*. Fast et al. (1978) detected that the ATP content of toxin damaged cells decreased rapidly. Geiser (1979) used the changes of the Cf-1 cell line (Sohi, 1971) for the quantitative determination of δ-endotoxin, including the examination of different crystal protein fractions and a comparative study between the two varieties *thuringiensis* and *kurstaki*. The reduction of ATP in a toxin-treated cell culture is shown in Figure 9-14. The decrease of ATP starts after 1 min and reaches the minimal level after 5 min. However, the cell culture of the Cf-1 line retained at least 25% of its ATP which is an indication that here also a proportion of cells did not react with the δ-endotoxin. The final 25% ATP level could not be lowered even by increasing the concentration of the δ-endotoxin above 50 μg/ml. A comparison between the two varieties *kurstaki* and *thuringiensis* shows that the *kurstaki* toxin was approximately thirty times more active (Figure 9-15). When different crystal protein components were assayed, it was found that whole crystals and chemically dissolved crystal protein had no effect on the cells. Full activity was only found with the high molecular weight fractions obtained after proteolytic degradation with trypsin or with gut juice (see Chapter 8). The low molecular weight fragments which are produced as by-products during proteolysis did not react with the cell line. The action of the δ-endotoxin could be blocked with specific antisera.

It is important to note that the sensitivity of the *in vitro* assay is 50 to 100 times lower than a comparable *in vivo* test with *P. brassicae* or *B. mori* larvae. Amounts of toxin of just below 1 ng can be detected in the larval assay whereas the *in vitro* test requires a minimal dose of 100 ng of toxin.

Ebersold et al. (1980) selected two enzymes, lactate dehydrogenase (LDH) and acid phosphatase, as markers to detect alterations in the permeability of the membranes of the Cf-1 cells. LDH (MW 140,000) is present in the cytoplasm in an unbound state whereas acid phosphatase is located in the autolytic vacuoles and partially bound to membranes. LDH activity was detected outside Cf-1 cells as early as 10 min following incubation of the culture with δ-endotoxin. After 60 min, 80% of the total activity was present in the supernatant. This experiment confirmed the rapid loss of permeability control of the cells. The membranes become permeable even for high molecular weight compounds. Using acid phophatase as the other marker, it was shown that the intracellular activity of the cell-bound enzyme increased by 50% within 1 h because p-nitro-phenylphosphate was able to pass through the membrane of toxin-damaged cells while in control cultures the membrane acted as a barrier for the substrate.

Figure 9-12. Section from a Cf-1 tissue culture. N = nucleus, NE = nuclear envelope.

Figure 9-13. A cell from a δ-endotoxin treated Cf-1 tissue culture (10 μg/ml for 30 min) with the typical symptoms of a δ-endotoxin poisoning. Note the vacuolization of the nuclear envelope (NE). M = empty former mitochondria.

Figure 9-14. Plasmatic fracture phase (PF) of a cell membrane from a Cf-1 control cell.

Figure 9-15. Plasmatic fracture phase of a δ-endotoxin treated Cf-1 cell membrane. The tissue culture was exposed to the δ-endotoxin (var. kurstaki, 14 μg/ml) for 30 min at 23°C.

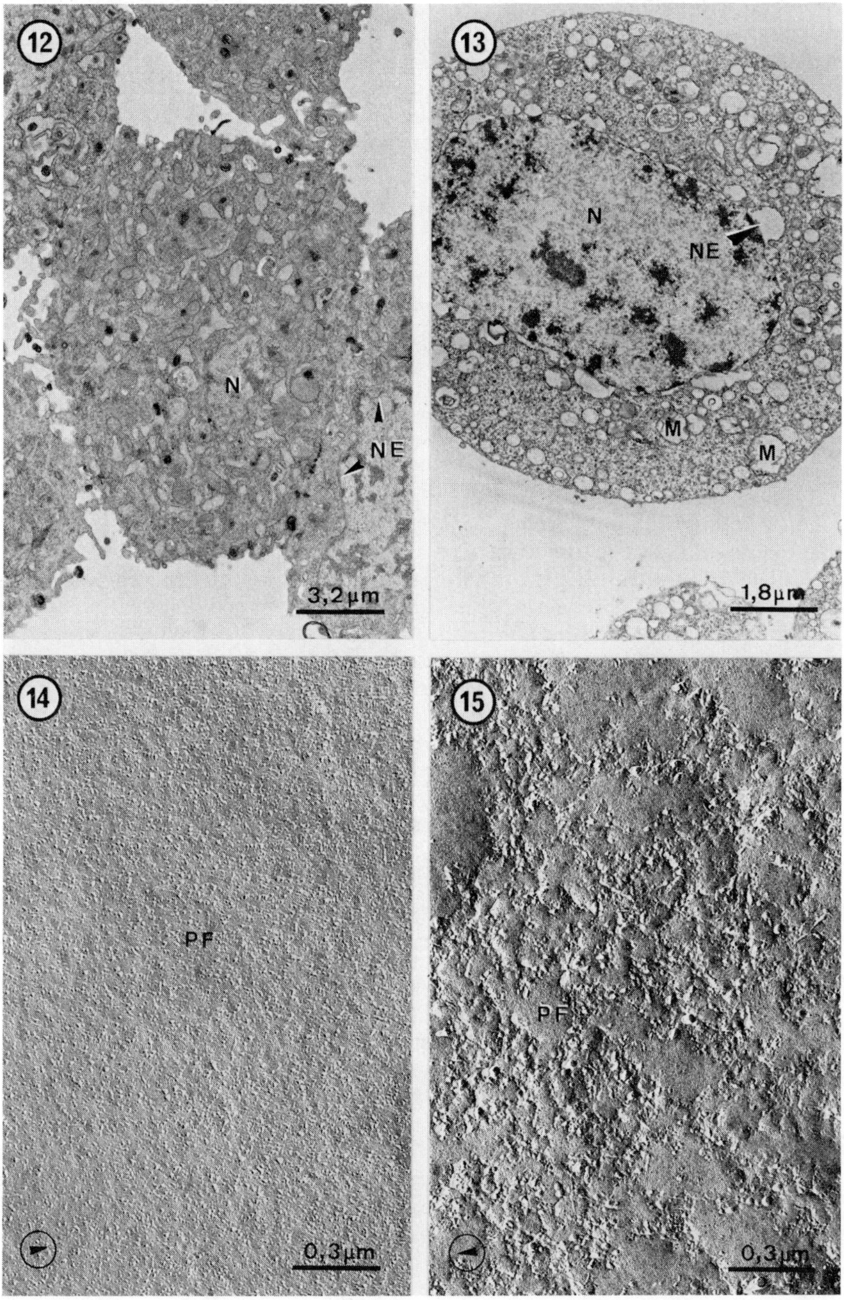
12
N
NE
3,2 µm
13
N
NE
M
M
1,8 µm
14
PF
0,3 µm
15
PF
0,3 µm

In another series of experiments, Ebersold et al. (1980) investigated structural changes in membranes of Cf-1 cells induced by the δ-endotoxin. When the cells (10^6/ml) were incubated with δ-endotoxin (10 μg/ml) for 30 min, the molecular structure of the membranes changed greatly. The regular distribution of the protein complexes within the membrane disappeared. Figure 9-14 represents the plasmatic fracture phase of a membrane derived from a control cell. A freeze-fractured cell membrane which was exposed to δ-endotoxin is shown in Figure 9-15. Freeze fracturing of healthy membranes resulted in a homogenous plane fraction through the hydrophobic part of the membrane. Toxin-treated specimens lost this property and the levels of the fractures fluctuated strongly. These investigations were completed by including isolated and partially purified brush border membranes (Luthy et al., 1978). Again the number of protein complexes within the brush border membranes was greatly reduced following incubation with δ-endotoxin.

A direct action of the δ-endotoxin on the cell membranes was postulated by Fast et al. (1978). Their assumption, as we already mentioned, was based on an experiment in which the δ-endotoxin was bound to Sephadex G-25 beads which should have prevented the penetration of the active moiety into the cells. The authors reported that the ATP levels of the cells were reduced at the same rate as in the control where unbound δ-endotoxin was applied.

The above described experiments, carried out with the C. fumiferana cell line, show a direct and rapid action of the δ-endotoxin on the cell membranes. Similar membrane-action toxins are produced by strains of Staphylococcus aureus, e.g. the δ-toxin. Thelestam and Mollby (1975a) studied the changes in membrane structures caused by the δ-toxin of S. aureus. Comparing their results with the effects of the δ-endotoxin of B. thuringiensis, no obvious differences can be detected. The action of the staphylococcal toxin was described as detergent-like, since Triton X-100 is known to induce similar changes in the molecular structure of membranes. There exists however one major difference between the action of Triton X-100 and the δ-toxin of S. aureus. The toxin leads to a gradual increase in membrane permeability while Triton X-100 is responsible for an immediate general loss of permeability control (Thelestam and Moellby, 1975b). It is now of interest to find out if the δ-endotoxin brings out a gradual change in the permeability properties of membranes or if the alterations occur in a single step.

Travers et al. (1976) assume that the mitochondria play a key role as primary target for the δ-endotoxin. Using mitochondria isolated from the midgut epithelium of B. mori, they measured an uncoupling of oxidative phosphorylation, leading to the depletion of the energy pool. The authors were working with an active toxin fraction of a molecular weight of

33,000. It is still unexplained, however, how a molecule of this size is able to reach the site of action at the inner mitochondrial membrane. A lag phase should be present as in the case of diphtheria toxin, and there should also exist a specific binding mechanism between the δ-endotoxin and the membranes which must be passed. In our laboratory (Ebersold, unpublished) similar experiments utilizing purified rat liver mitochondria did not lead to the detection of an uncoupling reaction induced by the δ-endotoxin. The different origin of the mitochondria could of course have been responsible for the negative result.

Cooksey et al. (1969) demonstrated that the δ-endotoxin blocked nerve conduction in a desheathed spinal ganglion preparation of *Periplaneta americana* (cockroach). It was suggested that the δ-endotoxin acted on membrane permeability and that potassium ion regulation was disrupted.

The uncoupling of oxidative phosphorylation as well as the neurotoxic effect could be explained by the demonstrated membrane activity of the δ-endotoxin. In both cases a breakdown of the permeability barriers must necessarily lead to collapse of the membrane potential which is required for normal functioning of the oxidative phosphorylation and for nerve conduction. The negative response of rat liver mitochondria could be explained by the specificity of the δ-endotoxin.

It was mentioned above that Angus (1968) showed that valinomycin caused the same symptoms as the δ-endotoxin in larvae *B. mori*. Valinomycin acts as a ionophore increasing ion flux, especially of potassium ions through membranes. The potassium ions passing from the gut to the hemocoel cause a rise in pH of the hemolymph which is manifested by paralysis. Geiser (1979), using black lipid membranes, presented evidence that a distinct difference exists in the mode of action between valinomycin and δ-endotoxin. While valinomycin increased the flux of potassium ions through the black lipid membranes, the δ-endotoxin did not influence the membrane potential. This result does not exclude ionophoric properties of the δ-endotoxin but it demonstrates at least a basic difference among the two agents and underlines once more the specificity of the δ-endotoxin.

The species of the crystal protein, the mode of activation and the type of target cells are the three factors which determine the spectrum of activity. Under natural conditions the activation of the crystal protein occurs only in the intestine of a relatively small number of insect species. The activated δ-endotoxin can be regarded as an intermediate product of crystal proteolysis. The only tissue which comes in direct contact with activated δ-endotoxin is the gut epithelium. Susceptibility of cells exposed to the δ-endotoxin varies from one insect species to the other.

In vitro activation of the crystal protein by gut juice proteases or by trypsin made it possible to test numerous different cell types for susceptibility. It is therefore not surprising that besides insect cell cultures, some types of vertebrate cells have also responded to the δ-endotoxin. Up to the end of 1979 two groups of authors reported activity of the δ-endotoxin against vertebrate cells; in both cases these are tumor cells. Prasad and Shetna (1974) were able to suppress the development of Yoshida ascites sarcoma cells in mice by peritoneal injections of crystal protein. In 1976, the same authors working with short term *in vitro* cultures of the tumor cells, showed that the δ-endotoxin was responsible for a general breakdown of metabolism, accompanied by a significant loss of high and low molecular weight constituents. It was proposed that the δ-endotoxin affects the integrity and function of the cellular membrane. Seki et al. (1978) obtained similar results when they studied the activity of δ-endotoxin against mouse sarcoma 180 ascites cells. Their cell system also lost UV-absorbent compounds and phospholipids during incubation with δ-endotoxin. Their control consisted of trypsin-treated midgut cells of *B. mori* which reacted in the same way to the presence of δ-endotoxin. The activities of the δ-endotoxin against tumor cells have to be reconfirmed and work in this direction has to be intensified before any conclusions can be drawn. It is questionable whether tumor cells can be regarded as a representative type of vertebrate cells for assays with membrane toxins. It is known that membranes of tumor cells are modified in their structure, their composition and their function. It is therefore not surprising that *in vitro* activated toxin interacts with the permeability of cell types which have lost their original identity. It would be of course very interesting from the medical point of view if such abnormal cells respond to the δ-endotoxin.

In addition, Prasad and Shetna (1976) used for *in vitro* experiments concentrations of δ-endotoxin which were at least twenty times higher (250 to 1,000 μg/ml) than the amounts which were applied by Geiser (1979) for the experiments with the *C. fumiferana* cell line (10-20 μg per ml). The corresponding dose of δ-endotoxin which is needed for the destruction of the gut epithelium of L5 *P. brassicae* is one hundred times lower again than the concentrations used by Geiser against insect tissue culture cells. Geiser (1979) investigated the influence of the δ-endotoxin upon chicken fibroblasts. He was not able to demonstrate any morphological alterations of the culture or inhibition in the development, despite the high concentration of 67 μg/ml.

SEROTYPE 14; BACILLUS THURINGIENSIS VAR. ISRAELENSIS

The parasporal inclusions formed during the sporulation process are responsible for the insecticidal activity of B. thuringiensis var. israelensis. This can be demonstrated with purified parasporal bodies, using the separation method developed by Delafield et al. (1968). In our laboratory the minimal lethal dose to A. aegypti in the second instar was 2.5 ng of crystal protein/ml. The low amount of parasporal protein which has to be present in the water, compares favorably with chemical insecticides.

Although the symptoms induced by the δ-endotoxin of B. thuringiensis var. israelensis are difficult to compare with the classical poisoning caused by B. thuringiensis in lepidopterous larvae, there seem to be some similarities. The first symptoms appear rapidly, within minutes, if high doses are applied. In our laboratory L_2 of A. aegypti were exposed to a spore/crystal suspension containing 10^7 spores/ml for different periods of time. An exposure time of 5 min caused 100% mortality within 21 min. An exposure time of 1 min led to 80% mortality within 3 h and 100% within 24 h. The LD_{50} was reached with a few hundred spores/ml. The symptoms are manifested by a slowing of movements with prolonged resting periods. The larvae become unable to control the direction of their movements. Paralysed and dead larvae accumulate at the water surface.

De Barjac (1978b) presented evidence in a histopathological study that the gut epithelium is the primary target. When sections of larvae which were exposed to the toxin were examined in the light microscope, the gut epithelia showed the typical swelling followed by a general destruction of this tissue.

No reports have yet been published on experiments on the histopathology and mode of action in vitro. It would be of interest to know if mosquito cell lines respond to the δ-endotoxin of B. thuringiensis var. israelensis. This would be a convenient system to compare the histopathology and molecular mode of action.

Serotype 14 has been classified as a variety of B. thuringiensis, based on the presence of a parasporal body with insecticidal activity and based on far-reaching similarities in biochemical properties. There are however important differences between the other serotypes and this new one, opening a considerable systematic gap. The spectrum of activity does not overlap at all; serotype 14 does not share crystal antigens with the other serotypes, the parasporal inclusions are irregular in shape and if they are present in a crystalline form, their lattice is unique (see Chapter 8). On the other hand, what is known of the mode of action seems to be the same as for the other serotypes.

CONCLUSIONS

Histopathological results and the few investigations on the level of the molecular mode of action indicate that the membrane is the primary target for δ-endotoxin. The breakdown of permeability control by the cell membrane induces general autolysis within the cell which is expressed, for example, by the vacuolization of the endoplasmic reticulum or the disintegration of the mitochondria. The increase in the number of lytic vacuoles seems to indicate that part of the events occurring within the cell is caused by autolysis. In the healthy gut epithelium, a small proportion of the cells are always in a state of autolysis, due to the continuous renewal process. These events are very similar to the histopathological changes produced by δ-endotoxin. Stimulation of the production of lytic vacuoles can be particularly well demonstrated if the gut epithelium is exposed to sublethal doses (Ebersold, unpublished).

The nature of the first contact between the δ-endotoxin and the cell membranes (which must take place following degradation of the crystal and activation of the protein by gut juice proteases) is open for discussion and speculation. The specificity of the δ-endotoxin suggests that receptors or other components of the cell membrane are involved which are exclusive for the target organisms. Receptors are known to play a key role, for example, in the mode of action of cholera and diphtheria toxins. In both cases, however, a lag-phase is found between binding of the B-fraction of the molecule to the receptor and the introduction of the A-fraction into the cell. Histopathological changes do not occur during the early phases of the action of cholera and diphtheria toxins.

The absence of a lag phase and the rapid destruction of affected cells justifies an assignment of the δ-endotoxin to the group of the cytolytic toxins (Bernheimer, 1968). Cytolytic toxins interact directly with the membranes. According to their mode of action the cytolytic toxins can be divided into three groups (Alouf, 1977):

i) General surfactant activity involving reaction with bilayer lipids or proteins and solubilization of these components (e.g. δ-toxin of *Staphylococcus aureus*)

ii) Enzymatic activity involving the hydrolysis of lipids or proteins (β-toxins of *Staphylococcus aureus* or α-toxin of *Clostridium perfringens*.

iii) Binding to specific molecules of the bilayer causing a transient or permanent rearrangement of the membrane framework (streptolysin O produced by *Streptococcus pyogenes* and leukocidin produced by *Staphylococcus aureus*).

The high degree of specificity of the δ-endotoxin of *B. thuringiensis* probably excludes a detergent-like mode of action. The δ-toxin of *S. aureus*, which acts in this way, is rather unspecific and affects many different cell types. The second possibility, that the δ-endotoxin would act as an enzyme, is still open and has to be considered in future investigations. Bulla et al. (1977) detected proteolytic activity associated with the crystal protein. The authors suggested that the proteases were responsible for the activation of the δ-endotoxin. In contrast, Chestukhina et al. (1978) regarded these proteases as impurities which are embedded during the synthesis of the crystals. If the δ-endotoxin happens to be an enzyme, a protease or a lipase, it must be highly specific since common substrates are not attacked. The low proteolytic and lipolytic activities which are found associated with the crystal protein cannot account for the mode of action at the level of the target cells (Huber, pers. comm.).

The behaviour of the δ-endotoxin can best be explained by the third proposed mode of action for cytolytic toxins. The binding of the toxic moiety to a component of the cell membrane would explain the high specificity. The rapid destruction of osmotic barriers has been demonstrated and the physical changes of the membrane structure induced by the δ-endotoxin have been clearly shown. Further evidence has been presented by Griego et al. (1979) that active potassium transport from the hemolymph side to the lumen side of the brush border membrane in isolated midguts of *Manduca sexta* was blocked by δ-endotoxin.

Future research with the δ-endotoxin of *B. thuringiensis* will have to emphasize the molecular mode of action. Histopathology can be regarded as substantially finished. Experiments with brush border membranes of susceptible insects would be a suitable basis to proceed in the elucidation of the mode of action. Brush border vesicles can be used for influx and efflux measurements, for binding studies and for the determination of changes in the membrane composition. Insect cell cultures exhibiting susceptibility to the δ-endotoxin will also be of importance for future investigations. They will allow determination of whether permeability of the cell membrane increases gradually or changes abruptly. A further tool in the investigation of the δ-endotoxin are artificial membranes, for example liposomes, in which interactions between membranous lipid components and the δ-endotoxin could possibly be demonstrated. Liposome preparations are easily obtained following lipid extraction of different types of cell membranes.

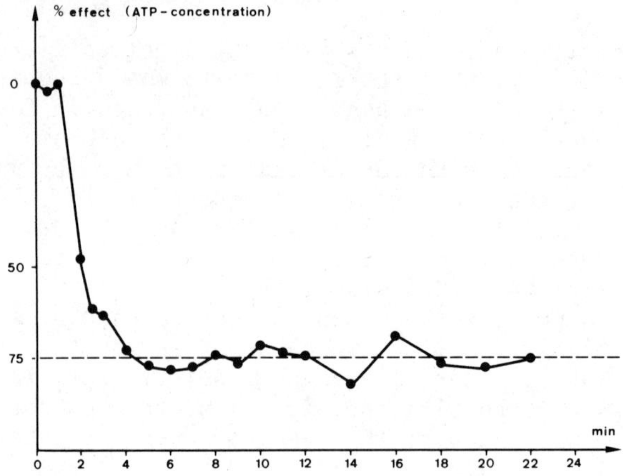

Figure 9-16. The loss of ATP after addition of δ-endotoxin (var. kurstaki, 50 μg/ml) to a Cf-1 culture. The total ATP-reduction does not exceed 75%, indicating that a proportion of the cells does not respond.

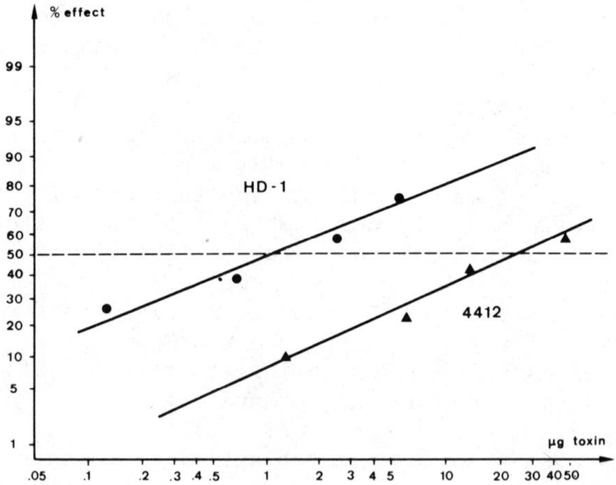

Figure 9-17. Comparison of the susceptibility of the Cf-1 cells against two varieties of B. thuringiensis, namely kurstaki (HD-1) and thuringiensis (4412). Incubation for 30 min at 23°C. The effect is expressed as the loss of ATP referred to the susceptible portion of the cells.

ACKNOWLEDGMENTS

The authors express their gratitude to Prof. Dr. L. Ettlinger for his helpful advice and criticism and to Franziska Klausener for her technical assistance. Our Bacillus thuringiensis research project is supported by the Swiss National Science Foundation, Project Nr. 3.357-0.78.

REFERENCES

Alouf, J. E. 1977. Cell membrane and cytolytic bacterial toxins. In The Specificity and Action of Animal, Bacterial and Plant Toxins, P. Cuatrecasas, ed., Chapmann and Hall, London. pp. 220-270.

Angus, T. A. 1968. Similarity of effect of valinomycin and Bacillus thuringiensis parasporal protein in larvae of Bombyx mori. J. Invertebr. Pathol., 11, 145-146.

Angus, T. A. 1970. Implications of some recent studies of Bacillus thuringiensis - A personal purview. Proc. IV Int. Coll. Insect Pathol., College Park, Maryland. USA.

de Barjac, H. 1978a. Une nouvelle variété de Bacillus thuringiensis très toxique pour les Moustiques: Bacillus thuringiensis var. israelensis sérotype 14. C. R. Acad. Sc. Paris, Série D, 286, 797-800.

de Barjac, H. 1978b. Etude cytologique de l'action de Bacillus thuringiensis var. israelensis sur larvae de moustiques. C. R. Acad. Sc. Paris, Serie D, 286, 1629-1632.

de Barjac, H. and Bonnefoi, A. 1973. Mise au point sur la classification des Bacillus thuringiensis. Entomophaga, 18, 5-17.

de Barjac, H. and Coz, J. 1979. Sensibilite compareé de six especes differentes de moustiques a Bacillus thuringiensis var. israelensis. Bull. Wld. Hlth. Org., 57, 139-141.

Benz, G. 1966. Die Pufferkapazitat von Blut und Verdauungssaft von Pieris brassicae und der Einfluss von Anoxie und Bacillus thuringiensis auf die Permeabilitat der Darmwand. J. Insect Physiol., 12, 137-151.

Bernheimer, A. W. 1968. Cytolytic toxins of bacterial origin. Science, 159, 847-851.

Bulla, L. A. Jr., Kramer, K. J. and Davidson, L. I. 1977. Characterization of the entomocidal parasporal crystal of Bacillus thuringiensis. J. Bact., 130, 375-383.

Burgerjon, A. 1962. Relation entre l'intoxication provoquee par Bacillus thuringiensis Berliner et la consommation chez Pieris brassicae. L. Ann. Epiph., 13, 59-72.

Burgerjon, A. and Martouret, D. 1971. Determination and significance of the host spectrum of *Bacillus thuringiensis*. *In* Microbial Control of Insects and Mites, H. D. Burges and N. W. Hussey, eds., Academic Press, London. pp. 305-322.

Chestukhina, G. G., Zalunin, I. A., Kostina, L. I., Kotova, T. S., Katrukha, S. P., Kusnetsov, Y. S. and Stepanov, V. M. 1978. Proteinase bound to crystals of *Bacillus thuringiensis*. Biokhimiya, 43, 857-864.

Cooksey, K. E., Donninger, C., Norris, J. R. and Shankland, D. 1969. Nerve-blocking effect of *Bacillus thuringiensis* protein toxin. J. Invertebr. Pathol., 13, 461-462.

Delafield, F. P., Sommerville, H. J. and Rittenberg, S. C. 1968. Immunological homology between crystal and spore protein of *Bacillus thuringiensis*. J. Bact., 96, 713-720.

Delucca, A. J., Simonson, J. and Larson, A. 1979. Two new serovars of *Bacillus thuringiensis*: Serovars *dakota* and *indiana* (serovars 15 and 16). J. Invertebr. Pathol. 34, 323-324.

Ebersold, H. R., Luethy, P. and Mueller, M. 1977. Changes in the fine structure of the gut epithelium of *Pieris brassicae* induced by the δ-endotoxin of *Bacillus thuringiensis*. Bull. Soc. Ent. Suisse, 50, 269-276.

Ebersold, H. R., Luethy, P., Geiser, P. and Ettlinger, L. 1978. The action of the δ-endotoxin of *Bacillus thuringiensis*: An electron microscope study. Experientia, 34, 1672.

Ebersold, H. R., Luethy, P. and Huber, H. E. 1980. Membrane damaging effect of the δ-endotoxin of *Bacillus thuringiensis*. Experientia, 36, 495.

Endo, Y. and Nishiitsutsuji-Uwo, J. 1980. Mode of action of *Bacillus thuringiensis* δ-endotoxin: Histopathological changes in the silkworm midgut. J. Invertebr. Pathol., 36, 90-103.

Estes, Z. E. and Faust, R. M. 1964. Studies on the mucopolysaccharides of the greater wax moth, *Galleria mellonella* (Linnaeus). Comp. Biochem. Physiol. 13, 443-452.

Fast, P. G. and Donaghue, I. P. 1971. The δ-endotoxin of *Bacillus thuringiensis*. II. On the mode of action. J. Invertebr. Pathol., 18, 135-138.

Fast, P. G. and Morrison, I. K. 1972. The δ-endotoxin of *Bacillus thuringiensis*. IV. The effect of δ-endotoxin on ion regulation by midgut tissue of *Bombyx mori* larvae. J. Invertebr. Pathol., 20, 208-211.

Fast, P. G. and Videnova, E. 1974. The δ-endotoxin of *Bacillus thuringiensis*. V. On the occurrence of the endotoxin fragments in hemolymph. J. Invertebr. Pathol., 23, 280-284.

Fast, P. G., Sohi, S. S. and Murphy, D. W. 1978. On Bacillus thuringiensis δ-endotoxin: Evidence that the toxin acts at the surface of susceptible cells. Experientia, 34, 762-763.

Geiser, P. 1979. Versuche in vitro zum Nachweis einer Wirkung des δ-endotoxins von Bacillus thuringiensis. Diss. ETH Zurich, Switzerland, Nr. 6411.

Goldberg, L. and Margalit, J. 1977. Bacterial spore demonstrating rapid larvicidal activity against Anopheles sergentii, Uranotaenia unguiculata, Culex univittatus, Aedes aegypti and Culex pipiens. Mosquito News, 37, 355-458.

Griego, V. M., Moffet, D. and Spence, K. D. 1979. Inhibition of active K^+ transport in the tobacco hornworm Manduca sexta midgut after ingestion of Bacillus thuringiensis endotoxin. J. Insect Physiol., 25, 283-288.

Guillet, P. and Escaffre, H. 1979. Evaluation de Bacillus thuringiensis de Barjac pour la lutte contre les larves de Simulium damnosum s. l. II. Efficacité comparée de trois formulations expérimentales. WHO/VBC/79.730, Geneva, Switzerland.

Hall, I. M., Arakawa, K. Y., Dulmage, H. T. and Correa, J. A. 1977. The pathogenicity of strains of Bacillus thuringiensis to larvae of Aedes and to Culex mosquitoes. Mosquito News, 37, 246-251.

Heimpel, A. M. and Angus, T. A. 1959. The site of action of crystalliferous bacteria in Lepidoptera larvae. J. Insect Pathol., 1, 152-170.

Heimpel, A. M. and Angus, T. A. 1960. Bacterial insecticides. Bacterial. Rev., 24, 266-288.

Lecadet, M. M. 1970. Bacillus thuringiensis toxins - the proteinaceous crystal. In Microbial Toxins, T. C. Montie, S. Kadis and S. J. Ajl, eds., Vol. III, pp. 437-471. Academic Press, N.Y.

Lecadet, M. M. and Martouret, D. 1967. Hydrolyse enzymatic de l'inclusion parasporale de Bacillus thuringiensis par les protéases de Bombyx mori L. C. R. Acad. Sc. Paris, Série D, 265, 1543-1546.

Luethy, P. 1975. Zur bakteriologischen Schädlingsbekampfung: Die entomopathogenen Bacillus - Arten, Bacillus thuringiensis und Bacillus popilliae. Vjschr. naturf. Ges. Zurich, 120, 81-163.

Luethy, P., Huber, H., Ebersold, H. R. and Geiser, P. 1978. New aspects of the mode of action of the δ-endotoxin of Bacillus thuringiensis. Proceed. Int'l. Colloq. Invertebr. Pathol., Prague, pp. 115-116.

Luethy, P., Raboud, G., Delucchi, V. and Kuenzi, M. 1980. Field efficacy of Bacillus thuringiensis var. israelensis. Bull. Soc. Ent. Suisse, 53, 3-9.

Murphy, D. W., Sohi, S. S. and Fast, P. G. 1976. Bacillus thuringiensis enzyme digested δ-endotoxin: Effect on cultured insect cells. Science, 194, 954-956.

Murphy, J. R. 1976. Structure activity relationships of diphteria toxin. In Mechanisms in Bacterial Toxinology, A. W. Bernheimer, ed., pp. 31-51, John Wiley and Sons, N.Y.

Nishiitsutsuji-Uwo, J., Endo, Y. and Himeno, M. 1980. Effects of Bacillus thuringiensis δ-endotoxin on insect and mammalian cells in vitro. Appl. Ent. Zool. 15, 133-139.

Prasad, S. S. V. and Shetna, Y. I. 1974. Purification, crystallization and partial characterization of the antitumor and insecticidal protein subunit from the toxin of Bacillus thuringiensis var. thuringiensis. Biochim. Biophys. Acta, 363, 568-570.

Prasad, S. S. V. and Shetna, Y. I. 1976. Mode of action of a purified antitumor protein from the proteinaceous crystal of Bacillus thuringiensis subsp. thuringiensis on Yoshida Ascites sarcoma cells. Antimicrob. Agents Chemother., 10, 293-298.

Schnetter, W., Engler, S., Morawcsik, J. and Becker, N. 1981. Efficacy of Bacillus thuringiensis var. israelensis against mosquito larvae and non-target organisms. Mitt. Dtsch. Ges. Allg. Angew. Ent., in press.

Seki, T., Nagamatsu, M., Nagamatsu, Y., Tsutsui, R., Ichimaru, T., Watanabe, T., Koga, K. and Hayashi, K. 1978. Injuring reaction of δ-endotoxin upon sarcoma 180 ascites cells and silkworm midgut cell in vitro. Sci. Bull. Fac. Agr., Kyushu Univ., 38, 19-24.

Singrè, G., Gaven, B. and Jullien, J. L. 1979. Sécurité d'emploi du sérotype H-14 de Bacillus thuringiensis pour la faune non-cible des gîtes à moustiques du littoral méditerranéen francais. WHO/VBC/79.742, Geneva, Switzerland.

Sohi, S. S. 1971. In vitro cultivation of larval tissue of Choristoneura fumiferana. Proc. III Int. Coll. Invertebr. Tissue Culture, Bratislava, CSSR.

Staehelin, L. A. and Hull, B. E. 1978. Junctions between living cells. Scientific American, 238, 140-153.

Strehler, B. L. 1968. Bioluminicence assay: Principles and practice. Methods Biochem. Analysis, 16, 99-180.

Thelestam, M. and Moellby, R. 1975a. Determination of toxin induced leakage at different size nucleotides through the plasma membrane of human diploid fibroblasts. Infect. Immun., 11, 640-648.

Thelestam, M. and Moellby, R. 1975b. Sensitive assay for detection of toxin induced damage to the cytoplasmic membrane of human diploid fibroblasts. Infect. Immun., 12, 225-232.

Travers, R. S., Faust, R. M. and Reichelderfer, C. F. 1976. Effects of _Bacillus thuringiensis_ var. _kurstaki_ δ-endotoxin on isolated Lepidopteran mitochondria. J. Invertebr. Pathol., _28_, 351-356.

Weiser, J. and Vankova, J. 1978. Toxicity of _Bacillus thuringiensis israelensis_ for blackflies and other freshwater invertebrates. Proc. Int'l. Colloq. Invertebr. Pathol., Prague, pp. 243-244.

Chapter 10

BACTERIAL DISEASES OF INSECTS CAUSED BY TOXIN-PRODUCING BACILLI OTHER THAN *BACILLUS THURINGIENSIS*

by Elizabeth W. Davidson*

INTRODUCTION

A wide variety of rod-shaped, sporeforming bacteria are associated with insect disease. Some of these produce one or more toxins which are of primary importance in their pathology. These pathogens all have in common the failure to produce hemocoelic septicemia until near or after death of the host. In most cases, the host exhibits symptoms of intoxication while the bacteria are confined within the peritrophic membrane of the midgut. Some of these bacilli carry their toxins with them when ingested, having produced them elsewhere (a "food poisoning" in Lysenko's terminology; see Chapter 6), while others produce the toxins *de novo* while multiplying in the host gut. Among these bacteria, the best studied species is the commercially-important *Bacillus thuringiensis*; other toxin producing bacilli include the aerobes *Bacillus cereus*, *B. sphaericus*, and *B. alvei*, and the anaerobes *Clostridium brevifaciens* and *Cl. malacosomae*.

BACILLUS CEREUS FRANKLAND AND FRANKLAND

Most readers are aware of the controversy concerning the taxonomic position of entomocidal bacterial strains related to *Bacillus cereus*. Many microbial taxonomists (e.g. Gordon et al., 1978; Seki et al., 1978) place the crystalliferous entomocidal strains in a variety of *B. cereus*, "*B. cereus* var. *thuringiensis*", while insect pathologists have traditionally referred to the crystalliferous entomopathogens as a separate species, "*B. thuringiensis*" (e.g. Heimpel and Angus, 1958): the 8th edition of Bergey's Manual (Buchanan

*Department of Zoology, Arizona State University, Tempe, AZ 85281

PATHOGENESIS OF INVERTEBRATE MICROBIAL DISEASES. Davidson, ed. © Allanheld, Osmun (Totowa, N.J.)

and Gibbons, 1974) likewise places B. thuringiensis in a separate species. In this book, the latter terminology is used. The pathogenesis of B. thuringiensis toxins is discussed in Chapters 8, 9, and 11. Some non-crystalliferous strains of B. cereus which produce an apparently different pathology from B. thuringiensis are also known (reviewed by Heimpel and Angus, 1963), and these strains are discussed here.

The symptoms of B. cereus infection vary with the host and the method of introduction. Codling moth, Laspeyresia (Carpocapsa) pomonella, larvae injected with B. cereus spores become immobile and dark, and die within 24-48 hours. Dead larvae are black, soft, and shrunken. Codling moth larvae fed spores become sluggish by 24-72 hours, and brown spots appear on the integument. The brown color then spreads to cover the whole integument. Dead larvae are soft and flaccid, internal organs are broken down, and viscous fluid seeps from the body wall. This fluid yields almost pure cultures of B. cereus. The virulence of B. cereus to codling moth larvae is increased by insect passage (Stephens, 1952).

When B. cereus spores are fed to larvae of the larch sawfly, Pristiphora ehrichsonii, diarrhea and vomiting occur within 10-18 hours, and larvae cease feeding; infected larvae appear normal except for slight yellowing of the ventral body in some cases. When spores are injected into sawfly larvae, the body changes color, and larvae become shrunken and flaccid within 8-10 hours. Histological sections of larvae fed spores show that tissue damage preceeds bacterial septicemia in the hemocoel. The bacteria multiply within the gut, whereupon the peritrophic membrane disappears, midgut nuclei become pycnotic, and basophylic granules accumulate in the distal ends of the midgut cells. Abdominal nerve ganglia, salivary glands, and muscles may also be altered (Heimpel, 1955a).

Pathogenicity of B. cereus strains is strongly correlated with the ability of the strains to produce lecithinase. Injection of purified phospholipase from B. cereus causes mortality of larch sawfly larvae, and the pH of the larval gut is within the pH range of lecithinase activity. Resistant host species have more alkaline midgut contents (Heimpel 1955a, b; Heimpel and Angus, 1956). From this evidence it was concluded that lecithinase produced by B. cereus growing in the larval gut is the primary cause of the observed symptoms. Lysenko (Chapter 6) discusses the isolation and properties of phospholipase from B. cereus in greater detail.

BACILLUS SPHAERICUS NEIDE

The first mosquito larvicidal strain of Bacillus sphaericus was isolated from Culiseta incidens larvae collected in

California. This strain, later designated the "K" strain, was infective for larvae of three species of Culex, five species of Aedes, and Orthopodomyia spp., as well as for the original host (Kellen et al., 1965), and also killed house flies (Rogoff et al., 1969). The activity of the K strain however, was not sufficient to encourage further interest. In 1973 Singer reported the isolation of B. sphaericus strain SSII-1 from Culex fatigans larvae collected in Delhi, India by employees of the World Health Organization (WHO). Strain SSII-1 was 10,000 times as insecticidal as the K strain to C. quinquefasciatus[1] (= C. pipiens quinquefasciatus; C. pipiens fatigans; C. fatigans) larvae (Singer, 1973). Isolation of SSII-1 stimulated renewed interest in B. sphaericus, and within a few years Singer isolated several more strains from WHO collections; each of these strains bears its WHO accession number. Strains of B. sphaericus have also been isolated by other workers from mosquitoes in Japan and Sri Lanka (Table 10-1).

As shown in Table 10-1, insecticidal strains of B. sphaericus have world-wide distribution. The species itself is cosmopolitan, but only strains isolated from mosquito larvae are larvicidal (Krych et al., 1980). Krych et al. (1980) have shown that the mosquito-pathogenic strains have a high degree of DNA/DNA homology to each other, but less than 50% homology to most of the noninsecticidal B. sphaericus strains, and only 19% homology to the type strain. DNA/DNA homology suggests that the insecticidal strains may eventually be placed in a new species.

When seven of the insecticidal strains were phage-typed and serotyped, three groups were formed: 1. SSII-1 and 1404; 2. 1593, 1881, and 1691; and 3. Kellen K and Q. In addition, strain M R-4 did not fall into any of these three groups. (Yousten, A. A. et al., in press).

The insecticidal strains of B. sphaericus exhibit a varied spectrum of activity toward larvae of the three major genera of mosquitoes, Culex, Anopheles, and Aedes (Table 10-1). Among the better-studied strains, strain 1593 shows the best activity toward all tested species and is also the most stable (Myers et al., 1979). For these reasons, strain 1593 is the current strain of choice for commercialization. Other, more recently isolated strains may also be commercially useful (Singer, 1980; Wickeremsinghe and Mendis, 1980).

The pathology of three strains, K, SSII-1, and 1593, has been studied in some depth. Because these three strains are different in phage susceptibility, serotype, and in some aspects of their pathology, they will be discussed individually.

[1]Common Names of Insects and Related Organisms, 1978 Revision. Entomological Society of America.

Table 10-1

ORIGIN AND INSECTICIDAL ACTIVITY OF *BACILLUS SPHAERICUS* INSECTICIDAL STRAINS

Strain	Original Host	Country of Origin	*Culex* species	LC_{50}[1]
K	*Culiseta incidens*	USA	quinquefasciatus	10^6-10^7
Q	derived from K		quinquefasciatus	10^4
SSII-1 (1321)	*Culex fatigans*	India	quinquefasciatus tarsalis nigripalpus	10^2-10^3 10^2-10^3 10^3
1593	*Culex fatigans*	Indonesia	quinquefasciatus tarsalis nigripalpus	10^1-10^2 10^1 10^2
1404	*Culex pipiens quinquefasciatus*	Phillippines	quinquefasciatus nigripalpus	10^2-10^3 10^4
1537	*Anopheles gambiae*	Nigeria	quinquefasciatus	inactive
1595	*Anopheles gambiae*	Nigeria	quinquefasciatus	active
1691	*Anopheles albimanus*	El Salvador	quinquefasciatus	10^2
1881	*Anopheles albimanus*	El Salvador		
1883 through 1895	?	Israel	quinquefasciatus	10^3-10^5
2013	*Culex* spp.	Roumania	quinquefasciatus	10^2-10^3
MR-4 (2297)	*Culex quinquefasciatus*	Sri Lanka	quinquefasciatus	10^1-10^2
SC 1713	*Culex pipiens*	Japan	pipiens	10^4-10^6

[1]Approximate LC_{50} in bacterial cells/ml

Table 10-1 (cont.)

Strain	*Anopheles* species	*Anopheles* LC_{50}[1]	*Aedes* species	*Aedes* LC_{50}[1]	Reference
K			aegypti	10^8	Kellen, 1965; Singer, 1973, 1975a, b
Q			aegypti	10^7	Singer, 1973, 1975a, b
SSII-1 (1321)	gambiae albimanus	10^5-10^6 10^6-10^7	aegypti taeniorynchus	10^4-10^6 10^8	Singer, 1973, 1974, 1975a, b 1980; Ramoska et al., 1977
1593	gambiae albimanus	10^5 10^6	aegypti nigromaculis taeniorynchus	10^3-10^4 10^2 10^7	Singer, 1975b, 1980; Myers et al., 1979; Mulligan et al., 1978; Ramoska, et al., 1977
1404	gambiae albimanus	10^8 10^6-10^7	aegypti taeniorynchus	10^3-10^4 10^8	Singer, 1975b, 1980; Ramoska et al., 1977
1537					Singer, 1975b
1595					Singer, 1975b
1691 1881					Singer, 1980; Davidson and Myers, *in press*
1883 through 1895					Singer, 1980
2013					Singer, 1980, Davidson and Myers, *in press*
MR-4 (2297)					Wickremsinghe and Mendis, 1980; Davidson and Myers, *in press*
SC 1713					Takarazuka, 1980

[1]Approximate LC_{50} in bacterial cells/ml

K:

Mosquito larvae infected with the K strain lose normal turgor and mobility within three days, becoming sluggish and turning brown. Death often requires seven days; dead larvae characteristically hang from the surface film. There is progressive degeneration of the posterior midgut, which is filled with vegetative *B. sphaericus* cells. Midgut cells become vacuolated and pycnotic, and are eventually sloughed into the lumen, permitting bacterial access to the hemocoel when the host is near death (Kellen et al., 1965). These authors concluded that this bacterium is a facultative parasite which must dominate the gut flora in order to kill the larva. Singer et al. (1966) felt that the effectiveness of the K strain depended on infection rather than on a toxin.

SSII-1:

Insecticidal activity in strain SSII-1 cultures increases sharply at the onset of sporulation, peaks midway through the sporulation cycle, and then drops off to 1/3 to 1/2 the maximum (Singer, 1973, 1974, 1975a). Synthetic medium produces cultures which are one to two logs more active than those produced on complex media (Singer, 1974). Singer (1975a) postulated that activity was produced during secondary metabolism, prior to mature spore formation. Activity was shown to be due to toxic material which is closely associated with the bacterial cell, and not to toxin secreted into the medium, to a crystal, or to invasion of larval gut cells or hemocoel (Davidson and Singer, 1973; Singer, 1974, 1975a; Davidson, 1979).

Myers and Yousten (1977, 1978a) verified that SSII-1 infection involves a toxin, as chloroform- or ultraviolet-killed cells retain their insecticidal activity. By the use of carefully synchronized cultures, Myers and Yousten showed that SSII-1 cells of all ages are equally insecticidal whether grown on complex or defined medium, with or without Mn+, in sporogenous or oligosporogenous condition. Since SSII-1 spores are less insecticidal than sporulating cells, activity drops as the culture completes sporulation. Myers and Yousten also confirmed by electron microscopy that there is no parasporal body of the sort found in *B. thuringiensis* in sporulating SSII-1 cells.

The gross symptoms of SSII-1 infections in highly susceptible *C. quinquefasciatus* larvae are similar to those of the K strain, but occur much more quickly. Thirty minutes to one hour after ingestion of a large dose, the midgut swells against the body wall, allowing the peritrophic membrane to fall into zigzag folds. Larvae undergo shuddering, tremors, and many fail to moult. Death occurs as early as 12 hours; all mortality is expressed by 48 to 72 hours (Singer, 1974;

Davidson et al., 1975; Davidson, 1975, 1979; Ramoska et al., 1977).

Numbers of viable B. sphaericus and normal bacterial flora in C. quinquefaciatus larvae were determined during the course of SSII-1 infections. Bacillus sphaericus numbers dropped dramatically from the approximately 10^5 cells per larva ingested to only two cells per larva by six hours, rising again at the onset of mortality at 12 to 24 hours. Normal flora numbers similarly rose at the onset of mortality (Figure 10-1). When chloroform-treated B. sphaericus cells were fed, no viable B. sphaericus cells were found in the larvae, though 98% mortality was produced (Figure 10-2) (Davidson et al., 1975). The authors concluded that B. sphaericus numbers dropped rapidly due to digestion during the first hour, and later due to defecation. They proposed that digestion is the toxin-releasing mechanism. Multiplication of B. sphaericus in the larval gut is not necessary in the pathogenesis, since larvae are killed by chloroform- or UV-killed cells. The role of increasing normal flora numbers was not determined by Davidson et al. (1975). However, when larvae were separated into groups of live, moribund, and dead larvae 24 hours after feeding, only dead larvae contained elevated numbers of normal flora and B. sphaericus, indicating that multiplication of normal flora and B. sphaericus is a result of, not a cause of larval death (Davidson, 1975). Replication of B. sphaericus in the assay cups is also not important, as LC_{50} values are not changed by the addition of inhibitory levels of antibiotics (Myers and Yousten, 1978a). When B. sphaericus numbers in Ae. aegypti larvae were similarly determined, it was found that these more resistant larvae ingest and digest B. sphaericus cells just as Culex larvae. Resistance in Ae. aegypti is therefore due to some other factor (Davidson, 1975).

An ultrastructural study of larvae fixed at various times after feeding on SSII-1 vegetative cells confirmed that these cells are digested very rapidly (Davidson, 1975, 1979). The cell wall structure of B. sphaericus, B. polymyxa, and a few related species is unusually complicated for gram-positive organisms. By electron microscopy, the cell wall is seen to consist of an outer, densely-staining protein layer, an intermediate unstained layer, and an inner densely-stained peptidoglycan layer (Holt et al., 1975; Myers and Yousten, 1978b) (Figure 10-3). The outer cell wall layer is tetragonally arranged in many B. sphaericus strains, and has been called the "T layer" (Howard and Tipper, 1973). However, in all insecticidal strains and in 9 of 31 examined nontoxic strains, the outer cell wall layer is not tetragonally arranged (Myers and Yousten, 1978b; Myers, 1979).

Bacillus sphaericus SSII-1 cells begin to lose integrity immediately upon entering the anterior midgut (Figure 10-4).

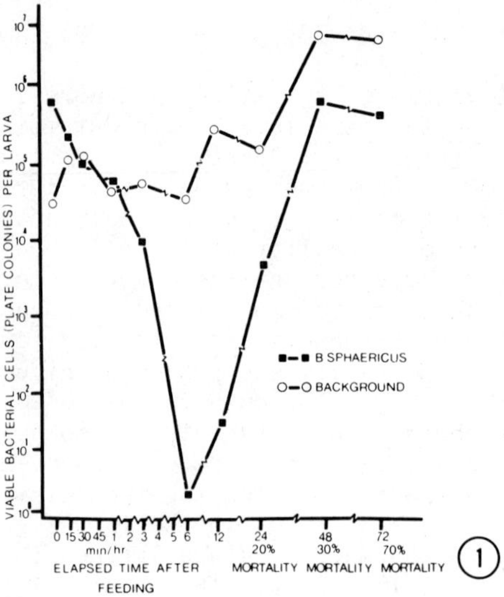

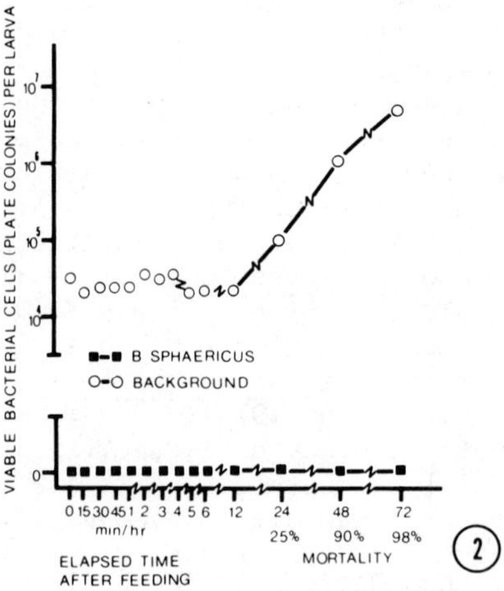

Figure 10-1. Bacterial populations in second-instar Culex pipiens quinquefasciatus larvae fed B. sphaericus/SSII-1 final whole cultures at 1/100 dilution for 10 min.

Figure 10-2. Bacterial populations in second-instar Culex pipiens quinquefasciatus larvae fed chloroform-treated B. sphaericus/SSII-1 final whole culture at 1/100 dilution for 10 min.

Figures 10-1 and 10-2, from Davidson et al., 1975. Reprinted by permission of Academic Press, Inc.

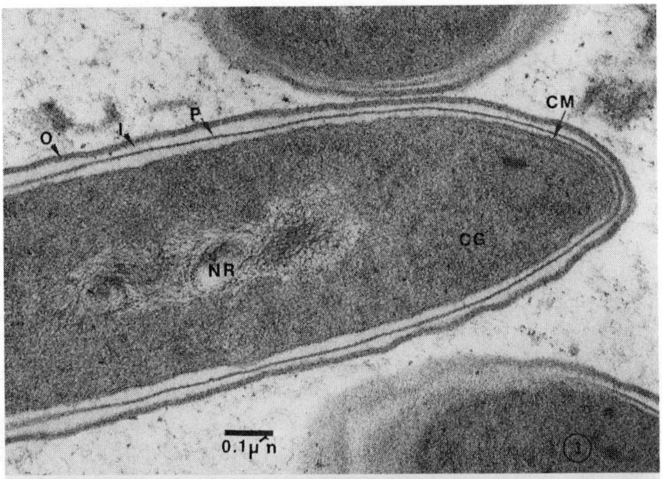

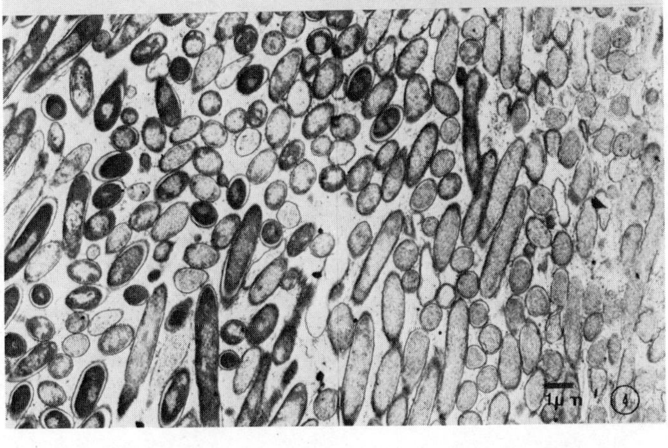

Figure 10-3. *Bacillus sphaericus* SSII-1 vegetative cell with outer cell wall layer (O), unstained intermediate cell wall layer (I), inner peptidoglycan cell wall layer (P), cytoplasmic membrane (CM), nuclear region (NR), and dense cytoplasmic ground substance (CG).

Figure 10-4. Lumen of anterior midgut, demonstrating rapid degradation of *B. sphaericus* SSII-1 vegetative cells immediately after feeding. Stomodeal valve to the left.

The outer cell wall layer disappears first, followed by loss of cytoplasmic ground substance and breakage of some cell walls (Figure 10-5). In the central midgut of larvae fed 30 minutes earlier, empty, broken cell walls containing ruffled cell membranes are found (Figure 10-6). It was again shown that defecation at about 1 hour removes nearly all bacteria from the gut. Apparently B. sphaericus and normal flora numbers increase during the next six hours by multiplication of the small number of cells which escape both digestion and defecation.

In examining tissues of infected larvae, Davidson (1979) found that only midgut cells show visible changes until near death of the larva. Midgut swelling is manifested ultrastructurally by separation of midgut cells from one another at their bases; only the tight junctions near the lumen surface remain closed. By 10 to 12 hours after ingestion of B. sphaericus SSII-1 cytolysosomes in midgut cells are markedly larger and more numerous than in controls (Figure 10-7, 10-8). Sloughing of posterior midgut cells is frequently seen, though bacteria do not breach the peritrophic membrane or enter the hemocoel.

The anterior midgut of C. pipiens larvae has a high pH (9 to 10), and is the site of most proteolytic activity (Dadd, 1975). The posterior midgut is the site of amylase activity (Dadd, 1975), and absorption of amino acids and sugars (Wigglesworth, 1942). Release of the toxin in the anterior midgut and absorption and subsequent toxic effects in the posterior midgut agree with the hypothesis (Davidson et al., 1975) that the toxin is proteinaceous or polypeptide in nature.

As activity in SSII-1 is greatly decreased by breaking the cells sonically, by French pressure cell, or by freeze-thawing, the location of insecticidal activity in strain SSII-1 is not known (Myers and Yousten, 1978a). In contrast, activity in strain 1593 is stable to cell breakage, and successful attempts to fractionate cells have utilized this strain.

1593:

The symptoms of strain 1593 infections are very similar to those of SSII-1; however death may occur as early as four hours. Strain 1593 is more insecticidal than strain SSII-1 to most species of mosquitoes (Table 10-1), and its toxin is more stable than is SSII-1 toxin. Also in contrast to strain SSII-1, toxin production in strain 1593 is a sporulation event, and spores are highly insecticidal (Myers et al., 1979). These characteristics, along with phage susceptibility and serotype (see above), indicate that strains SSII-1 and 1593 are quite different, and that their toxins may also be different (Davidson et al., 1978; Myers et al., 1979).

The fate of ingested 1593 spores in the larval gut has been observed with the electron microscope. In contrast to the rapid degradation of SSII-1 vegetatives (Davidson, 1979), 1593 spores were not visibly affected by larval digestion, and germinated at 4-6 hr after ingestion (Figure 10-9). Though no obvious structural changes were seen in the spores, toxin apparently left the spores rapidly as larvae showed symptoms of intoxication within 30 min and many were dead within 4 to 6 hr. Host response to intoxication was similar to that seen in SSII-1 infections; however spores and other bacteria were not removed from the gut by peristaltic action before death of the host (Davidson, 1981a).

Fractionation of 1593 sporulating cells revealed that most toxin is located in the cell wall, with much smaller amounts in the cell membrane and soluble cytoplasmic fractions (Myers 1979; Myers and Yousten 1980a, b) (Table 10-2). Extraction of cell walls with 8M urea for 30 minutes removed the outer cell wall layer but did not lower the activity of the walls; hence, at least in strain 1593, the outer cell wall layer is not the location of the toxin. Serial extraction with chloroform:methanol:water, Triton X-100, and 8M urea does not reduce activity, nor does oxidation of sugars by periodate. These data indicate that neither lipid nor carbohydrate is a critical component of the toxin. In addition to the four amino acids making up the cell wall peptidoglycan in urea-extracted cell walls, 13 other amino acids make up the 25% protein component of the extracted walls. If the cell wall toxin is a protein or polypeptide, it is included in this protein component; it may be related to spore coat protein (Myers and Yousten, 1980a).

In a parallel study, spheroplasts were formed by treating 1593 cells with lysozyme. After osmotic lysis of spheroplasts, cell membrane and cytoplasmic fractions were assayed. Density of the cell membrane fraction, its activity compared to cell membranes prepared by sonication or French pressure cell treatment (Myers and Yousten, 1980a), and the presence of peptidoglycan, showed that the cell membranes contained adhering cell wall. Strain 1593 cell walls are apparently quite resistant to digestion by lysozyme. Most insecticidal activity was again found in the cell wall/cell membrane fraction, and very little in the cytoplasm (Davidson et al., 1978).

Production of large amounts of soluble cell wall toxin is a critical step in characterization of the *B. sphaericus* toxin(s). Attempts to solubilize active toxin from cell walls have so far met with little success. Treatment of cell walls with lipid solvents or enzymes neither solubilizes nor destroys activity. Treatment with 8M urea for 30 minutes removes a small amount of toxin (Table 10-2), while treatment for one hour with 8M urea, 8M urea plus mercaptoethanol, or 3M guanidine HCl solubilizes but inactivates toxin. Activity is destroyed by 0.01 M NaOH (30 min at 22° C) or by boiling (10 min) (Myers and Yousten, 1980a, b).

Figure 10-5. *Bacillus sphaericus* SSII-1 cells in anterior midgut immediately after feeding. Outer cell wall layer (O) is being removed (arrows); other cellular structures are unchanged.

Figure 10-6. *Bacillus sphaericus* SSII-1 cells 30 minutes after feeding. Cytoplasmic ground substance is nearly absent, internal membranes are whorled (W) and ruffled (CM). Only the peptidoglycan cell wall layer remains; in 6b this layer has broken.

Figure 10-7. Anterior midgut cells fixed 10 hr after feeding on *B. sphaericus* SSII-1, showing separation of cells at bases (arrows) and increase in number and size of cytolysosomes (CL).

Figure 10-8. Enlarged cytolysosome (CL) in midgut cell 10 hr after feeding on *B. sphaericus* SSII-1. Apparent fusion of cytolysosome with mitochondrion.

Figures 10-3, 10-5, 10-6 and 10-7 from Davidson (1979); reprinted by permission of National Research Council of Canada

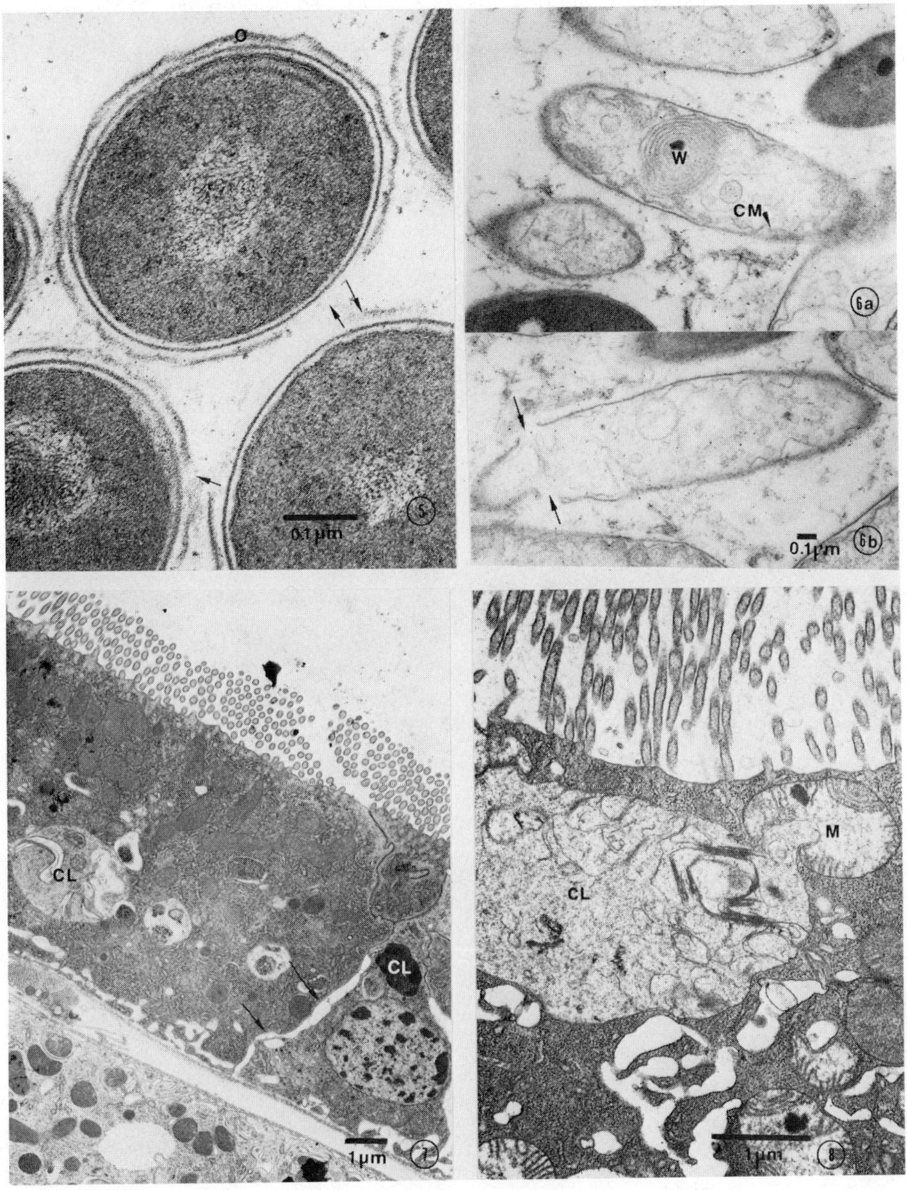
O
0.1 μm
5
W
CM
6a
0.1 μm
6b
CL
CL
1 μm
7
M
CL
1 μm
8

Table 10-2

Purification of Toxic Activity from Bacillus sphaericus 1593 cells[a]

FRACTION	ACTIVITY[b]
Original cells	0.104 μg/ml
Broken cells	0.524 μg/ml
Cell wall	5.9×10^{-3} μg/ml
Cell membrane	28.6 μg/ml
Soluble or cytoplasmic	58.0 μg/ml
Urea soluble wall protein	71.0 μg/ml

[a]Cells were grown in NYSM broth for 12 hr on a New Brunswick model G25 shaker operating at 33°C and 150 rpm.

[b]Toxic activity is expressed as LC_{50} values to Culex quinquefasciatus larvae based on dry weight of fractions. Each value represents the mean of 3 separate fractionation studies.

From Myers, 1979; reprinted by permission of the author.

The only source of active soluble *B. sphaericus* 1593 toxin at present, therefore, is the small amount present in the soluble cytoplasmic fraction of cells broken by sonication or French Press. When soluble cytoplasmic fractions were subjected to gel filtration on Sephadex G-25, G-100, or G-200, toxin was eluted in the void volume on G-25, but was retarded on G-100 and G-200. On Sephadex G-100, toxin emerges in a single narrow region of the elution curve shortly after the void volume, indicating that there may be a single toxin present with a molecular weight near the exclusion limit of G-100, 150,000 daltons. Toxin is readily precipitated by ammonium sulfate, and all activity is retained in a fraction containing less than 20% of the protein in the original cytoplasm. Semipurified cytoplasmic toxin is active in a pH range of 4 to 10, but is destroyed at pH 11, is active at 40°C but is gradually inactivated at 60°C, is resistant to trypsin, α-chymotrypsin, protease K, β-glucuronidase, α-amylase, and thermolysin, but is destroyed by the broad-range proteolytic action of pronase and subtilisin (Davidson, 1981b). The relationship of cytoplasmic toxin to cell membrane and cell wall toxins is not yet known. Indirect evidence indicates that cell wall and cytoplasmic toxins may be the same or closely related, as chemical and physical properties are similar, and their effects on mosquito larvae are identical.

No bodies resembling the parasporal crystals of *B. thuringiensis* have previously been seen in sporulating nontoxic *B. sphaericus* (Holt et al., 1975) nor in toxic strains SSII-1 and 1593 (Myers and Yousten, 1978a; Myers et al., 1979). However, during a recent ultrastructural study of larvae fed 1593 spores, densely-staining bodies were seen near the 1593 spore (Davidson, 1981a). These bodies appeared to be of two types: small eliptical bodies located between the exosporium and the mother cell remmant near the apical end of the cell; and larger, oval to triangular bodies located between exosporium and the mother cell remnant at the non-apical end of the cell. Myers (1979) and Myers and Yousten (1980b) have also observed densely staining bodies accompanying 1593 spores, and Wickremsinghe and Mendis (1980) observed inclusions accompanying strain MR-4 spores. Further ultrastructural examination of toxic and nontoxic strains revealed that strains 1593, MR-4, 2013-4, and 1691 produce, in addition to the densely staining bodies described above, crystal-like polyhedra within which crystalline lattice can occasionally be seen. The polyhedra are four- or six-sided in section, and may be bounded by a membrane (Figure 10-10). One or more polyhedra appear in each cell. All strains bearing polyhedra are also highly insecticidal in the spore stage. Toxic strains which are not highly insecticidal in the spore stage (e.g. strain SSII-1) do not bear polyhedra. No polyhedra were seen in nontoxic strains (Davidson, E. W. and Myers, P., *in press*).

Figure 10-9. *Bacillus sphaericus* 1593 spore germinating in larval gut 6 hr after ingestion. Outer cell wall layer (OCW) is appearing on germinant. From Davidson (1981a), reprinted by permission of the Society for Industrial Microbiology.

Figure 10-10. *Bacillus sphaericus* strain MR-4 spore (SP) with large accompanying polyhedral bodies (P). Spore shape is atypical.

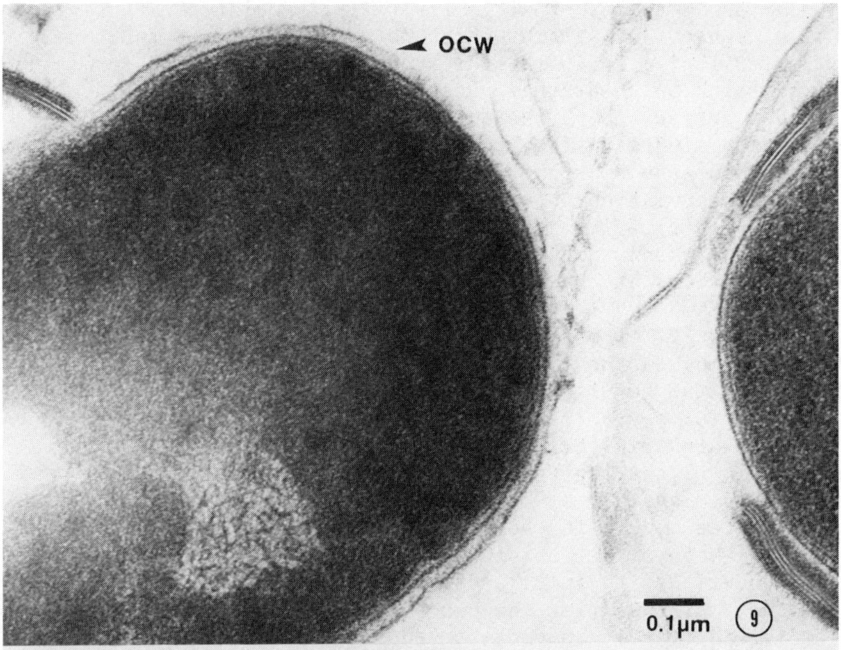
OCW
0.1µm
9

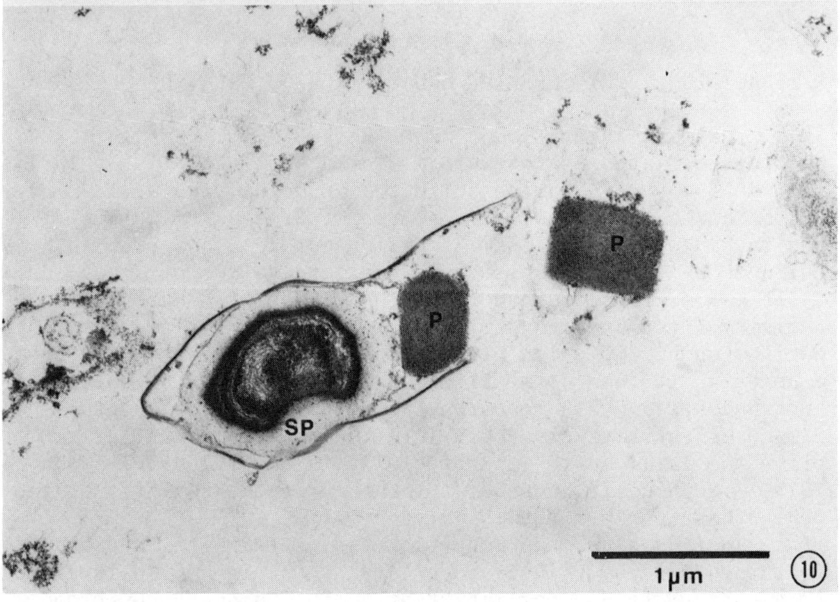
P
P
SP
1µm
10

The role of the polyhedra in the B. sphaericus pathology is not yet known; however they are not likely to be the sole location of toxin. Toxic activity is associated with spores of B. sphaericus strains which do not bear polyhedra (e.g. SSII-1), and with spores of polyhedra-bearing strains from which polyhedra have been removed or are not at the moment present (e.g. the sonicated 1593 spores studied by Davidson, 1981a) . Also, toxicity is present at varying levels throughout the B. sphaericus cell cycle, and is not restricted to the sporulation phase (Myers and Yousten, 1978a; Myers et al., 1979).

At present, nothing is known of the mode of action of B. sphaericus at the molecular and cellular levels. Histological evidence implicates the midgut as the target tissue, though neurotoxicity or other effects are possible. Rapid swelling of the midgut, the first visible symptom of intoxication, is probably the result of influx of fluid into the gut lumen. This influx may be caused by disruption of the cells' ability to regulate transport, perhaps by membrane damage as may occur in B. thuringiensis δ-endotoxin intoxication (see Chapters 8 and 9), or by stimulation of the adenylate cyclase system, as occurs in cholera infections (reviewed by Holmgren, 1978). Mode of action studies await purification of the major B. sphaericus toxin(s) and development of a sensitive assay system. At that time the causes for the marked resistance of larvae of some species of Aedes and Anopheles can also be investigated.

BACILLUS ALVEI CHESHIRE AND CHEYNE AND RELATED SPECIES

Organisms identified as B. alvei, B. circulans, B. brevis, and intermediate strains were isolated by Singer (1973, 1974) from dead mosquito larvae collected in Delhi, India (WHO Accession No. 1321; also the source of B. sphaericus SSII-1) and in Rangoon, Burma (WHO Accession No. 1242 and 1246). Cultures of these organisms produce LC_{50} values of about 10^6 cells per ml to C. quinquefasciatus larvae. Insecticidal activity first appears during early sporulation, and does not decline upon completion of sporulation. These strains produce a heat-labile, non-crystalline soluble toxin in the culture supernatants. Death requires two to seven days. No gross damage of gut tissues occurs in C. quinquefasciatus larvae prior to larval death (Singer, 1973, 1975a, 1975b; Davidson, 1975). Nothing is known of the nature of the toxin(s) or the mode of action in mosquito larvae.

Bacillus alvei has long been associated with European Foulbrood disease of honey bee larvae. It is now believed to be a saprophyte and not important to the European Foulbrood infection cycle (reviewed by Shimanuki, 1978; see Chapter 7).

CLOSTRIDIUM BREVIFACIENS BUCHER AND CLOSTRIDIUM MALACOSOMAE BUCHER

Bucher (1957) described a disease of tent caterpillars, *Malacosoma* spp., which he called "brachytosis" from the shortened condition of infected larvae. Though Bucher originally described the causative organisms as *Bacillus* spp., he later learned that they were anaerobic and named them *Clostridium brevifaciens* and *Cl. malacosomae* (Bucher, 1961). A similar infection was reported in Essex Skippers (*Thymelicus lineola*) (Bucher and Arthur, 1961).

Visible symptoms of brachytosis disease include regurgitation, diarrhea of rusty-brown color, sluggishness, and the characteristic shortening of the body. Larvae become moribund on the fifth or sixth day, having emptied the gut of fluid which contains large numbers of spores. Death occurs 1-4 days later. Dead larvae are dry, mummified, and very short (Bucher, 1957).

Vegetative rods appear in the gut 16-24 hours after ingestion of spores, and the bacteria multiply rapidly during the next 12 hours. Bacterial rods in the gut arrange themselves around leaf cells to form characteristic "pin-cushions". By 36-48 hours, free spores are produced and secondary bacterial infections may occur. Masses of vegetative rods occur in the folds of the gut produced by body shortening during the third day. By the fourth day, the gut is emptied of most Clostidia by regurgitation and defecation. Once the host has become moribund, there is little evidence of the primary pathogen, and the secondary invaders are predominant.

The causative organisms of brachytosis are confined to the gut and do not invade the hemocoel. The pH of the host foregut increases from the normal 6.5-6.8 to 9.0-9.2, then drops to 7.7-7.8. Bucher (1957) felt that the direct cause of larval death is dehydration from diarrhea. Shortening of the larva is due to contraction of both longitudinal intersegmental and longitudinal gut muscles. Bucher postulated that Clostridia in the gut produce a toxin during autolysis of the mature sporangia.

Though *Cl. brevifaciens* and *Cl. malacosomae* can be cultured on a complex anaerobic medium which includes leaf extract, sporulation does not occur (Bucher, 1961). Repeated transfer on artificial medium causes a loss of infectivity. There is not a dosage/mortality dependence, but rather a minimal lethal dose which sets up an infection. Some infected larvae are able to recover, and larvae become progressively more resistant to infection as they age. Fifth instar larvae are refractory to infection (Bucher, 1961, 1963).

Toxin production and symptoms in brachytosis are similar to those of Clostridial infections of mammals. The mammalian Clostidial pathogens produce a wide range of toxic substances (reviewed by Raskova and Masek, 1971); it is likely that the insect pathogens likewise produce several toxins. Isolation of these toxins and study of their modes of action would provide an important comparative study to the toxins affecting mammals.

ACKNOWLEDGEMENTS

Previously unpublished research by the author was supported by the following sources: UNDP/World Bank/WHO Special Programme for Research and Training in Tropical Diseases; and National Science Foundation Grant No. PCH 8004262.

REFERENCES

Buchanan, R. E. and Gibbons, N. E. 1974. Bergey's Manual of Determinative Bacteriology, 8th edition. Williams and Wilkins Co., Baltimore, MD 1246 pp.

Bucher, G. E. 1957. Disease of the larvae of tent caterpillars caused by a sporeforming bacterium. Canad. J. Microbiol., 3, 695-709.

Bucher, G. E. 1961. Artificial culture of *Clostridium brevifaciens* n. sp. and *C. malacosomae* n. sp., the causes of brachytosis of tent caterpillars. Canad. J. Microbiol., 7, 641-655.

Bucher, G. E. and Arthur, A. P. 1961. Disease in a field population of the introduced Essex Skipper, *Thymelicus lineola* (Ochs.) (Lepidoptera:Hesperiidae). Canad. Entomologist, 93, 1048-1049.

Bucher, G. E. 1963. Ultrasonic cleaning of spores of an obligate bacterial pathogen. J. Insect Pathol., 5, 427-433.

Dadd, R. H. 1975. Alkalinity within the midgut of mosquito larvae with alkaline-active digestive enzymes. J. Insect Physiol., 21, 1847-1853.

Davidson, E. W. 1975. Pathogenesis of bacterial diseases of vectors. In Biological Regulation of Vectors, J. D. Briggs, ed., U.S. Dept. Health, Educ. and Welfare Publ. No. (NIH) 77-1180, pp. 19-30.

Davidson, E. W. 1979. Ultrastructure of midgut events in the pathogenesis of *Bacillus sphaericus* strain SSII-1 infections of *Culex pipiens quinquefasciatus* larvae. Canad. J. Microbiol., 25, 178-184.

Davidson, E. W. 1981a. A review of the pathology of bacilli infecting mosquitoes, including an ultrastructural study of larvae fed *Bacillus sphaericus* 1593 spores. Devel. in Industr. Microbiol., 22: in press.

Davidson, E. W. 1981b. Purification and properties of soluble cytoplasmic toxin from the mosquito pathogen, Bacillus sphaericus strain 1593. J. Invertebr. Pathol., in press.

Davidson, E. W. and Singer, S. 1973. Pathogenesis of Bacillus sphaericus infections in mosquito larvae. Proceed. Int'l Colloquium Insect Pathol. and Microb. Control., Oxford, England. p. 95.

Davidson, E. W., Singer, S. and Briggs, J. D. 1975. Pathogenesis of Bacillus sphaericus strain SSII-1 infections in Culex pipiens quinquefasciatus (= C. pipiens fatigans) larvae. J. Invertebr. Pathol., 25, 179-184.

Davidson, E. W., Myers, P. and Yousten, A. A. 1978. Cellular location of insecticidal activity in Bacillus sphaericus. Proceed. Int'l. Congress Invert. Pathol., Prague, Czech, p. 47.

Gordon, R. E., Hayes, W. C. and Pang, C. H-N. 1973. The Genus Bacillus. U.S. Dept. of Agric., Agric. Handbook No. 427.

Heimpel, A. M. 1955a. Investigations of the mode of action of strains of Bacillus cereus Fr. and Fr. pathogenic for the larch sawfly, Pristiphora ehrichsonii (Htg.), Canad. J. Zool., 33, 311-326.

Heimpel, A. M. 1955b. The pH in the gut and blood of the larch sawfly, Pristiphora ehrichsonii (Htg.), and other insects with reference to the pathogenicity of Bacillus cereus Fr. and Fr. Canad. J. Zool., 33, 99-106.

Heimpel, A. M. and Angus, T. A. 1956. Recent advances in the knowledge of some bacterial pathogens of insects. Proceed. 10th Int'l Congress of Entomol., 4, 711-722.

Heimpel, A. M. and Angus, T. A. 1958. The taxonomy of insect pathogens related to Bacillus cereus Frankland and Frankland. Canad. J. Microbiol., 4, 531-541.

Heimpel, A. M. and Angus, T. A. 1963. Diseases caused by certain sporeforming bacteria. In Insect Pathology, An Advanced Treatise, Vol. 2, E. A. Steinhaus, ed., Academic Press, N.Y., pp. 21-73.

Holmgren, J. 1978. Cholera toxin and the cell membrane. In Bacterial Toxins and Membranes, J. Jeljaszewicz and T. Wadstron, eds., Academic Press, N.Y., pp. 333-336.

Holt, S. C., Gauthier, J. J. and Tipper, D. J. 1975. Ultrastructural studies of sporulation in Bacillus sphaericus. J. Bacteriol., 122, 1322-1338.

Howard, L., and Tipper, D. J. 1973. A polypeptide bacteriophage receptor: Modified cell wall protein subunits in bacteriophage-resistant mutants of Bacillus sphaericus strain P-1. J. Bacteriol., 113, 1491-1504.

Kellen, W. R., Clark, T. B., Lindegren, J. E., Ho, B. C., Rogoff, M. H. and Singer, S. 1965. Bacillus sphaericus Neide as a pathogen of mosquitoes. J. Invertebr. Pathol., 7, 442-448.

Krych, V. K., Johnson, J. L. and Yousten, A. A. 1980. Deoxyribonucleic acid homologies among strains of Bacillus sphaericus. Int'l. J. Systematic Bacteriol., 30, 476-484.

Myers-Keith, P. 1979. Toxic activity of Bacillus sphaericus for mosquito larvae. Ph.D. Dissertation, Virginia Polytechnic Inst. and State Univ., Blacksburg, Va. 102 pp.

Myers, P. and Yousten, A. A. 1977. Mosquito pathogenicity of Bacillus sphaericus SSII-1. Abstr. Ann. Mtg. Amer. Soc. Microbiol. p. 162.

Myers, P. and Yousten, A. A. 1978a. Toxic activity of Bacillus sphaericus SSII-1 for mosquito larvae. Infect. and Immun., 19, 1047-1053.

Myers, P. and Yousten, A. A. 1978b. Cell wall structure of an entomocidal strain of Bacillus sphaericus. Spores VII, pp. 31-33.

Myers, P., Yousten, A. A. and Davidson, E. W. 1979. Comparative studies of the mosquito-larval toxin of Bacillus sphaericus SSII-1 and 1593. Canad. J. Microbiol., 25, 1227-1231.

Myers, P. and Yousten, A. A. 1980a. Localization of the mosquito-larval toxin of Bacillus sphaericus 1593. Appl. and Environ. Microbiol., 39, 1205-1211.

Myers, P. and Yousten, A. 1980b. Toxic activity of Bacillus sphaericus for mosquito larvae. Devel. in Industr. Microbiol., 22, in press.

Mulligan, F. S. III, Schafer, C. H. and Miura, T. 1978. Laboratory and field evaluation of Bacillus sphaericus as a mosquito control agent. J. Econ. Entomol., 71, 774-777.

Ramoska, W. A., Singer, S. and Levy, R. 1977. Bioassay of three strains of Bacillus sphaericus on field-collected mosquito larvae. J. Invertebr. Pathol., 30, 151-154.

Raskova, H. and Masek, K. 1971. Microbial toxin. In Pharmacology and Toxicology and Naturally Occurring Toxins, H. Taskova, ed., Vol. 1H. Pergamon Press, N.Y., pp. 3-212.

Rogoff, M. H., Ignoffo, C. M., Singer, S., Gard, I. and Prieto, A. P. 1969. Insecticidal activity of thirty-one strains of Bacillus against five insect species. J. Invertebr. Pathol., 14, 122-129.

Seki, T., Chung, C-K., Mikami, H. and Oshima, Y., 1978. Deoxyribonucleic acid homology and taxonomy of the genus Bacillus. Int'l. J. Systematic Bacteriol., 28, 182-189.

Shimanuki, H. 1978. Bacteria. In Honey Bee Pests, Predators, and Diseases, R. A. Morse, ed., Cornell Univ. Press, pp. 44-61.

Singer, S. 1973. Insecticidal activity of recent bacterial isolates and their toxins against mosquito larvae. Nature (London), 244, 110-111.

Singer, S. 1974. Entomogenous bacilli against mosquito larvae. Devel. in Indust. Microbiol., 15, 187-194.

Singer, S. 1975a. Use of bacteria for control of aquatic insect pests. In Impact of the Use of Microorganisms on the Aquatic Environment, A. W. Bourquin, D. G. Ahearn, and S. P. Myers, eds., U.S. Environ. Protect. Agcy. Ecol. Res. Series, EPA 660/3-75-001, pp. 5-22.

Singer, S. 1975b. Isolation and development of bacterial pathogens of vectors. In Biological Regulation of Vectors, J. D. Briggs, ed., U.S. Dept. of Health, Educ. and Welfare Public No. (NIH) 77-1180, pp. 3-17.

Singer, S., 1980. *Bacillus sphaericus* for the control of mosquitoes. Biotechnol. and Bioengineer., 22, 1325-1355.

Singer, S., Goodman, N. and Rogoff, M. H. 1966. Defined media for the study of bacilli pathogenic to insects. Ann. N.Y. Acad. Sci., 139, 16-23.

Takarazuka, H. 1980. Pathogenicity of *Bacillus sphaericus* to larvae of the mosquito, *Culex pipiens*. Jap. J. Appl. Ent. Zool., 24, 93-97 (In Japanese, Engl. summ)

Wickremsinghe, R. S. B. and Mendis, C. L. 1980. *Bacillus sphaericus* spore from Sri Lanka demonstrating rapid larvicidal activity on *Culex quinquefasciatus*. Mosquito News, 40, 387-389.

Wigglesworth, V. B. 1942. The storage of protein, fat, glycogen, and uric acid in the fat body and other tissues of mosquito larvae. J. Exp. Biol., 19, 56-57.

Chapter 11

BACILLUS THURINGIENSIS BETA-EXOTOXIN

by MARGUERITE-M. LECADET* and HUGUETTE DE BARJAC†

INTRODUCTION

The thermostable exotoxin of *Bacillus thuringiensis*, which is also referred to as β-exotoxin, fly factor or thuringiensin can be recognized from the other *Bacillus thuringiensis* toxins by its chemical nature, thermostability, and wide range of action. The first report of a substance stable at 120°C for 15 minutes and toxic by injection for wax moth larvae, *Galleria mellonella*, was made by MacConnell and Richards (1959). Then, Burgerjon and de Barjac (1960, a,b) demonstrated the activity *per os* of a soluble heat-stable toxin which was supposed to be the fly factor described by different authors as being toxic for *Musca domestica* larvae (Hall and Arakawa, 1959; Dunn, 1960). Further studies have shown the nucleotidic nature of this extracellular toxin (de Barjac and Dedonder, 1965; Benz, 1966) which is excreted into the culture medium by several strains belonging to different varieties or serotypes of *B. thuringiensis* (Burgerjon and de Barjac, 1964; Cantwell et al., 1966; de Barjac et al., 1966). It should be noted that the accumulation of exotoxin is not a general property of *B. thuringiensis*, as is the crystalline δ-endotoxin which is considered as a basic taxonomic criterion for the species. Furthermore, in contrast to the specific action of the crystal toxin, which is limited to Lepidoptera and a few Diptera (mosquito) larvae (de Barjac, 1978), the β-exotoxin is highly toxic for a wide range of species in many insect orders

*Institut de Recherche en Biologie Moléculaire, C.N.R.S., Universite Paris VII, France
†Institut Pasteur, Paris, France

including: Lepidoptera, Diptera, Hymenoptera, Coleoptera, Orthoptera, Hemiptera and Isoptera and even for nematodes (Prasad et al., 1972) and mites (Krieg, 1968; Hall et al., 1971). When given to larvae the effects of the exotoxin are delayed in time and usually appear at moulting or during metamorphosis, thus preventing normal insect development (Burgerjon et al., 1964). Teratological effects also occur as a response to the ingestion of sublethal doses of exotoxin by many insect species, especially Lepidoptera (Burgerjon and Biache, 1967a) and Coleoptera (Burgerjon et al., 1969).

Such original and interesting properties as well as the nucleotidic nature of the toxin, account for the extensive literature devoted to the structure of the β-exotoxin and to its effects at the molecular level. Other aspects including host/pathogen interactions, histopathology or physiological responses which appear essential to an insect pathologist, have been poorly documented.

This chapter will deal mainly with the following closely related points: the structure, the molecular mode of action, and the teratological effects of the β-exotoxin.

CELLULAR INTERACTION OF HOST AND PATHOGEN

In the usual terms of insect pathology, this approach would include interactions between the pathogen taken as a whole and its host. In the case of _B. thuringiensis_ β-exotoxin, the toxin is an extracellular compound accumulated by the bacteria during some stage of the growth cycle, under defined cultural conditions and preferentially in complex media. Experimental studies of the basic effects of the toxin use partially purified or pure preparations of exotoxin, free of spores and crystals. In diseased larvae having ingested a total sporulated culture of an exotoxin-producing strain, however, the effects of the nucleotidic toxin would be added to the specific toxemia induced by the δ-endotoxin (Burgerjon and de Barjac, 1960a). For example, Benz (1975) has shown that β-exotoxin enhanced the action of commercial bacterial preparations against larvae of the larch bud moth (_Zeiraphera diniana_), thus allowing a synergy of the toxins in species with low sensitivity to each toxin alone. Several other important aspects are the route of intoxication, the amount of exotoxin and the larval age. When high doses are administered _per os_ during a given instar, the toxic effects appear only at the next moult, inducing mortality which is delayed by several days without precise symptoms.

There are few data describing the pathological effects at the organ or tissue level. Vankova and Leskova (1972) injected the exotoxin (culture supernatant or purified exotoxin from _B. thuringiensis_ serotype H1) into the hemocoel of 6th instar G. mellonella larvae. Within 24 hours, the micronucleocytes and macronucleocytes were severely damaged, with vacuolation and disintegration of the cytoplasm, and loosening of the cytoplasmic and nuclear membranes. More recently Galani and Beratlief (1977) have reported the histopathological changes induced in larvae of the Colorado potatoe beetle Leptinotarsa decemlineata, by the oral application of a concentrated supernatant of B. thuringiensis serotype H8, leading to 90% mortality in 5 days. Histopathological changes in the midgut appeared 20 hours after ingestion of the toxin and consisted of an "elongation of the epithelial cells toward the gut". Rupture of the gut wall was observed 72 hours after the larvae were treated, thus preventing the normal nutritional process. These effects appear to be specific; however they would be more convincing if obtained after administration of pure or partially purified toxin.

Using in vitro cultures of cardiac cell line from Leucophea maderae (Orthoptera), and 5 to 100 mg/ml of purified exotoxin prepared by one of us from B. thuringiensis serotype H1, Quiot (1975) obtained a two-step response. First, cell multiplication stopped. Second, after 3 or 4 days, vacuolation of the cytoplasm occurred with lipid accumulation and chromatolysis of the nuclei.

Simpson et al. (1971) used the same purified exotoxin to treat 100 hour-old imaginal discs of Drosophila melanogaster. The authors treated these discs in vitro for 5 minutes with varying doses of exotoxin (1 to 100 μg/ml) and re-implanted the discs in 80-90 hour-old host larvae, which then underwent metamorphosis. They observed general necrosis of the genital discs even with 1 μg/ml of exotoxin in some cases but more striking was the differential sensitivity of discs to the exotoxin concentrations. The discs of halteres were the most suceptible, then in decreasing order of sensitivity, were the discs of antennae, eyes, legs and wings. These results are comparable to the observations of Burgerjon and Biache (1967a) on the abnormalities of Pieris brassicae pupae or adults produced from larvae treated per os with sublethal doses of exotoxin (see p. 307). Head and thorax appendages were found to be lacking but the wings remained. These authors also found the relative "protective effect" of high doses of adenine (100 fold the exotoxin concentration); adenosine triphosphate (ATP) is a competitive inhibitor of β-exotoxin for the Eschericia coli RNA-polymerase reaction (see p. 299).

Tremblay et al. (1972) have shown that the exotoxin penetrates the chorion of Acheta domesticus (cricket) eggs between the 30th and 110th hour of incubation, thus disturbing normal development of embryos. The uptake of exotoxin, as measured by the percentage of embryonic mortality, increases at the same time as the water permeability of the chorion increases. The mechanism for penetration was not elucidated. These examples demonstrate that interactions between host and pathogen can occur at various stages of insect development and induce general perturbations, but the cause of these abnormalities remains to be explained.

In parallel studies conducted on mammalian cell cultures *in vitro*, comparable pathological effects have been found. Riou (1968) was among the first authors to describe morphological alterations of hamster kidney cell lines treated with relatively high doses of purified exotoxin (500 to 2500 μg/ml). Laurent (1978) confirmed the work on various other vertebrate cell lines. In addition (personal communication) he demonstrated the very high sensitivity of embryonic chicken myoblasts and fibroblasts. The cell fusion of myoblasts is inhibited by 10 to 100 μg/ml of exotoxin, and, moreoever, the acetylcholine receptor is depressed after 5 days from 16 to 66% according to the dose of exotoxin (1 to 50 μg/ml). Ten μg/ml of exotoxin causes 97 to 100% inhibition of DNA synthesis after 24 hours, as determined by incorporation of ^{3}H thymidine. In contrast the lowest dosage, 0.1 μg/ml, induces a stimulation, which could be related to the mitogenic property of the toxin (see p. 313). RNA and protein synthesis, as respectively measured by incorporation of ^{3}H-uridine and ^{3}H-leucine, react in the same way to the different doses of exotoxin. These data point out the variability of toxin action on mammalian cells according to the dose.

MOLECULAR INTERACTION OF HOST AND PATHOGEN

The general effects of the β-exotoxin on a wide range of insect species have been previously reviewed (Heimpel, 1967; Bond et al., 1971). They are mainly characterized by a slow intoxication leading to death after moulting without precise symptoms, and thus point to an "anti-metabolite-like" action of the toxin.

Nature and molecular structure of the exotoxin

The chemical nature and the precise structure of the toxin were simultaneously elucidated in different laboratories. The first indication of the nucleotidic nature of the toxin was reported by de Barjac and Dedonder (1965) who isolated, from the active supernatant of *B. thuringiensis* serotype H1 cultures, a nucleotide containing adenine, ribose and phosphate. This nucleotide induced the same toxic effects as the culture filtrates when tested on *Mamestra brassicae*

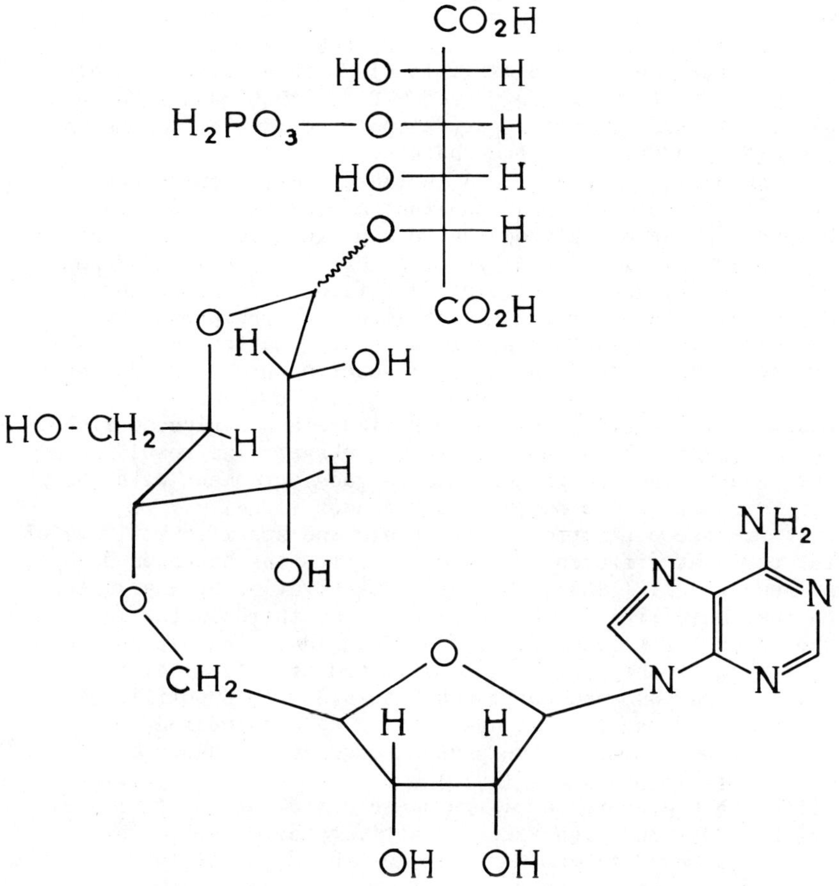

Figure 11-1. Molecular structure of the β-exotoxin (Farkas et al., 1969).

larvae; the LD_{50} for this compound, expressed as "AMP equivalent"/g of larvae was 2.42 μg. In 1968, the same authors described a complete purification procedure using column chromatography and provided a more detailed characterization of the compound; the precise molecular ratio of the constituents adenine, ribose and phosphate was 1:1:1. A molecular weight in the range 700-800 daltons was found and the substitution of the AMP molecule with a radical involving a glycol function was suggested. Benz (1966) and Sebesta et al. (1967) also reported partial purification of a compound containing adenine and phosphate.

Soon after, Bond (1969) indicated that adenine was substituted in position 9 and that a glucose residue was present in the α-D-glycopyranose configuration. The final structure was elucidated by Farkas et al. (1969) and Bond et al. (1969) and definitely identified as an adenosine derivative linked in the 5' position to a phosphoallaryl glucose. The structure according to the pattern established by the Farkas group, in the configuration proposed by Bond et al. (1969) is given in Figure 11-1. It shows a dissaccharide moiety formed by D-ribose and D-glucose linked together through an unusual ether linkage. The results obtained by the two groups were in good agreement with the first reports by de Barjac and Dedonder (1968).

Precise estimation of the yield and specific activity of the toxin at different steps of purification has been detailed by Bond et al. (1969); they used quantitative bioassays based on the inhibition of the development of third instar larvae of the house fly M. domestica. The ED_{50} (dose inhibiting 50% emergence of the adults) was estimated as 3 μg/ml of test solution per os, and corresponds to a 300-fold purification with a yield range of 35-40%. Toxicity disappeared by hydrolytic removal of the phosphate group, as shown by other authors. Kim and Huang (1970) have reported a slightly different procedure for large scale purification, giving a 15% recovery and high specific activity as measured in bioassays based on the mortality of third instar larvae of the house fly. These authors called the isolated toxin "thuringiensin" and after the last purification step they separated two interconvertible forms: thuringiensin A and B. The less active form thuringiensin B was characterized as the γ-lactone of thuringiensin A. The molecular weight estimated by chromatography on Sephadex G10 was 701 daltons which is in agreement with the lower estimate proposed by de Barjac and Dedonder (1968) and with the final molecular structure. A few years later this complex structure was fully confirmed by the total chemical synthesis of the exotoxin (Kalvoda et al., 1973).

Molecular mode of action

It became clear that β-exotoxin could be considered as a structural analogue of ATP and as a potential inhibitor in reactions involving a structural role of ATP. The same groups who elucidated the structure of the toxin also demonstrated its inhibitory effect on RNA synthesis in vitro, specifically on the polymerization reaction catalyzed by DNA-dependent RNA-polymerase. Sebesta and Horska (1968 ; 1970) reported detailed in vitro experiments using purified exotoxin from B. thuringiensis (strain gelechiade) and RNA polymerase from E. coli. Their results led to the following conclusions: i. the exotoxin binds by its adenine moiety to the ATP-specific site of the DNA-enzyme complex, thus competing exclusively with ATP and preventing the polymerization step of the enzyme reaction; ii. it is possible to reverse the inhibition by addition of ATP to five-to ten-fold the exotoxin concentration; iii. the initiation step of the polymerase reaction does not appear substantially influenced by exotoxin; iv. N-oxidation of adenosine in the exotoxin molecule leads to a considerable loss of inhibitory activity. In other words the exotoxin acts as an analogue of ATP in "slowing down the polymerization reaction", thus inducing a depression in RNA synthesis.

Klier et al. (1973) confirmed the above interpretation using another in vitro bacterial system in which both template DNA and RNA polymerases were purified from B. thuringiensis. By the method of kinetic analysis the authors demonstrated the competitive nature of the inhibition induced by pure exotoxin and determined the value of 0.6×10^{-4} M for the inhibition constant K_i. The inhibition was of the same order when RNA-polymerase was extracted from vegetative or from sporulating cells (Johnson et al., 1975). As a practical consequence of the inhibitory effect of β-exotoxin on prokaryotic RNA-polymerase, de Barjac and Lecadet (1976) have proposed an enzymatic titration for estimating the amounts of exotoxin in an autoclaved culture supernatant, using RNA polymerase from B. thuringiensis or the commercial enzyme from E. coli and calf thymus DNA as a substrate. This method appears much more sensitive than bioassays and gives a concentration of exotoxin in crude supernatant reaching 200-250 μg/ml under optimal culture conditions, thus allowing the detection of toxin in low-producing strains. This method may also be used to determine the timing of exotoxin excretion during the bacterial growth cycles. The toxin is detected exclusively during the stationary phase about two hours after the end of exponential phase, and its concentration in the medium increases progressively during sporogenesis until complete cell lysis. The accumulation of a nucleotide-like compound during the course of a differentiation process raises many exciting questions about regulatory mechanisms and finality of the toxin production as previously posed by Rogoff and Yousten (1969).

Though the inhibition of RNA-polymerase may be a general mechanism by which exotoxin can prevent RNA synthesis in prokaryotic *in vitro* systems, we must now examine whether such mechanisms can work *in vivo* at the molecular level in diseased insects.

Eukaryotic mammalian RNA-polymerases are very sensitive to β-exotoxin which is very toxic when injected into mice (Sebesta et al., 1969; de Barjac and Riou, 1969). Beebee et al. (1972) have examined its effects on the RNA polymerases from rat liver; they have shown that polymerase activity in intact nuclei, and in partially purified enzymes corresponding respectively to the nucleolar and the nucleoplasmid forms, are strongly inhibited by the toxin (cf. Table 11-1) and appear more sensitive than bacterial polymerases. The inhibition, competitive in nature, was reversed by ATP and suppressed when the toxin molecule was dephosphorylated. The activity which is resistant to α-amanitin and generally considered as the nucleolar enzyme was preferentially inhibited. Moreover, Smuckler and Hadjiolov (1972), studying the action of the exotoxin on mouse liver RNA-polymerases *in vivo* and *in vitro*, have shown a decreased RNA production in intact nuclei after intraperitoneal injection of exotoxin (20 μg/g body weight), as a result of decreased activity of both nucleolar and nucleoplasmid RNA-polymerases. However when the *in vitro* inhibition of polymerases purified from treated animals is considered, the nucleoplasmic enzyme (polymerase II) is preferentially inhibited. Differences between the results from the two groups clearly illustrate the difficulties encountered when comparing the *in vitro* and *in vivo* action in mammalian cells; in addition microheterogenities in RNA-polymerases from eukaryotes might explain such differences (Bond, personal communication).

Nonetheless, the mammalian RNA-polymerases were shown to be extremely sensitive to the inhibition by β-exotoxin; the K_i values for purified RNA polymerases from rat liver range between 0.1 to 2.0 x 10^{-6} M as deduced from data reported by the authors cited above. When examining overall effects on the nature of RNA synthesized *in vitro* Mackedonski et al. (1972) demonstrated the preferential inhibition of ribosomal RNA synthesis in mouse liver after injection of exotoxin, while the synthesis of heterogenous "DNA-like" RNAs was also depressed but to a lesser extent.

Let us examine now the case of insect RNA-polymerases. Beebee and Bond (1973a) considered the effects of exotoxin on normal and ecdysone-stimulated RNA-polymerase activities in intact nuclei extracted from the fat body of blowfly, *Sarcophaga bullata*, larvae. Two RNA polymerase activities,

TABLE 11-1

INHIBITION OF EUKARYOTIC RNA-POLYMERASES BY β-EXOTOXIN AS COMPARED WITH TOXICITIES

	DOSES OF EXOTOXIN WHICH INHIBIT RNA POLYMERASE ACTIVITIES IN INTACT NUCLEI (1)		TOXICITIES: LD_{50} VALUES IN μg/g BODY WEIGHT		ORIGIN
	α-amanitin resistant activity	α-amanitin sensitive activity	Injection	Ingestion (2)	
Insect systems	0.010 μM	5 μM	0.1-0.5 μg		Beebee and Bond (1973)
				0.5 μg/ml	Norris (1969)
				0.25-0.5 μg/ml	Kim et al. (1973)
				2.42 μg (3)	de Barjac and Dedonder (1965)
				3 μg (ED_{50})	Bond et al. (1969)
Mammalian systems	0.8 μM	100 μM	18 μg		Beebee and Bond (1973)
				400-800 γ/g	Norris (1969)
				930 γ/g	de Barjac (not published)

(1) RNA polymerase activities in intact nuclei were measured through incorporation of labeled trinucleotides in the presence or not of exotoxin as reported by Beebee and Bond (1973a).

In insect systems nuclei were prepared from *Sarcophaga bullata*; in mammalian systems nuclei were from rat liver.

(2) Doses for ingestion tests were given in μg/ml of the medium used for the alimentation of larvae.

(3) The value reported here was expressed as "AMP" equivalent.

not quite similar to the mammalian ones, but also differing in their response to α-amanitin are detected in this in vitro system. Both activities are strongly inhibited by low doses of exotoxin, especially the activity resistant to α-amanitin which is shown to be stimulated by ecdysone. It is very important to note the differences between insect and mammalian systems in response to exotoxin inhibition. Insect systems appear much more sensitive (by two orders of magnitude) than mammalian systems, thus explaining the differences in the toxic effects (Table 11-1). The involvement of RNA synthesis, sensitive to exotoxin and stimulated by moulting hormones during metamorphosis would explain some characteristics of exotoxin poisoning in insects.

But the in vivo response is another problem. An approach to understanding the nature of the in vivo effects of the toxin at the molecular level was given by Kim et al. (1972a) who examined the effects induced by injecting mature larvae of the cabbage looper, *Trichoplusia ni* with pure exotoxin. Doses > 90 μg induced malformed mouth parts in 100% of the adults. The experiments roughly demonstrated the inhibition of proteins and RNA synthesis (and to a lesser extent DNA synthesis) within two hours after injection of 50 μg of toxin, as measured by ^{14}C valine or ^{14}C uracil incorporation in whole tissue homogenates. Using an ovary cell line from adult cotton bollworm *Heliothis zea*, Kim et al. (1972b) also evaluated the in vitro effects of β-exotoxin on protein and nucleic acid synthesis in tissue cultures. Ninety-seven percent inhibition of cell growth after 11 days resulted from the addition of exotoxin to the culture medium (0.5 to 5 μg/ml) and was attributed to a specific inhibition of RNA synthesis. The incorporation of ^{14}C uracil into RNA was shown to decrease progressively during the hours following the addition of the toxin.

Taking together the inhibitory effects induced by β-exotoxin in the various insect systems examined and the basic effects at the level of eukaryotic RNA polymerases, it appears clear that a direct correlation exists between enzyme inhibition and toxicity to larvae of most holometabolous insects. Significant confirmation of this correlation was reported by Beebee and Bond (1973b) who reported the relative insensitivity of RNA polymerase from adult *S. bullata* which parallels their higher resistance to the toxin. The effects of exotoxin on the nucleolar RNA synthesis, specifically rRNA synthesis, and on ecdysone stimulated RNA synthesis, may have consequences on the system directing the synthesis of dopa decarboxylase involved in sclerotization of the larval cuticle (Sekeris, 1965). For all these reasons we agree with the interpretation proposed by Beebee and Bond. "The sequence of events (synthesis of RNA, synthesis of enzyme, sclerotization, pupation) explains why holometabolous insects when exposed to high doses of exotoxin, either fail to pupate, or at slightly lower doses, die as deformed pupae."

Interference by the toxin in the complex systems involved in metamorphosis is highly probable. At the present time no direct evidence is available regarding the primary impact of the toxin *in vivo* in insect larval systems; this may reflect the well-known discrepancies between *in vitro* and *in vivo* systems.

On the other hand, the possible action of exotoxin on other metabolic pathways cannot be ruled out. For instance, Grahame-Smith et al. (1975), reported competitive inhibitory effects of the toxin on various vertebrate adenyl cyclases, which would suggest an important impact at the level of hormone-mediated biosynthesis. Moreover possible effects on protein synthesis have been suggested by Somerville and Swain (1975) who found that high concentrations (2.0 and 6.0 mM) of exotoxin can inhibit polyphenylalanine formation in cell free system from *E. coli*. If demonstrated in other systems, such effects would support previous observations by Kim *et al.* (1972a) using *in vivo* insect systems. It is obvious that many unsolved problems of great interest remain with regard to the molecular mode of action of β-exotoxin.

PHYSIOLOGICAL RESPONSES OF HOST TO PATHOGEN

In most cases, studies on the toxic effects of β-exotoxin reported above involved third instar larvae. When considering the physiological aspects of the response, it is necessary to examine all the different stages of development from the embryo to the adult insect. It is also important to distinguish between lethal and sublethal doses, because many studies have focused on the teratological effects induced by low doses. In Table 1 we have assembled LD_{50} values reported for different insect systems using purified toxin preparations. It is more difficult to precisely estimate sublethal doses when crude heated supernatants or partially purified solutions were used, as is the case in most studies. Burgerjon and Biache (1967b) have suggested that the difference between the minimum lethal dose and the minimum dose inducing teratologies may be at least a hundred fold.

Metabolic changes

There is very little information available about the classical aspects of the physiological response of host to toxin: metabolic changes, respiratory metabolism, immune response or other essential functions. An example is the previously mentioned study by Tremblay et al. (1973) on the effects of the toxin on oxygen uptake by embryos of the cricket, A. domesticus. Embryonic metabolism was greatly disturbed when eggs at 0-3 hours were treated with partially purified toxin; changes in oxygen uptake were observed over 5-6 days of incubation. High doses of exotoxin killed the embryos between the katatrepsis and the beginning of the dorsal closure, whereas at five- to ten-fold lower doses organogenesis was delayed by two days, thus suggesting inhibitory effects at the level of morphogenesis. Vankova and Horska (1975) studied the in vitro effects of gut homogenates from larvae of G. mellonella, M. domestica and adults of the honeybee, Apis mellifera on the exotoxin. Gut acid and alkaline phosphatases were shown to detoxify the exotoxin.

Teratological effects of the toxin: Timing of the response

Before describing the teratological effects induced by sublethal doses of exotoxin, we would like to examine the changes in diseased larvae according to the instar chosen for intoxication and the time of intoxication within the period between two moults. In all cases the toxin was administered per os.

The overall effects in larvae of various species were reviewed by Burgerjon (1974). In many species of holometabolous insects, Lepidoptera, Coleoptera, Diptera, Hymenoptera, and to a lesser extent in some heterometabolous species, the response will appear at the next moult causing either the death of the larvae in a premoulting stage or a cessation of feeding after the moulting is accomplished, followed by death spread out over a few days. With sublethal doses, the larvae can get through one or two larval moults and generate atrophied pupae after the pupal moult. According to the degree of atrophy, adults either fail to emerge from the pupae or emerge with deformities.

Within the general response common to holometabolous insects slight variations may be observed which are characteristic of a given species. In Lepidoptera detailed studies have been reported by Burgerjon and Biache (1967 a, b) and Mayas (1969) on P. brassicae, M. brassicae, and Laspeyresia pomonella. A schematic representation of the results is a given in Figure 11-2. Young L_3 larvae are more sensitive to the toxin (70% mortality) than old L_3 larvae (20% mortality); mortality is observed at the moult or immediately after, during young L_4, whereas teratological effects in the surviving population are not severe. Intoxication during the last larval instar L_5 also produces the most severe effect when applied at the beginning of this

Table 11-2

MORPHOLOGICAL ABNORMALITIES INDUCED BY SUBLETHAL DOSES OF β-EXOTOXIN

ORDER/SPECIES		
Coleoptera	- Atrophied or modified antennae: antennae with club extremities bearing claws	Burgerjon et al. (1969)
	- Atrophy of labial or maxillary palps	
	- Eyes deformed	
Lepidoptera	- Atrophy of the mouthparts, principally of the proboscis	Burgerjon and Biache (1967a,b)
	- Absence of cephalic or thoracic appendages	Burges (1975)
	- Occasionally atrophy of legs and wings	
Diptera		
Musca domestica	- Incomplete pupae-slight atrophy of the mouthparts Partially atropied wings	Burgerjon and Biache (1967b) Cantwell et al. (1964)
Drosophila melanogaster	- No teratologies detectable: delayed moulting	Van Herrewege (1969)
Hymenoptera	- No teratologies detectable: delayed moulting	Burgerjon and Biache (1964) Martouret and Euverte (1964)
Orthoptera	- No teratologies detectable: delayed or incomplete moulting	Burgerjon et al. (1964) Gry (1971)

stage; in addition, the prepupal period is increased in length. The main factors determining the nature and the degree of the response seems to be the larval instar and the precise time within a given instar at which toxin is applied.

In Coleoptera, typically represented by the Colorado potato beetle, *L. decemlineata*, the physiological response is roughly similar (Burgerjon, 1974). Sublethal doses applied at the L_3 or L_4 period induce teratological effects in pupae and emerging adults; Coleoptera differ from Lepidoptera in the nature of these effects and the considerably lengthened prepupal period when the toxin is ingested during the last stage L_4 (Figure 11-2).

In both cases, the percentage of deformed insects is related to the dose of toxin within the sublethal range; in this respect, *L. decemlineata* larvae respond to smaller doses than do Lepidoptera.

Nature of the morphological abnormalities

Burgerjon *et al*. (1969) provided striking examples of the specific effects of the toxin. Generally the malformations affect buccal parts, the eyes, and the antennae and consist of atrophy of these organs and/or morphological modifications. The most common abnormalities in different species are summarized in Table 11-2. In *L. decemlineata* the abnormalities are principally in the antennae (99% of insects examined); malformations include abbreviated antennae and antennae with clubbed extremities bearing a pair of claws which are similar to those normally found on the tarsi of the legs. Other alterations induced with a high frequency (60 to 70%) are abnormalities in the mouth parts; the paired parts of the labial palps are transformed into a reduced unpaired part, and maxillary palps are deformed or reduced. Protuberances on the eyes, or eyes deformed and separated into two parts are also seen but less frequently (Figure 11-3).

In Lepidoptera the minimum effect is atrophy of the mouth parts especially of the proboscis as seen in Figure 11-4, while maximum effects can include the absence of cephalic or thoracic appendages and occasionally atrophy of the legs. The mouthparts are typically the first structures to be affected, with changes in the shape of segments, shortening, bloating or fusion of neighbouring appendages. In most species the abnormalities are easily observed in pupae and are the same nature as those encountered in adults.

Burges (1975), describing the symptoms of sublethal β-exotoxin doses in *G. mellonella*, suggested that the deformities in the mouth parts could be used as "a qualitative diagnostic effect of exotoxin" and possibly as a quantitative assay.

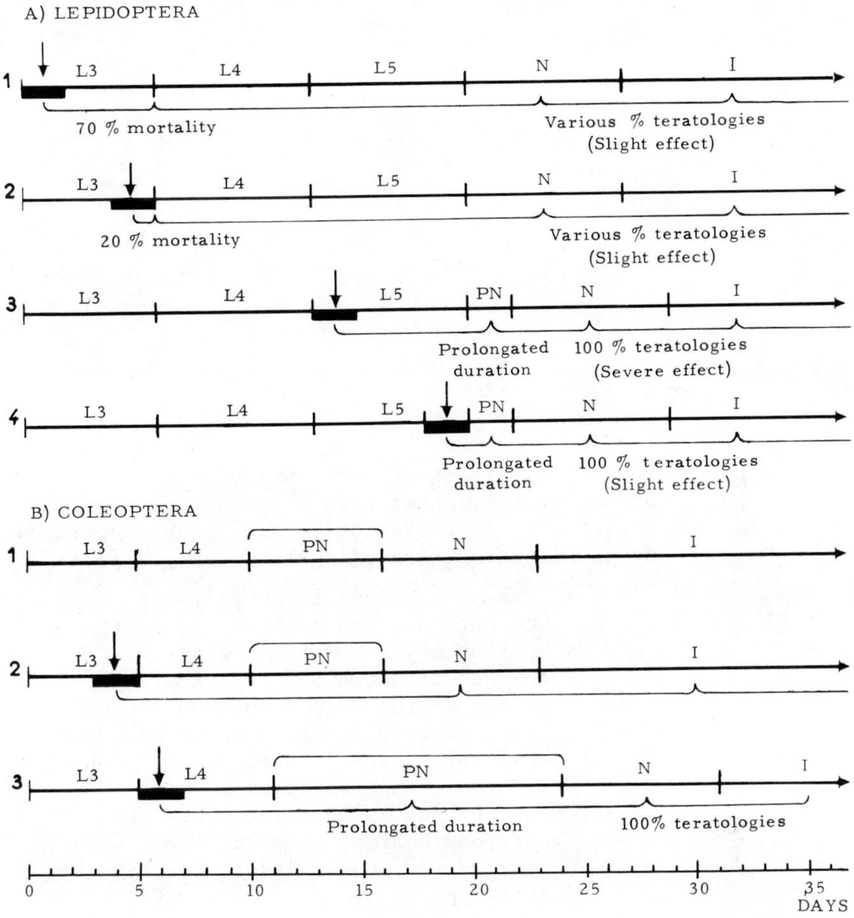

Figure 11-2. Schematic representation of the effects of sublethal doses of β-exotoxin in: Lepidoptera and Coleoptera (In Burgerjon, 1974; reprinted by permission of Masson, S. A., Paris)

L3-L4-L5 = 3rd, 4th and 5th instar larvae; PN = prenymphs (prepupae); N = nymphs (pupae); I = imagos; full bars and arrows indicate the application of the toxin (In B, no. 2, various % of teratologies are observed, but not indicated on the figure).

Figure 11-3. Teratological aspects of *Leptinotarsa decemlineata* treated with sublethal doses of β-exotoxin (In Burgerjon et al., 1969; reprinted by permission of Academic Press Inc., N.Y.).

A - Head of a normal *L. decemlineata*.
B - One antenna with clubbed extremity bearing a pair of small claws.
C - Absence of antennae; the unpaired part of the labial palps appeared reduced.
D - Antennae very deformed or atrophied. Maxillary palps deformed. Labial palps reduced to one unpaired part.
E - One antenna absent; the other antenna deformed bearing claws at the extremity. Abnormalities seen in the left eye.

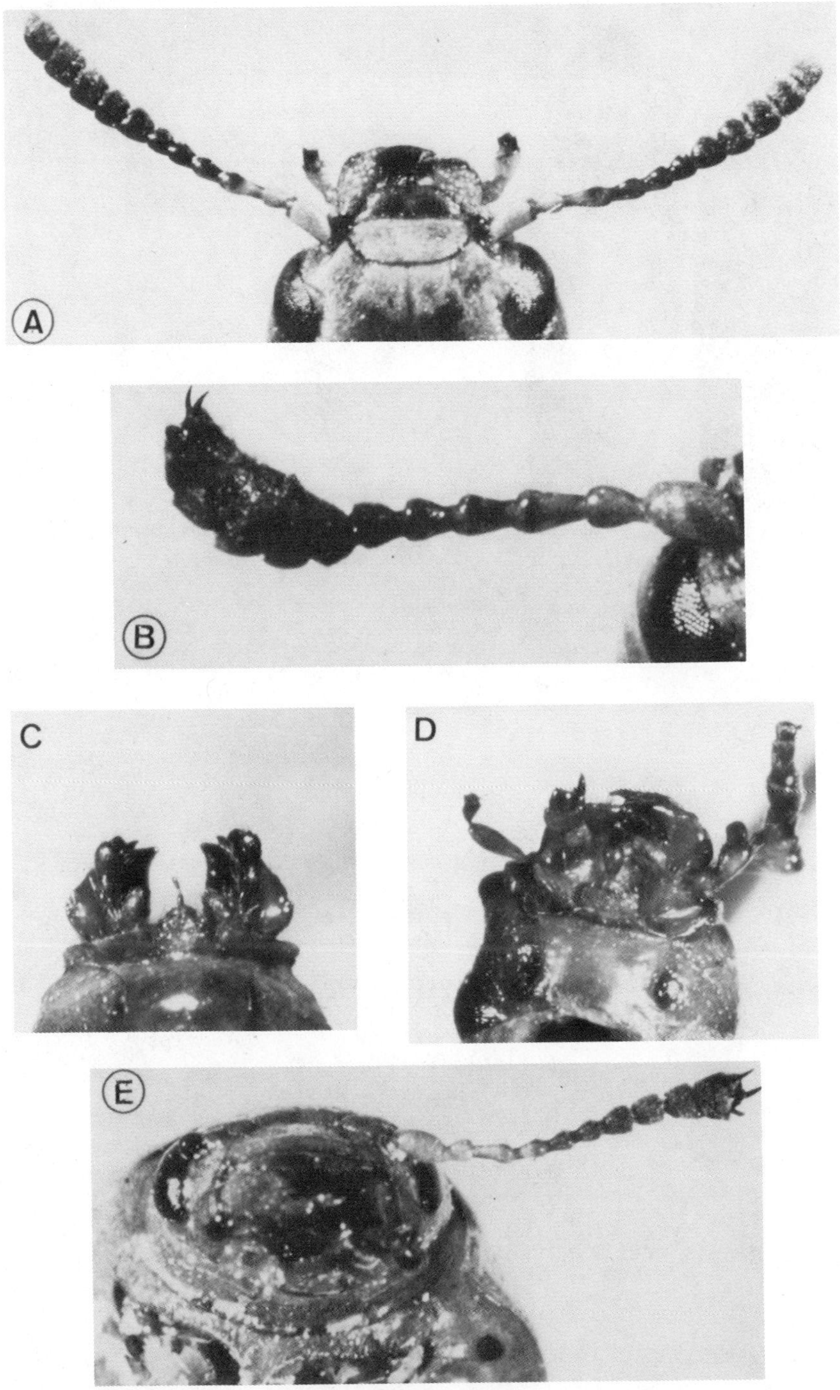
A
B
C
D
E

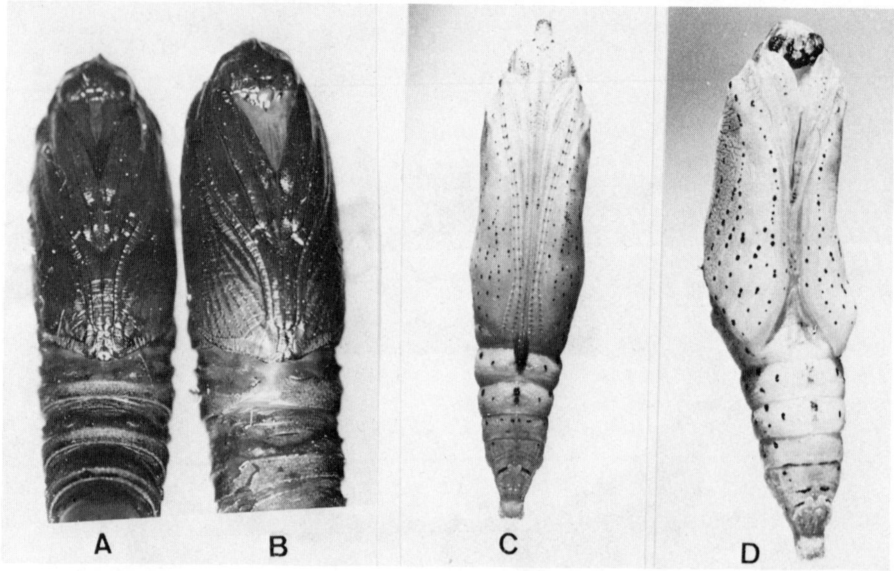

Figure 11-4. Some typical abnormalities induced by sublethal doses of β-exotoxin in Lepidoptera pupae.

A - normal *Tortrix viridans*.
B - *T. viridans* with atrophied mouthparts (absence of proboscis and palps). Slight teratological effect.
C - normal *Pieris brassicae*.
D - *Pieris brassicae* with severe teratologies: atrophied mouthparts, partial necrosis of eyes, depression of the thorax.

(Photographs were kindly provided by A. Burgerjon).

In Diptera, observing teratologies in pupae is more difficult. When sublethal doses are administered to *M. domestica*, incomplete pupae or abnormal adults with reduced wings are seen; in most cases the minimum effect in pupae as well as in adults is atrophy of the epipharynx and labrum mouthparts. In *Drosophila melanogaster*, delayed larval development leading to death (especially of females) was observed by Van Herrewege (1970); whereas David and Vago (1967) reported abnormalities of ovogenesis when the toxin was administered to adults. In this species no teratological effects were seen in pupae or adults emerging from treated larvae. More recently Linnamainmaa *et al.* (1977) observed effects on sex ratio of emerging *D. melanogaster* adults when doses lower than the LD_{50} were administered during the entire development period; males were more resistant to the toxin.

On the other hand, Prkopjev et al. (1976) demonstrated deformities and changes in the function of organs and tissues induced by the endotoxin in flea larvae (*Xenopsylla cheopsis*, *Xenopsylla gerbilli minax* or *Ceratophyllus laeviceps*).

Thus it appears that with some species variation, the emergence of deformed individuals is one of the basic responses of insects to sublethal doses of β-exotoxin.

Mutagenic effects of the toxin

An important and unexpected aspect of the teratologies described above was their ability to be transmitted to the offspring, thus indicating that exotoxin had acted as a mutagen. Burgerjon (1974) discovered this phenomenon in the Colorado potato beetle, *L. decemlineata*; in this species the deformed adults remain able to reproduce normally. The progeny resulting from such individuals was examined for at least two generations. As seen in Figure 11-5, the offspring exhibit antennae whose abnormal claws have disappeared but shortened legs with atrophied claws are still seen. The abnormalities in the mouthparts are transmitted without further modifications (not shown in Figure 11-5). Thus it appears that transmissible characters can be reproduced with some differences in the case of antennae and legs; the results observed in progeny point to the tarsal origin of the claws located at the extremities of antennae in adults emerging from intoxicated larvae. These observations may lead to very promising investigations and indicate to the great complexity of the morphogenetic processes in holometabolous insects. It is not clear whether these hereditary characters are limited to Coleoptera. Preliminary results with Lepidoptera did not suggest the same effects. However in *Drosophila*, Linnamainmaa et al. (1977) found that crude

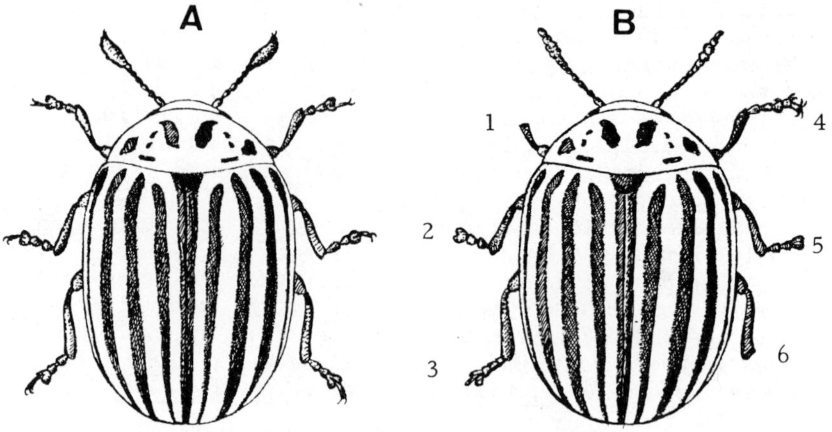

Figure 11-5. Schematic representation of the mutagenic effects of the β-exotoxin in Leptinotarsa decemlineata Say (dessin Prechae, I.N.R.A.) in Burgerjon (1974); reprinted by permission of Masson, S. A., Paris.

A - teratological adult with modified antennae.

B - F_1 progeny from teratological adults with atrophied or modified legs. Claws have disappeared from antennae.

supernatants seem to have a slight effect in increasing sex-linked recessive mutants; in addition, they showed that a significant proportion of the F_1 progeny of males treated with sublethal doses failed to emerge from the pupae.

Obviously any hypothesis explaining such phenomena at the molecular level is rather difficult. Steinhaus and Zeikus (1968) have described morphological abnormalities occuring spontaneously in homogenous populations of the beetle *Tenebrio molitor*; in this insect abnormalities involving shortened, divergent or crumpled elytra, deformed or incomplete legs, and less frequently, antennae are seen, while mouthparts are never involved. However these modifications did not appear similar in nature and in frequency with those induced by exotoxin. The authors did not explain the origin of the spontaneous teratologies observed, but wondered if they are "the result of interaction of various genetic and environmental factors". Metamorphosis-inhibiting compounds with juvenile hormone-like activity are known, however, such as the species-specific "paper factor" isolated by Saxena and Williams (1966) which is toxic for *Dysdercus koenigii*, the red cotton bug, and induces lethal effects or abnormalities in development. Similarly, David (1966) demonstrated that aminopterin, an antagonist for folic acid, produces teratological effects in adult *Drosophila*. So it is obvious that "antimetabolite-like" substances are able to induce morphological abnormalities comparable with those induced by the β-exotoxin, by interrupting some biochemical process with subsequent cytoplasmic changes. Cellular activity is thus disturbed at critical stages during development. However, in agreement with Burges (1976), we would like to emphasize that the action of exotoxin in deforming only the mouthparts and possibly the wings is unique.

Mitogenic action

Using the blast transformation test on mice splenocytes *in vitro*, Truffa-Bachi et al. (1977) have shown that the β-exotoxin has a mitogenic action. This mitogenic stimulation is optimal at an exotoxin concentration of 1 to 4 μg/ml.

BEHAVIORAL AND REPRODUCTIVE RESPONSES OF HOST TO TOXIN

In the course of this review we have frequently mentioned the lack of precise symptoms in the intoxication process leading either to death or to the emergence of deformed pupae or adults. As in the case of other aspects of intoxication discussed above, the behavioral and reproductive responses depend on the period of development and on the doses chosen for intoxication, and are also a function of the species.

As previously seen, larvae treated with sublethal doses can get through one or two larval stages and generate atrophied pupae from which teratological adults will emerge. The behavior of such insects provides interesting examples of the effects of exotoxin in adults which may differ according to the insect species. Generally these adults are characterized by strongly reduced fecundity and longevity, especially in the case of Lepidoptera. In P. brassicae the butterflies have normal behavior when flying, but have difficulties in feeding as a consequence of the proboscis atrophy, so premature death of individuals in the population is observed at the 8th day (instead of 30th day in the control). Their ability to copulate is reduced and fewer eggs are laid (1 to 2% of the control) (Burgerjon and Biache, 1967b). In Coleoptera the response appears rather different (Burgerjon et al. 1969); the process of feeding is less affected, probably because their mandibles are not essentially deformed. Malformations of antennae or labial palps do not prevent feeding nor fecundity. The insects can copulate and generate new teratological individuals as discussed above. However rare adults showing abnormal wings and elytra are not viable.

In D. melanogaster teratologies were not observed; however treating imagos with exotoxin (Van Herrewege, 1970) produced reduced fecundity and a decrease in hatching which are attributed to physiological alterations in the males. Earlier, David and Vago (1967) showed a reduction of longevity and fecundity, and a decrease in the size of eggs which were interpreted as a direct action on the ovaries.

Before ending this discussion, it is interesting to note that some species of Arthropods other than insects are sensitive to β-exotoxin. Krieg (1968) reported the toxicity of heated culture supernatants for Tetranychus telarius larvae and proposed utilization of the toxin against these mites. Similarly Hall et al. (1971) have shown the effectiveness of the exotoxin against the citrus red mite, Panonychus citri, larvae and adults. When high doses of toxin were administered orally to adults mortality in the whole population occurred within ten days; with decreasing doses reduced fertility was observed, whereas recently-laid eggs were directly intoxicated on the treated leaves.

The general effects of β-exotoxin extend to a large variety of species; however there is specificity of response to sublethal doses, as a result of the various processes involved in metamorphosis. We have shown the very high sensitivity of insect systems to the toxin in vivo and in vitro. In eukaryotic systems, while it is true that toxic effects can be induced by injecting exotoxin into mice, several studies have shown its harmlessness when administered orally to mammals. More recently we have shown (de Barjac and Lecadet, 1975) that radioactive toxin ingested by mice was recovered within 24 hours into feces without any fixation at the level of the examined organs: liver, heart, kidneys, spleen and adrenal bodies. On the other hand cytotoxic effects in plants have been reported by Linnamainmaa et al., (1977) but the unusually high doses of exotoxin used (up to 30% in crude supernatant) render questionable the nature of these abnormalities. So the specific effects of the β-exotoxin on insects would make it valuable tool in the future if its innocuity for mammals and plants under field conditions could be definitely confirmed.

CONCLUSIONS

A number of studies have demonstrated the inhibitory effects of the *B. thuringiensis* β-exotoxin on RNA synthesis mediated by DNA-dependent RNA polymerase and the high sensitivity of insect systems. However unsolved problems remain with regard to the exact nature of the inhibitory mechanisms involved in intoxicated larvae. It is obvious that essential biosynthetic pathways are affected thus explaining the general effects of the toxin, possibly amplified by ecdysone-mediated synthesis. Another important problem is elucidation of the specific effects inducing aberrant morphogenesis at precise sites in the course of development and probably involving the imaginal discs. The mutagenic properties are also poorly explained. It is clear that abnormalities during ovogenesis, as well as a possible affinity of the toxin for DNA would constitute a base for studying the effects at the molecular level.

So, due to its mode of action the β-exotoxin raises problems of fundamental interest justifying further studies to which geneticists and molecular biologists as well as entomologists can contribute.

ACKNOWLEDGEMENTS

We are grateful to Dr. A. Burgerjon for giving us documents, for his valuable advice and for helpful discussions. We thank Dr. R. P. M. Bond, Dr. H. D. Burges, and Dr. K. Sebesta, who provided reprints or new information.

REFERENCES

de Barjac, H., 1978. Toxicité de Bacillus thuringiensis var. israelensis pour les larves d'Aedes aegypti et d'Anopheles stephensis. C. R. Acad. Sci. Paris (Série D), 286, 1175-1178.

de Barjac, H. and Dedonder, R., 1965. Isolement d'un nucleotide identifiable a la toxine thermostable de Bacillus thuringiensis var. Berliner, C. R. Acad. Sci. Paris, 260, 7050-7053.

de Barjac, H., Burgerjon, A. and Bonnefoi, A., 1966. The production of heat-stable toxin by nine serotypes of Bacillus thuringiensis. J. Invertebr. Pathol., 4, 537-538.

de Barjac, H. and Dedonder, R., 1968. Purification de la toxine thermostable de Bacillus thuringiensis var. thuringiensis et analyses complémentaires. Bull. Soc. Chim. Chiol., 50, 941-944.

de Barjac, H. and Rious, J. Y., 1969. Action de la toxine thermostable de Bacillus thuringiensis var. thuringiensis administrée a des souris. Rev. Pathol. Comp. Med. Exp., 6, 19-26.

de Barjac, H. and Lecadet, M-M., 1975. Elimination de l'exotoxine thermostable de Bacillus thuringiensis apres ingestion chez la souris. C. R. Acad. Sci. Paris (Série D), 280, 677-679.

de Barjac, H. and Lecadet, M-M., 1976. Dosage biochèmique de l'exotoxine thermostable de Bacillus thuringiensis d'aprés l'inhibition d'ARN-polymérases bactériennes. C. R. Acad. Sci. Paris (Série D), 282, 2119-2122.

Beebee, T. J. C., Korner, A. and Bond, R. P. M., 1972. Differential inhibition of mammalian ribonucleic acid polymerases by an exotoxin from Bacillus thuringiensis. Biochem. J., 127, 619-624.

Beebee, T. J. C. and Bond, R. P. M., 1973a. Effect of the exotoxin of Bacillus thuringiensis on normal and ecdysone stimulated ribonucleic acid polymerase activity in intact nuclei from the fat body of Sarcophaga bullata larvae, Biochem. J., 136, 1-7.

Beebee, T. J. C. and Bond, R. P. M., 1973b. Effect of an exotoxin from Bacillus thuringiensis on deoxyribonucleic-acid dependent ribonucleic acid polymerase in nuclei from adult Sarcophaga bullata. Unusual behavior of eukaryotic polymerase to inhibitors. Biochem. J., 136, 9-13.

Benz, G., 1966. On the chemical nature of heat stable toxin of Bacillus thuringiensis Berliner in Locusta migratoria. J. Invertebr. Pathol., 6, 381-383.

Benz, G., 1975. Action of Bacillus thuringiensis preparation against Larch Bud Moth, Zeiraphora diniana Gu. enhanced by β-exotoxin and DTT. Experientia, 31, 1288-1290.

Bond, R. P. M., 1969. Chem. Comm., 338.

Bond, R. P. M., Boyce, C. B. C. and French, J., 1969. A purification and some properties of an insecticide exotoxin from Bacillus thuringiensis Berliner. Biochem. J., 114, 477-488.

Bond, R. P. M., Boyce, C. B. C., Rogoff, M. H. and Shieh, T. P. 1971. The thermostable exotoxin of Bacillus thuringiensis. In Microbial Control of Insects and Mites, H. D. Burges and N. W. Hussey, eds. Academic Press, London. pp. 275-303.

Burgerjon, A., 1972. Quelques effets physiologiques de la toxine thermostable de Bacillus thuringiensis sur le doryphore Leptinotarsa decemlineata. Ent. Exp. et Appl., 15, 112-117.

Burgerjon, A., 1974. Effets physiologiques et mutagènes sur les insectes de la toxine thermostable de Bacillus thuringiensis Berliner. Annales de Parasitologie (Paris) 48, 835-844.

Burgerjon, A. and de Barjac, H., 1960a. Nouvelles données sur le role de la toxine soluble thermostable produite par Bacillus thuringiensis Berliner. C. R. Acad. Sci. Paris (Serie D), 251, 911-912.

Burgerjon, A. and de Barjac, H., 1960b. Essais préliminaires sur le role insecticide de la toxine thermostable produite par Bacillus thuringiensis Berliner. Proc. Int'l. Congr. Entomol. Vienna, pp. 835-839.

Burgerjon, A. and de Barjac, H., 1964. Etude de la toxine thermostable chez différentes souches de Bacillus thuringiensis Berliner. Coll. Int. Pathol. Insect. Lutte Microbienne Paris, 1962. Entomophaga (hors series) 2, 222-226.

Burgerjon, A., Grison, P. and Kachouli, A., 1964. Activity of the heat stable toxin of Bacillus thuringiensis Berliner in Locusta migratoria. J. Insect Pathol., 6, 381-383.

Burgerjon, A. and Biache, G., 1967a. Effets tératologiques chez les nymphes et les adultes d'insectes dont les larves ont ingèré des doses subléthales de toxine thermostable de Bacillus thuringiensis Berliner. C. R. Acad. Sci. Paris (Serie D), 264, 2423-2425.

Burgerjon, A. and Biache, G., 1967b. Divers effets speciaux et symptomes teratologiques de la toxine thermostable de Bacillus thuringiensis en fonction de l'age physiologique des insectes. Ann. Soc. Ent. Fr. (N. S.), 3, 929-952.

Burgerjon, A., Biache, G. and Cals, P., 1969. Teratology of the Colorado potato beetle Leptinotarsa decemlineata as provoked by larval administration of the thermostable toxin of Bacillus thuringiensis. J. Invertebr. Pathol., 14, 274-278.

Burges, H. D., 1975. Teratogenicity of the thermostable β-exotoxin of Bacillus thuringiensis in Galleria mellonella. J. Invertebr. Pathol., 26, 419-420.

Cantwell, G. H., Heimpel, A. M. and Thompson, M. J., 1964. The production of an exotoxin by various crystal forming bacteria related to Bacillus thuringiensis var. thuringiensis Berliner. J. Insect Pathol., 6, 466-480.

David, P. H., 1966. Role physiologique de l'acide folique chez la Drosophile etudie au moyen d'un antagoniste specifique. Ann. Nut. et Aliment, 4, 339-347.

David, P. H. and Vago, C., 1967. Influence des toxines de Bacillus thuringiensis sur divers caractères physiologiques des Drosophiles adultes. Entomophaga, 12(2), 153-159.

Dunn, P. H., 1960. Control of house flies in bovine feces by a feed additive containing Bacillus thuringiensis var. thuringiensis Berliner. J. Insect Pathol., 2, 13-16.

Farkas, J., Sebesta, K., Horkas, K., Samek, Z., Dolijs, J. and Sorm, F., 1969. The structure of exotoxin of Bacillus thuringiensis var. gelechiae. Coll. Czech. Chem. Comm., 34, 1118-1120.

Galani, G. and Berartlief, Z., 1977. Studies on the histophatological changes induced in the Colorado beetle by the β-exotoxin of Bacillus thuringiensis Berliner. Ann. Inst. Cer. Prot. Plant Rom., 13, 163-176.

Grahame-Smith, D. G., Isaac, P., Heal, D. J. and Bond, R. P. M., 1975. Inhibition of adenyl cyclase by an exotoxin of Bacillus thuringiensis, Nature, 253, 58-60.

Gry, J., 1971. Action de la toxine soluble thermostable de Bacillus thuringiensis sur la croissance et le développement du criquet migrateur africain Locusta migratoria migratoriodes. Orthoptera: Acrididae. Thèse Paris-Sud Orsay 3e cycle Biol. Anim.

Hall, I. M. and Arakawa, K. Y., 1959. The susceptibility of the housefly Musca domestica Linnaeus to Bacillus thuringiensis var. thuringiensis Berliner, J. Insect Pathol., 1, 351-355.

Hall, I. M., Hunter, D. K. and Arakawa, K. Y., 1971. The effect of the β-exotoxin fraction of Bacillus thuringiensis on the citrus red mite. J. Invertebr. Pathol., 18, 359-362.

Heimpel, A. M., 1967. A critical review of Bacillus thuringiensis var. thuringiensis Berliner and other crystalliferous bacteria. Ann. Rev. Entomol., 12, 287-322.

Van Herrewege, J., 1970. Effets de la toxine thermostable de Bacillus thuringiensis sur les larves et les adultes de Drosophila melanogaster. Entomophaga, 15, 209-222.

Johnson, D. E., Bulla, L. A. Jr. and Nickerson, K. W., 1975. Differential inhibition by β-exotoxin of vegetative and sporulation specific ribonucleic acid polymerases from Bacillus thuringiensis cells. In Spores VI. American Society for Microbiology, pp. 248-254.

Kalvoda, L., Prystas, M. and Sorm, F., 1973. Synthesis of exotoxin produced by Bacillus thuringiensis. Tetrahedron Letters, 47, 4671-4674.

Kim, Y. T. and Huang, H. T., 1970. The β-exotoxin of Bacillus thuringiensis I. Isolation and characterization. J. Invertebr. Pathol., 15, 100-108.

Kim, Y. T., Gregory, B. G. and Ignoffo, C. M., 1972a. The β-exotoxin of Bacillus thuringiensis. III. Effects on in vivo synthesis of macromolecules in an insect system. J. Invertebr. Pathol., 20, 46-50.

Kim, Y. T., Gregory, B. G., and Ignoffo, C. M., 1972b. The β-exotoxin of Bacillus thuringiensis. VI. Effects on in vitro synthesis of macromolecules in an insect cell line. J. Invertebr. Pathol., 20, 284-287.

Klier, A. F., Lecadet, M-M. and de Barjac, H., 1973. Inhibition des ARN-polymerases ADN-dependantes de Bacillus thuringiensis par l'exotoxine de B. thuringiensis. C. R. Acad. Sci. Paris, Serie D, 277, 2805-2808.

Krieg, A., 1968. Effectiveness of Bacillus thuringiensis exotoxin on Tetranychus telarius (Acarina: Tetranychidae) J. Invertebr. Pathol., 12, 478-480.

Laurent, P., 1978. Sensibilité relative de diverses cultures cellulaires vis-a-vis de l'exotoxine thermostable de Bacillus thuringiensis. Rev. Can. Biol., 37, 127-130.

Linnamainmaa, K., Sorsa, M., Carlberg, G., Gripenberg, U. and Meretoja, T., 1977. Mutagenicity of Bacillus thuringiensis exotoxin. II. Submammalian tests. Hereditas, 85, 112-122.

Mac Connel, E. and Richards, A. G., 1959. The production by Bacillus thuringiensis Berliner of a heat stable substance toxic for insects. Can. J. Microbiol., 5, 161-168.

Mackedonski, V. V., Nilolaev, N., Sebesta, K. and Hadjiolov, A. A., 1972. Inhibition of ribonucleic acid biosynthesis in mice liver by the exotoxin of Bacillus thuringiensis. Biochem. Biophys. Acta. 272, 56-66.

Martouret, D. and Euverte, G., 1964. The effect of Bacillus thuringiensis Berliner preparations on the honey bee under conditions of forced feeding. J. Insect Pathol., 6, 198-203.

Mayas, I., 1969. Contribution a l'étude du mode d'action de Bacillus thuringiensis Berliner sur Laspeyresia pomonella (Lepidoptera: Tortricidae) Thèse de Troisieme Cycle, Paris, Biologie Animale Entomologie.

Norris, J. R., 1969. Sporeformers as insecticides. In The Bacterial Spore, C. Gould and A. Hurst, eds. Academic Press London and New York. pp. 485-516.

Prasad, S. S. V., Tilak, K. V. B. R. and Gollakota, K. G., 1972. Role of Bacillus thuringiensis var. thuringiensis on the larval survivability and egg hatching of Meloidogyne sp., the causative agent of root knot disease. J. Invertebr. Pathol., 20, 377-378.

Prokopjev, V. N., Jakunin, B. M., Chervjakov, V. D. and Dibickiy, A. M., 1976. Pathological changes in fleas caused by the exotoxin of Bacillus thuringiensis. Parazitologija S. S. S. R., 10(3) 222-226.

Quiot, J. M., 1975. Recherches sur la culture in vitro de cellules d'insectes et l'action de germes entomopathogênes en culture cellulaire. Doct. Sciences Naturelles, Académie de Montpellier.

Riou, J. Y., 1968. Etude in vitro de l'action de la toxine thermostable de Bacillus thuringiensis var. Berliner sur la souche cellulaire BHK21C13. Diplôme d'Etudes Approfondies Biologie Cellulaire, Université Paris.

Rogoff, M. and Yousten, A., 1969. Microbial considerations on Bacillus thuringiensis. Ann. Review of Microbiology, 23, 357-386.

Saxena, K. N. and Williams, C. M., 1966. "Paper factor" as an inhibitor of the metamorphosis of the red cotton bug Dysdercus koenigii. Nature, 210, 441-442.

Sebesta, K., Horska, K. and Vankova, J., 1967. Isolation, purification and toxicity of a thermostable exotoxin from the strain of Bacillus gelechiae (Ault) In: Proc. Int. Colloq. Insect Pathol. and Microbial. Control, Wageningen. North Holland Publishing Co., Amsterdam. p. 238.

Sebesta, K., Horska, K. and Vankova, J., 1969. Inhibition of de novo RNA synthesis by the insecticidal exotoxin of Bacillus thuringiensis var. gelechiae. Coll. Czech. Chem. Comm., 34, 1786-1791.

Sebesta, K. and Horska, K., 1968. Inhibition of DNA-dependant RNA polymerase by the exotoxin of Bacillus thuringiensis var. gelechiae. Biochem. Biophys. Acta. 169, 281-282.

Sebesta, K. and Horska, K., 1970. Mechanisms of inhibition of DNA dependant RNA polymerase by exotoxin of Bacillus thuringiensis. Biochem. Biophys. Acta. 209, 357-376.

Sekeris, C. E., 1965. In Mechanisms of Hormone Action, D. Karlson, ed. Academic Press, New York. pp. 149-164.

Simpson, P., Ziller, C. and de Barjac, H., 1971. Effets de la toxine thermostable de Bacillus thuringiensis sur le développement des disques imaginaux de la Drosophile. C. R. Acad. Sci. Paris, Série D, 272, 3158-3162.

Smuckler, E. and Hadjiolov, 1972. Inhibition of hepatic deoxyribonucleic acid-dependant ribonucleic acid polymerases by the exotoxin of Bacillus thuringiensis in comparison with the effect of D-amanitin and cordycepin. Biochem. J., 129, 163-166.

Somerville, H. J. and Swain, H. M., 1975. Temperature dependant inhibition of polyphenylalanine formation by the exotoxin of Bacillus thuringiensis. F. E. B. S. Letters, 54, 330-333.

Steinhaus, E. A. and Zeikus, R. D., 1968. Teratology of the beetle Tenebrio molitor I. Gross morphology of certain abnormality types. J. Invertebr. Pathol., 10, 190-210.

Tremblay, F. L. J., Huot, L. and Perron, J. M., 1972. Pénétration de l'exotoxine thermostable de Bacillus thuringiensis a travers le chorion d'Acheta domesticus. Entomol. Exp. App., 15, 397-398.

Tremblay, F. L. J., Huot, L. and Perron, J. M., 1973. Effets de l'exotoxine thermostable de Bacillus thuringiensis Berliner sur la consommation d'oxygène par les embryons d'Acheta domesticus L. Canad. J. Zool., 51, 433-435.

Truffa-Bachi, P., Motta, I., de Barjac, H. and Laurent, P., 1977. Mise en évidence d'un pouvoir mitogène chez l'exotoxine thermostable de Bacillus thuringiensis, C. R. Acad. Sci. Paris, Série D, 285, 455-457.

Vankova, J. and Leskova, A. J., 1972. Changes in the homolymphes of Galleria mellonella caterpillars infected by the exotoxin of Bacillus thuringiensis. Acta Entomol. Bohemoslovaca, 69, 297-304.

Vankova, J. and Horská, K., 1975. The activity of gut phosphatases in some insect species and the effects of gut homogenates on the exotoxin of Bacillus thuringiensis in vitro. Acta. Entomol. Bohemoslovaca, 72, 7-12.

Chapter 12

BACTERIAL DISEASES OF INVERTEBRATES OTHER THAN INSECTS

by ALBERT K. SPARKS*

INTRODUCTION

In a review of the fungal diseases of Crustacea, Unestam (1973) noted the literature indicates that bacterial diseases are far more important in vertebrates than fungal diseases, but that the reverse is true among the invertebrates. The reasons for the apparent relatively minor role of bacterial diseases in invertebrates are unknown, but Unestam speculated on several possibilities. He discounted the likelihood that the fungi are more interesting to scientists or that they are more obvious than bacteria and postulated that the difference probably was the result of a more efficient host defense mechanism against fungi in the vertebrates (see also Chapter 17).

Although not disputing Unestam's basic premise that invertebrates appear to have more fungal than bacterial diseases, there are some points worth considering. To the non-specialist, fungi are usually more obvious than bacteria, especially in tissue sections prepared by standard methods. Furthermore, fungi are usually more easily identified as a primary pathogen rather than a secondary invader of a moribund animal. Finally, I submit that the fungi are probably more interesting to the general invertebrate pathologist who is not a specialist on either group.

In the matter of host defense, I feel that Unestam is probably correct. Since the primary host response of invertebrates to foreign material lies in the hemocytes, specifically by phagocytosis and encapsulation, invertebrates are intrinsically more competent to deal with invaders of

*National Marine Fisheries Service, National Oceanic and Atmospheric Administration, 2725 Montlake Boulevard East, Seattle, Washington 98112

small size, including bacteria, than the larger fungi. It is interesting to note that bacterial diseases in invertebrates are almost always caused by species of bacteria that have developed a mechanism to escape phagocytosis, are able to survive phagocytosis, or are enabled to overwhelm the host because of some extrinsic stress to the host.

In recent years, bacterial diseases in invertebrates have been recognized and reported in increasing numbers. Most of these have been associated with captive populations, in species of commercial importance being held prior to sale or being grown under aquaculture conditions. While it is true that the stress of unfavorable environmental conditions, along with the enhanced potential for transmission associated with crowding, enhance the potential for epizootics, at least some of these diseases have been reported to occur in nature. However, the role of bacterial diseases in epizootics of feral invertebrates is not well documented and is difficult to establish. A review of the current knowledge of the bacterial diseases of invertebrates other than insects, in both captive and feral populations, follows.

ROOT-KNOT NEMATODE BACILLOSIS

Syn. Nemapox Disease

Bacillus penetrans, the causative organism was originally described as a microsporidian, *Duboscquia penetrans* (Thorne, 1940), was later (Williams, 1960) thought to be a haplosporidian, then a fungus (Canning, 1973), before Mankan (1975) transferred it to the genus *Bacillus*. Root-knot nematode bacillosis is a fatal disease and *B. penetrans* is an excellent candidate as a biological control agent of these economically important plant parasites (Mankan, 1975).

The nematode becomes infected by coming in contact with spores present in the soil or root debris. In *Meloidogyne* spp., and other root-knot nematodes over a wide geographic range, the spores become firmly attached to the cuticle and, when present in large numbers, cause a distinctive appearance that led Mankan to propose "Nemapox Disease" as a descriptive name for the syndrome. Infected female worms taken from galls on host plant roots are somewhat smaller than normal and usually do not produce eggs.

The spore enzymatically invades the cuticle and apparently germinates to become a vegetative bacterial cell that may undergo a typical logarithmic growth phase, but this stage of the life-cycle is difficult to observe. Numerous minute rod-shaped bodies are occasionally present in infected nematodes, but they have not been unequivocally linked to the life cycle of *B. penetrans*. At some undetermined point, the vegetative cells cease dividing and elongate to form a dichotomously branched filamentous structure, with the terminal cells of the branches enlarging so that the entire structure resembles a cluster of grapes.

The terminal cells are developing sporangia that contain spores which, with phase contrast or Nomarski optics, have the highly refractile appearance characteristic of the endospore of "milky disease" of the Japanese beetle and other scarabeid grubs caused by *B. popilliae* and *B. lentimorbis* (see Chapter 7).

All stages of development are present in early infections, but in terminal stages of the disease most host tissues have disappeared and the nematode becomes essentially a container for the spores. Infective spores are released into the soil when the nematode's cuticle ruptures.

Bacillus penetrans spores are highly resistant to dessication and remain viable for several years. Larvae of several species of *Meloidogyne* and *Pratylenchus* are easily infected when exposed to the spores and under experimental conditions the disease is highly effective in controlling root-knot nematode populations.

ASCARID CUTICULAR LESION BACTEREMIA

Cuticular lesions of probable bacterial etiology have been frequently reported in a number of parasitic ascarid nematodes. These reports were reviewed by Anderson et al. (1971) and include the genera *Ascaris*, *Parascaris*, *Toxascaris*, *Heterakis*, *Oxyurus*, and *Strongylus*. The lesions in the different genera are indistinguishable grossly, even though several bacterial genera have been implicated as causative organisms. Both the lesions and their etiology have been studied most thoroughly in the swine ascarid, *Ascaris suum*, and serve as an example of this disease or syndrome.

The lesions are grossly recognizable as yellow-brown or white circular spots scattered over the entire cuticle (Figure 12-1) (Stewart and Godwin, 1963; Anderson et al., 1971) or elongate, somewhat irregular, yellowish-brown lesions that are not widely distributed (Anderson et al., 1971). The round lesions apparently begin as white pinhead spots that enlarge to become ring-like and finally, as they continue to enlarge, to become yellow-brown, circular, circumscribed spots (Stewart and Godwin, 1963). The older lesions are slightly elevated peripherally and depressed centrally, with the outer regions somewhat darker than the centers.

Histologically, the early, pinpoint lesion consists of a thickened cuticle (Figure 12-2) that becomes cratered as the disease progresses (Figure 12-3). Although appearing grossly

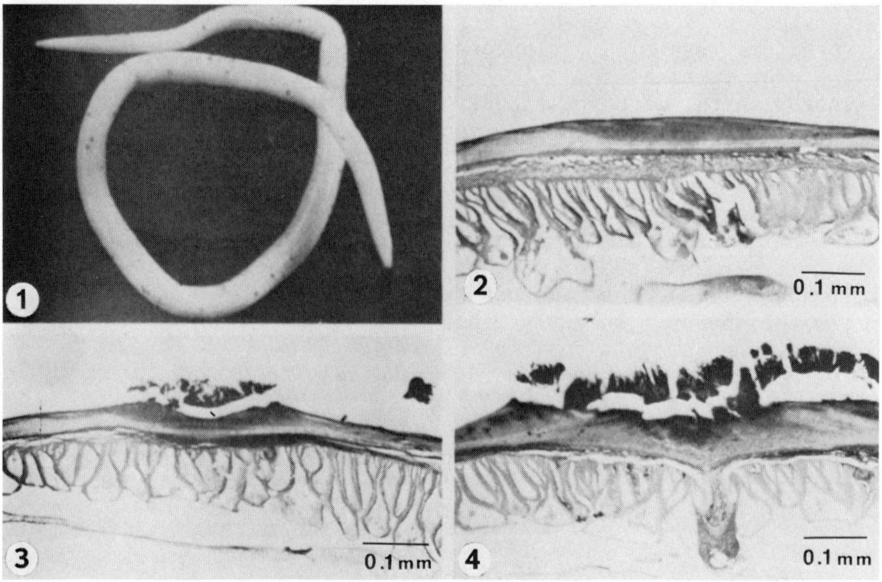

Figure 12-1. *Ascaris suum* with grossly visible cuticular lesions.

Figure 12-2. Early cuticular lesion. Note thickened cuticle. Would appear grossly as a pinpoint lesion. H&E.

Figure 12-3. Crater lesion. Would appear grossly as a dark ring with a lighter central zone. H&E.

Figure 12-4. Older lesion. Note prominent erosion of cuticle and perforation of body wall. H&E.

Figures 12-1 through 12-4 from Stewart and Godwin, 1963; reprinted by permission of Journal of Parasitology.

to be confined to the cuticle, older lesions perforate the body wall (Figure 12-4), particularly when present on the lateral line of the worm (Stewart and Godwin, 1963). Scanning electron microscopy reveals that, in even the smallest lesions, the cuticular architecture is markedly altered (Anderson et al., 1971). The more extensive cuticular damage in the larger lesions consists of destruction of the external and internal cortical layers of the cuticle and cavitation of the underlying homogenous layers. In tissue sections, the underlying muscle is often separated from the sub-cuticle, a possible artifact but probably associated with the lesion.

Although Escherichia coli, Pseudomonas sp. and a yeast, Candida sp. were cultured from the lesions, Stewart and Godwin believed that the Pseudomonas sp. was responsible for the lesions because they were able to induce the lesions experimentally with cultures of the pseudomonad. Although it is possible that the lesions result from invasion of injured cuticle by normal gut flora of the pig, it is more likely that they are caused by primary microbial infection of intact cuticle.

EARTHWORM BACTERIAL SEPTICEMIA

Syn. Earthworm Blister Disease

Earthworms, Lumbricus terrestris, were reported to be highly susceptible to experimental challenges by cultured Bacillus thuringiensis (Smirnoff and Heimple, 1961). There was a mortality rate of 100 per cent over a two month period of earthworms exposed to even the lowest dosage of crude cultures of B. thuringiensis, while no control worms died over a four month period.

From preliminary studies of the histopathology of the dead worms, a massive septicemia was reported. It appeared that bacterial invasion occurred in the foregut, but all tissues, including blood vessels, were invaded. The bacteria sporulated and formed crystals in the earthworms, a phenomenon of rare occurrence in insects.

Subsequently, Heimple (1966), isolated two strains of B. thuringiensis from another species of earthworm, Eisenia foetida, suffering from "blister disease", a syndrome present in serious epizootics in worm farms in Tennessee and other parts of southeastern United States. Affected earthworms developed blisters (see Figure 12-5 and 12-6), containing yellow cloudy material, at random locations on the external surface of the body. Soon after the blisters formed, the lesions ulcerated, the worms fragmented and died. Histological examination revealed that all the blisters contained bacteria

Figure 12-5. Four specimens of *Eisenia foetida* with blisters on anterior regions of bodies.

Figure 12-6. A close up of the blister on anterior portion of second worm from right.

Figure 12-7. Photomicrograph of sporeformer with crystal-shaped body in the cell.

From Heimpel, 1966; reprinted by permission of Academic Press, Inc.

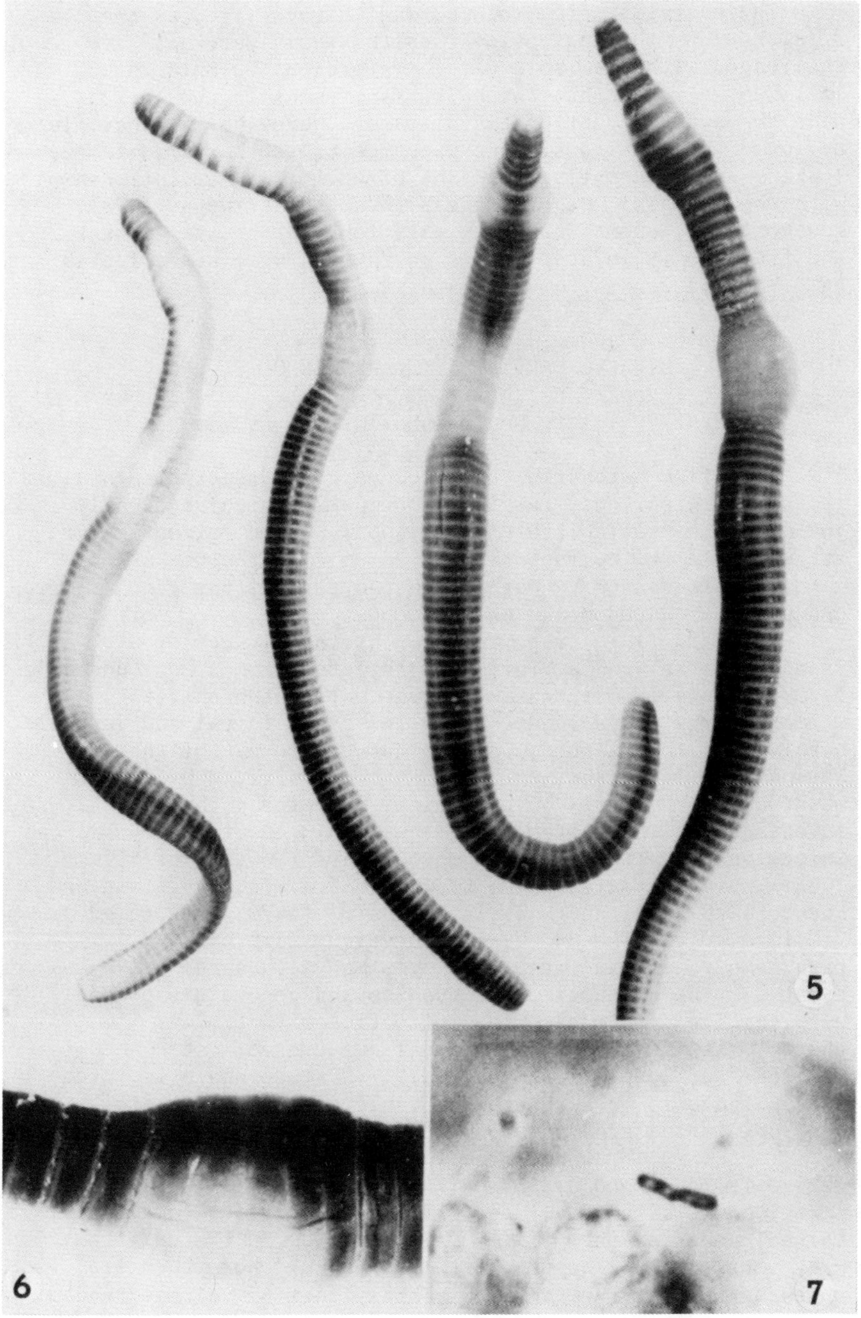
5
6
7

resembling *B. thuringiensis*, while adjacent tissue was free of bacteria. In one case the bacteria had sporulated and contained one or two diamond shaped crystals (Figure 12-7).

The bacteria were isolated and characterized as two strains of *B. thuringiensis*. Healthy worms were not challenged with the isolates, therefore, as Heimple noted, it is not certain that the bacteria present in the lesions were the causal agent of the disease. Nevertheless, Heimple believed that the consistent presence of the bacteria in the lesions was presumptive evidence of an intimate relationship with the disease. Other workers (Benz and Altwegg, 1975), however, have questioned this relationship because laboratory and field tests indicated that earthworms were not affected by *B. thuringiensis*.

BIVALVE LARVAL BACILLARY NECROSIS

Syn. Bacillary Necrosis of Oyster Larvae

Shellfish hatcheries, both commercial and experimental, frequently experience fatal epizootics in larval and early juvenile stocks (Tubiash et al., 1965). Some epizootics are attributable to the fungus *Sirolpidium zoophthorum* but the majority are caused by pathogenic bacteria of the genus *Vibrio* and possibly *Pseudomonas* and *Aeromonas*.

Guillard (1959) reported that broth cultures of a *Pseudomonas* sp. and a *Vibrio* sp. isolated from a moribund hard clam (*Mercenaria mercenaria*) larvae were pathogenic to *M. mercenaria* larvae. Recurring losses of larval and juvenile bivalves at the Milford, Connecticut Laboratory of the National Marine Fisheries Service, characterized by sudden onset and high levels of mortality, prompted a thorough investigation that established the role of bacterial infections in bivalve larval mortalities (Tubiash et al., 1965; Tubiash et al., 1970; Brown, 1973, 1974; and Brown and Losee, 1978). Although Tubiash et al., (1970) identified *Vibrio anguillarium* and *V. alginolyticus* (listed as *V. parahaemolyticus* in latest Bergey's Manual) as causative agents of the disease, other species and genera are probably also involved. For example, epizootics with similar characteristics occurred in larval cultures of *Ostrea edulis* in 1968 and 1969 (Sindermann, 1977). Pseudomonads isolated were pathogenic to larvae of *O. edulis*, *Mytilus edulis*, *Venerupis decussata* and *Crassostrea gigas*.

The course of the epizootic in experimentally infected larvae is rapid and dramatic (Tubiash et al., 1970). Within 4 to 5 hours after challenge with a virulent strain, there is marked reduction in motility of the larvae in the culture and many settle to the bottom of the container where they lie quiescent with either the rudimentary foot or velum extended.

"Swarms" of bacteria, originating from discrete foci on the margins of a few larvae, appear and become progressively more numerous.

Larval tissues often appear to pulsate because of the activity of the proliferating bacteria. In a heavily infected culture, all larvae may succumb within an 18 hour period, but ciliary activity on loosely attached or detached portions of the velum continues after all other soft tissues have been destroyed. At the peak of the epizootic, bacteria can be seen swarming around the vela of moribund larvae and secondary invasion by opportunistic ciliated protozoans frequently occurs (Brown and Losee, 1978).

The progression of the disease in spontaneous and experimental infections is quite similar, differing only in timing, undoubtedly because the higher concentrations of larvae in the experimental challenges and the large bacterial inoculum combine to accelerate the progression of the epizootics.

Oyster embryos are more susceptible to infection than the veliger stage (Brown, 1973) and veligers, in turn are more susceptible than presetting larvae. It is not known whether the increasing resistance with age is due to the added shell of older larvae or to actual increased resistance with increasing age. Some larvae surviving the epizootics develop abnormally while others, though normal in appearance, have decreased growth rates. Although heat-killed bacteria are harmless to bivalve larvae, the filtrates of cultures pathogenic to larvae are toxic to larvae and are also believed to be teratogenic.

AMERICAN OYSTER *VIBRIO* CARDIAC EDEMA

In contrast to the marked susceptibility of larval oysters to *Vibrio* infections, adult *C. virginica* are relatively resistant to bacterial infections, ingesting or otherwise clearing from the medium high concentrations of vibrios and other reportedly pathogenic bacteria (Tubiash et al., 1973). However, four of more than 10,000 adult oysters in a routine survey from the Maryland portion of Chesapeake Bay had grossly enlarged hearts and pericardial chambers that appeared to be related to infection with *Vibrio anguillarium*.

In all cases, the heart and pericardial chamber were greatly enlarged, swollen and distended with fluid which, in one case, had a jelly-like consistency. Histologic examination of the four oysters revealed no tissue abnormalities other than the cardiac involvement.

Aspirated pericardial fluid was estimated to contain concentrations of 10^6 to 10^7 per ml of small gram-negative rods, which, on culture, were identified and confirmed to be

V. anguillarium. When isolates from three of the oysters were used to challenge week-old oyster larvae, average larvae mortalities ranged from 87 to 96 per cent. However, adult oysters challenged by intracardiac injection of the isolates and maintained in flowing seawater for five months, remained in good condition with normal hearts.

PACIFIC OYSTER MASS MORTALITIES OF POSSIBLE BACTERIAL ETIOLOGY

Syn. Adult Oyster Vibriosis, Focal Necrosis

Mass mortalities of Pacific Oysters, Crassostrea gigas, both in Japan (Kanno, et al., 1965) and the Pacific coast of the United States (Grischkowsky and Liston, 1974) have occurred for many years. Despite much research in both the United States and Japan, the etiology of these mortalities has not been unequivocally established. The epizootics occur in summer when temperatures are elevated and phytoplankton production is high. Dying oysters are typically fat and sexually mature or spawning, leading the Japanese biologists to conclude that the mortalities result from physiological stress of heavy gonadal production and incomplete spawning despite the fact that in all their studies high percentages of dying oysters had bacterial infections and associated tissue lesions which they called "multiple abcesses". The same or a similar disease was termed "focal necrosis" by researchers at the Bureau of Commercial Fisheries (now National Marine Fisheries Service) laboratory at Oxford, Maryland (Sindermann, 1970).

Colwell and Sparks (1967) isolated Pseudomonas enalia from dying oysters in Washington waters and were able to produce the disease in other oysters exposed to the bacterium under laboratory conditions. Histopathological studies of dying oysters were suggestive of bacterial disease and P. enalia was recovered from the challenged oysters. However, the route of infection was artificial, i. e. by injection of large numbers of bacteria into either the body or shell cavity of the oyster, and the numbers of oysters challenged were small.

Lipovsky and Chew (1971) transferred oysters from commercial beds to the laboratory and held them at either 10°C or 20°C. Over a period of 20-25 days, oysters kept at 20°C for 24 h per day had a cumulative mortality of 100 per cent; oysters at 20°C for 16 h and 10°C for 8 h had a 70 per cent mortality; those held at 20°C for 8 h and 10°C for 16 h suffered a 53 per cent mortality; and those held at 10°C for 24 h a day had zero mortality. Vibrio parahemolyticus was isolated from the water and the oysters maintained at both

10°C and 20°C. However, bacterial counts in both the water and the oysters were higher by several orders of magnitude at the higher temperatures.

In another laboratory experiment, Lipovsky and Chew (1972) studied the histopathology in oysters infected with the pathogenic bacteria. They reported hemocyte infiltration into the gonads and phagocytosis of sex products; degeneration, fragmentation, and sloughing of the digestive tubule epithelium; and massive hemocytic infiltration of the connective tissue around the stomach, gut, and blood sinuses (Figure 12-8). There were also focal lesions and hemocytic infiltrates in the connective tissue of the mantle and irregular bands of infiltrated hemocytes beneath the mantle epithelium. "Brown" cells, commonly associated with a variety of diseased conditions in oysters, were common throughout the connective tissue. Although scattered foci of bacteria were present in various locations, the greatest concentration was in the gastric shield (Figure 12-9) and the adjacent stomach wall. The areas of the gastric shield containing bacteria were necrotic and the stomach epithelium was necrotic and appeared autolytic (Figure 12-10). Numerous hemocytes containing phagocytosed bacteria were present in the lumen of the stomach (Figure 12-11), apparently as a result of diapedesis. The massive hemocytic infiltration beneath the mantle epithelium was recognizable grossly by a greenish discoloration of the body surface due to the copper content of the massed hemocytes.

Grischkowsky and Liston (1974) also investigated the interrelationships of temperature and bacterial pathogenicity of adult *C. gigas* under laboratory conditions, using strains of bacteria isolated by culture from dead or moribund oysters and added directly to the water in tanks. They found mortality to be highest in high (20°C) temperature bacteria-inoculated tanks and lowest in low-temperature control tanks. Statistical analysis showed that high temperature was a substantial contributing factor to mortalities and the moribund or dead oysters held in the high temperature consistently had higher bacterial counts than those held in low temperature control tanks. Further, they stated that *Vibrio anguillarium* and *V. alginolyticus* (*V. parahemolyticus* biotype *alginolyticus*) were implicated as faculative pathogens of Pacific oysters at elevated temperatures.

Although it appears quite clear that several species of bacteria are capable of causing disease and death in adult Pacific oysters in the laboratory under elevated temperature conditions, the possible interrelationship of physiological stress and bacterial infection under natural conditions remains unclear. My personal belief is that the summer mortalities are probably of bacterial etiology, with the physiological stress of gonadal maturation and spawning of, at best, minor importance.

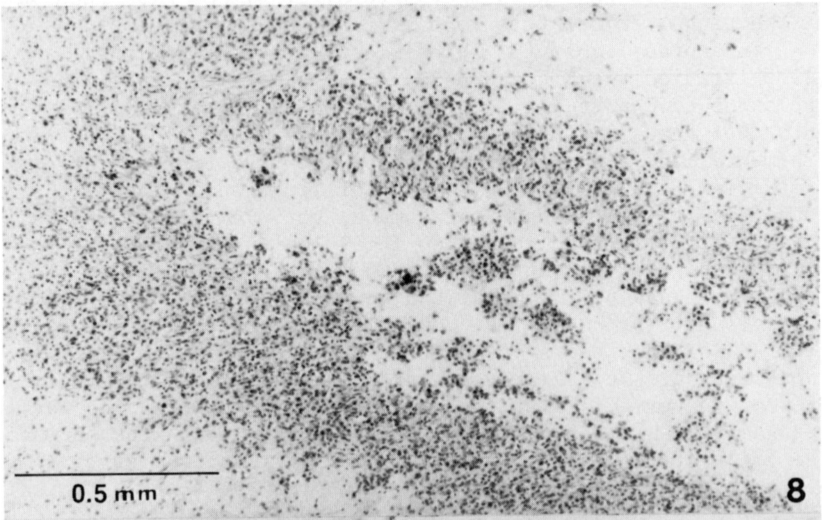

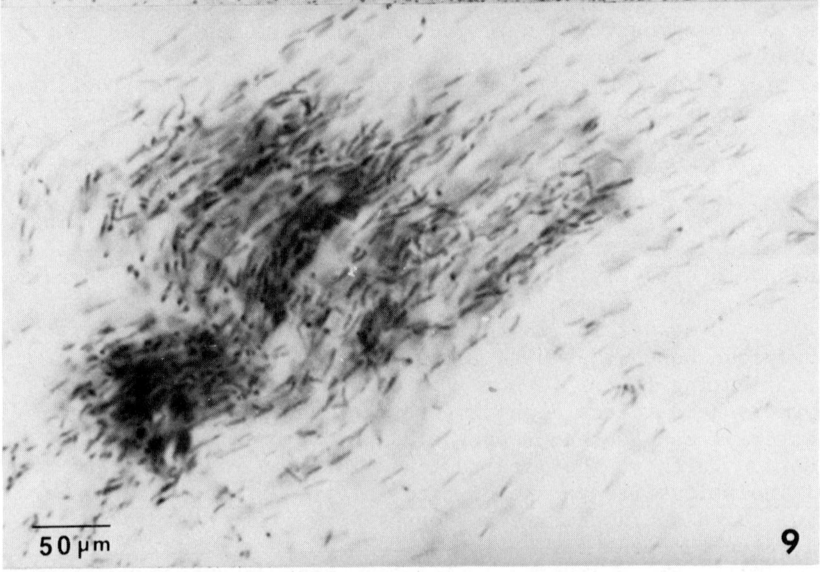

Figure 12-8. Hemocytic infiltration around blood sinus of a Pacific oyster infected with pathogenic Vibrio. H&E.

Figure 12-9. Bacteria in gastric shield. H&E.

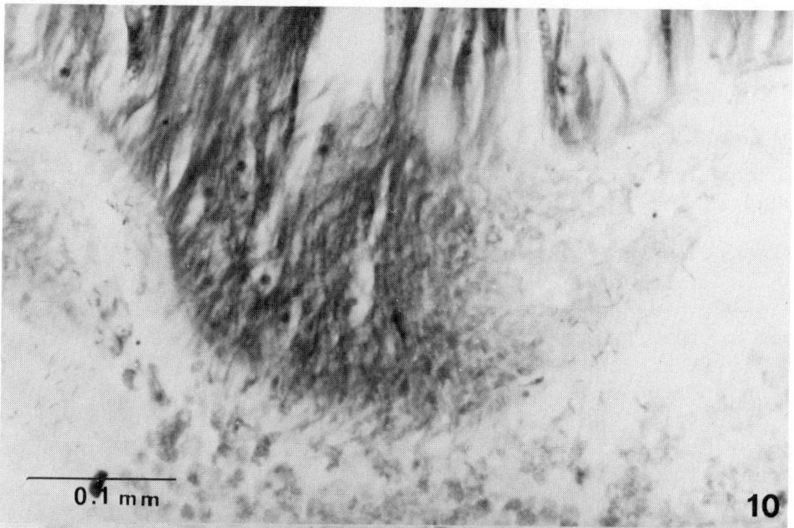

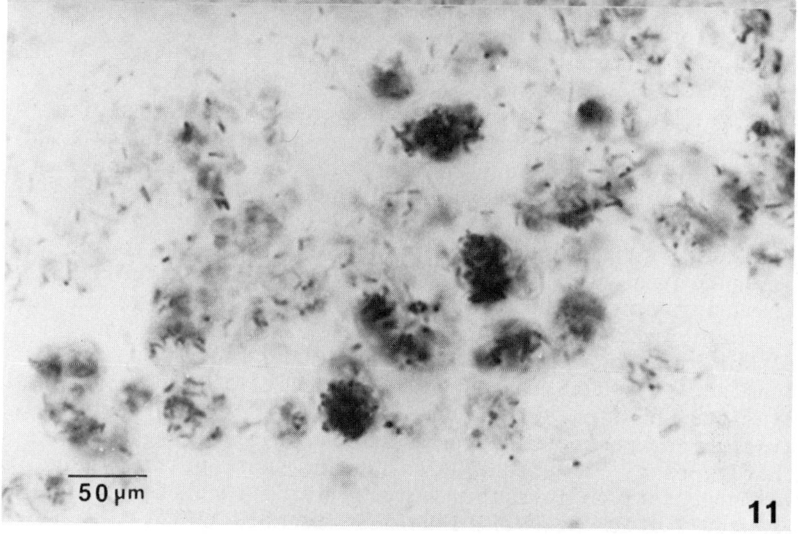

Figure 12-10. Stomach epithelium of Vibrio challenged oyster. Note necrosis (NEC), autolysis (AU), and bacteria (Bac).

Figure 12-11. Necrotic phagocytes laden with bacteria.

Figures 12-8 through 12-11 from Lipovsky and Chew, 1972; reprinted by permission of the National Shellfisheries Association and the authors. Printed from negatives provided by authors.

GASTROPOD ACID-FAST BACTEREMIA

The presence of acid-fast bacilli in an aquatic snail was first reported in Australorbis glabratus (Pan, 1956). Subsequently, a similar, if not identical, organism was found in laboratory colonies of Heliosoma anceps (Michelson, 1961) of unknown origin that had been maintained for a number of years in the Harvard University Department of Tropical Public Health. The bacillus was isolated, cultured and identified as a species of Mycobacterium, the first reported from an invertebrate. Although cultures were sent to several specialists, its relationship to known species of Mycobacterium remains unknown.

Grossly, the earliest manifestation of the disease is a conspicuous clubbing of the tips of the tentacles, usually accompanied by localized increase in pigmentation. Small spherical yellow bodies develop on the tentacles or body that can be observed through the shell.

Histopathologically, the tips of the tentacles are markedly aplastic and contain large vacuolated areas (Figure 12-12). The grossly recognizable increased pigmentation is seen microscopically as aggregations of pigment particles in the swollen tips of the tentacles or closely adjacent tissue. The normally ciliated columnar outer epithelium appears compressed and squamose with frequent loss of cilia.

Microscopic study of sections containing the yellow bodies reveals that they are discrete tubercle-like lesions containing numerous acid-fast bacilli (Figures 12-12 and 12-13). In older infections, tubercles develop on the surface epithelium concurrent with atrophy of the tentacles. Michelson stated that the bacteria are intracellular parasites, particularly in amoebocytes, and noted that parasitized amoebocytes accumulate in small aggregates with other types of infected cells that are subsequently walled off by fibroblasts to form the characteristic tubercles (Figure 12-14). However, I believe it more likely that the bacilli are phagocytosed by the amoebocytes and that the aggregates are encapsulated by fibroblasts in the typical gastropod response to foreign material.

In either case, immature tubercles are composed primarily of masses of cells containing the bacteria. As the tubercles mature, the cells become necrotic and eventually are resorbed or destroyed, leaving vacuoles containing enormous numbers of bacilli (Figure 12-15). Subsequently, the bacilli appear to invade the fibroblasts forming the wall of the tubercle and spread the infection to surrounding tissue. Sattelite lesions occur most frequently in the kidney, tentacles, rectal ridge and foot, but have been observed in the tissues in all the major organs.

Spontaneous transmission of the disease occurred not only in *H. anceps* but also in *Australobis glabratus* and *Biomphalaria pfeifferi* in the laboratory. It was also experimentally transmitted by inoculation or contamination of the water to 8 of 13 strains of planorbid snails, including a total of 6 species of *Heliosoma*, *Australorbis*, *Biomphalaria* and *Bulinus*. There is evidence of host specificity by the mycobacterium, because Michelson was unable to produce infection in two species of land snails, *Otala lactea* and *Helix aspara* by inoculation.

GIANT AFRICAN SNAIL AEROMONASIAS

Syn. Giant African Snail Bacteriemia

Leucodermic Disease Syndrome

Mead's Syndrome

The giant African snail, *Achatina fulica* was introduced into Hawaii by residents for aesthetic and medicinal purposes in 1936 (Pemberton, 1938, cited by Dean et al., 1970). The species became established and quickly developed into a major agricultural pest in Hawaii and many other areas of the Indo-Pacific Region and various chemical, biological, legislative, and manual efforts were initiated in efforts to control it (Mead, 1961). Subsequently, the species became established in southeastern Florida.

An epizootic of then undetermined etiology was reported as causing a significant decline in populations of this pest in Ceylon in the mid-1950's (Mead, 1956). The disease syndrome was characterized by leucodermic lesions on the tentacles, anterior portion of the head and other surface areas of the body. A similar or identical epizootic with a parallel population decline, accompanied by greater incidence and severity of leucodermic lesions, occurred in Oahu in Hawaii (Mead, 1960). As the epizootic intensified, the percentage of the population exhibiting the characteristic lesions increased from 14.5% in January, 1955 to 83.0% in September, 1959.

An attempt by Mead in 1954 to determine the etiology of the epizootic in Ceylon absolved protozoans, fungi, yeasts and spirochetes as causative agents. Mead (1963) reported that pseudomonad bacteria were present in the lesions in snails from the island of Mahiniu, Hawaii, but, at that time, believed the bacteria were probably secondary invaders and suspected a virus to be the causative agent. He noted that snail populations had declined to 15 per cent of their 1955 levels with incidences of the disease attaining 94 per cent of

Figure 12-12. Longitudinal section through a portion of the anterior head region of *Helisoma anceps* infected with acid-fast bacillus. Yellow lesions (grossly) are the tubercles "a", "b" and "c" at the base of the tentacle. Note aplasia at the tip of the tentacle. Hemalum and Eosin.

Figure 12-13. Tubercle "a" at higher magnification. Note fibroblasts surrounding the tubercle and inner mass of necrotic cells. Hemalum and Eosin.

Figure 12-14. Aggregate of amebocytes containing bacilli in the connective tissue surrounding the digestive gland. Carbol fuchsin and acid hemalum.

From Michelson, 1961; reprinted by permission of American Journal of Tropical Medicine and Hygiene and the authors.

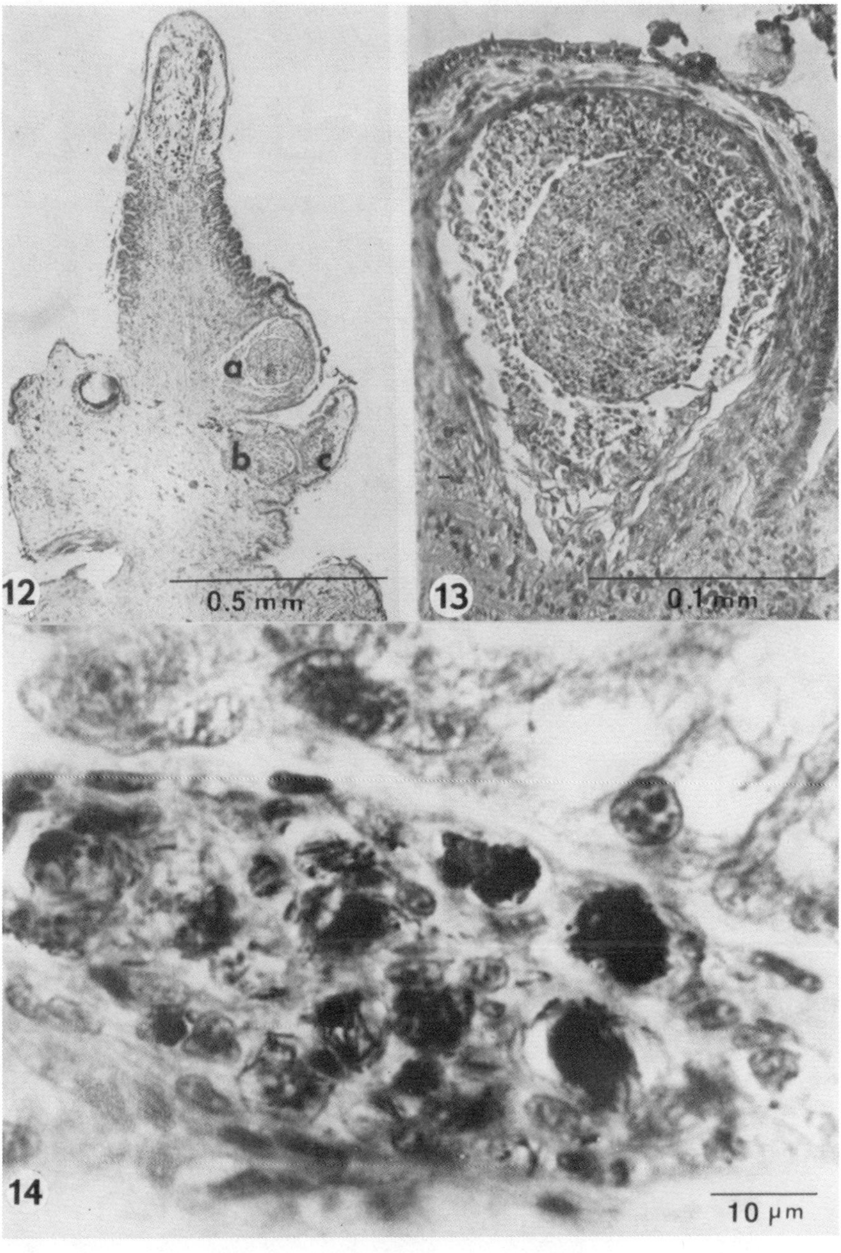
12
a
b
c
0.5 mm
13
0.1 mm
14
10 μm

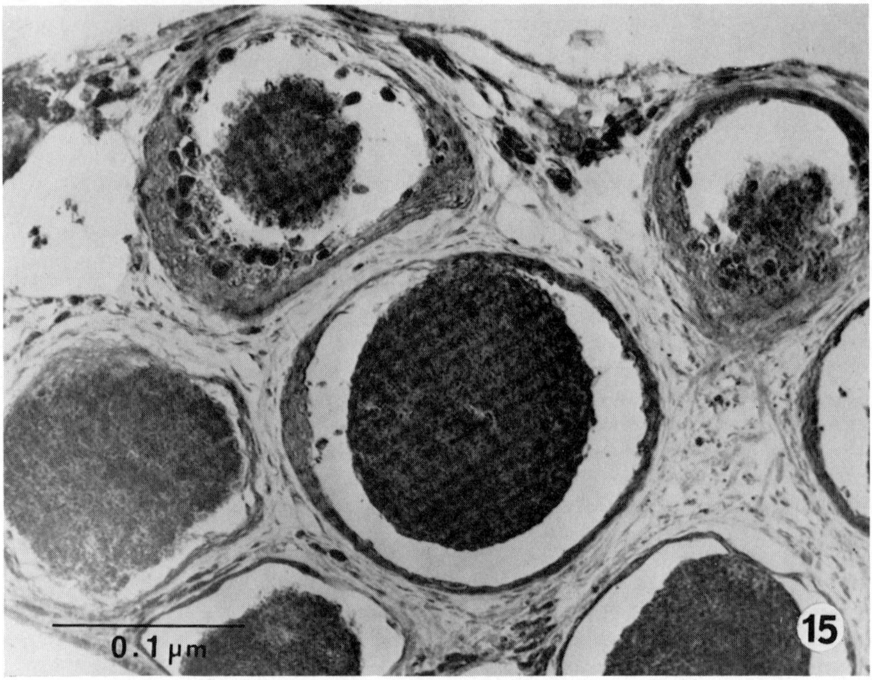

Figure 12-15. Mature tubercles. Note necrotic cells and huge numbers of acid-fast bacteria. Carbol fuchsin and acid hemalum. From Michelson, 1961; reprinted by permission of American Journal of Tropical Medicine and Hygiene and the authors.

the population. The peak of the cyclical disease was in June and appeared to be related to a variety of stress factors, including an increase in rainfall and greater metabolic activity of the snails. However, repeated isolation of an *Aeromonas* sp. in affected snails by Kojima (1965) focused additional attention to the possible role of bacteria in the epizootics. This led to further study (Dean et al., 1970) of the possible bacterial etiology of the epizootic. Cultures were obtained from affected snails collected from 5 locations on the island of Oahu, Hawaii. Cultures were attempted from symptomless, apparently uninfected, snails from one of the areas as controls. Since it was virtually impossible to obtain adequate samples of apparently uninfected older snails in the infested areas, control samples were obtained from a relatively young population.

A species of *Aeromonas*, identified as *A. liquefaciens*, was repeatedly isolated from affected snails, but was rare in control snails. More recent work on the taxonomy of the genus *Aeromonas* (Sturtevant, 1969) indicated that the correct name of the bacterium infecting the giant African snail is *A. hydrophila*, and *A. liquefaciens* is no longer recognized.

In Hawaiian populations of the giant African snail, heavy rainfall and consequent high humidity have been shown to result in a consistent pattern of increased snail activity, heavy feeding, increased reproductive activity, and a subsequent marked population decline. Thus, it appears that the epizootics are caused by *A. hydrophila* compounded by stress factors (referred to by Kojima as Mead's Syndrome).

Preliminary experiments (Srivastava and Srivastava, 1968), using an extract of diseased snails to increase the incidence of the leucodermic disease syndrome in laboratory and field populations of the snail, are promising. However, further studies are necessary before *A. hydrophila* can be used in biological control of this serious pest.

LIMULUS BACTEREMIA

Syn. Horseshoe Crab Bacterial Disease

A progressive, fatal bacterial disease of the horseshoe crab, *Limulus polyphemus*, has been produced experimentally in the laboratory (Bang, 1956). Bacteria randomly collected from sea water were injected into horseshoe crabs of various sizes. One individual became sluggish and appeared ill. Hemolymph drawn from its heart failed to clot, in contrast to the instant clotting response of normal *Limulus* hemolymph. Cultures from hemolymph of the diseased crab yielded numerous bacteria which, on injection into other crabs, produced the typical progressive disease marked by intravascular clotting and death.

Within a few minutes after injection into the vascular system of the horseshoe crab, the infectious bacteria were mechanically disseminated throughout the vascular system, as is the case with non-infectious bacteria. Initially, bacteria-injected individuals produced negative hemolymph cultures, but after they began to exhibit the typical signs of poor clotting and lethargy, the hemolymph cultures became positive and the affected animals soon died. In one case, a localized infection developed in one leg. Hemolymph drawn from the distended, immobile leg was watery, failed to clot and contained few hemocytes. Culture of hemolymph from the localized infection produced a confluent spreading bacterial growth, but a hemolymph culture from a leg on the opposite side of the body was negative. However, the animal became moribund 13 days later (24 days after the original infection) and the hemolymph produced a positive culture.

The disease-producing bacteria were transferred in culture and passaged through crabs numerous times. Heat killed bacterial suspensions, even in small amounts, quickly killed horseshoe crabs when injected and caused intravascular clots identical to those produced by the living bacteria. Shortly after injection, recipient crabs became tetanic and hemolymph was difficult to withdraw. When obtained, the hemolymph contained stringy masses of clotted hemocytes.

Phase microscopy of gill leaflets from living animals demonstrated internal clotting by degranulation and formation of filmentous processes connecting the small clumps of hemocytes, a process confirmed by study of fixed tissue sections.

Although the bacterial disease of _Limulus_ was an artificial disease, originating from the injection of free-living marine bacteria and transferred only through injection of crabs under laboratory conditions, it is important for several reasons. It was a progressive fatal disease rather than initial response to injected toxic bacteria. It clearly demonstrated the host response to the bacteria, the histopathological effects, and identified the cause of death to be from the toxin produced by the bacteria. Finally, the study illustrated the potential for bacteremias in marine invertebrates arising from secondary invasion of traumatized individuals by normally free-living bacteria.

BRINE SHRIMP SPIROCHETOSIS

The brine shrimp, _Artemia salina_, is a cosmopolitan inhabitant of saline waters. The dried eggs readily hatch when placed in saline water and the newly hatched larvae are of tremendous importance as live food for ornamental fish and in aquaculture in fish hatcheries and, especially, as food for the post-larval and early juvenile stages of cultured crustaceans.

Spirochetes were first reported (Tyson, 1970) in the maxillary glands of three of more than 100 brine shrimp examined. They were present in the cytoplasm of all three regions of the gland as well as in the cytoplasm of hypodermal cells, muscle fibers, and amebocytes. Spirochetes also occurred extracellularly and were especially abundant in hemocoelic spaces and in the lumen of the end-sac of the maxillary gland.

The three infected individuals were vigorous and apparently normal prior to fixation and the infections became apparent only on microscopic examination. Subsequently, four additional spirochete-infected brine shrimp were found (Tyson, 1974b). The basic fine structure of the spirochete was described (Tyson, 1974a and 1974b) and a distinctive renal lesion associated with the presence of the organisms was discovered in the region of the basal lamina of the efferent tubule epithelium (Tyson, 1974b).

At the site of the lesions, groups of spirochetes occupied a cave-like indentation of the epithelium (Figures 12-16 and 12-17). Also, discontinuities of the basal lamina were clearly evident (Figure 12-18). Although cave-like indentations occur occasionally in uninfected brine shrimp, the basal lamina discontinuities were found only in infected individuals. It is not known whether the spirochetes present in the indentations were responsible for their development or had entered previously existing caves. Tyson believed that the discontinuities of the basal lamina resulted from the release of a chemical agent or agents by the spirochetes.

The large, fat-containing "phagocytic storage cells" in the hemocoel contained large numbers of spirochetes, frequently in membrane-limited "digestive vacuoles" where they sometimes appeared to have undergone structural alterations. This led Tyson to believe that the phagocytes were capable of phagocytosing and degrading the spirochetes. However, other non-degenerated spirochetes were present in the cytoplasm of the phagocytes. Thus, it appears that while some spirochetes are phagocytosed others are capable of invading the phagocytes.

Although it is obvious that the spirochetes are infectious agents in the brine shrimp and the occurrence of the renal lesions are suggestive of pathogenicity, the role spirochetosis as a disease of brine shrimp remains enigmatic. Because of the importance of brine shrimp as food for cultured fish and invertebrates, further investigation of brine shrimp spirochetosis is warranted.

Figure 12-16. Low magnification micrograph showing lesions (arrow) in basal region efferent tubule of brine shrimp (BL) basal lamina, (H) hemocoel, (L) lumen of efferent tubule.

Figure 12-17. Higher magnification micrograph of lesion in basal region of efferent tubule.

Figure 12-18. Micrograph of efferent tubule lesion containing spirochetes, spherical (nonspiral) forms (SF) and small vesicles (V) of unknown origin. Arrows mark the extent of interruption or gap in the basal lamina (BL) of the epithelium.

From Tyson, 1974b, reprinted by permission of American Society of Microbiology and the author. Micrographs courtesy of the author.

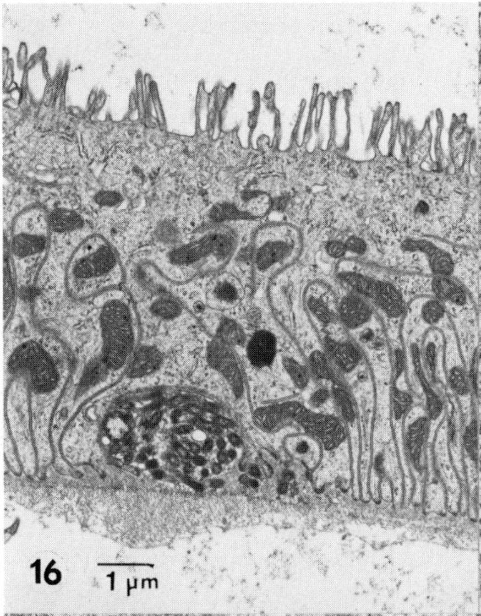
16
1 μm

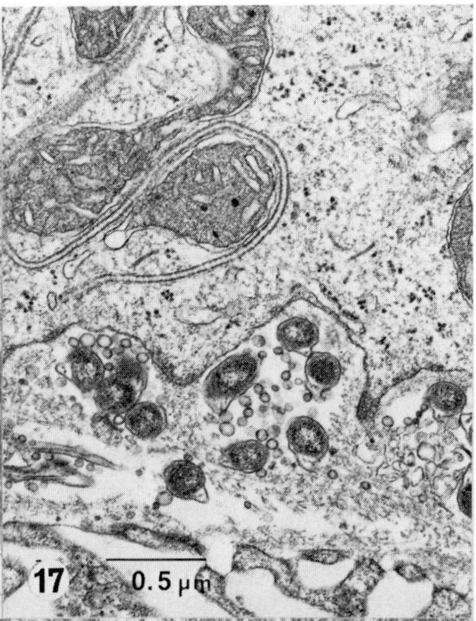
17
0.5 μm

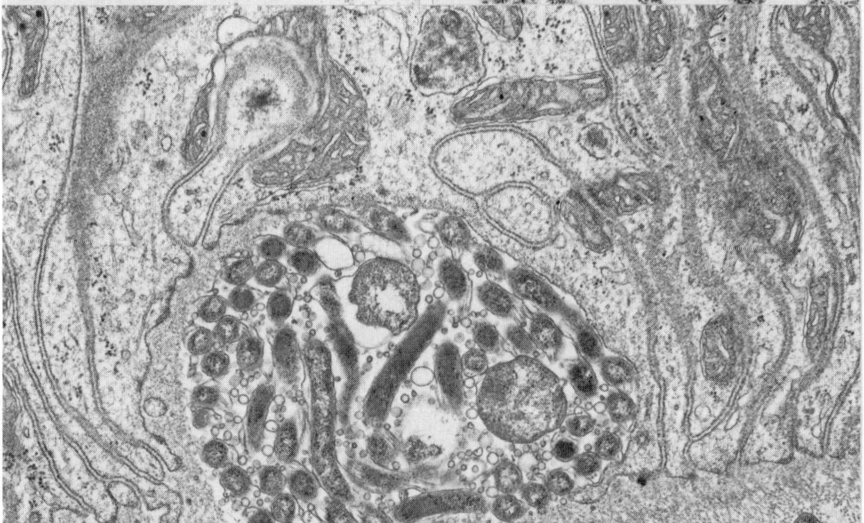
0.5 μm
18

LOBSTER GAFFKEMIA

Synonyms: Gaffkaemia, Lobster Aerococcosis,

Lobster Pedicoccosis, Red Tail

Gaffkemia is a highly virulent septicemic disease of American and European lobsters (*Homarus americanus* and *H. vulgaris*). The causative organism, a micrococcus, was originally named *Gaffkya homari* (Hitchner and Snieszko, 1947), subsequently renamed *Pedicoccus homari* (Kocur and Martinec, 1965) and finally *Aerococcus viridans* (var.) *homari* (Editorial Secretary, 1971; Kelly and Evans, 1974). In view of the rejection of the generic name Gaffkya, Lobster Aerococcosis would be a more correct descriptive name for the disease, but it continues to be called gaffkemia.

Although endemic in wild populations (Stewart et al., 1966), it was first isolated from and is most important in lobster pounds where the animals are held under crowded conditions with attendant wounding from fighting and pegging, accompanied by frequent cannibalism (Snieszko and Taylor, 1947; Stewart and McDonald, 1962). It will undoubtedly be a serious disease problem under aquaculture conditions. Its known range is Eastern Canada and Northeastern United States in *H. americanus* and Norway and the Netherlands (Roskam, 1957), Ireland (Gibson, 1961) and the United Kingdom (Wood, 1965a and 1965b) in *H. vulgaris*. Largely through the efforts of Stewart and his associates at the Halifax, N.S. Laboratory of the Canadian Department of the Environment, Fisheries and Marine Service, the physiopathology of gaffkemia has been more thoroughly investigated than any non-insect invertebrate disease. However, the histopathology of the disease has not been adequately described to date.

Gross Pathology and Symptomology

The ventral side of the abdomen of gaffkemic lobsters is sometimes pinkish, resulting in the common name "red tail". Usually, however, there are no external signs of the disease, even in terminal cases, until shortly before death at which time they lie quietly with their chelae extended. Hemolymph clotting time is progressively prolonged as the infection intensifies and clotting is completely eliminated in advanced infections. Withdrawn hemolymph is less viscid than normal and is pinkish in color.

Histopathology

Microscopic examination of smears of hemolymph from gaffkemic lobsters reveals the characteristic tetrads of encapsulated gram positive cocci and strikingly reduced numbers of circulating hemocytes. This hemocytopenia is believed to arise from the combination of two phenomena: (1) initially by the removal of significant numbers of hemocytes of circulating hemolymph as they phagocytose the bacteria and are sequestered in clumps in the connective tissue associated with the heart, hepatopancreas, and gills, and (2) a rapid decline in hemocytes because of the death of hemocytes in the hemolymph that have phagocytosed bacteria.

Physiopathology

Because of the acidity of the gastric fluid (pH 4.7 to 5.0), oral transmission of gaffkemia does not occur (Stewart et al., 1969b) and lobsters become infected only through breaks in the integument. The time interval of susceptibility is quite short because of the efficient wound sealing mechanism (Stewart et al., 1969b), but up to 80% of the lobsters wounded just prior to exposure to heavy concentrations of *A. viridans* var. *homari* contract the disease. Unwounded controls exposed to the same concentration of the organism do not become infected.

Lobsters are extremely susceptible to virulent strains of the bacteria, with animals injected with dosage as low as 1×10^1 organisms/Kg body weight (approximately 5 bacteria per lobster) dying as quickly as individuals injected with 6×10^8 organisms/Kg body weight. However, the ambient temperature is of great importance in the progression of the disease, with the mean death time decreasing with increasing temperatures above 1^oC (Stewart et al., 1969c). However, the bacteria are neither eliminated nor lose their virulence at low temperatures.

Although lobster hemolymph has bactericidal or bacteriostatic properties that inhibit the growth of a number of species of bacteria and a nonspecific agglutinin or aggluglutinins that agglutinate most bacteria studied, *A. viridans* var. *homari* is unaffected. Indeed, lobster hemolymph is an excellent culture medium for the bacterium.

The marked hemocytopenia in infected lobsters has a direct effect on the lobster's clotting mechanism, with clotting time progressively prolonged soon after lobsters become infected. Since a "true" clot is formed by the release, on rupturing, of "coagulin" by the hemocytes, the failure to clot results from the lack of coagulin because the concentration of plasma proteins, including fibrinogen, is essentially unaffected by gaffkemia (Stewart et al., 1969d).

Drastic declines in hepatopancreatic glycogen and hemolymph non-protein nitrogen occur as the disease progresses. These are specific effects rather than a general physiological decline, but the inability to replace these substances from hepatopancreatic lipid, hemolymph protein or muscle may result from impaired metabolism or normal conversion rates too slow to cope with an immediate massive requirement (Stewart and Rabin, 1970).

Experimentally infected lobsters that have been fed died much earlier and with higher bacterial counts than those that had been starved for 2 or 3 months prior to challenge. Lobsters become anorectic within one or two days after infection, thus eliminating the opportunity to replenish energy reserves from extraneous sources. The mean bacterial levels at death were 2 to 4 times higher and were attained 2 to 4 times earlier in the fed groups than in the starved groups. These data confirmed earlier findings that constituents of the hemolymph were adequate to support the growth of the bacteria and the rate of growth is controlled by the available nutrients. Therefore, the higher the nutrient level in the hemolymph, the more rapidly the bacteria are capable of attaining lethal population levels.

Cause of Death

As is the case in almost all invertebrate infectious diseases, the precise cause of death from gaffkemia is not known. However, because of the thoroughness with which the physiological effects have been studied, it is probably less speculative than in any other non-insect invertebrate disease. Progressive prolongation of clotting time resulting from hemocytopenia led Goggins and Hurst (1960) to suggest that gaffkemic lobsters bleed to death. While this is undoubtedly true when infected lobsters are wounded, the conclusion (Stewart et al., 1969a and 1969; Stewart and Cornick, 1972) that death from gaffkemia is a "wasting process" resulting from unsuccessful competition by the infected lobster for its own energy reserves is probably more accurate.

Epizootiology

Although the major epizootics occur in captive lobsters, almost 5 per cent of feral lobsters are gaffkemic (Stewart et al., 1966). The disease is unlikely to be epizootic, but the numbers of infected wild lobsters are probably underestimated because gaffkemic lobsters are anorectic and would, therefore, be unlikely to enter lobsters pots in search of food.

Bacteria consistent in morphology and growth characteristics with A. viridans var. homari can be readily isolated from the mud of tidal ponds (Goggins and Hurst, 1960) and from several species of crabs. Further evidence of the ubiquitousness of the organism is indicated by the wide range of hosts that can be infected experimentally (Stewart and Rabin, 1970). However, the bacterium was less virulent in all the crabs tested, even though passage through the crabs does not affect the virulence of the pathogen for lobsters. As Stewart and Rabin (1970) pointed out, A. viridans var. homari is capable of infecting a wide range of crustaceans, is maintained for a long periods of time within them (up to 80 days in Cancer irroratus) and is able to survive outside a host, all of which can enhance the survival potential of the organism. It is also quite possible that one or more species of crabs may serve as reservoirs of the disease.

Commercial lobster fishermen's use of wooden pegs to immobilize the large chelae produces wounds that are effective portals of entry for the pathogen for at least 24 hours. In lobster pounds, crowded conditions that promote fighting and cannibalism result in the immediate release of large numbers of infective bacteria from gaffkemic lobsters rather than their being contained for 48 hours or more after the death of an undisturbed moribund lobster. Thus, the presence of massive numbers of bacteria, combined with available portals of entry through wounds caused by pegging and fighting, leads to the epizootics common in captive lobsters.

PENAEID BACTERIAL SEPTICEMIA

Syn. Penaeid Vibriosis

Bacterial diseases have been implicated as major causes of mortality of juvenile penaeid shrimp in aquaculture (Sindermann, 1971; Lewis, 1973; Vanderzant et al., 1971). Juvenile shrimp being held in tanks prior to sale as live bait also frequently experience high mortalities that are attributable to bacterial septicemia (Lightner and Lewis, 1975). Bacterial disease of penaeid shrimp has thus far been shown to be caused by V. parahaemolyticus biotype alginolyticus and V. anguillarium, but there is evidence that an Aeromonas sp. may also be implicated.

Lightner and Lewis (1975) summarized the clinical signs of bacterial septicemia, noting that the syndrome appears to be indistinguishable in brown, white and pink shrimp (Penaeus aztecus, P. setiferus, and P. duorarum). The first recognizable sign is a gradual change in the appearance of the abdominal masculature from colorless and translucent to whitish and opaque (Figure 12-19). Melanization of gill filaments and ventrolateral edges of the carapace and cuticular lesions develop in some individuals. Apparent expansion of integumental melanophores causes slight darkening of the dorsal integument in some cases and apparent expansion of integumental erythropores produces a reddening of the periopods of moribund or recently dead shrimp. Moribund individuals exhibit a characteristic dorsal flexure of the abdomen, with the second and third abdominal segements at the apex of the flexure (Figure 12-20).

Pronounced, even diagnostic, behavioral symptoms develop as the disease progresses. Swimming is retarded and disoriented, with affected individuals swimming on their sides rather than in the normal upright position. Subsequently, the near-moribund shrimp becomes lethargic and rests on the bottom, often upright and supported by the periopods, pleopods and telson. At this stage some individuals respond to tactile stimulus by brief flurries of swimming. Death occurs 2 to 4 hours after onset of lethargy. Some individuals die and remain in an upright position, virtually diagnostic of the disease since shrimp have not been observed to remain upright after death from any other causes (Figure 12-21).

Hemolymph clotting time is markedly retarded in moribund shrimp suffering from bacterial septicemia, ranging from 5 minutes to more than an hour rather than clotting immediately on withdrawal. Hemolymph from bacteremic shrimp is slightly turbid and lacks the blue coloration typical of healthy shrimp. Hemocytopenia is pronounced and the hemolymph contains numerous gram-negative rods.

Pure cultures of bacteria can usually be obtained when hemolymph is drawn directly from the heart of moribund shrimp. *Vibrio parahemolyticus* biotype *alginolyticus* and *V. anguillarium* are the most prevalent organisms isolated from shrimp with clinical signs of a bacteremia in wild populations from Galveston Bay and in hatchery-reared shrimp.

Pathogenicity

Bacterial isolates from obviously bacteremic shrimp are invariably pathogenic to healthy juvenile white, brown and pink shrimp when injected into the abdominal musculature. Although the virulence varies, injection of the standard inoculum usually causes death of challenged healthy shrimp within 24 to 48 hours. Injection of 10-fold dilutions of the standard bacterial suspension results in progressively delayed onset of clinical signs of bacteremia and subsequent mortality; dilutions greater than 1:100 produce no clinical

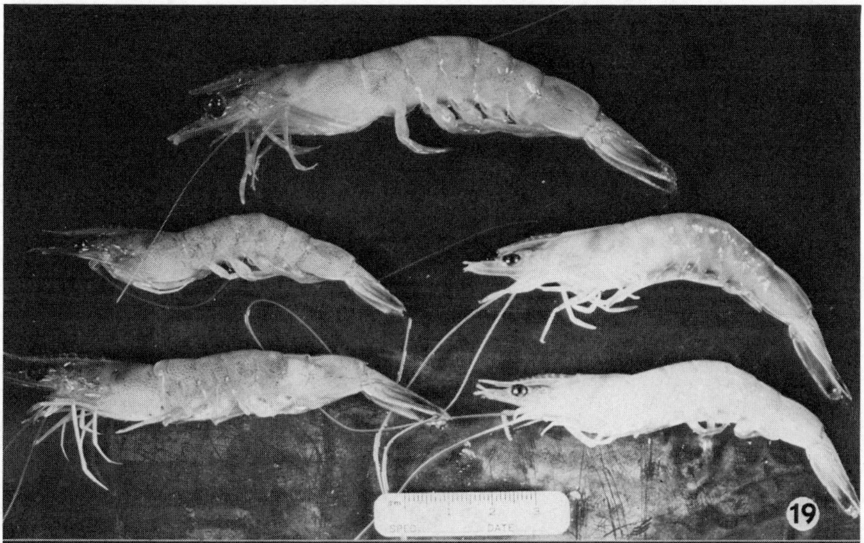

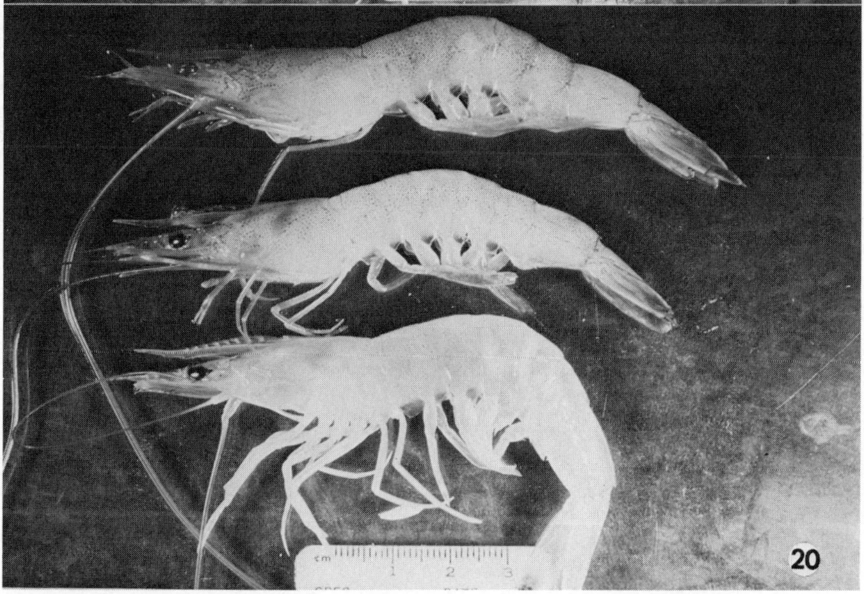

Figure 12-19. Uninfected shrimp on right. Shrimp with bacterial septicemia, left and top.

Figure 12-20. Normal shrimp bottom. Septicemic shrimp, middle and top. Note increased opacity, darkening melanophores and flexure of abdomen in infected shrimp. Photos courtesy of D. Lightner.

Figure 12-21. Morbund and dead septicemic shrimp. Note abdominal flexure and upright position of the three upper shrimp. Photo courtesy of D. Lightner.

signs of disease or mortality.

Feeding of bacteria isolated from obviously bacteremic shrimp seldom results in clinical disease, indicating that penaeid shrimp are resistant to infection by *Vibrio* via the oral route. The capture and retention of wild penaeid shrimp frequently initiates catastrophic epizootics caused by vibriosis. Less frequently, handling of healthy hatchery-reared shrimp has the same result. Wild shrimp are often traumatized while being captured in shrimp trawls and it is virtually certain that *Vibrio* infections develop through bacterial invasion of cuticular wounds. Small wounds undoubtedly occur in the handling and crowding of hatchery-reared shrimp, thus producing a portal of entry for potentially pathogenic bacteria that are part of the normal microbial flora of shrimp (Vanderzant et al., 1970, 1971).

The gross symptoms present in moribund shrimp suffering from bacterial disease may also occur in shrimp moribund from other causes. Non-specific gross clinical signs common to a bacteremia and one or more other disease conditions include lethargy, white discoloration of abdominal musculature, dorsal flexure of the abdomen and redness of the pleopods and periopods. In the absence of a bacteremia, shrimp exhibiting these symptoms are typically suffering from hypoxia or anoxia. Differential diagnosis of a bacteremia is, therefore, dependent on the demonstration of bacteria in the hemolymph.

BLUE CRAB BACTEREMIA

Because soft-shelled crabs command a premium price and can be identified when in premolt condition, a soft crab industry involving holding premolt crabs in shallow floats until they have molted have existed for many years in the Chesapeake Bay region. Although mortality is usually quite low in nature during the molting process, high mortality rates in the shedding floats have been known for many years. Physical factors, including the capture, handling, transport (frequently out of water), exposure to elevated temperatures, crowding and wounding by fellow captives, stress the crabs but bacterial disease is a major factor in the death of freshly caught blue crabs. Indeed, these stress factors are primarily responsible for the development of the bacteremia.

In the major study of blue crab bacteremia, Johnson (1976) used crabs from a variety of sources and handled in various ways to work out the course of infection and histopathology of the disease. She found that, while the amount of stress greatly influences the percentage of crabs developing fatal bacterial infections, the syndrome is consistent in all individuals contracting the disease.

Gross Pathology

Heavy mortalities frequently occur during the first week crabs are held in sheddling floats or laboratory aquaria. Moribund animals may develop extensive premortem hemolymph clots in the anteriodorsal and frontal blood sinuses that are visible grossly as large jelly-like masses, while blood freshly drawn from the heart clots incompletely. Many newly dead crabs also have extensive anterior clotting, a symptom never reported in crab without a bacterial infection.

Crabs with a bacteremia develop aggregations of hemocytes, visible grossly as cloudy areas, in the semitransparent distal segment of the posterior pair of appendages, the swimming legs or "paddles". The "cloudy" areas appear in most infections on the third day post-capture and approximately 24 hours premortem, infected crabs become markedly lethargic (Johnson, personal communication).

Histopathology

The acute infection begins during the first day of captivity and runs its course by the sixth day. Foci of hemocytes, consisting of a small center of cells with pycnotic or karyorrhectic nuclei surrounded by normal hemocytes, develop in the first day. These nodules, sometimes ensheathed by flattened granular or hyaline hemocytes, are most common in the heart (Figure 12-22). By the third day, nodules develop in gill lamellae and stems, along with increasingly large aggregations of hemocytes that distend the lamellae and cause sloughing of lamellar epithelium and stasis of hemolymph flow by the formation of thrombo-emboli. The numerous podocytes (nephrocytes) in the gill stem and scattered through the lamellae are unaffected and the thickened gill epithelium, the so-called salt gland involved in the ion transport, ranges from normal to degenerate.

Nodules in hemal spaces and blood vessels consist of aggregates of necrotic hemocytes surrounded by an areolar layer of degenerating hemocytes with normal, rounded hemocytes usually present peripherally (Figure 12-23) but no marked encapsulation. Bacteria are present both extracellularly, in the nodules and in nearby, often pycnotic, hyaline hemocytes. The latter are often so bacteria-filled that the nucleus is flattened against the cell membrane. Nodules gradually decrease in the heart concomitant with a general increase in the gills, antennal gland, Y-organ and, as previously noted, in the paddles.

Specific, but variable, pathological changes occur in many organs as a result of systemic bacteremia. In the hepatopancreas there is massive sloughing and metaplastic-like diminution in the height of the acinar epithelium, marked decrease or cessation of mitotic activity in the apical acinar epithelium and varying numbers of karyolytic acinar epithelial cells. The rosettes of fixed phagocytes (Figure 12-24) present in the connective tissue between the acini become swollen, vacuolated and necrotic, with all stages of necrosis (pycnosis, karyorrhexis and karyolysis) common (Figure 12-24). In heavy infections, phagocytosed bacteria may be present in the fixed phagocytes and the interacinar spaces are often distended by large aggregations of hemocytes.

The Y-organ is affected by the fourth day, with aggregations of hemocytes in arterioles and hemal spaces impinging on the glandular tissue, causing local infarcts and subsequent extensive focal necrosis of Y-organ tissue. Concurrently, the normal architecture of the antennal gland (Figure 12-25) may be altered by degeneration or completely destroyed by massive necrosis (Figure 12-26). Hemocytic aggregations in the hemal spaces between the involutions of the labyrinthal epithelium and between the interdigitated labyrinthal and end-sac epithelium result in degeneration of the end-sac and hemocytic infiltration of the disorganized tissue. The labyrinthal epithelium may become thinner and more acidophilic with pycnotic or karyorrhetic nuclei, suggestive of coagulation necrosis. The degeneration is generally multifocal, but entire glands may become necrotic.

Virtually all organs and tissues may be affected in systemic infections. Pycnotic hemocytes often occur in the small blood vessels and sinuses of the brain and thoracic ganglion and in the spongy glial tissue just beneath the neural lamellae. In central nervous tissue involvement, granulocyte infiltration of nerves is common and small nodules are formed sporadically.

The most probable source of infections are potentially pathogenic bacteria on the surface of the crab that are able to breach the protective barriers of captured crabs via traumatic wounds or other, unknown, routes rendered vulnerable by the stresses of capture and transport (Johnson, 1976). Several species of bacteria are involved, especially _Vibrio_ sp. (including _V. parahaemolyticus_). Differences in virulence among the different _Vibrio_ species probably affect the host response and course of infection.

STARFISH BACTEREMIA

Syn. Sea Star Bacteremia

Starfish coelomic fluid is normally bacteria-free (Bang and Chaet, 1959), but aseptically withdrawn fluid contains bacteria if the animal is undergoing autotomy, has been traumatized or is collected from stagnant waters (Bang and

Figure 12-22. Groups of aggregating hemocytes in heart of blue crab with bacterial infection.

Figure 12-23. Large aggregations of hemocytes in clotted blood. Note degenerate centers, surrounding areolar zone and peripheral normal hemocytes.

Figure 12-24. Abnormally swollen rosettes of fixed phagocytes between hepatopancreatic acini.

Figure 12-25. Normal antennal gland.

Figure 12-26. Massive necrosis of antennal gland in crab with bacterial infection. Note complete loss of normal architecture.

From Johnson, 1976; reprinted by permission of Academic Press, Inc. and the author. Photomicrographs courtesy of P. Johnson.

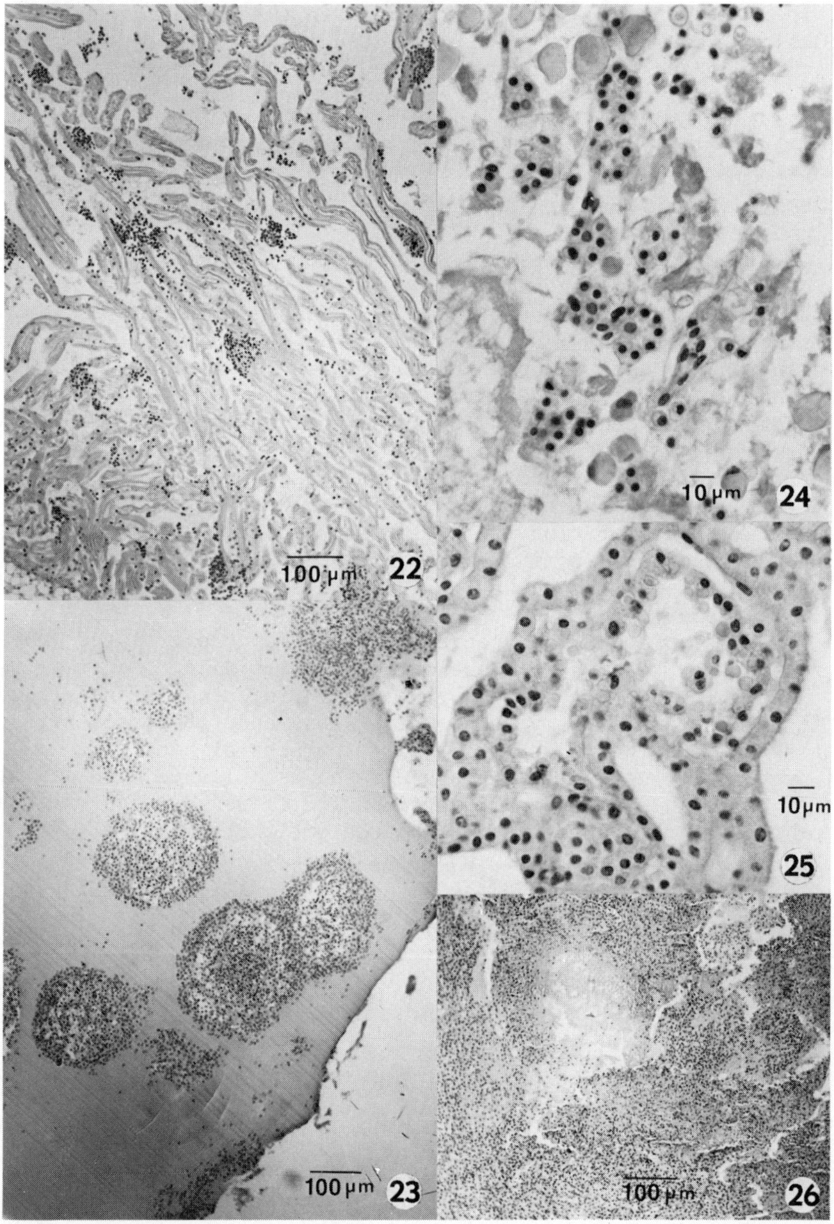
100 µm
22
10 µm
24
10 µm
25
100 µm
23
100 µm
26

Lemma, 1962). Many starfish collected by the commonly used starfish mop or dredge, consisting of a tangled mass of cotton fibers attached to a framework of chain that is dragged from a slowly moving boat, are traumatized to varying degrees.

Bang and Lemma (1962) reported that all dredged starfish that were "sick", as indicated by weak muscular activity and sluggish movement, had heavy bacterial infections and died shortly after the symptoms appeared. Traumatized individuals that did not exhibit these symptoms had mild infections and became negative after a few days in aquaria provided with running sea water.

These observations led Bang and Lemma to explore the possibility that, in addition to trauma, autotomy and/or stagnant water might enable bacteria to penetrate the coelom. Autotomy of a ray was invoked by the injection of a 1:5 dilution of heat-produced starfish toxin (Chaet and Albert, 1958). Cultures of coelomic fluid taken 2 hours after autotomy produced bacterial growth in 11 of 12 cases, but the infections were of mild intensity. Individuals maintained in stagnant water provided with air-bubble aeration for 12 hours developed heavy infections (9 of 10) as did those in stagnant water without aeration (3 or 4). Since starfish often remain in the tide pools during low tide, bacterial infections may develop in starfish in nature while in stagnant water following autotomy.

Five unidentified species of bacteria were isolated from symptomatic starfish. These and other strains of bacteria pathogenic to other invertebrates, when injected into healthy starfish, consistently cause weight loss and the losses are greater with the injection of large numbers of bacteria. However, a progressive disease was not established with any of the bacteria. Following injection of different concentrations of each of the bacteria with the coelom, the numbers of bacteria gradually declined and disappeared within 3 days, presumably by phagocytosis and diapedesis by the amebocytes.

REFERENCES

Anderson, W. R., Madden, P. A., and Tromba, F. G. 1971. Histopathologic and bacteriologic examination of cuticular lesions of *Ascaris suum*. J. Parasitol., 57, 1010-1014.

Bang, F. B. 1956. A bacterial disease of *Limulus polyphemus*. Bull. Johns Hopkins Hosp., 98, 325-351.

Bang, F. B. and Chaet, A. B. 1959. The effect of starfish toxin on amebocytes. Biol. Bull., 117, 403-404.

Bang, F. B. and Lemma, A. 1962. Bacterial infection and reaction to injury in some echinoderms. J. Insect Pathol., 3, 401-414.

Benz, G. and Altwegg, A. 1975. Safety of *Bacillus thuringiensis* for earthworms. J. Invertebr. Pathol., 26, 125-126.

Brown, C. 1973. The effects of some selected bacteria on embryos and larvae of the American oyster, *Crassostrea virginica*. J. Invertebr. Pathol., 21, 215-223.

Brown, C. 1974. A pigment producing pseudonomad which discolors culture containers of a bivalve mollusk. Chesap. Sci., 15, 17-21.

Brown, C. and Losee, E. 1978. Observations on natural and induced epizootics of vibriosis in *Crassostrea virginica* larvae. J. Invertebr. Pathol., 31, 41-47.

Canning, E. U. 1973. Protozool parasites as agents for biological control of plant-parasitic nematodes. Nematologica, 19, 342-348.

Chaet, A. B. and Albert, M. 1958. Evidence for the toxic theory. Fed. Proc., 17, 24.

Colwell, R. R. and Sparks, A. K. 1967. Properties of *Pseudomonas enalia*, a marine bacterium pathogenic for the invertebrate *Crassostrea gigas* (Thurnberg). Applied Microbiol., 15, 980-986.

Dean, W. W., Mead, A. R. and Northey, W. T. 1970. *Aeromonas liquifaciens* in the African snail, *Achatina fulica*. J. Invertebr. Pathol., 16, 346-351.

Editorial Secretary (for the Judicial Commission of the International Committee of Systematic Bacteriology) 1971. Opinion 39. Rejection of the generic name *Gaffkya* Trevisan. Int. J. Syst. Bacteriol., 21, 104-105.

Gibson, F. A. 1961. Gaffkaemia in stored lobsters. Cons. Perma. Int. Explor. Mer. Shellfish Committee Rept. 58, 1 p. (Mimeo.).

Goggins, P. L. and Hurst, J. W., Jr. 1960. Progress report on lobster gaffkyaremia (red tail). Maine Dept. Sea Shore Fish. Mss., 9 pp.

Grischkowsky, R. S. and Liston, J. 1974. Bacterial pathogenicity in laboratory-induced mortality of the Pacific oyster (*Crassostrea gigas*, Thunberg). Proc. Natl. Shellfish. Assoc., 64, 82-91.

Guillard, R. R. L. 1959. Further evidence of the destruction of bivalve larvae by bacteria. Biol. Bull., 117, 258-266.

Heimple, A. M. 1966. A crystalliferous bacterium associated with a "Blister Disease" in the earthworm, Eisenia foetida (Savigny). J. Invertebr. Pathol., 8, 295-298.

Hitchner, E. R. and Snieszko, S. F. 1947. A study of a microorganism causing a bacterial disease of lobsters. J. Bacteriol., 54, 48. (Abstract).

Johnson, P. T. 1976. Bacterial infection in the blue crab, Callinectes sapidus: Course of infection and histopathology. J. Invertebr. Pathol., 28, 25-36.

Kanno, H., Sasaki, M., Sakurai, Y., Watanabe, T. and Suzuki, K. 1965. General aspects of the mass mortality of the oyster in Matsushima Bay and its environmental conditions. Studies on the mass mortality of the oyster in Matsushima Bay. Bull. Tohoku Reg. Fish. Res. Lab., 25, 1-26. In Japanese. Translated by U.S. Joint Pub. Res. Ser. for Translation Program. Bureau of Commercial Fisheries, 1966.

Kelly, K. F. and Evans, J. B. 1974. Deoxyribonucleic acid homology among strains of the lobster pathogen Gaffkya homari and Aerococcus viridans. J. Gen. Microbiol., 81, 257-260.

Kocur, M. and Martinec, T. 1965. Proposal for the rejection of the bacterial generic name Gaffkya. Int. Bull. Bacteriol. Nomencl. Taxon, 15, 177-179.

Kojima, E. S. 1965. Bacteriological studies on leucodermic lesions on the giant African snail. M.S. Thesis, Univ. Hawaii, 40 pp.

Lewis, D. H. 1973. Response of brown shrimp to infection with Vibrio sp. Proc. Fourth Ann. Workshop Mariculture Soc., pp. 333-338.

Lightner, D. V. and Lewis, D. H. 1975. A septicemic bacterial disease syndrome of penaeid shrimp. Diseases of Crustaceans. Soc. Invert. Pathol. Symp. Mar. Fish. Rev. 37 (Nos. 5-6) 25-28.

Lipovsky, V. P. and Chew, K. K. 1971. A preliminary report on Pacific oyster (Crassostrea gigas) mortality after transfer from a natural bed into 10°C and 20°C water in the laboratory. Proceed. Natl. Shellfish. Assoc., 61, 9.

Lipovsky, V. P. and Chew, K. K. 1972. Mortality of Pacific oysters (Crassostrea gigas): The influences of temperature and enriched seawater on oyster survival. Proc. Natl. Shellfish Assoc., 62, 73-82.

Mankan, R. 1975. Bacillus penetrans n. comb. causing a virulent disease of plant-parasitic nematodes. J. Invertebr. Pathol., 26, 333-339.

Mead, A. R. 1956. Disease in the giant African snail (Achatina fulica Bowdich). Science, 123, 1130-1131.

Mead, A. R. 1960. An epizootic in the giant African snail. Ann. Rept. Am. Malacol. Union, 1961, 40.

Mead, A. R. 1961. A prognosis in the problem of the giant African snail (Achatina fulica Bowdich). In "Symposium: Ecology and Pacific Distribution of the Giant African Snail with Special Reference to the Measures That Are Being Taken for its Control". Proc. Pacific Sci. Congress, 9th Thailand, 1957, 7-10.

Mead, A. R. 1963. Disease, decline and predation in the giant snail populations of Hawaii (abstract). Ann. Rept. Am. Malacol. Union, p. 22.

Michelson, E. H. 1961. An acid-fast pathogen of fresh-water snails. Amer. Jour. Trop. Med. Hyg, 10, 423-433.

Pan, C-T. 1956. Studies on the biological control of schistosome-bearing snails: A preliminary report on pathogenic microorganisms found in Australobis glabratus. J. Parsitol., 42 (Sect. 2), 33.

Pemberton, C. E. 1938. The giant African snail Achatina fulica (Fer.) discovered in Hawaii. Hawaii Plant Rec., 42, 135-140.

Roskam, R. T. 1957. Gaffkaemia, a contagious disease in Homarus vulgaris. Cons. Perma. Int. Explor. Mer. Shellfish Committee Rept. (unnumbered), 4 pp. (Mimeo).

Sindermann, C. J. 1970. Principle Diseases of Marine Fish and Shellfish. Academic Press, N.Y. 369 pp.

Sindermann, C. J. 1971. Disease-caused mortalities in mariculture, status and predictions. Proc. Second Annual Workshop World World Mariculture Society, pp. 69-74.

Sindermann, C. J. 1977. Bacillary necrosis of oyster larvae. Disease Diagnosis and Control in North American Marine Aquaculture (C. T. Sinderman, Ed.). Developments in Aquaculture and Fisheries Science, Vol. 6. Elsevier Sci. Pub. Co. Amsterdam, 1977).

Smirnoff, W. A. and Heimple, A. M. 1961. Notes on the pathogenicity of Bacillus thuringiensis var. thuringiensis Berliner for the earthworm, Lumbricus terrestris Linnaeus. J. Insect Pathol., 3, 403-408.

Snieszko, S. F. and Taylor, C. C. 1947. A bacterial disease of the lobster (Homarus americanus). Science, 105, 500.

Srivastava, P. E. and Srivastava, Y. N. 1968. Role of snails' diseases in the biological control of Achatina fulica Bowdich, 1822 in the Andamans. Veliger., 10, 320-321.

Stewart, J. E. and MacDonald, J. F. 1962. A report to the fishing industry regarding lobster disease (gaffkaemia). Fish. Res. Bd. Can., Technol. Sta., Halifax, New Series Circ. No. 9, 2 pp.

Stewart, T. B. and Godwin, N. J. 1963. Cuticular lesions of Ascaris suum caused by Pseudomonas sp. Jour. Parasitol. 49, 231-234.

Stewart, J. E., Cornick, J. W., Spears, D. I. and McLeese, D. W. 1966. Incidence of Gaffkya homari in natural lobster (Homarus americanus) populations of the Atlantic. region of Canada. J. Fish. Res. Bd. Can., 23, 1325-1330.

Stewart, J. E., Foley, D. M. and Ackerman, R. G. 1969a. Characteristics of Gaffkya homari, the causative agent of the lobster disease, gaffkemia. J. Fish. Res. Bd. Can., 26, 1385-1389.

Stewart, J. E., Dockrill, A. and Cornick, J. W. 1969b. Effectiveness of the integument and gastric fluid as barriers against transmission of Gaffkya homari to the lobster Homarus americanus. J. Fish. Res. Bd. Can., 26, 1-14.

Stewart, J. E., Cornick, J. W., and Zwicker, B. M. 1969c. Influence of temperature on Gaffkemia, a bacterial disease of the lobster, Homarus americanus. J. Fish. Res. Bd. Can. 26, 2503-2510.

Stewart, J. E., Arie, B., Zwicker, B. M. and Dingle, J. R. 1969d. Gaffkemia, a bacterial disease of the lobster, Homarus americanus: Effects of the pathogen, Gaffkya homari, on the physiology of the host. Can. J. Microbiol., 15, 925-932.

Stewart, J. E. and Rabin, H. 1970. Gaffkemia, a bacterial disease of lobsters (Genus Homarus) In: A Symposium on Diseases of Fishes and Shellfishes. Amer. Fish. Soc. Special Pub. No. 5, 431-439.

Stewart, J. E. and Cornick, J. W. 1972. Effects of Gaffkya homari on glucose, total carbohydrates, and lactic acid of the hemolymph of the lobster (Homarus americanus). Can. J. Microbiol., 18, 1511-1513.

Stewart, T. B. and Godwin, N. J. 1963. Cuticular lesions of Ascaris suum caused by Pseudomonas sp. J. Parasitol., 49, 231-234.

Sturtevant, A. B., Jr. 1969. A taxonomic study of the genus Aeromonas. M. S. Thesis, Univ. Alabama, 58 pp.

Thorne, G. 1940. Duboscquia penetrans n. sp. (Sporozoa: Microsporidia Nosematidae), a parasite of the nematode Pratylenchus pratensis (deMan) Flipjev. Proc. Helmin. Soc. Wash., 7, 51-53.

Tubiash, H. S., Chanley, P. E., and Leifson, E. 1965. Bacillary necrosis, a disease of larval and juvenile bivalve molluscs. I. Etiology and epizootiology. J. Bact., 90, 1036-1044.

Tubiash, H. S., Colwell, R. R. and Sakazaki, R. 1970. Marine vibrios associated with bacillary necrosis, a disease of larval and juvenile bivalve mollusks. J. Bacteriol., 103, 272-273.

Tubiash, H. S., Otto, S. V. and Hugh, R. 1973. Cardiac edema associated with Vibrio anguillarium in the American oyster. Proc. Natl. Shellfish. Assoc., 63, 39-42.

Tyson, G. E. 1970. The occurrence of a spirochete-like organism in tissues of the brine shrimp Artemia salina. Arch. Microbiol., 99, 281-294.

Tyson, G. E. 1974a. Ultrastructure of a spirochete found in the tissues of the brine shrimp, Artemia salina. Arch. Microbiol., 99, 281-294.

Tyson, G. E. 1974b. A distinctive renal lesion of spirochete-infected brine shrimp. J. Bacteriol., 199, 629-631.

Unestam, T. 1973. Fungal diseases of crustacea. Rev. Med. and Vet. Mycology, 8, 1-20.

Vanderzant, C., Mroz, E. and Nickelson, R. 1970. Microbial flora of Gulf of Mexico and pond shrimp. J. Milk Food Technol., 33, 346-350.

Vanderzant, C., Nickelson, R. N. and Judkins, P. W. 1971. Microbial flora of pond-reared brown shrimp (Penaeus aztecus). Appl. Microbiol., 21, 916-921.

Williams, J. R. 1960. Studies on the nematode soil fauna of sugarcane fields in Mauritius 5: Notes upon a parasite of root-knot nematodes. Nematologica, 5, 37-42.

Wood, P. C. 1965a. A preliminary note on gaffkaemia investigations in England. Cons. Perma. Int. Explor. Mer. Rapp. Proc.-Verb. Reun., 156, 30-34.

Wood, P. C. 1965b. Gaffkaemia, the blood disease of lobsters. J. Gen. Microbiol., 41(3), 28 (Abstract).

SECTION III

FUNGI

Chapter 13

FUNGAL DISEASES OF TERRESTRIAL INSECTS

by R. Y. Zacharuk*

INTRODUCTION

Many fungal species have been recovered from a variety of diseased or dead terrestrial insects (MacLeod, 1963; Madelin, 1963; McEwen, 1963). For a large number of these fungi it remains to be proved whether they are parasitic or saprophytic. Of the relatively small number of fungi that have been shown to be primary pathogens, detailed information on the pathogenesis of their infection is limited to but a few of the most significant ones. These are in the genera *Metarrhizium*, *Beauveria* and *Aspergillus* in the deuteromycetous order Moniliales, and *Entomophthora* in the zygomycetous order Entomophthorales. Although these are given primary emphasis here, the limited information available for other genera in the above two orders and in the ascomycetous order Laboulbeniales is discussed as well.

This information is based primarily on selected findings reported since the 1963 treatises noted above. The ascomycetous genus *Cordyceps* is omitted for lack of significant pathogenetic information beyond that summarized by McEwen (1963). Other more recent general reviews of fungal entomopathogens are by Ferron (1978) and Madelin (1966).

*Department of Biology, University of Regina, Regina, Saskatchewan, Canada S4S 0A2

PATHOGENESIS OF INVERTEBRATE MICROBIAL DISEASES. Davidson, ed. © Allanheld, Osmun (Totowa, N.J.)

MONILIALES (HYPHOMYCETES)

Metarrhizium

The only significant entomopathogenic species in this genus is *M. anisopliae*. It is the cause of the green muscardine disease in many terrestrial insects. The information available on the pathogenesis of this disease is more detailed than that for any other fungal disease of terrestrial insects, and is summarized here first as a base of reference for the fungal diseases that will be dealt with subsequently.

Symptoms of disease: The symptoms of the green muscardine disease are basically similar in all the terrestrial insects that are susceptible to it, and may be exemplified by those noted in elaterid larvae (McCauley et al., 1968). The first symptom is usually a loss of appetite. This is followed by an apparent sluggishness, weakness and decreased sensitivity, and a subsequent progressive loss of function or paralysis. Color changes to transparency, yellow or brown, but not usually to black, may occur in integumental patches or in the whole body in some but not in all host species. At death the body of the host is usually soft, pliable and distended. It progressively hardens and mummifies, eventually becoming very brittle and more opaque. The cadaver may remain in this state for a prolonged period in a dry environment. Given moisture, however, a white mycelial growth appears over the body, proliferating particularly from body openings and membranous areas. The white mycelium gradually becomes covered by the dark green conidia produced during surface sporulation. The progression of symptoms and the time of death after infection varies between species of host insect and between individuals within a species.

Histopathology: The infective unit is the conidiospore (Madelin, 1963). It contains a single nucleus and, in the electron-dense cytoplasm, abundant nutrient stores in conspicuous lipoprotein inclusions and numerous lipoidal droplets (Fig. 13-1) (Hammil, 1972; Zacharuk, 1970a). They germinate only in contact with moisture, and epicuticular lipids of insect integument may influence their rate of germination (Veen, 1968). Germination on the host may be inhibited by antibiosis from other fungal and bacterial surface contaminants (Schabel, 1976b, 1978).

The primary site of infection in most terrestrial insects is through the integument (Madelin, 1963). In the beetle *Hylobius pales* the fungus penetrates into the hemocoel solely through membranes of the body wall or in the buccal cavity (Schabel, 1976a). It usually penetrates intersegmental membranes but can also penetrate sclerites in several species of elaterid larvae (McCauley et al., 1968; Zacharuk, 1970c). Sclerites are penetrated as easily and as often as membranes on the body of locust nymphs but, when fed spore contaminated food, penetration usually occurs at the base of the cephalic appendages or in the buccal cavity (Veen, 1966, 1968). The rectal wall can also be penetrated when spores are injected into the rectum (Veen, 1968). There is no definitive proof that M. anisopliae can infect any terrestrial insect through an intact epithelial wall of the midgut (i.e., Gabriel, 1959; Schabel, 1976a). Occasionally it may enter the body cavity through spiracles and trachae (Madelin, 1963; McCauley et al., 1968), or through the thin socket membrane and along the canal of integumental sensilla (McCauley et al., 1968).

Germ tubes produced by germinating conidia grow over the surface of the host cuticle in a meandering and random fashion (McCauley et al., 1968; Schabel, 1976b), and may attain lengths of 150 μm (Schabel, 1976b). They may have a short range chemotactic affinity for the surface of the host cuticle (Zacharuk, 1970b), on which they form lateral and terminal enlargements termed appressoria (Fig. 13-2) (McCauley et al., 1968; Schabel, 1976b, 1978; Zacharuk, 1970b). Germ tubes and appressoria from different conidia are often entangled with the spores into tight cushions (Fig. 13-6); these and individual appressorial cells seem to be glued to the surface of the host by a mucoid when viewed by scanning (Schabel, 1976b, 1978) and transmission (Zacharuk, 1970b) electron microscopy. Appressorial-like structures were produced in vitro by germ tubes of several entomogenous fungi, including M. anisopliae, against hard surfaces such as glass, artificial wax membranes, and excised pieces of Tenebrio molitor integument (Madelin et al., 1967). These authors suggested that host penetration by these fungi may resemble that of the appressorium-forming fungal pathogens of plants. The conidial nutritive stores are usually depleted and the cytoplasmic contents emptied during the formation of germ tubes and appressoria. Nutrition for the further development of appressoria and their penetrant apparatus seems to be derived from the surface cuticular secretions produced by the host (Zacharuk, 1970b, 1970c).

The epicuticle of the host integument is entered by very thin penetrant pegs (McCauley et al., 1968; Robinson, 1966; Zacharuk, 1970c). These are produced from the wall of the appressorium that is appressed to the surface of the host. The penetrant pegs push into the epicuticle as they extend (Fig. 13-4), but this mechanical penetrant pressure seems to be preceded and accompanied by an enzymatic action of secretions from the fungus (Figs. 13-3 and 13-4) (Zacharuk, 1970c). The resulting penetration canal seems to form more by dissolution than by pressure fracture or puncture.

The penetrant hypha extends considerably after it has traversed the host epicuticle. In elaterid larvae it usually

Figure 13-1. Conidiospore of *Metarrhizium anisopliae* with abundant nutrient stores in large, very electron-dense lipoprotein globules flanking the nucleus (Nu) and peripheral electron-translucent lipoidal globules in an electron-dense ground plasma. From Zacharuk (1970a).

Figure 13-2. Germinated conidium (Co), its germ tube (gt) and appressorial cells (A) formed against the surface of the cuticle (arrowheads) of an elaterid larva. From Zacharuk (1970b).

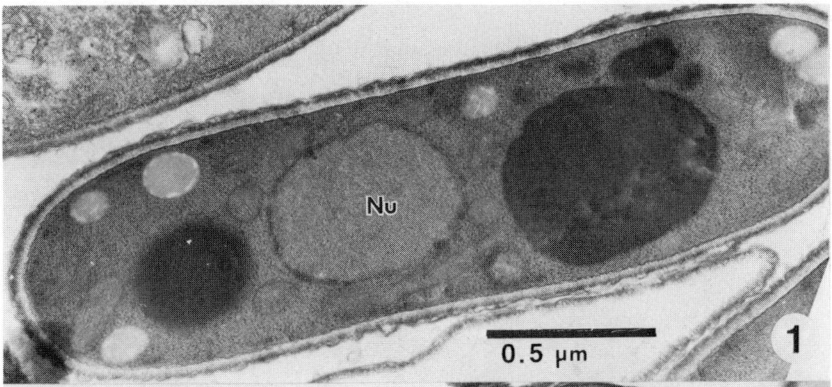
Nu
0.5 μm
1

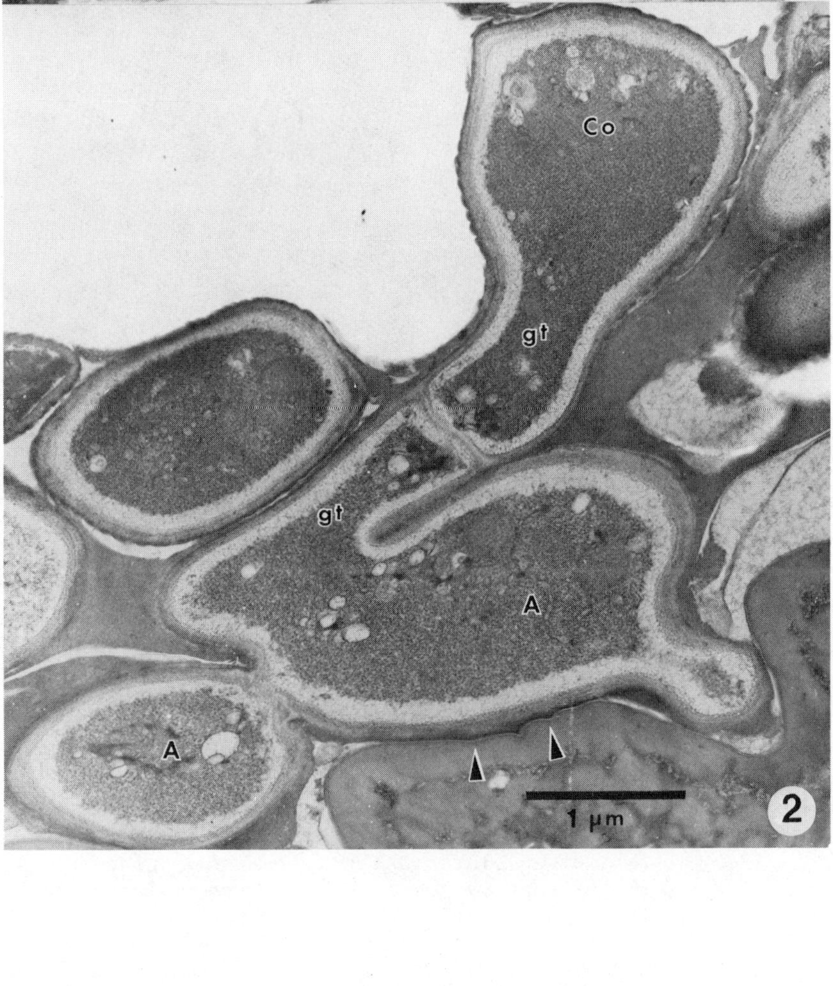
Co
gt
gt
A
A
1 μm
2

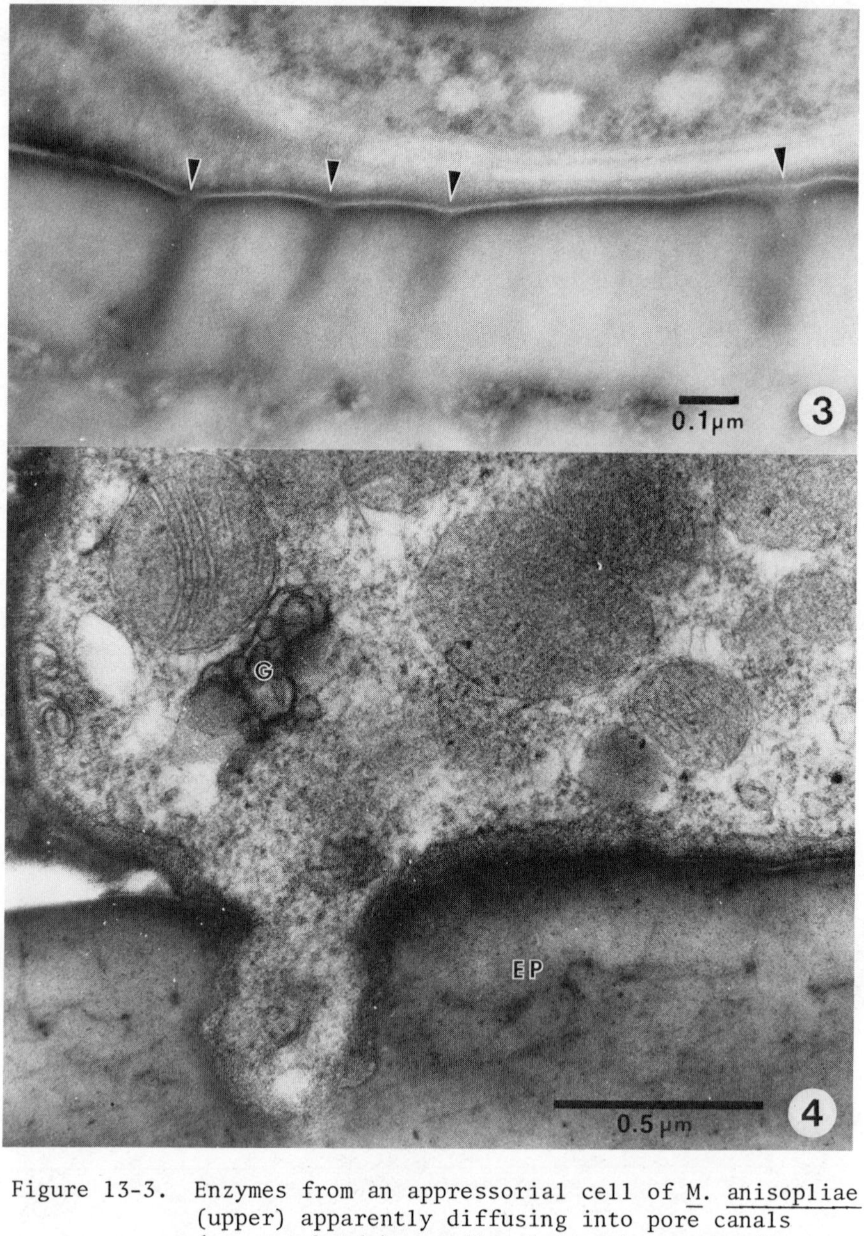

Figure 13-3. Enzymes from an appressorial cell of *M. anisopliae* (upper) apparently diffusing into pore canals (at arrowheads) and dissolving the surrounding epicuticle (lower) of an elaterid larva.

Figure 13-4. Penetrant peg from an appressorial cell of *M. anisopliae* (upper) invading the epicuticle (EP) of an elaterid larva, with a Golgi complex (G) presumably secreting the enzymes implicated in the penetration of the cuticle.

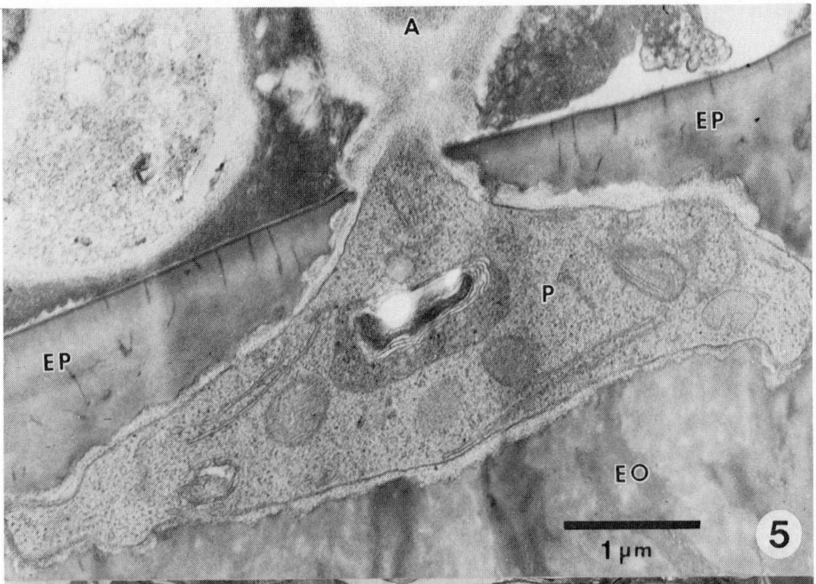

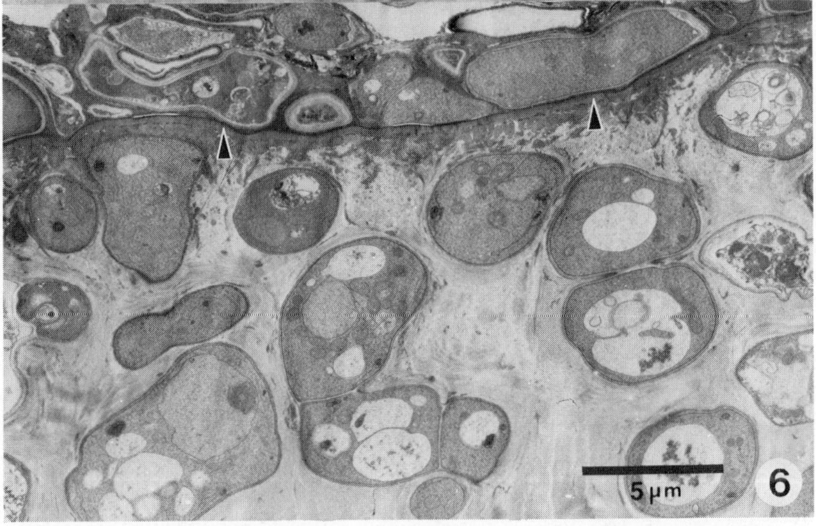

Figure 13-5. An expanded penetrant plant (P) formed by the appressorial cell of M. anisopliae (A) between the epicuticular (EP) and exocuticular (EO) layers of the cuticle of an elaterid larva. From Zacharuk, 1970c. Figures 13-1 through 13-5 reprinted by permission of Academic Press, Inc.

Figure 13-6. An appressorial pad (upper) against the surface of an epicuticle (at arrowheads) of an elaterid larva, and penetrant hyphal bodies and hyphae in the underlying procuticle.

forms a plate-like expansion just under the epicuticle (Fig. 13-5) (McCauley et al., 1968; Zacharuk, 1970c). This expansion extends thin hyphae laterally and inward, which produce numerous contiguous and scattered hyphal bodies between lamellae of the procuticle (Fig. 13-6). Hyphae from these hyphal bodies penetrate the underlying procuticular lamellae either directly, or in a stepwise fashion alternating between additional interlamellar hyphal bodies and translamellar hyphae (Fig. 13-7A). Direct hyphal penetration was noted most often in insects and body regions with thin and weakly sclerotized cuticles; stepwise penetration usually occurred in the thicker and more heavily sclerotized cuticles. There is both histological (McCauley et al., 1968) and ultrastructural (Zacharuk, 1970c) evidence that the fungus penetrates the procuticle by mechanical separation of the lamellae and by an accompanying enzymatic action (Figs. 13-5 and 13-17B). The latter is undoubtedly necessary to nourish the extensive fungal development in the host cuticle as well as to facilitate its penetration.

When fungal entry into the host occurred through spiracles of sensillar sockets, penetration seemed to be directly by germ tubes without the formation of appressoria (McCauley et al., 1968). In sensillar penetration the entrant hypha formed hyphal bodies in the sensillar canal just under the epicuticle, and fungal growth into the coelom continued from these (Zacharuk, 1970c). In the spiracular entries the germ tubes meandered within the tracheae but penetration of the tracheal wall was not observed.

In other host insects where penetration of the cuticle by M. anisopliae was observed, both mechanical pressure and enzymatic action are implicated (i.e. Madelin, 1963; Robinson, 1966). While the chitonolytic, lipolytic and proteolytic activities of this fungus have been demonstrated in vitro, there is not direct evidence available for their occurrence in vivo.

In elaterid larvae that are approaching a molt, the penetrant hyphae established in the exuvial cuticle can grow through the exuvial space, produce secondary appressoria on the surface of the new cuticle, and thus re-establish within the host (Zacharuk, 1973a). When this occurs, the insect cannot shed the exuvium and is subject to secondary bacterial infections through the ecdysial space into the midgut. Should the insect shed the exuvium before the fungus adheres to the new cuticle, the pathogen would be shed with the exuvium. This may be a common occurrence in insects that molt frequently, particularly in those with thick or heavily sclerotized cuticles which require a lengthy stepwise fungal penetration for infection.

Most of the host insects show behavioral symptoms of infection by the time the penetrant hyphae reach the epidermis or enter the coelom. In the coelom the hyphae produce hyphal bodies around the point of ingress. These are free cells usually less elongate and larger than conidia (Fig. 13-9). They circulate in the hemolymph (Prasertphon and Tanada, 1968) and produce secondary hyphae, thus spreading the pathogen throughout the body of the host. At the time of host death, however, coelomic fungal development is usually insignificant, and limited to the proliferation of penetrant hyphae and hyphal bodies around the point of ingress and to a few hyphal bodies scattered in other parts of the coelom (McCauley et al., 1968; Schabel, 1976b; Veen, 1968). This has been taken as evidence, albeit indirect, for the implication of toxins as primarily responsible for the behavioral symptoms of the disease and death of the host.

In elaterid larvae the penetrant hyphae traverse the epidermis and invade only the fat tissues in the coelom adjacent to the point of ingress before the host dies (Zacharuk, 1971a). However, the other tissues show degradative changes in ultrastructure before they are invaded by the fungus and before the host dies (as exemplified in muscle in Fig. 13-8). Postmortem there is further proliferation of hyphal bodies and of secondary hyphae throughout the body, and the hyphae invade and degrade all the body tissues (Prasertphon and Tanada, 1968; Zacharuk, 1971a, 1971b). As the cadaver begins to mummify, the fungus produces large resting spores with abundant nutritive reserves in the coelom (Fig. 13-10) (Zacharuk, 1971b). Concurrently or subsequently, depending on the moisture available, the mycelium egresses to the surface, proliferates, and produces the infective conidia on the cadaver (McCauley et al., 1968; Schabel, 1976b; Veen, 1968).

Fungal virulence and mycotoxins:

Based on the foregoing histopathology, the conidia of *M. anisopliae* must adhere to and germinate on some cuticular surface of a terrestrial insect host, and the resulting penetrant apparatus must traverse the integument and invade the coelom to initiate a lethal mycosis.

The germination of conidia of *M. anisopliae* is influenced by the composition of the medium used for sporulation (Veen, 1968). Strains of fungus isolated from the same or a closely related host species are more virulent to larvae of the Colorado potato beetle, *Leptinotarsa decemlineata*, or of the moth, *Galleria mellonella*, than are strains isolated from more divergent host species (Fargues et al., 1976; Ferron et al., 1972; Kmitowa et al., 1977). Integumental resistance to infection by non-specific strains is more pronounced than resistance to infection by intracoelic spore injection (Fargues et al., 1976). The results from several studies on different hosts show that virulence is directly related to the

Figure 13-7. Penetration of the procuticle of an elaterid larva. A. Interlamellar hyphal bodies (hb) and translemellar hyphae (at arrowheads). B. Penetrant hyphae (h) in the procuticle (c), showing degradation of the cuticle adjacent to the fungal growths (arrowheads) of *M. anisopliae*.

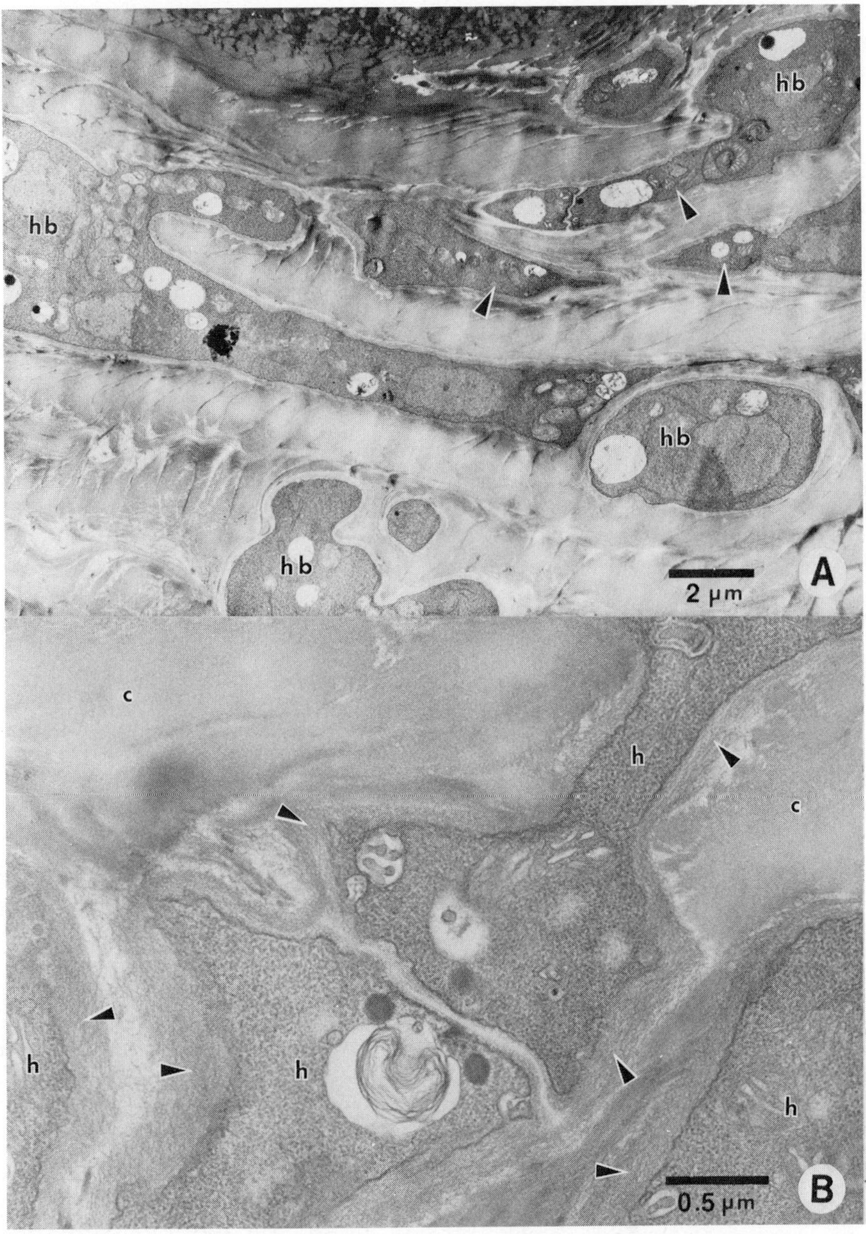
hb
hb
hb
hb
2 μm
A
c
h
c
h
h
h
0.5 μm
B

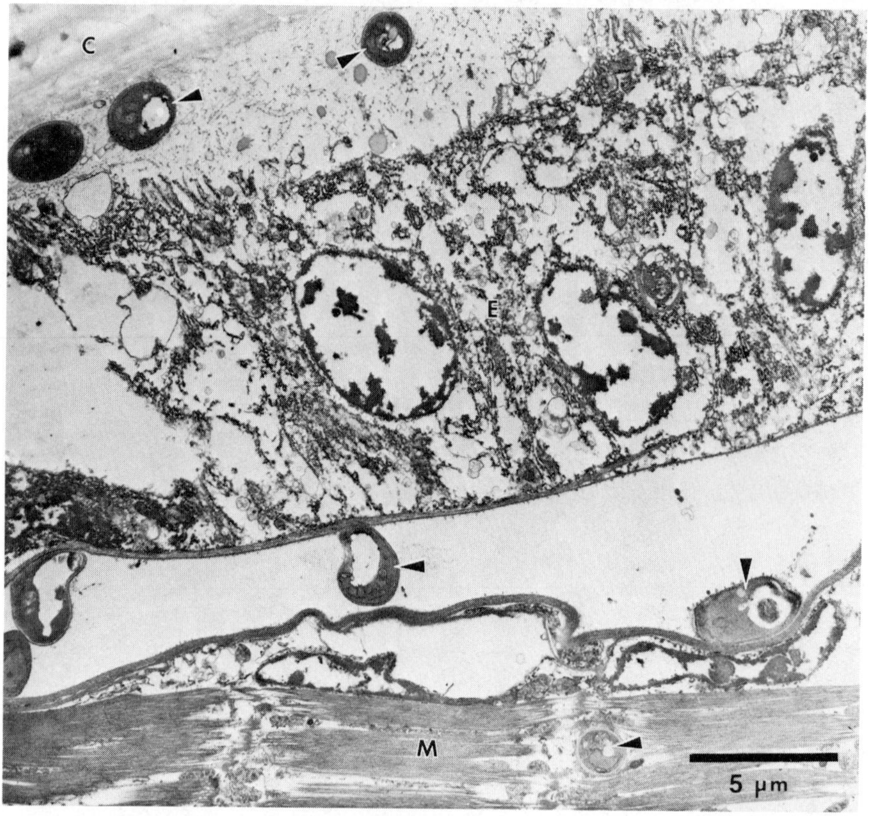

Figure 13-8. The cuticle (C), epidermis (E) and underlying coelom of an elaterid larva, the latter just invaded by penetrant hyphae (arrowheads) of M. anisopliae, showing degradation of the epidermis and muscle (M).

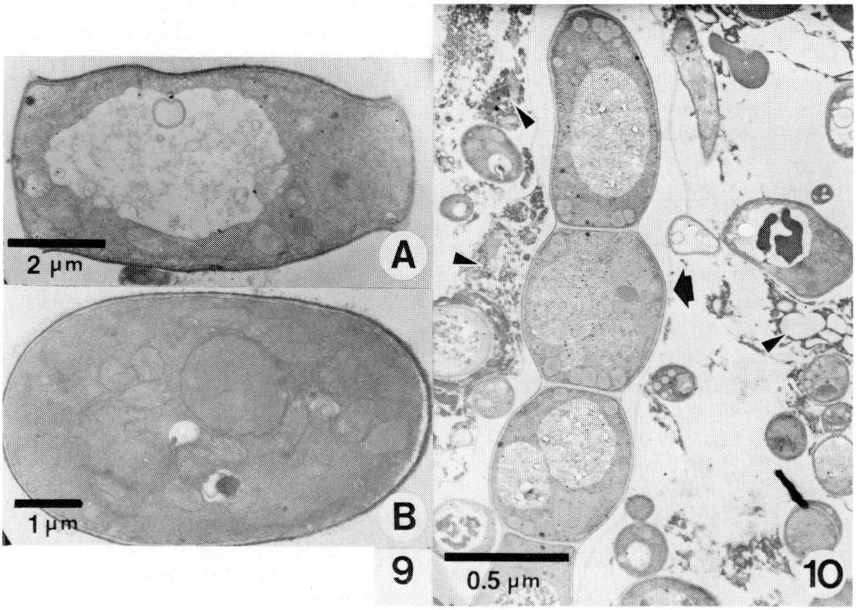

Figure 13-9. Coelomic hyphal bodies of M. anisopliae in an elaterid larva. A. Immature, just separated from a hypha. B. Mature. From Zacharuk, 1971b.

Figure 13-10. Chain of resting spores of M. anisopliae in an elaterid larva (at arrow), forming postmortem; arrowheads denote degraded coelomic tissue of the host. From Zacharuk,1971b. Figs.13-9 and 10 reprinted by permission of the National Research Council of Canada.

conidial dosage applied to the integument or injected into the hemocoel. Conidial germination is inhibited by other microbial contaminants on the surface of the host (Schabel, 1976b, 1978) and perhaps also in its environment (Zacharuk and Tinline, 1968). Mutants of this fungus that have an enhanced ability to degrade starch also have an enhanced conidial germination rate and were hypervirulent to mosquito larvae, but only when cultured on a medium that contained starch (Al-Aidroos and Seifert, 1980). However, alterations in their ability to degrade protein did not correlate with changes in virulence or in rate of conidial germination. This may apply similarly to strains of this fungus that are pathogenic to terrestrial insects. Thus, the nutrition of the sporulating medium and the nature of the surface of the host, as well as the composition of its integument, may significantly influence the germination of conidia on the host and the subsequent production of the number of penetrations of the integument that are necessary to establish a mycosis in a terrestrial insect.

Earlier workers inferred that M. anisopliae produced toxic substances that cause or enhance mortality of host insects (Madelin, 1963). More recently Kodaira (1961, 1962) isolated two toxins from culture filtrates of this fungus, Destruxins A and B. Both were lethal to larvae of the silkworm, Bombyx mori, and to mice by injection. They were also toxic when ingested by Leptinotarsa larvae, particularly newly molted third instars, although their presence on food discouraged feeding. Toxic substances lethal to silkworms were also recovered from silkworms infected with M. anisopliae. Roberts (1966) noted that filtrates from cultures of this fungus caused tetanic paralysis in silkworms and Galleria larvae when injected intrahemocoelically, and blood from infected silkworms that were in tetanic paralysis just before death was toxic when administered intrahemocoelically to Galleria larvae. In addition to Destruxins A and B, submerged cultures of this fungus produced other unidentified compounds that also caused a tetanic paralysis in Bombyx and Galleria larvae (Kodaira, 1961; Roberts, 1969). Kucera (1980) isolated a protease produced in vitro that was toxic to Galleria larvae.

Conidia of M. anisopliae also contain toxins, which are released when spores are disrupted probably by digestion in the gut of mosquito larvae (Crisan, 1971). If both the conidia and the fungal growths in the hemocoel of the host produce toxins, it is unlikely that the intervening fungal germ tubes and integumental penetrant hyphae do not. Toxins could diffuse from the latter growths into the hemocoel, and thus intoxify terrestrial insects and start degrading their hemocoel tissues before physical invasion actually occurs, as was noted histopathologically.

Host reactions: Reports on reactions in the host that could be regarded as defensive against the progress of infection by *M. anisopliae* are very limited. Veen (1968) saw no evidence of phagocytosis of the fungus by hemocytes in locust nymphs. Similarly, there was no histological or ultrastructural evidence for any defensive mechanism in elaterid larvae (McCauley et al., 1968; Zacharuk, 1973b). A cellular defense mechanism was noted in larvae of *Oryctes rhinoceros* that were injected intrahemocoelically with conidia of specific and non-specific strains of the fungus (Fargues et al., 1976). Within 24 hr the fungal cells of both strains were surrounded by a layer of hemocytes. After 4 days, however, the specific fungus grew out of the hemocytic envelope, but the non-specific fungus remained completely encapsulated within a heavily melanized granuloma. The lack of such observations in more terrestrial insects is probably the result of workers concentrating their histopathological studies on host insects that are infected through the integument with those strains of fungi that are most specific or virulent to that host.

There is no direct evidence for the presence of a humoral defense mechanism against *M. anisopliae* in terrestrial insects. The fact that pathogenicity is usually related to spore dosage suggests that such a defense mechanism may be operative when the number of spores that germinate and the resulting number of penetrations of the integument is low. Field populations of elaterid larvae are less susceptible to infection by *M. anisopliae* than are laboratory reared populations (Bell and Hamalle, 1971), perhaps due to a differential in an internal defense mechanism or in surface antibiosis.

Beauveria

Two species of this genus are known entomopathogens, *B. bassiana* and *B. tenella*. Significant pathogenetic information is available only for the white muscardine disease caused by the former species.

Symptoms of disease: The host symptoms to infection by *Beauveria* are largely similar to those caused by *M. anisopliae* in terrestrial insects. However, the conidia of *Beauveria* are white, and so the mycelial mat on the cadaver of the host continues to be white after sporulation. Also, there are more frequent reports of a color change to dark brown or black melanotic patches at the sites of penetration by *B. bassiana* (Fargues and Vey, 1974; Machowicz-Stefaniak, 1979). This fungus can synthesize pigments, apparently enzymatically, from aryl amines *in vitro* (Zajic, 1963). However, it did not darken the integument greatly in silkworms (Aoki and Yanase, 1970). The phenol oxidase activity in the integument decreased appreciably as infection progressed, and the activity remained at higher levels in blackened than in non-blackened integuments. Presumably the fungus disturbed the phenol oxidase activity of the host integument more severely in instances where integumental darkening was less intense.

Histopathology: Penetrations into the hemocoel can occur through any part of the host's body integument (Fargues and Vey, 1974; Pekrul and Grula, 1979), but may be more concentrated in intersegmental folds where spores are more easily trapped (Wasti and Hartmann, 1975). Penetrations may also occur through the oral apparatus (Bajan and Bilewicz-Pawinska, 1971). Microfed (Gabriel, 1959) and ingested (Wasti and Hartmann, 1975) spores can germinate in the midguts of some insects and the fungus thus penetrate into the hemocoel through the gut wall, but this does not occur in all insects. Conidia may also enter spiracles, and germinate within and penetrate the walls of trachae (Pekrul and Grula, 1979). This seems to be the preferred mode of infection in the alfalfa weevil, *Hypera postica* (Hedlund and Pass, 1968).

The germ tubes from germinating conidia seem to have an affinity for the surface of the host cuticle, resulting in growth directly towards it (Pekrul and Grula, 1979). Germinal hyphae produced appressorial-like structures against hard surfaces *in vitro* (Madelin et al., 1967; Robinson, 1966), but penetrant hyphae were not seen to emerge from these structures. In larvae of the corn earworm, *Heliothis zea*, direct entry of the germ tubes into the host cuticle was observed by scanning electron microscopy, apparently through enzyme-degraded holes (Pekrul and Grula, 1979). A similar direct penetration of the body wall by germinal hyphae seems to occur also in larvae of the gypsy moth, *Lymanthia* (*Porthetria*) *dispar*, (Wasti and Hartmann, 1975) and in the alfalfa weevil (Hedlund and Pass, 1968). However, Vey and Fargues (1977) report appressorial formation by *B. bassiana* in the penetration of the integument of *Leptinotarsa* larvae. This fungus produces lipase, protease and chitinase *in vitro* (Pekrul and Grula, 1979; Samsináková et al., 1971). The maximum amount of chitinase was produced earlier than were the maxima of the other two enzymes, and the amount produced seemed to be directly related to the ability of the germinal hyphae to penetrate the host cuticle.

Hyphal body production and their circulation in the hemolymph resemble that noted for *M. anisopliae*, and there is a similar rapid proliferation of the mycelium throughout the body. The host's hemocoel is usually filled with hyphal bodies and hyphae at the time of death, but the extent of fungal invasion of tissues varies. It may be minimal (Pekrul and Grula, 1979; Prasertphon and Tanada, 1968) or extensive, permeating most or all of the tissues (Hedlund and Pass, 1968; Wasti and Hartmann, 1975). Degradative changes in tissues occur before they are invaded by hyphae (Hedlund and Pass, 1968). The fat bodies are the first of the tissues to be affected or invaded. Alterations in levels of hemolymph proteins occur during infection; these vary with the host species and routes of infection (Gardner et al., 1979). The hemocoel mycelium grows out to the body surface where sporulation occurs. No recent reports were seen on the formation of resting spores.

Molting *Leptinotarsa* larvae can be infected through both cuticles and the intervening ecdysial space, with the formation of blastospores in the fluid-filled space (Fargues and Vey, 1974; Vey and Fargues, 1977). Some larvae shed the fungal contamination with the exuvium, and thus escape infection. Others develop integumental wounds through which the fungus or bacteria gain entry into the coelom after the exuvium is shed (Vey and Fargues, 1977).

Fungal virulence and mycotoxins: There is a direct relationship between the dosage of conidia of *B. bassiana* and pathogenicity in the population of the host (Bajan and Bilewicz-Pawinska, 1971; Zacharuk and Tinline, 1968). Strains of this fungus also tend to be fairly host specific (Kmitowa et al., 1977). Soil extracts inhibited germination of conidia and growth of germ tubes. This inhibitory activity was greater in the surface than in the deeper layers of soil, and was reduced in autoclaved soil (Clerk, 1969). Dextrose and extracts from silkworm cocoons enhanced conidial germination and growth of germ tubes. The pathogenicity to elaterid larvae was similarly greater in sterilized than in non-sterilized spore-infested soils (Zacharuk and Tinline, 1968). Thus, antibiosis and nutrition may have a significant influence on the germination of the conidia, the development of the penetrant apparatus and, consequently, the pathogenicity of *B. bassiana* to terrestrial insects.

The pathogenicity of strains of *B. bassiana* to *Leptinotarsa* larvae was directly related to their chitinolytic and proteolytic activities (Samsináková et al., 1977). Paris and Segretain (1975) noted a similar direct relationship between the lipolytic activity of *B. tenella* and its pathogenicity to larvae of the cockchafer, *Melolontha melolontha*, but the relationship with its proteolytic activity was inverse. While virulence may be related in part to the production of enzymes for cuticular penetration, such enzymes should also have a "virulent" degradative effect on the hemocoel tissues and vital processes of the host during fungal invasion.

Zajic (1963) demonstrated an inverse relationship between the pathogenicity of strains of *B. bassiana* for the salt marsh caterpillar, *Estigmene acrea*, and their ability to synthesize pigments. Aoki and Yanase (1970) similarly noted that melanotic blackening of the integument of silkworms was more intense in moderate than in acute infections. Whether the phenoloxidases involved are fungal, as interpreted by Zajic, or in the integument of the host and disrupted to a greater extent by the more virulent fungi, as interpreted by Aoki and Yanase, remains to be clarified.

This fungus produces substances *in vitro* that are toxic to *Galleria* larvae (Kucera and Samsináková, 1968; West and Briggs, 1968) and to silkworms (Kodaira, 1961). Kucera (1971) identified a proteolytic complex, the activity of which was enhanced by the presence and influenced by the composition of nutrient nitrogen sources *in vitro*. Toxin production was shown also *in vivo*; blood from infected silkworms was toxic to healthy silkworms when injected intrahemocoelically (Kodaira, 1961). Organic acids were also produced in infected silkworms, but these only appeared postmortem and were not implicated in cause of death.

The antibiotic "beauvericin" produced by *B. bassiana* is toxic to brine shrimp and mosquito larvae (Hamill et al., 1969), but not to *Heliothis* larvae (Champlin and Grula, 1979). In the latter instance, it could not be detected in the hemolymph of infected larvae, and intrahemocoelic injections failed to induce toxic symptoms in healthy larvae. Kodaira (1961) had earlier demonstrated the antibacterial activity of this fungus *in vitro* and *in vivo* in infected silkworms. The water loss that resulted in mummification of the silkworm cadavers occurred 3 to 5 days after death, yet there was no postmortem bacterial degradation. Extracts of the mummified cadavers were highly antibiotic, which would explain the preservation of mycosed cadavers for extended periods.

Host reactions: Salt (1970), on the basis of earlier studies, believed it unlikely that phagocytosis could restrain any fungus that was able to penetrate the integument of an insect. However, a highly pathogenic strain of *B. bassiana* was subjected to a very active cellular reaction in the hemocoel of *Leptinotarsa* larvae; Vey and Fargues (1977) concluded that this, combined with the exuvial shedding of the invading fungus at a molt, was a fairly effective defense mechanism. Such a defense mechanism may also be effective against this fungus when there are only a few localized hyphal penetrations into the hemocoel, or against strains that are low in enzymatic or toxic activity (Vey and Vago, 1971).

A hidden phase of a *B. bassiana* infection may persist in *Leptinotarsa* larvae after a molt, with no symptoms evident, and could be activated at a later stage in some manner as yet unknown (Vey and Fargues, 1977). The pathogenicity of this fungus and of *M. anisopliae* to flies was not affected by the age of the host (Rizzo, 1977).

Aspergillus

Species of Aspergillus are more predominantly saprophytic (Madelin, 1963), but several species are also pathogenic to terrestrial insects (Burnside, 1930). In general, however, they seem to be less significant entomopathogens than are Metarrhizium or Beauveria species.

Disease symptoms and histopathology: The symptoms in terrestrial insects infected with Aspergillus resemble those of Beauveria infections in most aspects. The infective conidia of A. flavus germinate in the alimentary canal, and hyphae penetrate the wall of the gut into the hemocoel of bees (Burnside, 1930) and of some lepidopterous larvae (Sussman, 1952). This seems to be a favored mode of infection. Germinal hyphae can penetrate the skin of bee larvae but not of adult bees (Burnside, 1930), and can infect lepidopterous larvae by penetrating intersegmental membranes (Sussman, 1952; Aoki, 1961) and the anus (Aoki, 1961). Germinal hyphae can also penetrate the chorion and infect eggs of Galleria (Behnke and Yenodol, 1969). This fungus was not observed to form appressoria in vivo or on hard surfaces in vitro (Robinson, 1966). Penetrant hyphae lifted the epicuticle at the point of entry into the integument of the mound-building termite, Odototermes obesus. The endocuticular lamellae were pushed downwards as they were penetrated, suggesting a mechanical force. The endocuticle around penetrating hyphae became PAS positive, melanin appeared, polyphenols were liberated, and the ascorbic acid normally present in the epidermis disappeared. Lateral hyphal growth in the endocuticle was extensive before the coelom was invaded (Sannasi, 1969a).

When penetrant hyphae gained access to the host's hemocoel they proliferated throughout the body. They entered fat tissue first, then other tissues, degrading them physically and by presumed enzymatic action (Burnside, 1930; Sussman, 1952). When the body of the host was filled and its tissues riddled with hyphae, movements ceased. Outgrowths of mycelia then appeared on the surface of the cadaver where sporulation occurred. No recent reports were seen on the formation of hyphal bodies or resting spores.

In the mycosis of grasshoppers by A. parasiticus (Ogloblin and Jauch, 1943), conidia enter spiracles and germinate in the trachae, from where they penetrate into the hemocoel. The subsequent histopathology resembles that noted above. In a natural mycosis of dipterous pupae by A. tamarii (Vey et al., 1967), the fungus developed in the clots of hemorrhagic integumental wounds caused by sharp grains of sand in the substrate. The body was not penetrated by mycelia until after the host died.

Fungal virulence and mycotoxins: Entomogenous species of *Aspergillus* seem to be pathogenic to a limited number of insect species, and to be variably virulent to individuals within a species. Probably for this reason, studies related to specific aspects of their pathogenicity are also limited.

Insecticidal mycotoxins were produced *in vitro* by *A. flavus* (Beard and Walton, 1971; Kodaira, 1961) and by *A. ochraceus* and *A. oryzae* (Kodaira, 1961). Blood from silkworms infected by these fungi was toxic to healthy silkworms by injection (Kodaira, 1961). Toxins produced *in vitro* retarded development or killed larvae of the house fly, the milkweed bug and the confused flour beetle, but showed no effect on the earwig, cockroach, Indian meal moth, greater wax moth or termites (Beard and Walton, 1971).

The melanization observed in the integument and hemolymph of termites infected with *A. flavus* was not a requirement for their death (Sannasi, 1969a). Ascorbic acid inhibited melanosis. Its level in the hemolymph of healthy insects gradually decreased after infection as melanization appeared. Tyrosine was present in normal hemolymph, but polyphenols first appeared 1 day after infection. The latter may contribute to the toxicity symptoms in termites. This fungus could be the source of the phenol oxidases involved, as was shown by Zajic (1963) for other moderately entomopathogenic fungi.

Host reactions: The most detailed recent descriptions of the action of insect's cellular defense mechanisms against a fungal infection involve *Aspergillus*, perhaps because of its generally low and variable entomopathogenicity (Vey and Vago, 1969, 1971). In *Galleria* larvae infected with *A. niger* the antifungal hemocytes appear to be plasmatocytes (Vey, 1971). They form a tight, multilayered envelope around the fungal elements in the hemocoel, thus forming a granuloma. The interhemocytic spaces are narrow and filled with a dense material, and intercellular junctions are common. The innermost cells become modified in fine structure, suggestive of a partial lysis. The fungus within the granuloma becomes encased in a deposit of melanin, which may extend into the inner layer of hemocytes. This melanized zone increases with time. The granulomes may migrate against the integument where they become partially cuticularized and may be shed with the exuvium at the next molt. Ogloblin and Jauch (1943) describe a somewhat similar cellular defense mechanisms in grasshoppers and locusts against invading *A. parasiticus* elements. Primary granulomes formed against the walls of trachae to prevent fungal invasion of the hemocoel. Secondary granulomes enveloped fungal elements that had invaded the hemocoel.

The cuticular lipids of silkworms and of larvae of the rice stem borer, *Chilo simplex*, are fungistatic or fungicidal to A. flavus, depending on their concentration (Koidsumi, 1957). The antifungal activity is present in exuvial cuticle but is greater in lipids extracted from living cuticle. The most active were medium-chained fatty acids, presumably caprylic or capric acid. The susceptibility of both host species to infection by the fungus increased significantly after the cuticular lipids were removed mechanically or chemically.

Paecilomyces

The entomopathogenic Spicaria (Isaria) farinosa and S. fumoso-rosea were transferred recently to the genus Paecilomyces. While they are fairly common pathogens of a variety of terrestrial insects, there is little information beyond that reported by Madelin (1963) on the pathogenesis of their diseases. In pathogenicity studies (Bajan and Kmitowa, 1969), infection of host insects occurs readily after body contact with spores. The fungus forms appressorial-like structures in vitro and readily penetrates excised integuments of Tenebrio larvae (Madelin et al., 1967; Robinson, 1966). Mycelial growths form on the surface of the host while it is paralyzed but still alive, as well as postmortem (Bajan and Kmitowa, 1969). Melanization of the integument occurs with infection as in B. bassiana but may be more severe in some insects (Aoki and Yanase, 1970), to the extent of blackening the entire body at death (Bajan and Kmitowa, 1969), and less severe in other insects (Machowicz-Stefaniak, 1979).

Paecilomyces farinosus and P. fumoso-roseus produce toxic substances in vitro (Kodaira, 1961; Wojciechowska, 1973) and in vivo (Kodaira, 1961). Chitinolytic and proteolytic activity was demonstrated in the former species (Samsináková et al., 1977). Strains of both species tend to be host specific in pathogenicity (Kmitowa et al., 1977). No recent information is available on defense reactions by host insects to infection by Paecilomyces.

Other Hyphomycetous genera

Fusarium solani was recently reported to be a weak pathogen of the elm bark beetle, Scolytus scolytus, but the host-pathogen interaction is not known (Barson, 1976). The hyphae may cause asphyxiation by blocking trachae, or may weaken the host by non-lethal toxins. Toxic substances produced in vitro by this fungus and by F. larvarum were insecticidal to the blow fly Calliphora (Claydon et al., 1977, 1979).

Mucor hiemalis normally cannot penetrate an intact integument or intestinal wall, but invades the hemocoel through integumental wounds. The host often dies with little or no intrahemocoelic hyphal development (Heiter, 1962; Vey, 1968). The cellular defense mechanism is usually overpowered, presumably by toxins, and there is a subsequent rapid proliferation of hyphae. Some infected Galleria larvae show no symptoms of disease; the fungal elements in these larvae are encapsulated within granulomes and eliminated at the next molt much as described for A. flavus (Vey, 1968; Vey and Vago, 1971). Mucor hiemalis produced substances that were toxic in vitro to dictyopteran cells in culture. This technique using insect cell cultures shows promise for the bioassay of entomopathogenic mycotoxins (Vey and Quiot, 1976).

Nomuraea rileyi was formerly in the genus Spicaria or Botrytis. It causes epizootics in field populations of several species of caterpillars. The conidia germinate on the integument of Heliothis larvae in two days, and the germ tubes penetrate the cuticle by an apparent enzymatic activity in the epi- and exocuticles and a lateral hyphal growth into the epidermis through the endocuticle (Mohamed et al., 1978). Hyphae proliferate in the hemocoel and attack the internal organs in the following order: blood cells, fat bodies, Malpighian tubules, muscles and mesenteron. The tissues disintegrate before death, and hyphae start growing out of the body at death. This fungus secretes chitinase, protease and lipase in vitro; these enzymes are probably involved in vivo in cuticular penetration and tissue degradation. It can also affect the development of a braconid parasite in infected boll-worm larvae (King and Bell, 1978). Caterpillars can also be infected by feeding on spore-infested plants (Ignoffo et al., 1976), but it is not known whether penetration is through the cuticle of the buccal apparatus or through the wall of the midgut. Pathogenicity is directly related to the conidial dosage applied to the integument (Mohamed et al., 1977).

Torula nigra invaded tissues of the parasitoid Exeristes comstockii through leg wounds (Bucher and Bracken, 1966). The resulting lesions caused loss of leg function, which interfered with motility and oviposition. A vigorous hemocytic reaction encapsulated and melanized fungal elements at or near the point of ingress into the coelom. The infection was thus localized, and mortality was light.

Verticillium lecanii infects the aphid Macrosiphoniella sanborni by surface contact with conidia (Hall, 1976). The fungus sporulates on the appendages of the host within a day after infection. This early sporulation on live aphids is prolific and, with apparently unimpaired locomotion, these individuals can rapidly spread the infection in an aphid colony. Hyphal bodies appear in the hemocoel 3 days after infection, and proliferate rapidly to engorge the hemolymph 1 day before death. Aphids died 4 to 6 days infection, and the fungus sporulated over the entire body postmortem. The aphids were fecund in the early stages of the disease, and produced uninfected nymphs in the later stages of mycosis. Aphids that were infected with very high spore dosage died within two days with an abundant surface fungal growth but with no invasion of the hemocoel. Death was believed to be due to a physical impairment of function rather than to the action of a mycotoxin.

ENTOMOPHTHORALES

Twenty years ago only two genera in this order were considered to be entomogenous: Entomophthora and Massospora (MacLeod, 1963). Since then a number of new entomopathogenic genera were recognized, and the generic placement of some of the species was revised (Ferron, 1978). The taxonomy used here will follow that of the original references, with the exception of E. coronata which will be discussed under the genus Conidiobolus.

Entomophthora and Conidiobolus

Disease symptoms and histopathology: The symptoms and progress of the disease in terrestrial insects infected with Entomophthora are similar in many aspects among the various fungal and host species (MacLeod, 1963). In summary, germ tubes from primary, secondary or in some cases subsequent condia that adhere to the surface of the host penetrate the integument probably in the thinner membranous regions, and the penetrant hyphae thus invade the hemocoel. There are no known reports of infections through the alimentary tracts. In the hemocoel the penetrant hyphae produce hyphal bodies which, with few exceptions, proliferate the fungus throughout the body of the host. Subsequently the hyphae and hyphal bodies produce conidiophores which egress to the surface of the host and produce there the primary conidia. Under certain conditions, however, the vegetative fungal elements may produce resting spores (azygospores, zygospores or chlamydospores) in the host's hemocoel instead of or in addition to conidiophores and conidia. Under favorable conditions the resting spores may produce infective hyphae or conidia.

Symptoms of the disease usually appear late in the mycosis. The host is usually restless for a short period, then movements become sluggish and soon cease. A yellowish discoloration of the body often appears with the sluggish movements, and this intensifies as the disease progresses. Just before death many host insects either drop to the ground or assume characteristic positions and poses on their host plants, which their cadavers maintain after death. The formation of conidiophores or resting spores usually begins at about the time the host dies. Postmortem the cadaver is moist, turgid and extended, but becomes shrivelled, mummified and brittle when the conidiophores are formed and the conidia released. Cadavers with resting spores remain mummified, and show no signs of fungal egress to the surface.

In more recent studies Gabriel (1968a) demonstrated *in vitro* that *E. apiculata*, *E. thaxteriana* and *E. virulenta* can produce chitinolytic, lipolytic and proteolytic enzymes, as can *C. coronatus*. Such enzymatic activities were demonstrated histochemically in localized secretions from penetrant hyphae of the latter fungus *in vivo* in cuticles of *Bombyx* and *Galleria* larvae (Gabriel, 1968b). Melanin was also present in the blackened spots at the points of integumental penetration. These findings support the conclusions based on histological evidence that penetration of the host integument is by chemical as well as physical means (Klein and Coppel, 1973; Yendol and Paschke, 1965). In the penetration of the integument of larvae of the cabbage looper *Trichoplusia ni*, however, the ultrastructural evidence suggests that mechanical pressures are more important than chemical dissolution in cuticular penetration by *E. apiculata* (Lambiase and Yendol, 1977) or than it is in penetration of elaterid larvae by *Metarrhizium* (Zacharuk, 1973b). In the former there is a much greater physical disruption of the epicuticle and of the procuticular lamellae.

Conidiobolus coronatus showed no preference for site of penetration of the integument of termites; it traversed sclerites as often and as successfully as membranes (Yendol and Paschke, 1965). Conidia also germinated in and penetrated the walls of the oral cavity, crop and esophagus, but did not infect through the wall of the midgut or hindgut. The large quantity of stored lipids and of other organelles noted ultrastructurally in conidia of *E. apiculata* indicates a capacity for great metabolic activity and the growth required for the penetration of the integument of the host. Appressoria were not formed in the sequence of penetration of the host cuticle by this fungus (Lambiase and Yendol, 1977).

When *E. apiculata* and *C. coronatus* were cultured *in vitro*, the exhaustion of nutrients and accumulation of waste products did not affect significantly the production of hyphal bodies (Prasertphon and Tanada, 1968). In the same study, hyphal bodies in infected larvae of *Galleria* did not circulate in the hemolymph as did those of *Metarrhizium* and *Beauveria*, presumably due to their larger size and to their formation primarily in solid tissues that were invaded preferentially by the penetrant hyphae rather than in the hemolymph. Egress of the fungus to the surface of the host was by hyphae produced by the hyphal bodies in the case of *C. coronatus*. In contrast this fungus did not produce hyphal bodies in termites; the vegetative hyphae proliferated the fungus throughout the body of the host and egressed through the integument for sporulation at the surface. The termites often died with little proliferation of fungus in the hemocoel, hemocoel tissues or any vital organ (Yendol and Paschke, 1965).

As in the muscardine diseases discussed in the preceding section, fat tissue is one of the first tissues to be invaded by the penetrant hyphae in the coelom. Other coelomic tissues are invaded, riddled and degraded by the vegetative hyphae to varying extents before or after the death of the different insect hosts as the mycoses progress (Klein and Coppel, 1973; Prasertphon and Tanada, 1968; Yendol and Paschke, 1965).

A protoplast stage in the life cycle of E. egressa was first observed in tissue culture (Tyrrell and MacLeod, 1972), and was later determined to be part of the natural life cycle when seen in the hemolymph of infected spruce budworm larvae (Tyrrell, 1977). The process of regeneration to a cell-walled form has been determined in vitro (Dunphy and Nolan, 1977) but not in vivo.

Fungal virulence and mycotoxins: The information is very limited on the bases for the pathogenicity of Entomophthora and Conidiobolus to their terrestrial insect hosts. Krejzová (1975) concluded that virulence of C. coronatus depends on its precultivation in some living organism, that the same or a related insect is more suitable than a distant one for enhanced virulence and, from a later study (1977), that multiple precultivation for enhanced pathogenicity is dependent more on the individual properties of the specimens than on the species of insect host.

Yendol and Paschke (1965) implicated toxins as an apparent cause of death in termites infected with C. coronatus, on the basis of the histopathology observed. Substances toxic intrahemocoelically to Galleria larvae were produced in vitro by C. coronatus and E. apiculata but not by E. thaxteriana or E. virulenta (Prasertphon, 1967). However, those produced by the latter were toxic to adult face flies, Musca autumnalis, (Yendol et al., 1968) and to blow flies (Claydon, 1978). Filtrates from larvae infected with C. coronatus were toxic intrahemocoelically to healthy larvae of Galleria (Yendol et al., 1968).

Host reactions: In larvae of the sawfly Diprion similis infected with E. tenthredinis numerous blood cells congregated at several loci in the body (Klein and Coppel, 1973). They often encapsulated the fungus after it traversed the cuticle, and thus curtailed the ability of the fungus to establish in the hemocoel. Hemocytes of the eastern hemlock looper did not adhere to protoplasts, but spherule cells strongly adhered to hyphae and hyphal bodies of E. egressa (Dunphy and Nolan, 1980). The protoplasts may secrete hemocyte inhibitors, or the hemocytes and protoplasts may be held apart by an electrostatic repulsion.

Other Entomophthorales genera

Massospora has two valid entomogenous species, M. levispora and M. cicadina, both limited to cicadas as their host. The former only infects adults, that is, above ground populations (Soper et al., 1976a, b). Emerging adults are infected by resting spores that produce conidia. These adults usually produce only conidia, and only within the abdomen to which mycelial growth is also confined. The conidia are released as the dry, brittle abdominal segments disintegrate. Adults infected by these conidia produce resting spores, without conidia or a mycelial mass, and these are also confined to the abdomen. The resting spores are also released and disseminated as the abdomen crumbles. Infected insects are often left moving about or flying with head and thorax only. The pathogenesis of M. cicadina infections is basically similar to the above (MacLeod, 1963).

Tarichium megaspermum, a pathogen of several species of cutworms, produces only resting spores within the host; a conidial stage has not been confirmed. Bucher and MacLeod (1974) "rediscovered" resting spores of this fungus along with hyphal bodies and conidia of E. virescens in several mycosed cutworms. They suggest that these may be two states of the same fungus which, if proven, would resemble Entomophthora in development and pathogenesis. The hyphal bodies in the hemocoel of the host fan out towards the cuticle from a central core, a mode of growth that, although reported also for E. aphidis, is atypical for Entomophthora in general.

Basidiobolus ranarum isolated from a mosquito is also pathogenic to termites and Galleria larvae (Krejzová, 1972, 1973, 1978). The course of infection resembles that of an Entomophthora mycosis. Basidiobolus fungi were previously considered to be primarily saprophytic on frog and lizard excrement (MacLeod, 1963).

LABOULBENIALES

Several fungi in this order parasitize terrestrial insects. They were generally considered to be ectoparasites, causing minor or no pathological effects on the host. There is now more information on the histopathology of several of these infections.

A number of species of Herpomyces infect a variety of cockroaches. These are highly but not completely host specific (Richards and Smith, 1954). Infective spores are firmly attached to the host cuticle by holdfasts; they attach to all parts of the body, but mature fungi occur primarily on antennae, sometimes on palpi, and only rarely elsewhere (Richards and Smith, 1955a). Penetration of the cuticle appears to be enzymatic rather than by pressure, and three types of haustoria have been described entering the cuticle and the underlying layer of epidermal cells (Richards and Smith, 1956). The volume of the fungus increases many thousands of times from that of the spore that produced it, indicating that it derives considerable nourishment from the insect host. Adults remain infected until they die. Nymphs may lose the fungus at a molt, but are easily reinfected subsequently, with no immunity apparent from the preceding infection. While the infection causes a dermatitis in cockroaches, there is no evidence of pathogenicity and no invasion of the hemocoel. *Herpomyces stylopygae* is dioecious (Richards and Smith, 1955b). Cell lineage is different for male and female plants; this and the gross structure of female plants differs between setal and body wall sites on the host, indicating a developmental plasticity in the fungus that is releated to the geometric limitations of the host's surface.

Hesperomyces virescens, unlike *Herpomyces*, produces a rhizoidal apparatus (haustoria) that penetrates into the body cavity of the host beetle *Chilocorus bipustulatus* that is predatory on citrus scales in Israel (Kamburov et al., 1967). Infections occur through virtually any part of the body surface. The disease leads to premature death of the beetles; it reaches epizootic proportions in nature, and is implicated in observed sharp declines in populations of the host insect.

Trenomyces histophthorus also invades the body cavity of its host, the chicken louse (Mallophaga). It produces extensive rhizomycelia in the hemocoel that results in loss of body and skeletal muscles, and is believed to cause significant host mortality (Meola and DeVaney, 1976). In a recently completed study on the ultrastructural histopathology of the disease (Meola and Tavares, 1981), penetration of the host cuticle into the hemocoel and effects on underlying tissues were noted. The protoplast of the basal cell of the thallus produced from a germinated spore narrows on penetration into the cuticle but expands into a primary haustorial bulb beneath the integument. The cuticular lamellae are pushed inward and disrupted extensively by this bulb. Extensions from it form secondary and tertiary bulbs and tubular hyphae in the hemocoel. This is accompanied by pathological changes in the ultrastructure of some of the epidermal, fat, and muscle cells that are adjacent to and in advance of the fungal growths.

Laboulbeniopsis termitarius appears to be a true ectoparasite on subterranean termites (Kimbrough and Gouger, 1970). Its development is limited to the surface of the host cuticle and is nourished presumably by surface secretions; no penetration pegs or haustoria were evident in the cuticle, and there were not deleterious effects to the host.

The recent studies on *Hesperomyces* and *Trenomyces*, which are obviously endoparasitic, contradict the generalization that the Laboulbeniales fungi on terrestrial insects are merely obligate ectoparasites. The degradation of host tissues prior to their physical invasion and disruption by fungal growths is indicative of some toxic or degradative fungal secretion in the hemocoel of the host. It is also inconceivable that there would not be some effect on the host through integumental perforation and sapping of nutrients in the case of simultaneous multiple integumental penetrations by fungi such as Herpomyces, even though they do not invade the hemocoel.

The above reports give no evidence for any cellular defensive reactions in the host to infection by the Laboulbeniales noted.

CONCLUSIONS

Knowledge on the interactions between several entomogenous fungi and their terrestrial insect hosts has been advanced considerably in the past 20 years. Except for a few minor pathogens (i.e., Mucor and Torula), all infect through intact cuticle-covered ectoderm. Enzymes play a significant role in the penetration of the cuticle by fungi in all three orders; these and other toxic substances appear to contribute significantly to the degradation of hemocoel tissues, disruption of vital functions, and the consequent mortality in host populations. Fungal virulence seems to be multi-faceted. It may be related to the germination rate of the infective conidia, or to the production of enzymes and other toxins by the invading hyphae. These may be dependent on the previous nutrition of the fungus before and during sporulation, on the nature and composition of the cuticle and of the surface secretions of the host being invaded, and on antibiosis by other microbial flora on or around the host. The mobilization and effectiveness of the insect's cellular defense mechanism against an invading fungus, and melanization in the host, are inversely related to the virulence of the fungal strain or species.

The bulk of current evidence that supports the above generalized conclusions is based on mycoses primarily by hyphomycetous fungi in a few terrestrial insect species. The currently limited information for the Entomophthorales and endoparasitic Laboulbeniales basically supports such generalizations. However, the pathogenesis of their mycoses in terrestrial insects needs much further elaboration for more fungal and host species. For all the mycoses discussed here, much more biochemical, physiological, genetic and ultrastructural information is required to clarify the specific mechanisms involved.

REFERENCES

Al-Aidroos, K. and Seifert, A. M. 1980. Polysaccharide and protein degradation, germination, and virulence against mosquitoes in the entomopathogenic fungus Metarrhizium anisopliae. J. Invertebr. Pathol., 36, 29-34.

Aoki, J. 1961. Studies on the infection mechanism of Aspergillus disease in silkworm larvae, Bombyx mori. II. The invading loci of causal fungus. J. Sericult. Sci. Japan, 30, 43-48.

Aoki, J. and Yanase, K. 1970. Phenol oxidase activity in the integument of the silkworm Bombyx mori infected with Beauveria bassiana and Spicaria fumoso-rosea. J. Invertebr. Pathol., 16, 459-464.

Bajan, C. and Bilewicz-Pawińska, T. 1971. Preliminary studies on the role of Beauveria bassiana (Bals.) Vuill. in reduction of Lygus regulipennis Popp. Ekol. Pol., 19, 35-46.

Bajan, C. and Kmitowa, K. 1969. Pathogenicity of entomogenous fungi isolated from hibernating imagines of the Colorado beetle (Leptinotarsa decemlineata Say). Ekol. Pol., Seria A, 17, 409-420.

Barson, G. 1976. Fusarium solani, a weak pathogen of the larval stages of the large elm bark beetle Scolytus scolytus (Coleoptera: Scolytidae). J. Invertebr. Pathol., 27, 307-309.

Beard, R. L. and Walton, G. S. 1971. Insecticidal mycotoxins produced by Aspergillus flavus var. columnaris. Bull. Connecticut Agr. Exp. Sta. No. 725, New Haven. 26 pp.

Behnke, C. N. and Yendol, W. G. 1969. Pathogenesis of an Aspergillus flavus infection of Galleria mellonella eggs. Entomophaga, 14, 177-184.

Bell, J. V. and Hamalle, R. J. 1971. Comparative mortalities between field-collected and laboratory-reared wireworm larvae. J. Invertebr. Pathol., 18, 150-151.

Bucher, G. E. and Bracken, G. K. 1966. Fungus disease of adult parasitic insects caused by Torula nigra (Marpmann). J. Invertebr. Pathol., 8, 193-204.

Bucher, G. E. and MacLeod, D. M. 1974. Rediscovery of Tarichium megaspermum (Entomophthoraceae) in Canadian cutworms (Noctuidae). J. Invertebr. Pathol., 23, 157-163.

Burnside, C. E. 1930. Fungous diseases of the honeybee. USDA Tech. Bull. no. 140, Washington, 42 pp.

Champlin, F. R. and Grula, E. A. 1979. Noninvolvement of beauvericin in the entomopathogenicity of Beauveria bassiana. Appl. Environ. Microbiol., 37, 1122-1125.

Claydon, N. 1978. Insecticidal secondary metabolites from entomogenous fungi: Entomophthora virulenta. J. Invertebr. Pathol., 32, 319-324.

Claydon, N., Groves, J. F. and Pople, M. 1977. Insecticidal secondary metabolic products from the entomogenous fungus

Fusarium solani. J. Invertebr. Pathol., 30, 216-223.

Claydon, N., Grove, J. F. and Pople, M. 1979. Insecticidal secondary metabolic products from the entomogenous fungus *Fusarium larvarum*. J. Invertebr. Pathol., 33, 364-367.

Clerk, G. C. 1969. Influence of soil extracts on the germination of conidia of the fungi *Beauveria bassiana* and *Paecilomyces farinosus*. J. Invertebr. Pathol., 13, 120-124.

Crisan, E. V. 1971. Mechanism responsible for release of toxin by *Metarrhizium* spores in mosquito larvae. J. Invertebr. Pathol., 17, 260-264.

Dunphy, G. B. and Nolan, R. A. 1977. Regeneration of protoplasts of *Entomophthora egressa*, a fungal pathogen of the eastern hemlock looper. Can. J. Bot., 55, 107-113.

Dunphy, G. B. and Nolan, R. A. 1980. Response of eastern hemlock looper hemocytes to selected stages of *Entomophtora egressa* and other foreign particles. J. Invertebr. Pathol., 36, 71-84.

Fargues, J., Robert, P. and Vey, A. 1976. Rôle due tégument et de la défense cellulaire des Coléoptères hôtes dans la spécificité des souches entomopathogènes de *Metarrhizium anisopliae* (Fungi Imperfecti). C. R. Acad. Sc. Paris, Série D, 282, 2223-2226.

Fargues, J. and Vey, A. 1974. Modalités d'infection des larves de *Leptinotarsa decemlineata* par *Beauveria bassiana* au cours de la mue. Entomophaga, 19, 311-323.

Ferron, P. 1978. Biological control of insect pests by entomogenous fungi. Ann. Rev. Entomol., 23, 409-442.

Ferron, P., Hurpin, B. and Robert, P. H. 1972. Sur la spécificité de *Metarrhizium anisopliae* (Metsch.) Sorokin. Entomophaga, 17, 165-178.

Gabriel, B. P. 1959. Fungus infection of insects via the alimentary tract. J. Insect Pathol., 1, 319-330.

Gabriel, B. P. 1968a. Enzymatic activities of some entomophthorous fungi. J. Invertebr. Pathol., 11, 70-81.

Gabriel, B. P. 1968b. Histochemical study of the insect cuticle infected by the fungus *Entomophthora coronata*. J. Invertebr. Pathol., 11, 82-89.

Gardner, W. A., Sutton, R. M. and Noblet, R. 1979. Effects of infection by *Beauveria bassiana* on hemolymph proteins of noctuid larvae. Ann. Entomol. Soc. Am., 72, 224-228.

Hall, R. A. 1976. *Verticillium lecanii* on the aphid, *Macrosiphoniella sanborni*. J. Invertebr. Pathol., 28, 389-391.

Hamill, R. L., Higgens, C. E., Boaz, H. E., and Gorman, M. 1969. The structure of beauvericin, a new depsipeptide antibiotic toxic to *Artemia salina*. Tetrahedron Lett., 49, 4255-4258.

Hammill, T. M. 1972. Electron microscopy of phialoconidiogenesis in *Metarrhizium anisopliae*. Am. J. Bot., 59, 317-326.

Hedlund, R. C. and Pass, B. C. 1968. Infection of the alfalfa weevil, Hypera postica, by the fungus Beauveria bassiana. J. Invertebr. Pathol., 11, 25-34.

Heiter, F. 1962. Parasitisme de blessure par le champignon Mucor hiemalis Wehmer chez les insectes. Ann. Epiphyt., 13, 179-205.

Ignoffo, C. M., Puttler, B., Hostetter, D. L. and Dickerson, W. A. 1976. Susceptibility of the cabbage looper, Trichoplusia ni, and the velvetbean caterpillar, Anticarsia gemmatalis, to several isolates of the entomopathogenic fungus Nomuraea rileyi. J. Invertebr. Pathol., 28, 259-262.

Kamburov, S. S., Nadel, D. J. and Kenneth, R. 1967. Observations on Hesperomyces virescens Thaxter (Laboulbeniales), a fungus associated with premature mortality of Chilocorus bipustulatus L. in Israel. Israel J. Agric. Res., 17, 131-134.

Kimbrough, J. W. and Gouger, R. J. 1970. Structure and development of the fungus Laboulbeniopsis termitarius. J. Invertebr. Pathol., 16, 205-213.

King, E. G. and Bell, J. V. 1978. Interactions between a braconid, Microplitis croceipes, and a fungus, Nomuraea rileyi, in laboratory-reared bollworm larvae. J. Invertebr. Pathol., 31, 337-340.

Klein, M. G. and Coppel, H. C. 1973. Entomophthora tenthredinis, a fungal pathogen of the introduced pine sawfly in northwestern Wisconsin. Ann. Entomol. Soc. Am., 66, 1178-1180.

Kmitowa, K., Bajan, C. and Wojciechowska, M. 1977. 12. Differences in the pathogenicity of entomopathogenic fungi from France and Poland. Pol. Ecol. Stud., 3, 115-126.

Kodaira, Y. 1961. Biochemical studies on the muscardine fungi in the silkworms, Bombyx mori. J. Fac. Textile Sci., Technol., Shinshu Univ., No. 29; Ser. E. Agr., Sericult, 5, 1-68.

Kodaira, Y. 1962. Studies on the new toxic substances to insects, Destruxin A and B, produced by Oospora destructor. Agr. Biol. Chem., 26, 36-42.

Koidsumi, K. 1957. Antifungal action of cuticular lipids in insects. J. Insect Physiol., 1, 40-51.

Krejzová, R. 1972. The pathogenic effect of Basidiobolus sp. on Coptotermes formosanus (Shiraki) and Reticulitermes lucifugus (Rossi) (Isoptera). Vestn. Cesk. Spol. Zool., 36, 253-255.

Krejzová, R. 1973. The pathogenic effect of the fungus Basidiobolus sp. on the larvae of Galleria mellonella L. Vestn. Cesk. Spol. Zool., 37, 273-274.

Krejzová, R. 1975. Enhancement of pathogenicity of Conidiobolus coronatus for the termites Coptotermes formosanus and Reticulitermes lucifugus by precultivation on an insect host. Vestn. Cesk. Spol. Zool., 39, 12-22.

Krejzová, R. 1977. Enhancement of pathogenicity of the fungus Canidiobolus coronatus by means of multiple preculitvation on insect hosts. Vestn. Cesk. Spol. Zool., 41, 105-113.

Krejzová, R. 1978. Taxonomy, morphology and surface structure of Basidiobolus sp. isolate. J. Invertebr. Pathol., 31, 157-163.

Kucera, M. 1971. Toxins of the entomophagous fungus Beauveria bassiana II. Effect of nitrogen sources on formation of the toxic protease in submerged culture. J. Invertebr. Pathol., 17, 211-215.

Kucera, M. 1980. Proteases from the fungus Metarrhizium anisopliae toxic for Galleria mellonella larvae. J. Invertebr. Pathol., 35, 304-310.

Kucera, M. and Samsináková, A. 1968. Toxins of the entomophagous fungus Beauveria bassiana. J. Invertebr. Pathol., 12, 316-320.

Lambiase, J. T. and Yendol, W. G. 1977. The fine structure of Entomophthora apiculata and its penetration of Trichoplusia ni. Can. J. Microbiol., 23, 452-464.

Machowicz-Stefaniak, Z. 1979. Objawy chorobowe wystepujace u gasienic Malacosoma neustria L. (Lepidoptera) zakazonych przez niektóre grzyby. Acta. Mycol., 15, 145-149.

MacLeod, D. M. 1963. Entomophthorales infections. In Insect Pathology, An Advanced Treatise, E. A. Steinhaus, ed., Vol. 2. Academic Press, N.Y. pp. 189-231.

Madelin, M. F. 1963. Diseases caused by hyphomycetous fungi. In Insect Pathology, an Advanced Treatise, E. A. Steinhaus ed., Vol. 2. Academic Press, N.Y. pp. 233-271.

Madelin, M. F. 1966. Fungal parasites of insects. Ann. Rev. Entomol., 11, 423-448.

Madelin, M. F., Robinson, R. K. and Williams, R. J. 1967. Appressorium-like structures in insect-parasitizing deuteromycetes. J. Invertebr. Pathol., 9, 404-412.

McCauley, V. J. E., Zacharuk, R. Y. and Tinline, R. D. 1968. Histopathology of green muscardine in larvae of four species of Elateridae (Coleoptera). J. Invertebr. Pathol., 12, 444-459.

McEwen, F. L. 1963. Cordyceps infections. In Insect Pathology, An Advanced Treatise, E. A. Steinhaus, ed. Academic Press, N.Y. pp. 273-290.

Meola, S. and DeVaney, J. 1976. Parasitism of Mallophaga by Trenomyces histophtorus. J. Invertebr. Pathol., 28, 195-201.

Meola, S. and Tavares, I. 1981. Ultrastructure of the haustorium of Trenomyces histophthorus and adjacent host cells. J. Invertebr. Pathol., In press.

Mohamed, A. K. A., Sikorowski, P. P. and Bell, J. V. 1977. Susceptibility of Heliothis zea to Nomuraea rileyi at various temperatures. J. Invertebr. Pathol., 30, 414-417.

Mohamed, A. K. A., Sikorowski, P. P. and Bell, J. V. 1978. Histopathology of Nomuraea rileyi in larvae of Heliothus zea and in vitro enzymatic activity. J. Invertebr. Pathol., 31, 345-352.
Ogloblin, A. and Jauch, C. 1943. Reacciones patológicas de los acridios atacados por Aspergillus parasiticus. Revista Argentina de Agronomia, 10, 256-267.
Paris, S. and Segretain, G. 1975. Physiological characteristics of strains of Beauveria tenella related to virulence for cockchafer larvae, Melolontha melolontha. Entomophaga, 20, 135-138.
Pekrul, S. and Grula, E. A. 1979. Mode of infection of the corn earworm (Heliothis zea) by Beauveria bassiana as revealed by scanning electron microscopy. J. Invertebr. Pathol., 34, 238-247.
Prasertphon, S. 1967. Mycotoxin production by species of Entomophthora. J. Invertebr. Pathol., 9, 281-282.
Prasertphon, S. and Tanada, Y. 1968. The formation and circulation, in Galleria, of hyphal bodies of entomophthoraceous fungi. J. Invertebr. Pathol., 11, 260-280.
Richards, A. G. and Smith, M. N. 1954. Infection of cockroaches with Herpomyces (Laboulbeniales). III. Experimental studies on host specificity. Bot. Gaz., 116, 195-198.
Richards, A. G. and Smith, M. N. 1955a. Infection of cockroaches with Herpomyces (Laboulbeniales). I. Life history studies. Biol. Bull., 108, 206-218.
Richards, A. G. and Smith, M. N. 1955b. Infection of cockroaches with Herpomyces (Laboulbeniales). IV. Development of H. stylopygae Spegazzini. Biol. Bull., 109, 306-315.
Richards, A. G. and Smith, M. N. 1956. Infection of cockroaches with Herpomyces (Laboulbeniales). II. Histology and histopathology. Ann. Entomol. Soc. Am., 49, 85-93.
Rizzo, D. C. 1977. Age of three dipteran hosts as a factor governing the pathogencity of Beauveria bassiana and Metarrhizium anisopliae. J. Invertebr. Pathol., 30, 127-130.
Roberts, D. W. 1966. Toxins from the entomogenous fungus Metarrhizium anisopliae. II. Symptoms and detection in moribund hosts. J. Invertebr. Pathol., 8, 222-227.
Roberts, D. W. 1969. Toxins from the entomogenous fungus Metarrhizium anisopliae: Isolation of destruxins from submerged cultures. J. Invertebr. Pathol., 14, 82-88.
Robinson, R. K. 1966. Studies on penetration of insect integument by fungi. Pest Artic. News Summ., Sect. B, 12, 131-142.
Salt, G. 1970. The Cellular Defence Reactions of Insects. Cambridge Monogr. Exp. Biol., No. 16. Cambridge

University Press, London.
Samsináková, A., Bajan, C., Kàlalovà, S., Kmitowa, K. and Wojciechowska, M. 1977. The effect of some entomophagous fungi on the Colorado beetle and their enzyme activity. Bull. Acad. Sci. Pol., Série Sci. Biol., 25, 521-526.
Samsináková, A., Misiková, S. and Leopold, J. 1971. Action of enzymatic systems of Beauveria bassiana on the cuticle of the greater wax moth larvae (Galleria mellonella). J. Invertebr. Pathol., 18, 322-330.
Sannasi, A. 1969a. Studies of an insect mycosis. I. Histopathology of the integument of the infected queen of the mound-building termite Odontotermes obesus. J. Invertebr. Pathol., 13, 4-10.
Sannasi, A. 1969b. Studies of an insect mycosis. II. Biochemical changes in the blood of the queen of the mound-building termite Odontotermes obesus accompanying fungal infection. J. Invertebr. Pathol., 13, 11-14.
Schabel, H. G. 1976a. Oral infection of Hylobius pales by Metarrhizium anisopliae. J. Invertebr. Pathol., 27, 377-383.
Schabel, H. G. 1976b. Green muscardine disease of Hylobius pales (Herbst) (Coleoptera:Curculionidae). Z. Angew. Entomol., 81, 413-421.
Schabel, H. G. 1978. Percutaneous infection of Hylobius pales by Metarrhizium anisopliae. J. Invertebr. Pathol., 31, 180-187.
Soper, R. S., Delyzer, A. J., and Smith, L. F. R. 1976a. The genus Massospora entomopathogenic for cicadas. Part II. Biology of Massospora levispora and its host Okanagana rimosa, with notes on Massospora cicadina on the periodical cicadas. Ann. Entomol. Soc. Am., 69, 89-95.
Soper, R. S., Smith, L. F. R., and Delyzer, A. J. 1976b. Epizootiology of Massosopra levispora in an isolated population of Okanagana rimosa. Ann. Entmol. Soc. Am., 69, 275-283.
Sussman, A. S. 1952. Studies of an insect mycosis III. Histopathology of an Aspergillosis of Platysamia cecropia L. Ann. Entomol. Soc. Am., 45, 233-245.
Tyrrell, D. 1977. Occurrence of protoplasts in the natural life cycle of Entomophthora egressa. Exp. Mycol., 1, 259-263.
Tyrrell, D. and MacLeod, D. M. 1972. Spontaneous formation of protoplasts by a species of Entomophthora. J. Invertebr. Pathol., 19, 354-360.
Veen, K. H. 1966. Oral infection of second-instar nymphs of Schistocerca gregaria by Metarrhizium anisopliae. J. Invertebr. Pathol., 8, 254-256.
Veen, K. H. 1968. Recherches sur la maladie, due à Metarrhizium anisopliae chez le criquet pèlerin. Meded. Landbouwhogesch. Wageningen, 68, 1-117.

Vey, A. 1968. Réactions de défense cellulaire dans les infections de blessures à Mucor hiemalis Wehmer. Ann. Epiphyt., 19, 695-702.

Vey, A. 1971. Étude des réactions cellulaires anticryptogamiques chez Galleria mellonella L.: Structure et ultrastructure des granulomes a Aspergillus niger V. Tiegh. Ann. Zool. Écol. Anim., 3, 17-30.

Vey, A. and Fargues, J. 1977. Histological and ultrastructural studies of Beauveria bassiana infection in Leptinotarsa decemlineta larvae during ecdysis. J. Invertebr. Pathol., 30, 207-215.

Vey, A. and Quiot, J. M. 1976. Action toxique due champignon Mucor hiemalis sur les cellules d'insectes en culture in vitro. Entomophaga, 21, 275-279.

Vey, A. and Vago, C. 1969. Recherches sur la guérison dans les infections cryptogamiques d'insectes. Infection a Aspergillus niger V. Tiegh chez Galleria mellenolla L. Ann. Zool. Écol. Anim., 1, 121-126.

Vey, A. and Vago, C. 1971. Réaction anticryptogamique de type granulome chez les insectes. Ann. Inst. Pasteur, 121, 527-532.

Vey, A., Vago, C. and Delanoue, P. 1967. Mode d'action des Aspergillus parasites de blessures d'insectes. Rev. Mycol., 32, 300-305.

Wasti, S. S. and Hartmann, G. C. 1975. Experimental parasitization of larvae of the gypsy moth, Porthetria dispar (L.), with the entomogenous fungus, Beauveria bassiana (Balsamo) Vuill. Parasitology, 70, 341-346.

West, E. W. and Briggs, J. D. 1968. In vitro toxin production by the fungus Beauveria bassiana and bioassay in greater wax moth larvae. J. Econ. Entomol., 61, 684-687.

Wojciechowska, M. 1973. Preliminary observations on the production of toxins by the insect pathogenic fungi. Ekol. Pol., 21, 699-703.

Yendol, W. G., Miller, E. M. and Behnke, C. N. 1968. Toxic substances from Entomophthoraceous fungi. J. Invertebr. Pathol., 10, 313-319.

Yendol, W. G. and Paschke, J. D. 1965. Pathology of an Entomophthora infection in the eastern subterranean termite Recticulitermes flavipes (Kollar). J. Invertebr. Pathol., 7, 414-422.

Zacharuk, R. Y. 1970a. Fine structure of the fungus Metarrhizium anisopliae infecting three species of larval Elateridae (Coleoptera). I. Dormant and germinating conidia. J. Invertebr. Pathol., 15, 63-80.

Zacharuk, R. Y. 1970b. Fine structure of the fungus Metarrhizium anisopliae infecting three species of larval Elateridae (Coleoptera). II. Conidial germ tubes and appressoria. J. Invertebr. Pathol., 15, 81-91.

Zacharuk, R. Y. 1970c. Fine structure of the fungus *Metarrhizium anisopliae* infecting three species of larval *Elateridae (Coleoptera)*. III. Penetration of the host integument. J. Invertebr. Pathol., 15, 372-396.

Zacharuk, R. Y. 1971a. Ultrastructural changes in tissues of larval Elateridae (Coleoptera) infected with the fungus *Metarrhizium anisopliae*. Can. J. Microbiol., 17, 281-289.

Zacharuk, R. Y. 1971b. Fine structure of the fungus *Metarrhizium anisopliae* infecting three species of larval *Elateridae (Coleoptera)*. IV. Development within the host. Can. J. Microbiol., 17, 525-529.

Zachruk, R. Y. 1973a. Penetration of the cuticular layers of elaterid larvae (Coleoptera) by the fungus *Metarrhizium anisopliae*, and notes on a bacterial invasion. J. Invertebr. Pathol., 21, 101-106.

Zacharuk, R. Y. 1973b. Electron-microscope studies of the histopathology of fungal infections by *Metarrhizium anisopliae*. Misc. Publ. Entomol. Soc. Amer., 9, 112-119.

Zacharuk, R. Y. and Tinline, R. D. 1968. Pathogenicity of *Metarrhizium anisopliae*, and other fungi, for five *elaterids (Coleoptera)* in Saskatchewan. J. Invertebr. Pathol., 12, 294-309.

Zajic, J. E. 1963. The formation of pigments by *Beauveria bassiana* (Balsamo) Vuillemin in the presence of aryl amines. J. Insect Pathol., 5, 16-27.

Chapter 14

FUNGAL PATHOGENS OF MOSQUITO LARVAE

by A. W. Sweeney*

INTRODUCTION

Little is known about the pathology of fungal diseases of aquatic insects with the exception of those which infect mosquito larvae. For the most part these fungi have been investigated with a view of evaluating them as biological control agents of mosquito vectors of disease. This research has expanded in recent years because of the development of insecticide resistance in some mosquito populations and environmental problems associated with the widespread use of broad spectrum chemical insecticides for insect control. The pathogenic processes of candidate microbial control agents must be understood before these organisms can be seriously considered for practical use in the field; as a result of increased research effort, there have been a number of investigations of the pathology of mosquito fungal diseases. This review includes the natural mosquito larval pathogens with the exception of *Lagenidium giganteum* which is covered in Chapter 15. Two fungal pathogens of terrestrial insects, *Metarrhizium anisopliae* and *Beauveria bassiana*, are pathogenic to mosquito larvae and their pathogenesis in these insects is also included in this chapter. Recent reviews on the biocontrol potential of mosquito pathogenic fungi include those of Roberts (1970), Roberts (1974), and Federici (1980).

*Malaria Research Unit, Royal Australian Army Medical Corps, Ingleburn, New South Wales, 2174, Australia

PATHOGENESIS OF INVERTEBRATE MICROBIAL DISEASES. Davidson, ed. © Allanheld, Osmun (Totowa, N.J.)

PHYCOMYCETES

Coelomomyces

Coelomomyces spp. are the best known fungal pathogens of mosquito larvae. The genus was established by Keilin in 1921 but many aspects of its biology remained obscure until the discovery of an alternate crustacean host by Whisler et al. (1974). This discovery led to the complete elucidation of the life cycle and also to the maintenance of several species in the laboratory. Controlled experimental studies on the infection in both insect and crustacean hosts were then possible. This review will be restricted to the invasion and subsequent development of the infection in the mosquito host; the copepod cycle is discussed in Chapter 17.

Coelomomyces invades through the exterior cuticle and, even though degenerated zygotes are sometimes seen within the digestive tract, it does not appear to invade through the gut (Zebold et al., 1979). These authors showed that when the diploid zygotes were stained with methylene blue they could be seen attached to the intersegmental areas of the thorax and abdomen, to the bases of the anal papillae, and to the head capsule. This pattern of attachment was confirmed in this species by Travland (1979a) using the scanning electron microscope and in another species, C. opifexi, by Wong and Pillai (1980) who used the fluorescent antibody technique to stain the attached zygotes. In an ultrastructural study Travland (1979b) observed small vesicles in the cytoplasm of the zygote which migrated to the zygote walls and disappeared shortly after attachment to the larval cuticle. Travland (1979b) postulated that these vesicles may be involved in adhesion prior to penetration. This process of adhesion may be one of the factors which determine host specificity as it has been shown by Zebold et al. (1979) that mosquito species susceptible to C. psorophorae had many adherent zygotes whereas resistant mosquito species had very few.

A short bulb-shaped germ tube grows from the encysted zygote. The germ tube secretes a dense amorphous material at its tip which may act as an adhesive or may contain enzymes which assist penetration of the host cuticle (Travland, 1979a). A narrow penetrant hypha grows directly through the cuticle and maintains its size as it extends into the host epidermal cell. There is evidence that in C. psorophorae the parasite protoplasm is rapidly squeezed through this narrow tube by the expansion of a vacuole inside the cyst and then rounds up into a spherical fungal cell within the larval epidermal cell (Travland, 1979a). In C. opifexi the protoplasm from the zygote either enters the epidermis or is released directly into the hemocoel (Wong and Pillai, 1980). These spherical cells then develop protusions which grow into small, multibranched clumps of hyphae which later form characteristically sculptured sporangia at their tips (Martin, 1969). The mature sporangium develops a lateral bulge at a pre-formed germination slit (Couch, 1968) through which haploid zoospores, infective for the copepod host, are released.

Leptolegnia

A species of *Leptolegnia* was recently isolated from a larva of *Culex quinquefasciatus* collected from an experimental pond in Clemson, South Carolina (T. M. McInnis Jr. and W. R. Zattau, pers. comm.). This fungus was found to be highly virulent for early instar larvae of *Culex*, *Aedes*, and *Anopheles* mosquitoes. Sporangia formed at the apex of specialized hyphae release weakly swimming primary zoospores which encyst after swimming for several minutes. Highly active secondary zoospores which emerge from the primary cyst after several hours are capable of initiating infection in larvae. These secondary zoospores encyst on the exterior cuticle and a germ tube penetrates into the hemocoel. Ingested cysts have also been observed to germinate in the midgut and penetrate the body cavity through the gut wall. The fungus rapidly colonizes the host and hyphae then penetrate the exterior cuticle to produce sporangia and another generation of primary zoospores. Sexual fusion between specialized structures on the hyphae (female oogonia and male antheridia) leads to the development of thick walled zoospores which may function as a resting stage.

Phythium

A species of *Pythium* (near *Phythium adhaerens*) was found causing a high mortality of *Ae. sierrensis* larvae in an oak tree hole in Fresno County, California (Clark et al., 1966). The fungus was isolated into agar culture and laboratory infection experiments showed that the zoospores were attracted to damaged cuticle into the hemocoel. The injured larvae died within two days and sporangial vesicles formed on the exterior after death. This fungus is clearly an opportunistic pathogen as the mortality in these experiments depended on the proportion of injured larvae tested as well as the concentration of zoospores. Moreover, the zoospores were also attracted to other substrates such as nutrient agar and it is possible that the organism may normally exist as an aquatic saprophyte.

Entomophthora

Though there are several reports of Entomophthora infecting adult mosquitoes in the field (see Roberts and Strand, 1977) there appears to be only one species, E. aquatica, which has been found to be a natural pathogen of the aquatic stages (Anderson and Ringo, 1969). It was found infecting Ae. canadensis and Culiseta morsitans in woodland pools in Connecticut. Histological examination of infected larvae revealed hyphal bodies, which reproduced by budding, within the hemocoel and fat body. The fungus was also seen in eye tissue and cells of the nervous system. The early stages of the infection were not observed, but as hyphal bodies were found within the midgut cells of one specimen, the authors speculated that it may invade through this site. Larvae and newly moulted pupae killed by the fungus sank and external sporulation did not develop. Older pupae killed by the fungus remained floating. Hyphae penetrated the dorsal surface of the anterior abdominal intersegmental membranes, grew through the water into the air, and formed an aerial mat of conidiophores from which conidia were discharged.

Smittium

Trichomycetes of the genus Smittium live attached by holdfasts to the hindgut cuticle of aquatic dipterous larvae in the families Culicidae, Chironomidae, and Simuliidae. These fungi are usually not detrimental to their hosts as they are confined to the lumen of the gut and are shed with the cuticle during ecdysis (Lichtwardt, 1964). However, the death of a small proportion of early instar larvae of Ae. aegypti with S. culisetae growing from the anus was reported by Williams and Lichtwardt (1972). Coluzzi (1966) recorded a high mortality of laboratory reared An. gambiae larvae apparently caused by blockage of the rectum by a species of Smittium. Dubitskii (1978) reported a Russian strain of Smittium causing a high mortality among Ae. aegypti and Cx. pipiens molestus larvae under laboratory conditions. A blackened area, made up of a cluster of darkened hyphae, was seen under the cuticle of infected larvae. More recently Smittium morbosum was isolated from a colony of An. hilli in Australia (Sweeney, 1981). This fungus occupied the anterior part of the hindgut within the pyloric chamber and extended forward to the posterior midgut. Some of the hyphae penetrated the cells of the midgut and Malphighian tubules (Fig. 14-1). The melanization of these hyphae by the host defense reaction was seen as an abdominal black spot, visible to the naked eye, similar to that observed by Dubitskii (1978). The penetrating hyphae did not spread throughout the hemocoel

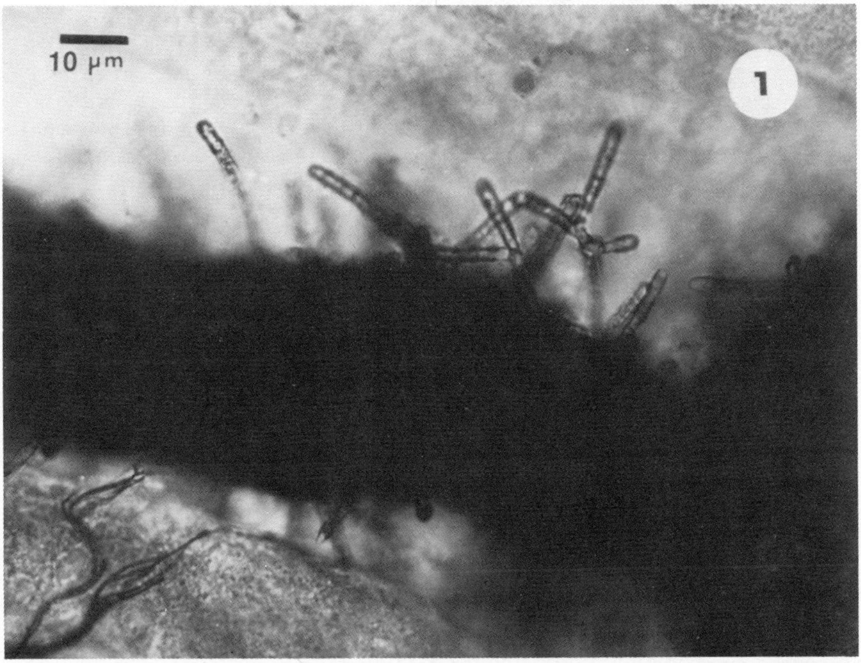

Figure 14-1. Malpighian tubule of Anopheles hilli larva, melanized hyphae of Smittium morbosum protruding into the hemocoel.

but appeared to act as anchorages which permitted the fungus to persist within the host through successive instars. Thalli continued to grow in the lumen of the hindgut until it was blocked and this led to the death of a significant number of infected larvae.

It should be pointed out that these records of larval mortality caused by Smittium have been observed in laboratory reared larvae and not in field specimens. It is possible that this phenomenon may be associated with growth of the fungus in unnatural hosts or it may occur in larvae reared under suboptimal conditions. However, penetration of the gut epithelium, as seen in the Russian and Australian observations, suggest that some strains of Smittium may have a pathogenic rather than commensal relationship with their arthropod hosts.

FUNGI IMPERFECTI

Culicinomyces

Fungi of this genus were discovered independently during 1972 in North Carolina, U.S.A. (Couch et al., 1974) and in Sydney, Australia (Sweeney et al., 1973). Coelomomyces is being evaluated in Australia for biological control of mosquito larvae. Comparative morphological observations have shown that both isolates belong to the same species, Culicinomyces clavosporus (A. W. Sweeney, C. Panter, and J. N. Couch, unpublished). This fungus normally invades the host through the cuticle lining the foregut and hindgut following ingestion of conidia (Figs. 14-2, 14-3). The most common sites of infection are the esophagus and pharynx but infection may also take place through the buccal cavity, ileum, colon, and rectum (Sweeney, 1975). Invasion through the midgut has not been seen. When larvae are exposed to large numbers of conidia (ca. 10^6 conidia/ml), massive invasion occurs through the foregut and hindgut, and conidia can be observed adhering to the cuticle of head, thorax, and abdomen, as well as the mouth brushes and setae, but actual invasion through the external integument has only been seen through the anal papillae (Sweeney, 1979).

It was originally suggested that chemical forces may be involved in adhesion of Culicinomyces to the host (Sweeney, 1975). However the conidia are slimy, and it has more recently been shown by ultrastructural observations that they are covered with an irregular layer of electron-dense material which is believed to be a mucilaginous substance responsible for adhesion (Sweeney et al., 1980). Following attachment to the gut cuticle, a short germ tube emerges from the apex of the conidium and presses against the epicuticle without enlarging in diameter. Penetration may be assisted by enzymes secreted by the narrow penetrant hypha as it enters the cuticle. Sometimes it grows directly through the gut epithelium (Fig. 14-4) but on other occasions, the cuticle surrounding the advancing hyphal tip is melanized by a host defense reaction. When this occurs the cuticle appears to be more resistant to penetration and the penetrant hypha is often deflected to grow parallel to the surface of the cuticle, forcing apart the cuticular layers by mechanical pressure, until it grows beyond the area of melanization. It then resumes its original course to the epithelial cells (Sweeney et al., 1980). The apex of the hypha then swells into an oval vesicle within the epithelial layer and a new growing point invades the hemocoel. Sometimes this early growth is enveloped by the humoral defense reaction of the host within melanized capsules (Fig. 14-4) but this is not sufficient to prevent invasion and the hyphae physically break out of the capsules. The hyphae then branch and spread from the infection sites; blastospores bud off laterally and circulate in the hemolymph. When larvae are exposed to large concentrations of conidia many of them die shortly after the penetration stage; it is possible that their rapid death may be due to the release of toxins into the hemocoel by massive numbers of invading hyphae (Sweeney, unpublished). However, most larvae which receive moderate infections are not killed until mycelium fills the thorax and abdomen. In these cases it seems likely that the larvae die of physiological starvation as do larvae infected with *Lagenidium giganteum* (Domnas et al., 1974) (see Chapter 15). After death, the cadaver is completely filled with *Culicinomyces* mycelium. Hyphae bulge against the outer integument and mechanically rupture the cuticle of the thorax and abdomen to form a dense mat of conidiophores. Sporulation occurs both aerially on floating dead larvae as well as under the water surface on larvae which are submerged.

Tolypocladium

The genus *Tolypocladium* was established by Gams (1971) for a group of hyphomycetes isolated from soil. One of these, *T. cylindrosporum*, has since been reported as a pathogen infecting mosquito larvae in Novato, California by Soares et al. (1979) and another isolate of the same species was discovered independently near Dunedin, New Zealand (J. S. Pillai, pers. comm.).

The pathogenesis of the California isolate of *Tolypocladium* has been studied by Soares (1979). Several isolates of this strain were recovered from field-collected *Ae. sierrensis* breeding in tree holes, and it was sometimes seen in association with *Beauveria bassiana* with mixed infections by both fungi in the one infected host. *Tolypocladium* forms aerial conidia on solid medium and produces

Figure 14-2. Dissected foregut of _Anopheles hilli_ showing adhesion of conidia of _Culicinomyces clavosporus_ (Australian strain) and penetration of gut cuticle. C = conidium; V = vesicle.

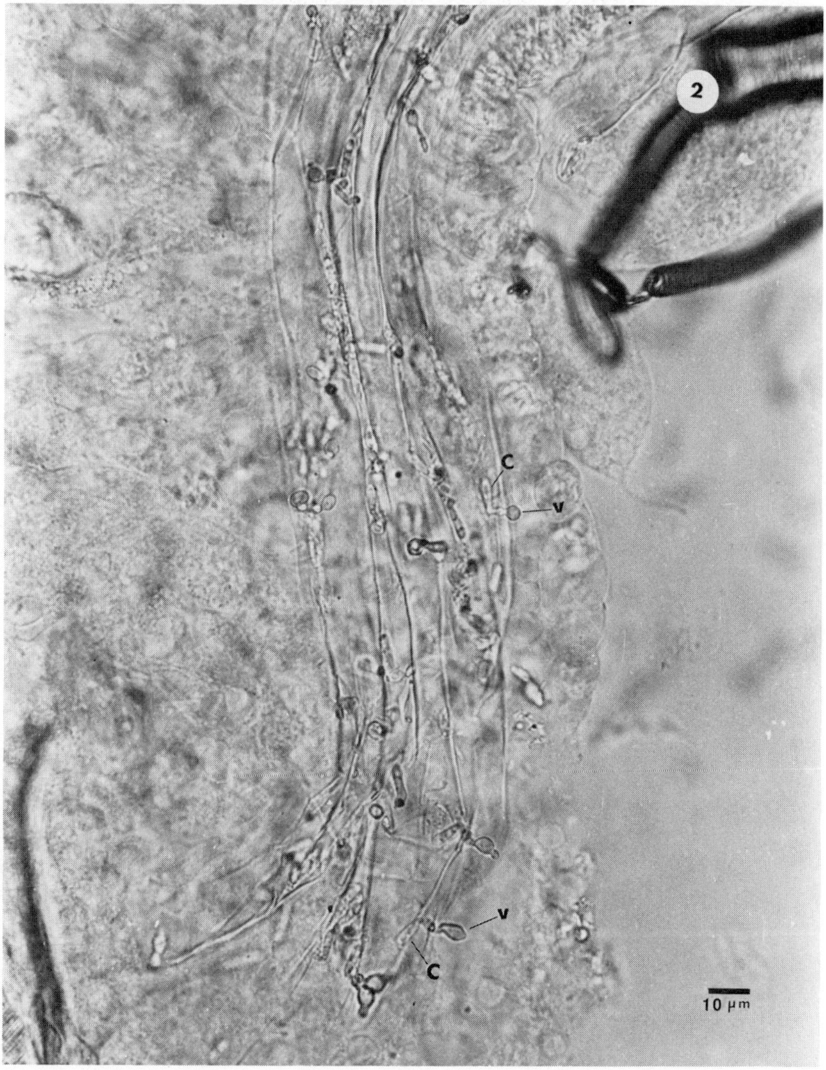
2
C
v
v
C
10 μm

Figure 14-3. Dissected foregut of Anopheles hilli showing hyphae of Culicinomyces clavosporus (Australian strain) growing into hemocoel.

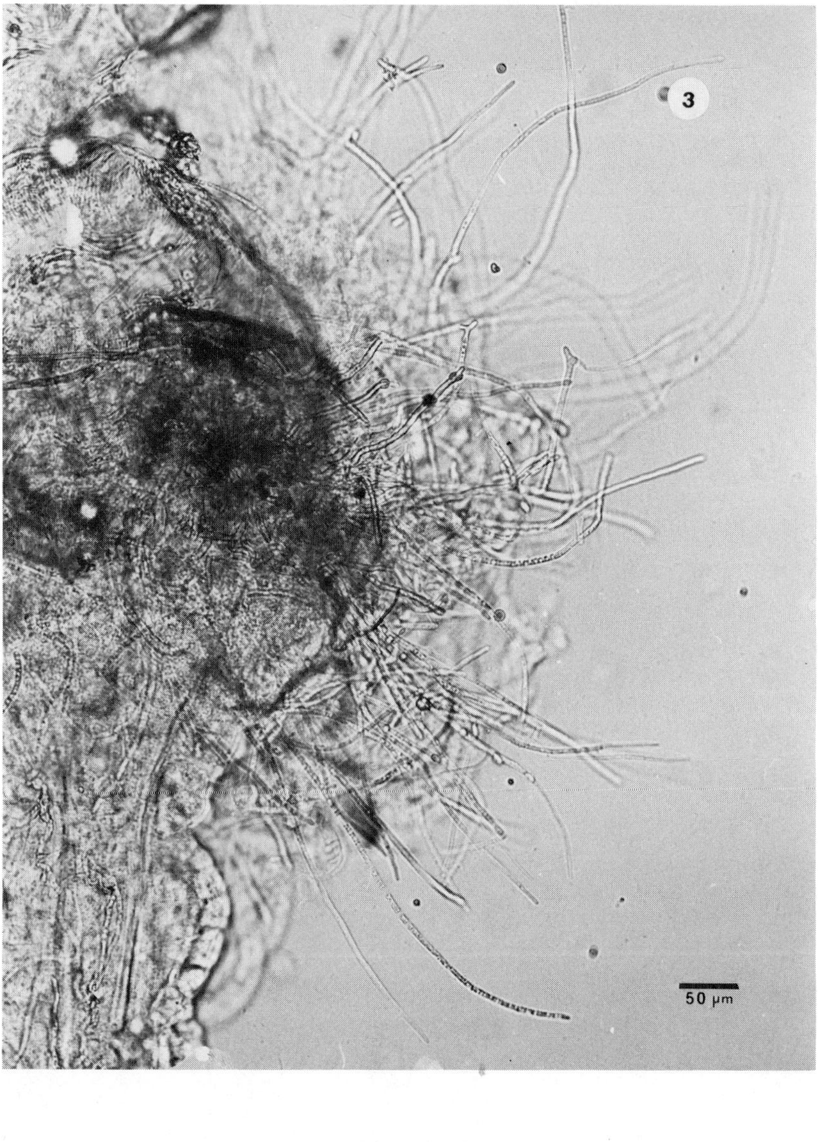
3
50 μm

Figure 14-4. Transmission electron micrograph of *Culicinomyces clavosporus* (Australian strain) invading foregut of *Culex quinquefasciatus* larva. Cs = capsule substance; Cu = cuticle of foregut; Gt = germ tube of conidium; H = hemocoel; Hyp = hypodermal cell; L = lumen of foregut; Ph = penetrant hypha extending from germ tube through cuticle into hypodermal cell.

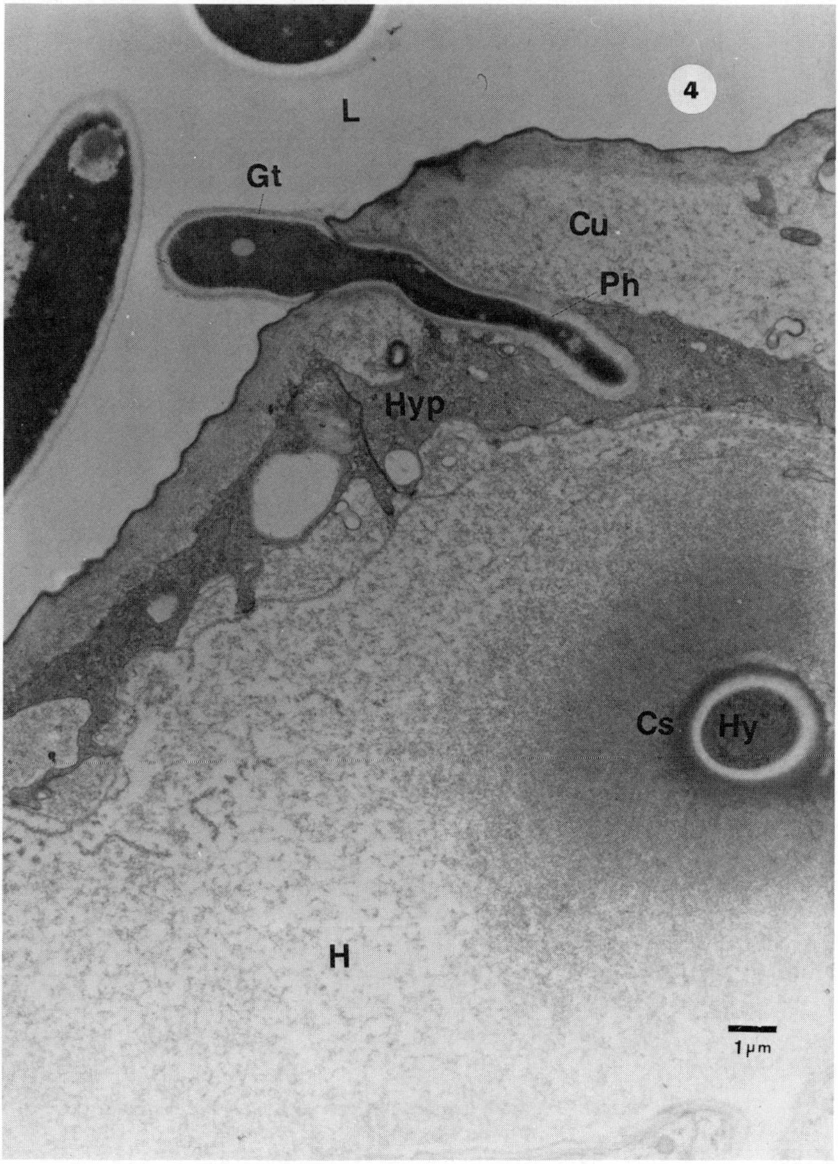
4
L
Gt
Cu
Ph
Hyp
Cs
Hy
H
1 μm

blastospores when grown in shake culture. Both types of spores are infectious to mosquito larvae. Invasion of the host occurs through the exterior cuticle; the head and thorax as well as the saddle area of the abdomen are common sites, and the fungus can also invade through the buccal cavity and midgut (Fig. 14-5). Both conidia and blastospores adhere to the cuticle (possibly aided by a mucilaginous secretion) with the long axis parallel to the cuticle surface. Both types of spores germinate at one end. The germ tube contacts the cuticle without forming an appressorium and a penetrant hyphae grows through into the hemocoel. Melanized encapsulation of invading hyphae was noticed in some cases but was not common.

The digestive tract is a secondary infection site of the California strain because, although penetration of the peritrophic membrane is frequently observed, the midgut epithelium is rarely penetrated and more extensive invasion is usually apparent from other sites on the exterior cuticle. Hyphae branch within the hemocoel and blastospores are budded off as in the case of *Culicinomyces*. Most larvae infected with *Tolypocladium* die after the hemocoel is heavily invaded with mycelium and hyphae then penetrate the exterior integument. Post-mortem enzymatic activity appears to be important in penetration of the cuticle as an appressorium is not formed and, unlike *Culicinomyces*, the cuticle does not appear to be ruptured by mechanical forces. Blastospores and conidia are not produced on hyphae which grow from submerged cadavers infected with the California strain, but tufts of aerial conidiophores and conidia form above the water surface on floating specimens.

The New Zealand strain of *T. cylindrosporum* was isolated from *Ae. australis* larvae in brackish pools at Quoin Point, 30 miles south of Dunedin (J. S. Pillai, pers. comm.). This fungus invades larvae through the foregut and hindgut. After colonizing the hemocoel, hyphae usually penetrate the posterior abdominal segments before death of the host, and, in contrast to the California strain, conidia are produced on submerged hyphae on the exterior cuticle of dead larvae (Weiser and Pillai, *in press*).

Metarrhizium

The well-known fungal pathogen of terrestrial insects, *Metarrhizium anisopliae*, is also lethal to mosquito larvae. Laboratory tests have demonstrated that one isolate (F84-1-1) is highly pathogenic for larvae of *Culex*, *Aedes*, and *Anopheles* (Roberts, 1967). The conidia of this fungus are hydrophobic and float on the water.

Invasions normally take place through the perispiracular valves of the siphon which are usually the only areas of the

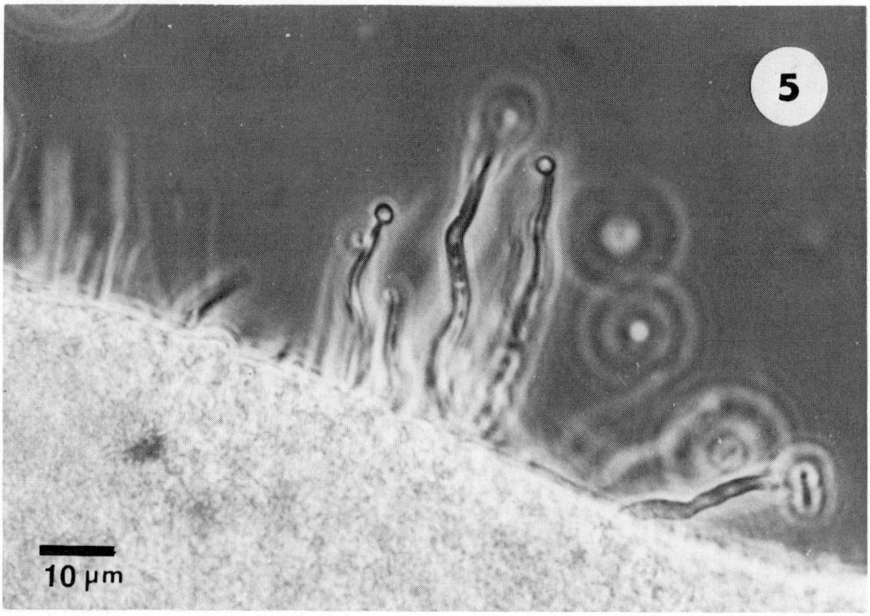

Figure 14-5. Dissected peritrophic membrane of Culex tarsalis larva showing penetration by hyphae of Tolypocladium cylindrosporum (California strain) (Photograph courtesy of G. Soares).

external cuticle in contact with the water surface. Details of the invasion process have not yet been fully investigated but it appears that an appressorium is seldom formed (Roberts, 1974) and that most specimens die with mycelium restricted to the siphon and anal segment. It is believed that blockage of the spiracles may lead to death by asphyxiation. Ingested spores may also release lethal toxins without germinating (Roberts, 1970; Crisan, 1971), though this occurs less frequently than perispiracular infection.

If the conidia are treated with detergents they submerge after application and do not apparently adhere to the cuticle but are ingested. They then germinate in the midgut (in *Aedes* larvae) or hindgut (in *Culex* larvae) and penetrate into the hemocoel, but death occurs before the insect is colonized by mycelium (Roberts, 1974). *Metarrhizium anisopliae* produces depsipeptides (Destruxins A and B) which are toxic to insects when it is grown in artificial culture (Roberts, 1969), but it is not yet known whether these toxins play a role in mosquito mortality following infection by conidia.

Beauveria

Pinnock et al. (1973) reported *Beauveria tenella* causing epizootics in treehole populations of *Ae. sierrensis* larvae in California but later studies by Soares et al. (1979) provided evidence that both *Tolypocaldium cylindrosporum* and *Beauveria bassiana*, rather than *B. tenella*, were involved in these fungal disease outbreaks. The pathogenicity of this strain of *Beauveria bassiana* does not yet appear to have been investigated.

Infection experiments using a different, terrestrial isolate of *B. bassiana* against laboratory reared mosquito larvae were performed by Clark et al. (1968). These experiments demonstrated that floating conidia were more effective against the larvae than those which were submerged. As with *M. anisopliae*, the sites of infection are the perispiracular lobes of the siphon. Penetration through these sites took place within 24 hrs and mortality usually occurred on the third and fourth days after exposure. At this time mycelial growth of the fungus was mostly confined within the siphon. The exact cause of death was not determined but appeared to be due to suffocation caused by mechanical blockage of the tracheal trunks within the siphon or by a toxin released into the hemocoel. Laboratory tests showed that this fungus was effective in killing *Cx. pipiens*, *Cx. tarsalis*, and *An. albimanus* but not *Ae. aegypti* and *Ae. sierrensis*. The lack of effect against the *Aedes* species may be explained by the fact that the fungus took longer to penetrate the perispiracular lobes in this genus.

HOST REACTIONS

Cellular encapsulation is a common defense reaction of insects to invasion by parasites and pathogens (Salt, 1970). Cells circulating within the hemolymph envelop the invading organisms in an effort to prevent their development; some of the inner cells of the capsule disintegrate, and a dense black substance coats the surface of the parasite (Poinar et al., 1968). However, melanized capsules can form in mosquitoes without the presence of blood cells. This was first noted in nematode parasitism of mosquito larvae by Bronskill (1962) and Esslinger (1962).

This type of reaction, termed "humoral encapsulation", has also been reported in larvae of Chironomidae invaded by fungi (Gotz and Vey, 1974). These authors observed melanization of invading hyphae of *Beauveria bassiana* by deposition of "capsule substance" within the cuticle, hypodermal cells and hemolymph of experimentally infected *Chironomus* larvae. Since histochemical tests of the capsule substance were all positive for melanin, this reaction is probably the result of phenoloxidase activity producing a capsule of melanin and other polyphenol-protein complexes. Further confirmation of this was provided by an *in vitro* study in which encapsulation by *Chironomus* hemolymph was prevented by glutathione, an inhibitor of phenolixidases (Vey and Gotz, 1975). Gotz and Vey (1974) also showed that when spores of *Aspergillus niger* were injected into the body cavity of *Chironomus* larvae the defense reaction was able to suppress the growth of mycelium. In contrast, although spores of *B. bassiana* were encapsulated soon after injection, they were able to germinate and break out of the capsule to invade the hemocoel. The pathogenicity of injected spores of *Mucor heimalis* was intermediate between the two. Though these experiments were performed under unnatural conditions they underline an important aspect of the host-pathogen relationship, as such defense reactions may slow down or prevent invasion by fungi. In an ultrastructural study of *C. psorophorae* infecting *Culiseta inornata*, Travland (1979a) observed that melanization of the cuticle sometimes leads to the degeneration of encysted zygotes before penetration, thereby demonstrating that reactions aroused in a susceptible host may directly affect the outcome of infection.

Melanization reactions have been mentioned in most of the published reports of fungal infections in mosquito larvae. In the case of *Culicinomyces* invading larvae of *Cx. fatigans* and *An. hilli*, ultrastructural observations (Fig. 14-4) showed that encapsulation of hyphae within the hemocoel takes place without the involvement of blood cells (A. W. Sweeney, and R. G. Wright, unpublished). Thus, humoral encapsulation, similar to that reported by Gotz and Vey (1974) in *Chironomus* larvae, may be a common host response of mosquitoes to fungal invasion.

CONCLUSIONS

Ferron (1978) has observed that fungal infections of insects are successful only when the organism is able to penetrate the cuticle and become established within the host between moults. Invasion of mosquito larvae at the intermoult period is particularly important as these larvae normally moult at short intervals of 2-4 days. If outgrowth from the spore occurs too slowly, or if the defense reactions of the host retard development of the fungus, then the infection may be cast off with the exuvium at the next moult.

The formation of appressoria (swellings of germ tubes or hyphae which provide an anchorage prior to penetration) are common in plant pathogenic fungi and in fungal pathogens of terrestrial insects, but they are not frequently observed in the fungal diseases of mosquito larvae. When *Metarrhizium anisopliae* invades wireworm larvae, large appressorial cells form in close contact with the exterior integument and penetration of the extensive cuticle proceeds through a series of steps taking between five and twenty days to reach the hemocoel (Zacharuk, 1973) (see Chapter 13). Appressoria have not been seen in *Culicinomyces* or *Tolypocladium* infections, and though Travland (1979a) found that *Coelomomyces psorophorae* forms bulbous appressoria prior to penetration, these appressoria do not appear to be as swollen or enlarged as those of *M. anisopliae* invading wireworms (Zacharuk, 1970) or those of *Beauveria bassiana* infecting Colorado potatoe beetles (Vey and Fargues, 1977). Moreover, appressoria are rarely formed by *M. anisopliae* invading mosquito larvae (Roberts, 1974). The cuticle of mosquito larvae is not as thick as in many terrestrial insects so is probable that the enzymatic and mechanical forces of the germinating spores of mosquito pathogenic fungi may be sufficient to affect rapid invasion without production of well developed appressorial structures and elaborate penetration processes.

The ability to invade through those areas of the integument where the host barriers are weakest would clearly be of selective advantage to a fungal pathogen. *Coelomomyces* zoospores commonly encyst on the intersegmental membranes, and *Culicinomyces* invades through the cuticle of the foregut, hindgut, and anal papillae. The unusually thin cuticle of these areas presumably offers less resistance to penetration than the more heavily sclerotized parts of the integument.

Some aspects of the infection processes of mosquito fungal diseases remain obscure. For example, the stimulus for spore germination following contact with the host is not known. Similarly, while it is clear that enzymes assist penetration of the cuticle, the nature of their activity on the chitin-protein complex is not understood. Though toxins appear to be involved in the pathogenesis of Metarrhizium and Beauveria in mosquito larvae it is not known whether toxic substances are released into the hemocoel following invasion by the natural mosquito larval pathogens. Investigations of fungal metabolic processes during invasion and growth within infected larvae are worthy areas for future research.

REFERENCES

Anderson, J. F. and Ringo, S. L. 1969. Entomophthora aquatica sp. n. infecting larvae and pupae of floodwater mosquitoes. J. Invertebr. Pathol., 13, 386-393.

Bronskill, J. F. 1962. Encapsulation of rhabditoid nematodes in mosquitoes. Can. J. Zool., 40, 1269-1275.

Clark, T. B., Kellen, W. R., Fukuda, T. and Lindegren, J. E. 1968. Field and laboratory studies on the pathogenicity of the fungus Beauveria bassiana to three genera of mosquitoes. J. Invertebr. Pathol., 11, 1-7.

Clark, T. B., Kellen, W. R., Lindegren, J. E. and Sanders, R. D. 1966. Pythium sp. (Phycomycetes: Pythiales) pathogenic to mosquito larvae. J. Invertebr. Pathol., 8, 351-354.

Coluzzi, M. 1966. Experimental infections with Rubetella fungi in Anopheles gambiae and other mosquitoes. Proc. 1st Int'l. Congr. Parasitol. (Rome) 1, 592-593.

Couch, J. N. 1968. Sporangial germination of Coelomomyces punctatus and the conditions favoring the infection of Anopheles quadrimaculatus under laboratory conditions. Proc. Joint U. S. - Japan Sem. Microb. Control Insect Pests (Fukuoka), 93-105.

Couch, J. N., Romney, S. V. and Rao, B. 1974. A new fungus which attacks mosquitoes and related Diptera. Mycologia, 66, 374-379.

Crisan, E. V. 1971. Mechanism responsible for release of toxin by Metarrhizium spores in mosquito larvae. J. Invertebr. Pathol., 17, 260-264.

Domnas, A., Giebel, P. E. and McInnis, T. M. Jr. 1974. Biochemistry of mosquito infection: Preliminary studies of biochemical change in Culex pipiens quinquefasciatus following infection with Lagenidium giganteum. J. Invertebr. Pathol., 24, 293-304.

Dubitskii, A. M. 1978. Biological Control of Blood Sucking Flies in the USSR. Institute of Biology, Kazakhstan Academy of Sciences, Alma Ata, 267 pp. (in Russian).

Esslinger, J. H. 1962. Behaviour of microfilariae of Brugia pahangi and Anopheles quadrimaculatus. Am. J. Trop. Med. Hyg., 11, 749-758.

Federici, B. A. 1980. Mosquito control by the fungi Culicinomyces, Lagenidium, and Coelomonyces. In Microbial Control of Pests and Plant Diseases 1970-1980, H. D. Burges, ed., in press.

Ferron, P. 1978. Biological control of insect pests by entomogenous fungi. Ann. Rev. Entomol., 23, 409-422.

Gams, W. 1971. Tolypocladium, eine Hyphomycetengattung mit geschwollenen phialiden. Persoonia, 6, 185-191.

Gotz, P. and Vey, A. 1974. Humoral encapsulation in Diptera (Insecta): Defense reactions of Chironomus larvae against fungi. Parasitology, 68, 193-205.

Lichtwardt, R. W. 1964. Axenic culture of two species of branched Trichomycetes. Am. J. Bot., 51, 836-842.

Martin, W. W. 1969. A morphological and cytological study of the development of Coelomomyces punctatus parasitic in Anopheles quadrimaculatus. J. Elisha Mitchell Sci. Soc., 85, 59-72.

Pinnock, D. E., Garcia, R. and Cubbin, C. M. 1973. Beauveria tenella as a control agent of mosquito larvae. J. Invertebr. Pathol., 22, 143-147.

Poinar, G. O., Jr., Leutenegger, R. and Gotz, P. 1968. Ultrastructure of the formation of a melanotic capsule in Diabrotica (Coleoptera) in response to a parasitic nematode (Mermithidae). J. Ultrastruct. Res., 25, 293-306.

Roberts, D. W. 1967. Some effects of Metarrhizium anisopliae and its toxins on mosquito larvae. In Insect Pathology and Microbial Control, P. A. Van der Laan, ed., North Holland Publ. Co., Amsterdam. pp. 243-246.

Roberts, D. W. 1969. Toxins from the entomogenous fungus Metarrhizium anisopliae: Isolation of destruxins from submerged cultures. J. Invertebr. Pathol., 14, 82-88.

Roberts, D. W. 1970. Coelomomyces, Entomophthora, Beauveria, and Metarrhizium as parasites of mosquitoes. Misc. Pub., Entomol. Soc. Amer., 7, 140-154.

Roberts, D. W. 1974. Fungal infections of mosquitoes. In Le Controle des Moustiques, A. Aubin, J. P. Bourassa, S. Belloncik, M. Pellisier, and E. Lacoursiere, eds. Univ. Quebec Press, Quebec. pp. 143-193.

Roberts, D. W. and Strand, M. A. 1977. Pathogens of medically important arthropods. Bull. World Heath Org. 55, Suppl. No. 1, 419 pp.

Salt, G. 1970. The cellular defense reactions of insects. Cambridge Monographs in Exper. Biol. No. 16. Cambridge University Press, 118 pp.

Soares, G. G. Jr. 1979. A study of Tolypocladium cylindrosporum Gams, a new, naturally occurring fungal pathogen of mosquitoes (with notes on Beauveria bassiana (Bals.) Vuill.). Ph.D. dissertation, University of California, Berkeley, 177 pp.

Soares, G. G. Jr., Pinnock, D. E. and Samson, R. A. 1979. Tolypocladium, a new fungal pathogen of mosquito larvae with promise for use in microbial control. Proc. 47th Ann. Conf. Calif. Mosq. Control Assn., pp. 51-54.

Sweeney, A. W. 1975. The mode of infection of the insect pathogenic fungus Culicinomyces in larvae of the mosquito Culex fatigans. Aust. J. Zool., 23, 49-57.

Sweeney, A. W. 1979. Infection of mosquito larvae by Culicinomyces sp. through anal papillae. J. Invertebr. Pathol., 33, 249-251.

Sweeney, A. W. 1981. An undescribed species of Smittium (Trichomycetes) pathogenic to mosquito larvae in Australia. Trans. Brit. Myc. Soc., in press.

Sweeney, A. W., Lee, D. J., Panter, C. and Burgess, L. W. 1973. A fungal pathogen for mosquito larvae with potential as a microbial insecticide. Search, 4, 344-345.

Sweeney, A. W., Wright, R. G. and van der Lubbe, L. 1980. Ultrastructural observations on the invasion of mosquito larvae by the fungus Culicinomyces. Micron, 11, 487-488.

Travland, L. B. 1979a. Initiation of infection of mosquito larvae (Culiseta inornata) by Coelomomyces psorophorae. J. Invertebr. Pathol., 33, 95-105.

Travland, L. B. 1979b. Structure of the motile cells of Coelomomyces psorophorae and function of the zygote in encystment on a host. Can. J. Bot., 57, 1021-1035.

Vey, A. and Fargues, J. 1977. Histological and ultrastructural studies of Beauveria bassiana infection in Leptinotarsa decemlineata larvae. J. Invertebr. Pathol., 30, 207-215.

Vey, A. and Gotz, P. 1975. Humoral encapsulation of Diptera (Insecta): Comparative studies in vitro. Parasitol., 70, 77-86.

Whisler, H. C., Zebold, S. L. and Shemanchuk, J. A. 1974. Alternative host for mosquito parasite Coelomomyces. Nature (London), 251, 715-716.

Williams, M. C. and Lichtwardt, R. A. 1972. Infection of Aedes aegypti larvae by axenic cultures of the fungal genus Smittium (Trichomycetes). Am. J. Bot., 59, 189-193.

Wong, T. L. and Pillai, J. S. 1980. Coelomomyces opifexi Pillai and Smith (Coelomomycetaceae: Blastocladiales) VI. Observations on the mode of entry into Aedes australis larvae. N.Z. J. Zool., 7, 135-139.

Zacharuk, R. Y. 1970. Fine structure of the fungus Metarrhizium anisopliae infecting three species of larval Elateridae (Coleoptera). J. Invertebr. Pathol., 15, 372-396.

Zacharuk, R. Y. 1973. Electron-microscope studies of the histopathology of fungal infections by Metarrhizium anisopliae. Misc. Pub. Entomol. Soc. Am., 9, 112-119.

Zebold, S. L., Whisler, H. C., Shemanchuk, J. A. and Travland, L. B. 1979. Host specificity and penetration in the mosquito pathogen Coelomomyces psorophorae. Can. J. Bot. 57, 2766-2770.

Chapter 15

BIOCHEMISTRY OF *LAGENIDIUM GIGANTEUM* INFECTION IN MOSQUITO LARVAE

by Aristotle J. Domnas*

INTRODUCTION

Several fungi such as *Coelomomyces* spp., *Metarrhizium anisopliae*, and *Culicinomyces clavosporus* attack mosquitoes, and a number of the Entomophthorales parasitize harmful insects (see Chapters 13 and 14). However, little is known about the biochemistry of these fungal infections. In some cases, the biochemistry is difficult because the organisms cannot be grown in axenic culture, e.g. *Coelomomyces*. At present, most biochemical knowledge of host-parasite relationships in diseases of insects is limited to the mosquito pathogen *Lagenidium giganteum*. What began as a cooperative venture among mycologists interested in biological control of mosquitoes, has evolved into a study of lipids and membranes, and eventually of enzymes that metabolize non-amino acid nitrogen compounds.

BIOLOGY OF *LAGENIDIUM GIGANTEUM* COUCH

The fungus is a water mold of the class Oomycetes which is divided into four orders, Saprolegniales, Leptomitales, Lagenidiales and Peronosporales. The last order includes the genera *Pythium* and *Phytophthora* that have physiological characteristics resembling those of *L. giganteum*. The established life cycle is outlined in Fig. 15-1 and has two phases (Sparrow, 1970).

*Department of Botany, University of North Carolina, Chapel Hill, North Carolina 27514

PATHOGENESIS OF INVERTEBRATE MICROBIAL DISEASES. Davidson, ed. © Allanheld, Osmun (Totowa, N.J.)

Figure 15-1. Life cycle of Lagenidium giganteum

A - zoospores.
B - spore cyst.
C - germination of zoospore cyst.
D - hypha.
E - hypha, with septum.
F - segmentation of hypha.
G - discharge papillum forming on hyphal segment (a sporangium).
H - discharge tube formed.
I - zoospores cleaved out in vesicle.
J - zoospore discharge.
K - initial infection in cephalic region, detail of head showing spore cysts and hyphae.
L - hyphae ramifying through head and thorax.
M - hyphae segmented in head and thorax; filaments extending into abdominal region.
N - larva completely filled with fungus; sporulation; detail of abdominal segments showing sporangium.

Saprophytic

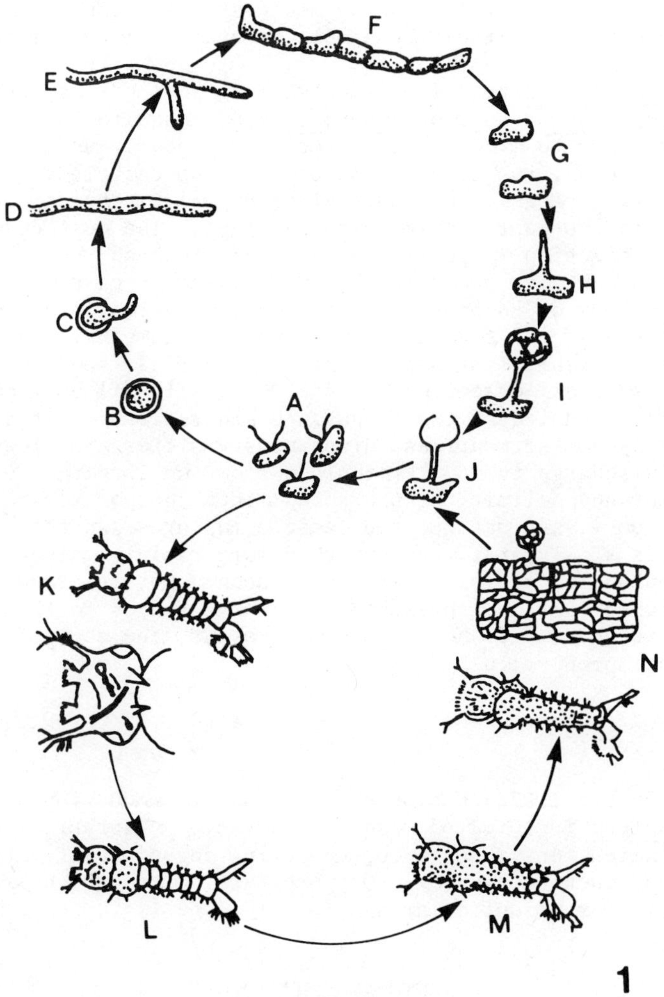

1

Parasitic

Life cycle

Saprophytic phase: After a few hours of restless swimming, the zoospore slows down, retracts and ingests its own flagella, then pushes out a hyphal extension. This hypha ramifies, segments, and eventually becomes a mycelial mat. Since the mycelia are septate, each segment is a potential sporangium, but all of these segments are not competent. The culture maintains itself by simple vegetative multiplication.

Parasitic phase: The free swimming zoospore attaches to a *Culex* spp., *Aedes* spp. or *Anopheles* spp. mosquito larva of 1st, 2nd, 3rd to middle 4th instar. By a poorly-understood boring process, the zoospore penetrates the cuticle and encysts (Fig. 15-2). A germ tube initial or hyphal extension appears, similar to the saprophytic germling stage. The most common site of infection is the distal area of the head, but infection can also occur in the siphon, and in cases where great numbers of zoospores are present, mosquito larvae can be attacked at all segments. Once the integumental barrier has been overcome, the hyphal initial ramifies, sending offshoots in all directions, until eventually all body spaces have been occupied by the fungus. Discharge tubes extend from certain hyphal segments and diaphanous vesicles are blown out of the discharge tubes. Protoplasm flows up into the vesicles and membranes delimit the protoplasm into zoospores. When the zoospores are mature, the vesicle ruptures and the zoospores slowly at first, but then more rapidly, disengage their entangled flagella, and enter upon a motile stage which, if suitable hosts are present, return the fungus to the parasitic phase. If hosts are not present, the zoospores develop saprophytically.

Sexual cycle

Umphlett (1973) has observed sexual reproduction of *L. giganteum* in infected mosquito larvae. An antheridium is formed which penetrates a corresponding oogonial-like structure with consequent karyogamy. Oospores are formed which are parasitic for mosquito larvae.

GROWTH AND PHYSIOLOGY

Nutrition

Mycological media: *Lagenidium giganteum* grows readily on many mycological media. Good mycelial yields can be obtained from Emerson YPSS, beef infusion, or peptone extracts, as well as media such as "V8" juice and potato extracts which

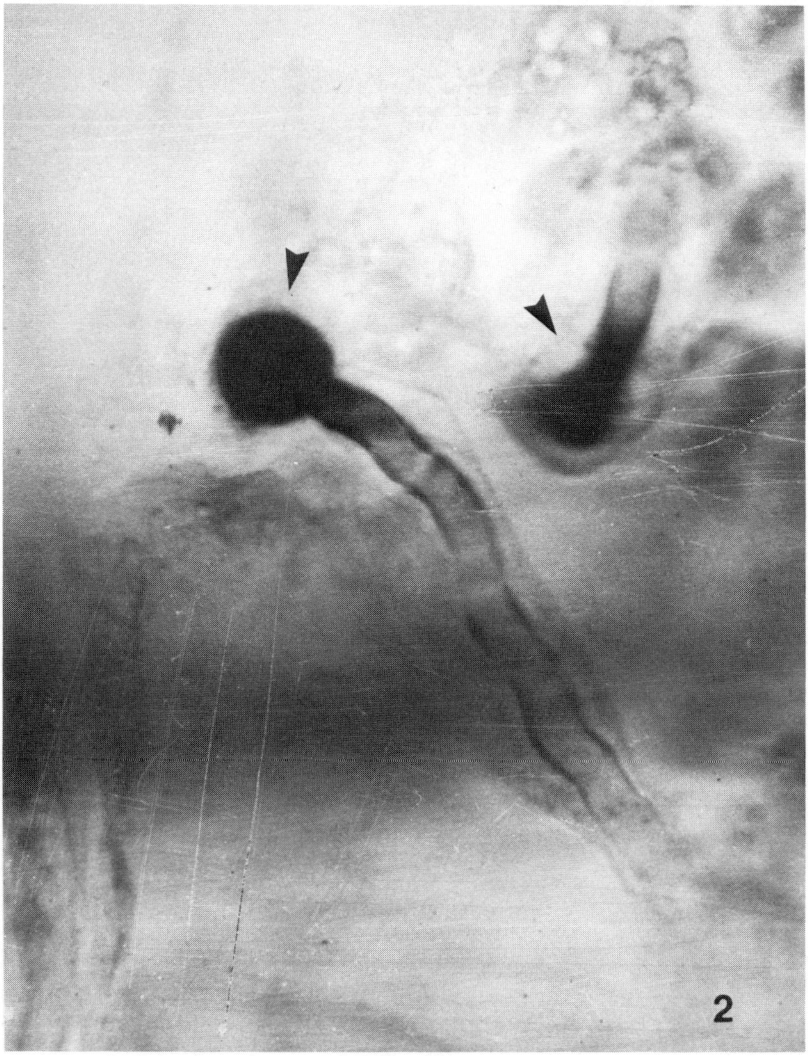

Figure 15-2. Zoospores (arrows) encysted on larval cuticle, heavily melanized and showing germ tube development.

TABLE 15-1. GROWTH OF *L. GIGANTEUM* ON SINGLE CARBON SOURCES

Carbon Source[1,2]	Dry Weight mg	Carbon Source	Dry Weight mg
Monosaccharides		Acids	
Glucose	177.2	*Lactate	54.9
*Fructose	163.5	Acetate	15.3
Mannose	80.8	Malate	Poor
*Galactose	21.9	*Pyruvate	42.9
*L-Arabinose	17.9	Succinate	44.8
Ribose	36.5	Citrate	42.6
		Butyrate	11.6
Disaccharides		Stearate	28.2
Trehalose	213.5	Palmitic	67.8
Sucrose	34.7	Oleic	29.3
Maltose	147.1	Tween 20	12.7
*Lactose	77.9		
		Miscellaneous	
Polysaccharides		*Glucosamine	16.5
CM cellulose	24.7		
Soluble starch	68.6	No sugar	34.9
Inuline	32.7		
Raffinose	35.2		
Chitin	None		
Alcohols			
*Ethanol	1.2		
Glycerol	162.7		
*Mannitol	61.9		

*Denotes filter sterilization

[1]Concentrations: Monosaccharides, disaccharides, alcohols, acids, raffinose, glucosamine: 0.5% (w/v). Tween 20, 0.2% (v/v). Polysaccharides, 01.% (w/v)

[2]All sugars are in the D form unless otherwise stated

(from McInnis, T. M. Jr., (1971) Ph.D. Dissertation, University of North Carolina; reprinted by permission of the author).

are used for pythiceous fungi. We have found that peptone: yeast extract:glucose (1.5 g:1.5 g:3.0 g per liter H_2O) (PYG) is the most useful undefined medium and many of our investigations have involved cultures grown in this medium. Supplements such as sterol and lipids are added to the broth which is autoclavable without detrimental or inhibitory effects on the fungus.

Adequate yields and growth times on defined medium can be obtained with Gleason's (1968) modification of the Machlis medium, which contains glucose, glutamic acid and thiamine, as well as trace metals and phosphate. This growth medium is used to determine carbon and nitrogen sources utilizable by *L. giganteum*.

Carbohydrate sources: Table 15-1 illustrates the sugars that *L. giganteum* can utilize for growth. It is apparent that trehalose is the best sugar, but glucose and maltose are also good carbon sources. Other substances such as cellobiose, chitobiose, and cellulose are very poor carbon sources. In this study, glutamate was employed as the nitrogen source.

Nitrogen requirements: The ability of the fungus to use various nitrogen sources was tested in the defined medium using glucose as the standard carbon source. Table 15-2 shows that amino acids of, or related to, the glutamate family are good single nitrogen sources. Glutamine is the best, and then the following amino acids are utilized in order: ornithine → glutamate → aspartate → alanine → asparagine. Other amino acids are essentially unusuable.

An extremely interesting aspect of nitrogen source utilization was observed. Cultures grown on PYG were innoculated into Gleason's defined medium, with ammonium sulfate as the nitrogen source. After an extremely long lag phase (14 days), the culture began to grow and attained yields in 5-6 days which were equal to those obtained with PYG. This growth effect was not consistent. Eventually, Moosa (1977) found that the addition of 0.1 ml of either olive, soy or peanut oil per 100 ml of Gleason's ammonium sulfate medium resulted in consistent growth of *L. giganteum*. This effect is not due to sterol requirements, but to the oils themselves, and may be due to a requirement for phospholipid such as lecithin. Oil-supplemented cultures will also utilize urea and nitrate but not nitrite.

Lipid requirements: As noted earlier, *L. giganteum* grows well in most media, but does not produce zoospores. Domnas et al. (1977) reported that sterols were required; this requirement is identical to that reported by Hendrix (1970) and Elliott (1977) for plant pathogenic *Phytophthora* spp. and *Pythia* spp.. *Lagenidium* has close affinities to these plant

TABLE 15-2. GROWTH OF *L. GIGANTEUM* ON VARIOUS SINGLE NITROGEN SOURCES

Compound[1]	Dry Weight mg	Compound	Dry Weight mg
Amino Acids[2]		Amino Sugars	
Glycine	10.5	*Glucosamine	9.9
Alanine	81.0	*N-acetylglucosamine	7.9
Leucine	11.5		
β-alanine	12.0	Inorganic Nitrogen	
Isoleucine	13.1	Sodium nitrate	10.3
Lysine	12.1	Ammonium sulfate	9.6
Valine	12.2	Ammonium sulfate[3]	170.7
Aspartate		Ammonium tartrate	10.9
(K salt)	131.1		
*Asparagine	41.0	Miscellaneous	
Glutamate		Chitin	Poor
(Na salt)	142.3	Keratin	None
*Glutamine	367.8	*L-leucine amide	11.2
*Arginine	11.2	*D-glutamine	11.2
*Citrulline	12.2	*Urea	11.0
*Ornithine	244.7		
Phenylalanine	12.3	No nitrogen	9.9
Tyrosine	Poor (insoluble)		
*Tryptophane	8.4		
Threonine	8.7		
Proline	10.2		
Serine	10.6		
*Histidine	7.1		
Cysteine	Poor (insoluble)		
*Methionine	10.6		
*Casamino acids	164.8		

*Denotes filter sterilization

[1]Concentrations based on amount of molecular nitrogen in 0.3% (w/v) sodium nitrate

[2]All amino acids are the *L* form unless otherwise stated

[3]Incubated 14 days

(from McInnis, T. M. Jr., (1971) Ph.D. Dissertation, University of North Carolina; reprinted by permission of the author).

pathogens. However, this is the first report of a sterol requirement in a fungus which infects an animal. In contrast, L. callinectes, which infects blue crab eggs (Couch, 1941) and shrimp (Lightner and Fontaine, 1973), does not need exogenous sterols for zoospore production. Lagenidium giganteum can use a large number of sterols as is shown in Table 15-3. The best compound was sitosterol (=campesterol) → ergosterol (=cholesterol), and the organism could use a number of unnatural compounds such as cholesteryl phosphate as well as sitosteryl glucoside.

Vitamin requirements: Thiamine was the only vitamin found to greatly improve mycelial yields in defined conditions, but it was not necessary for growth (Table 15-4).

Physiology

Temperature: Lagenidium giganteum grew best at 25-28° C in 5 days in both Gleason's medium and in PYG. Above and below this range considerable curtailment of yield was noted; no growth was recorded at 38° and 15°.

pH optimum: Optimal mycelial yields were obtained in defined medium with a pH range of 5.5 to 7.0. No growth below pH 5.0 or above 7.5 was noted indicating that the fungus has a narrow pH range and is closely adapted to its host, the mosquito.

Growth time: The growth time was found to be 5-7 days on the gyrotory shaker, in 100 ml broth cultures in PYG or defined medium.

HOST-PARASITE INTERACTIONS

Enzymes present in parasite and host

The parasite and host have a variety of enzymes (Table 15-5) and several enzymes are common to both, such as trehalases, proteases, L-ornithine transcarbamylases and alkaline phosphatases.

Infection mechanism

Infection nearly always begins in the periclypeal area of the head and is always initiated by zoospores. When abundant zoospores are present, the larva can be attacked anywhere on the body. It has been demonstrated in this laboratory and by T. McInnis (pers. comm.) that zoospores are apparently attracted to the larvae by some chemotactic agent emitted by the larvae. Once the zoospore finds the cuticle, it encysts,

TABLE 15-3. EFFECT OF VARIOUS STEROLS ON ZOOSPORE PRODUCTION BY *L. GIGANTEUM* AS MEASURED BY BIOASSAY

Nature of Medium[1]	No. of Trials	% Infection Range (3-day period)
Whole hemp seed (WHS)	20	>95
Sterols from WHS on peptone-yeast extract-glucose (PYG) agar	3	50-88
Sterol depleted WHS oil on PYG	3	0
Sterols from whole soy bean (WSB) on PYG	4	28-88
Lanosterol	1	30
Cholesterol	10	40-78
Desmosterol	1	45
Ergosterol	4	52-87
Campesterol	6	72-90
Sitosterol	6	70-90
Stigmasterol	3	5-10
Campesterol:stigmasterol 1:1	2	23-53
Stigmasterol:sitosterol	2	0
Sitosterol:campesterol:stigmasterol 3:1:1	2	32-38

[1]0.1% sterol depleted oils were added to PYG agar. Digitonin precipitated sterols were dissolved in either $CHCl_3$ or MeOH to give approximately 30 µg sterol/ml media. All authentic sterols were added at the levels of 30 µg/ml PYG media. Test larvae were *Aedes* sp.

(from Domnas et al., (1977); reprinted by permission of Mycological Society of America)

TABLE 15-4. EFFECT OF VITAMINS ON GROWTH OF *LAGENIDIUM GIGANTIUM*

Vitamin[1]	Dry Weight mg
None	105.7
Thiamine	325.7
Biotin	98.9
Pyridoxal-5'-phosphate	106.8
Riboflavin	114.2
All	288.9

[1]Concentration: 0.1 microgram per milliliter.

(from McInnis, T. M. Jr. (1971), Ph.D. Dissertation, University of North Carolina; reprinted by permission of the author).

TABLE 15-5. DISTRIBUTION OF ENZYMES IN *L. GIGANTEUM* AND IN UNINFECTED *CULEX QUINQUEFASCIATUS*

Enzyme (trival name)	Fungus[1]	Mosquito Larvae
β-galactosidase	-	+
α-glucosidase	tr.	
β-glucosidase	+++	-
N-acetyl-β-D-glucosaminidase	tr.	+++
Glucose 6-phosphate dehydrogenase	+++	tr.
Trehalase	+++	+++
L-aspartic acid transcarbamylase	+	+
Glutamate-oxalacetate transaminase	++	+++
Glutamate-pyruvate transaminase	++	-
L-ornithine transcarbamylase	+++	+
L-glutaminase	++	tr.
o-diphenol oxidase	-	+++
Lipase (non-specific)	++	++
Acid phosphatase	++	-
Alkaline phosphatase[2]	++	++
Trypsin[3]	+++	+++
Collagenase[3]	+++	
Chymotrypsin[3]	++	+++
Isocitrate dehydrogenase	++	?
Lactic acid dehydrogenase	+	-
Urease	+++	-
Uricase	+	-

[1]Fungus grown in PYG medium (+ = low; ++ = average; +++ = very active; tr. = trace)

[2]Membrane bound

[3]Extracellular enzymes

(from Domnas et al. (1974); reprinted by permission of Academic Press, Inc.)

perhaps at a specific recognized site on the cuticle. The zoospore retracts its flagellum and pushes out a germ tube, which apparently by mechanical force, penetrates the cuticle and enters the hemocoel. Details of the infection from this point on have been described by Domnas et al. (1974). Once cuticle penetration occurs, the fungus multiplies rapidly and the host organism has a limited capability to impede this multiplication. Many zoospores are involved in the initial infection and they apparently focus on the same target zone; the points of invasion are clearly seen as black "bore holes" which on closer examination reveal encystment of spores (Fig. 15-3). The black material surrounding the spore and the invading germ tube is probably a host melanization reaction.

Biochemistry of the mycosis

Some biochemical parameters: Domnas et al. (1974) separated parasite from host material and studied the impact of the fungus on the mosquito larvae. Parameters examined were protein, total free amino acids and sugars, as well as the enzymes chitobiase, trehalase, alkaline phosphatase, glutamate-oxalacetate transaminase and o-diphenol oxidase. As shown in Fig. 15-4, the protein content of infected larvae does not increase as in the controls but rather decreases in the dying animals. In Fig. 15-5 the amount of chitobiase available in infected larvae is shown to drop very rapidly up to the time of death; only the glutamate-oxaloacetate enzyme (Fig. 15-6) seems to maintain a relatively similar level to that of controls. In contrast, o-diphenol oxidase (Fig. 15-7) drops from levels at initial infection to essentially no activity at ca. 30 hours post-infection. In fact, dying infected larvae are quite bleached in contrast to the controls which may be due to a loss of ability to tan the cuticle. When the specific activity of the enzyme o-diphenol oxidase was plotted against time, a sharp increase was noted in contrast to controls. This was interpreted as an increased demand for melanin for encapsulation of the invading hyphal germ tubes (Fig. 15-8). From this combined evidence, it is clear that the host is overwhelmed and that physiological starvation occurs.

Amino acid levels in host during infection: Table 15-6 shows the effect of parasitism on the free amino acid levels in *C. quinquefasciatus* larvae (Geibel and Domnas, 1978). Glutamate levels remained high, which may be due to conversion of protein or amino acid reserves in an attempt to maintain a steady-state metabolism. This correlates with the rather high levels of glutamate-oxaloacetate transaminase observed (Fig. 15-6).

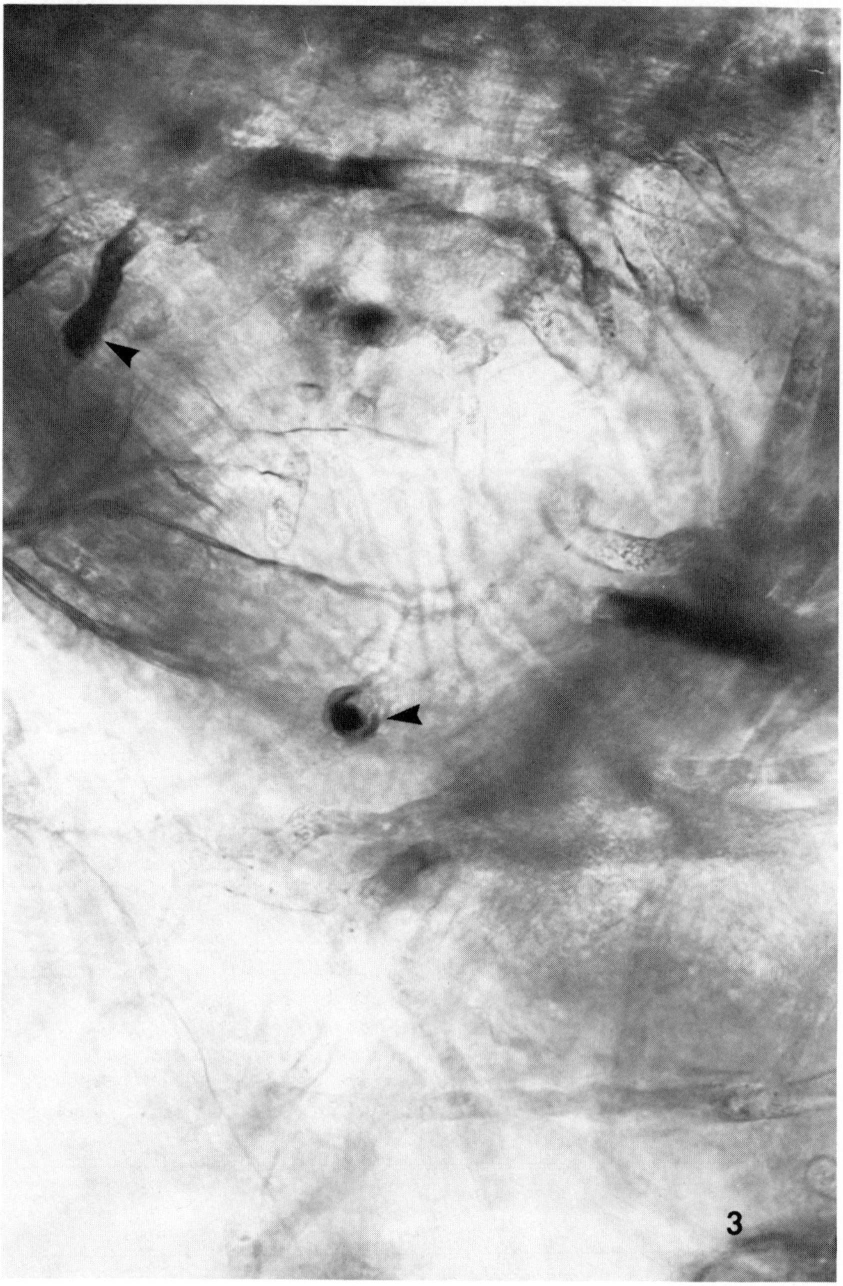

Figure 15-3. "Bore-hole" from zoospore encystment, arrow center. Melanized hypha, arrow upper left. Many non-melanized hyphae visible under larval cuticle.

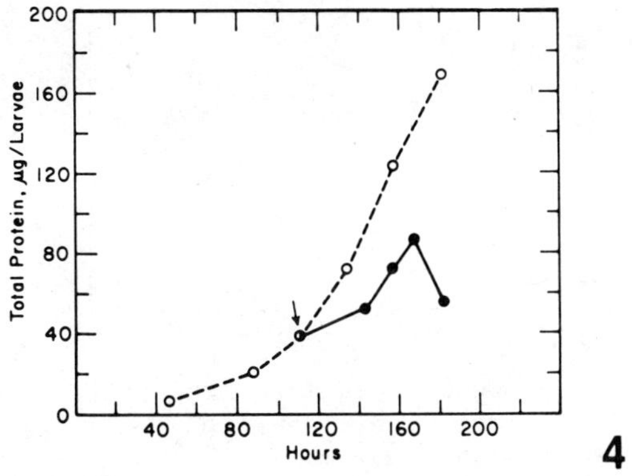

Figure 15-4. Development of total protein per larvae in healthy (○ -- ○) and infected _Culex_ sp. larvae populations (● — ●).

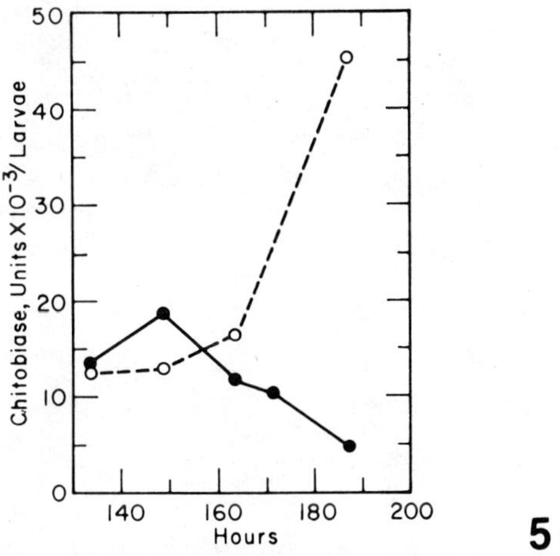

Figure 15-5. Time-course development of chitobiase in normally developing _Culex_ larvae (○ -- ○) and infected _Culex_ larvae (●—●).

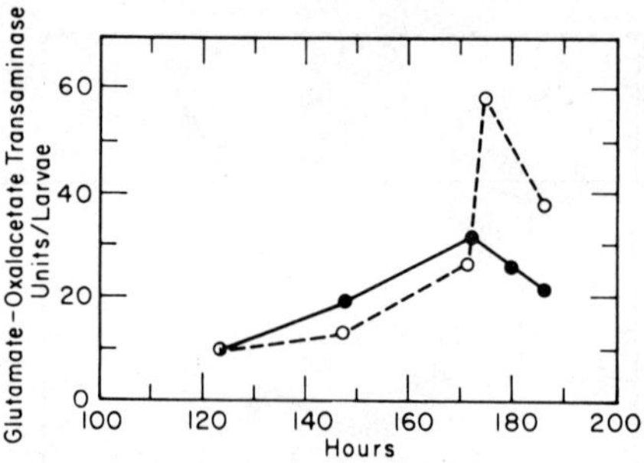

6

Figure 15-6. Response of glutamate-oxalacetate transaminase in normal Culex larvae (○---○) and infected larvae (● — ●).

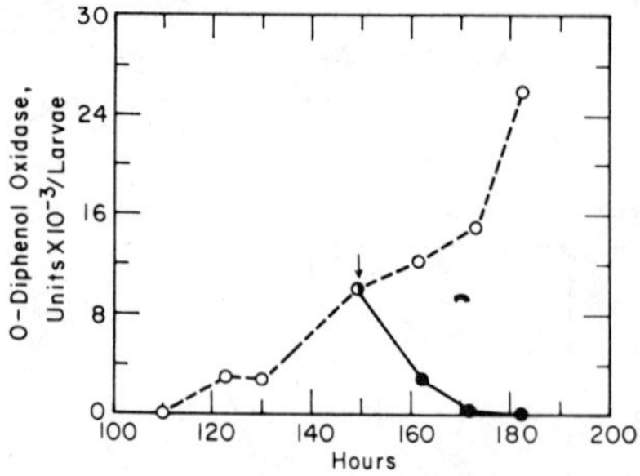

7

Figure 15-7. Behavior of o-diphenol oxidase in normal Culex larvae (○---○) and larvae infected with Lagenidium (● — ●).

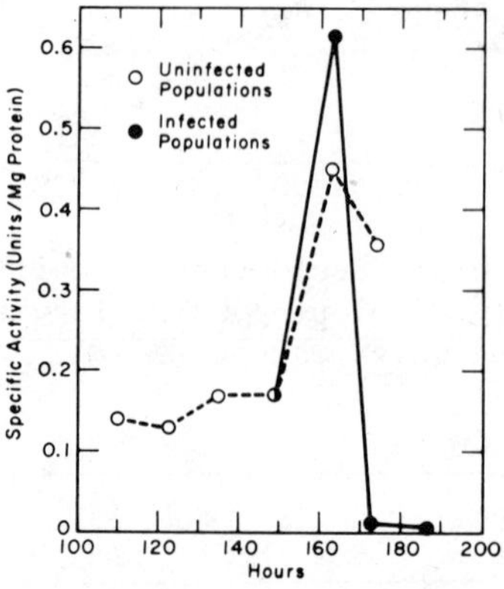

8

Figure 15-8. Specific activities of o-diphenol oxidase in normal and infected Culex larvae. Figures 15-2 through 15-8 from Domnas et al., 1974; reprinted by permission of Academic Press Inc.

TABLE 15-6. FREE AMINO ACIDS IN NORMAL AND INFECTED LARVAE OF *CULEX QUINQUEFASCIATUS*

Time, Hrs.[1]	0	120	120	168	168
Population[2]	C	C	I	C	I
Amino Acid (nmols/larva)					
Lysine	0.74	1.09	1.35	6.83	1.39
Histidine	0.77	2.63	2.27	4.92	3.04
Arginine	0.96	3.48	2.77	6.65	2.54
Aspartic acid	1.15	2.48	2.35	6.65	1.78
Threonine	0.53	1.02	0.31	4.74	1.14
Serine	0.87	1.38	1.68	5.27	1.68
Glutamic acid	2.16	5.00	5.52	6.31	4.53
Proline	0.78	1.42	1.26	4.69	0.97
Glycine	1.28	2.47	2.52	9.03	2.24
Alanine	1.44	2.38	2.52	8.56	2.49
Cysteine	0.06	0.21	0.03	0.08	0.00
Valine	0.72	1.14	0.01	5.63	1.10
Methionine	0.05	0.19	0.13	0.08	0.12
Isoleucine	0.41	0.73	0.84	4.04	0.91
Leucine	0.61	1.00	1.33	5.97	1.33
Tyrosine	0.135	0.23	0.43	1.36	0.21
Phenylalanine	0.31	0.67	0.058	3.35	0.51

[1] 0 hours - infection initiated
120 hours - 60% infection
168 hours - 100% infection

[2] C - control
I - infected

(from Geibel, P. E. (1974), Ph.D. Dissertation, University of North Carolina, reprinted by permission of the author).

Carbohydrate levels in the host: The carbohydrate level dropped in infected larvae (Table 15-7). Glucose, the principal free carbohydrate, was very low in C. quinquefasciatus larvae 168 hr post-infection. In contrast, glucosamine values were high, which could indicate the inability to synthesize chitin; infected larvae never ecdyse. Curiously, free trehalose was vanishingly low and probably reflects the extraction procedure for obtaining total sugars (Giebel and Domnas, 1978).

Comparative enzymology; Host and parasite trehalase: Trehalase from the C. quinquefasciatus was purified 187-fold and its properties studied and compared to the trehalase isolated from L. giganteum (Giebel and Domnas, 1976; McInnis and Domnas; 1973). Table 15-8 shows that the properties of both enzymes were quite similar with the exception of Arrhenius activation energies. A value of 28.9 Kcal/mole was measured for mosquito trehalase as against 11.4 Kcal/mole for the L. giganteum enzyme; this suggests that in competition for the common substrate trehalose, the fungal enzyme would require less energy for initiating catalysis and could out-compete the larval trehalase. Comparative inhibition studies showed some other interesting properties of these enzymes. The host trehalase is susceptible to inhibition by a number of common, normal metabolites; for example glucose produced 100% inhibition. Other inhibitors were sucrose (98%), glucosamine (30%), and the glucose phosphates (28%). Of these, only sucrose inhibited the fungal trehalase (34%) and all inhibitions were competitive. The study showed that the fundamental roles of the two enzymes were different. Thus the mosquito enzyme is subject to a number of regulatory controls while the fungal enzyme seems to be primarily poised for trehalose digestion. These comparative properties place the infected host at a disadvantage in competition for trehalose.

FURTHER ASPECTS OF FUNGAL BIOCHEMISTRY

Lipid requirements

Sterols: To enter the parasitic cycle, L. giganteum must produce zoospores, which depends on an exogenous supplement of sterols in the medium. The fungus cannot synthesize cholesterol de novo, but it can assimilate mevalonic acid which is converted to cholesterol (Warner and Domnas, unpublished). It appears that the major block in the biosynthetic pathway is absence of the enzyme β-OH-β, -β-methyl glutaryl CoA reductase, since the organism can convert mevalonic acid to sterols. The organism utilizes ^{14}C labeled glucose to synthesize lipid substances and can utilize ^{14}C labeled acetate but cannot

TABLE 15-7. FREE CARBOHYDRATES IN NORMAL AND INFECTED LARVAE OF *C. QUINQUEFASCIATUS*

Time, Hrs.[1]	0	120	120	168	168
Population[2]	C	C	I	C	I
Sugar (nmols/larva)					
Mannitol	0.0000	1.200	0.00	0.450	0.0000
Glucose	7.5000	4.100	1.13	3.100	1.9100
Glucosamine	0.8500	1.000	0.51	0.370	1.3000
Trehalose	0.0038	0.018	0.00	0.041	0.0005

[1] 0 hours - infection initiated
120 hours - 60% infection
168 hours - 100% infection; larvae dead

[2] C - control
I - infected

(from Giebel, P. E. (1974), Ph.D. Dissertation, University of North Carolina; reprinted by permission of the author)

TABLE 15-8. PROPERTIES OF TREHALASES FROM *C. QUINQUEFASCIATUS* AND *L. GIGANTEUM*

Parameter	Mosquito Trehalase	Fungal Trehalase
pH stability	5.5 - 9.0	4.0 - 7.0
pH optimum	5.5 - 6.5	5.5 - 6.5
K_m	1.21×10^{-3}M	1.4×10^{3} M
Thermal stability	up to 30° without substrate 40° with substrate	up to 50°
E_a	28.9 kcal/mole	11.4 kcal/mole
Specificity	specific for trehalose	specific for trehalose

(from Giebel and Domnas (1976); reprinted by permission of Pergamon Press) metabolize these substances to sterols. The crustacean parasite, L. callinectes, transforms glucose to cholesterol, and does not require exogenous sterols, or any other supplement to make zoospores. Lagenidium giganteum took up cholesterol in 3 hours from broth cultures whereas L. callinectes took it up in 3 days. When L. giganteum was cultured in the presence of sitosterol, ergosterol or campesterol, these substances were re-isolated from the mycelia essentially unchanged except for small amounts found as steryl esters. Domnas et al. (1977) reported that L. giganteum could use cholestanol, and indeed administration of β-cholestanol resulted in the recovery of cholesterol as was predicted; in addition, alpha cholestanol (or epicholestanol) was also recovered. Table 15-9 shows the metabolic fate of these Δ^0 analogs in both species of Lagenidium. The inversion of the β-OH to the α-OH cholestanol or to epicholestanol is rarely reported in the literature and the role of this substance is unknown. Its configuration makes it unsuitable for insertion into the lipid bilayer of the membrane and its uptake by both fungi is surprising.

Sterol esters: In common with many fungi (Elliott, 1977), L. giganteum converts cholesterol to esters with fatty acids whose nature were identified as myristic, palmitic, stearic and a C-20 acid, and anti-isoarachidate. Assimilation of sitosterol also leads to the formation of identical esters. Cholesteryl palmitate was also utilized by the fungus, although the uptake times were not measured. Free cholesterol was isolated from the mycelium and the cholesterol moiety was converted into other esters identical to those reported above (Warner and Domnas, unpublished).

Extracellular proteases

Gabriel (1974) discovered the presence of proteolytic activity in the broth of L. gigantium grown in PYG medium. The partially purified enzyme complex demonstrated many contradictory properties, particularly with respect to its inhibition. In 1979, Dean and Domnas (unpublished) demonstrated that the proteolytic activity was due to four types of enzymes - three or four trypsins, two collagenases (one of which hydrolyzed insoluble collagen and a second hydrolyzed acid-soluble collagen), a chymotrypsin and a moderate elastase activity.

These proteases could be induced in appropriate medium, by protein sources such as yeast extract or peptone. A certain amount of protease was secreted by the organism when grown in Gleason's medium, but never the levels produced in complex media. When mycelia of L. giganteum were washed clean of contaminating media and were resuspended in Gleason's salts without glucose but with peptone or yeast extract, proteases

TABLE 15-9. UPTAKE AND METABOLISM OF STEROLS BY TWO SPECIES OF LAGENIDIUM

	Sterol[1] added	T_r[2]	Sterol recovered
Lagenidium giganteum	Cholesterol	1.00	Cholesterol[3]
mixture {	Sitosterol (57.2%)	1.47	Sitosterol
	and		and
	Campesterol (42.8%)	1.24	Campesterol
	Campesterol	1.24	Campsterol
	Stigmasterol	1.29	Stigmasterol
L. callinectes	none	1.00	Cholesterol[3]
mixture {	Sitosterol (57.2%)	1.47	Sitosterol
	and		and
	Campesterol (42.8%)	1.24	Campesterol
	Cholesterol	1.00	Cholesterol[3]

[1]GLC-MS verified

[2]retention time relative to cholesterol - 1.0

[3]free, also recovered as the ester

were produced in ca. 12 hr. Addition of glucose to the medium reduced the lag period to ca. 9 hr. Fungus in Gleason's medium with glucose only produced a single inactive protein which could be activated by trypsin. It is not known whether the activation products were collagenase or trypsin, but increased total proteolytic activity was measured.

Sterols appear to be involved in production of proteases, for when cholesterol was added (as a suspension in 2% Tween 80) to growing cultures in PYG, considerable augmentation of total protease was observed (Table 15-10). Large amounts of single-band protein (as revealed by electrophoresis) were observed in the medium after at least 6 hr following broth inoculation. This protein again could be activated by trypsin to yield active fractions (Dean and Domnas, 1978). More recently Dean and Domnas (unpublished) showed that in PYG-cholesterol cultures, trypsin appeared first in substantial amounts after 36 hr of growth, whereas collagenase activity was absent or very low; 3 hr later, however, active collagenase was detected. Trypsin may activate a "procollagenase" to active collagenase, but this has not yet been proven. Broth cultures with cholesterol always secreted proteases before those without cholesterol, and preliminary results indicated that sterol esters were even more effective in increasing total protease output than free sterol.

CONCLUSIONS

Parasitic organisms may have developed because of environmental conditions which presented the parasite with some ecological niche relatively free from competition. This may be the case for Lagenidium giganteum since it has only been isolated from infected mosquito larvae in nature. In addition, certain nutritional deficiencies weigh heavily against this fungus which is unable to make sterols and take up ammonium ion from the aquatic domain. The fungus has the latent ability to synthesize sterols and to take up ammonium ion, but requires added lipids. Further, the sterol-synthesizing deficiency mimics superficially that of its hosts so that a dependency exists which compels the fungus to adopt the parasitic mode. This organism produces four different enzymes which act upon all major protein classes, including even collagen and elastin. If the enzymes are produced in the infected host, they give the parasite an enormous advantage over the host. Proteases probably are effective in the process of boring through the cuticle. Further investigations are designed to investigate the presence of extracellular phospholipases and glucanases, although no chitinase activity has ever been demonstrated.

TABLE 15-10. EFFECT OF STEROL IN TWEEN 80 AUTOCLAVED WITH PYG MEDIUM ON PRODUCTION OF EXOCELLULAR PROTEASE BY *L. GIGANTEUM* DURING FIVE DAYS GROWTH

Sterol Added	Specific Activity	No. of Trials
None - PYG Tween 80	74	
	72	4
	164	
	75	
Cholesterol	420	
	414	6
	110	
	968	
	492	
	507	
Ergosterol	287	
	330	
	217	5
	259	
	422	
Sitosterol	341	
	518	4
	510	
	626	
Campesterol/sitosterol[2]	573	
	512	4
	665	
	515	

2% Tween 80 with H_2O and 1 mg/m sterol. 1 ml added to 100 ml PYG (peptone-yeast extract-glucose broth).

Specific activity - units/mg protein. Unit equal to amount of protein that gives a ΔA_{520} of 0.1 in 30 min at 30° with Azocoll as substrate.

Aldrich NF - given as sitosterol. Contains sitosterol/campesterol 3:2.

Finally, this organism has been developed into a model for the study of the complicated relations of sterol-lipid bilayer interactions. There are many normally occurring reactions that are influenced by cholesterol, and this organism shows these reactions easily and in a measurable way.

REFERENCES

Couch, J. N. 1941. A new fungus on crab eggs. J. Elisha Mitchell Sci. Soc., 58, 158-162.

Dean, D. D. and Domnas, A. J. 1978. Influence of sterols on protease production by the mosquito pathogen Lagenidium giganteum. Proc. Int'l. Congr. Invertebr. Pathol., Prague, p. 27.

Domnas, A., Giebel, P. E. and McInnis, T. M. Jr. 1974. Biochemistry of mosquito infection: Preliminary studies of biochemical change in Culex pipiens quinquefasciatus following infection with Lagenidium giganteum. J. Invertebr. Pathol., 24, 293-304.

Domnas, A., Srebro, J. P. and Hicks, B. F. 1977. Sterol requirement in the mosquito-parasitizing fungus Lagenidium giganteum. Mycologia, 69, 875-886.

Elliott, C. G. 1977. Sterols in fungi: Their functions in growth and reproduction. Adv. Microbial. Physiol., 15, 121-173.

Gabriel, J. D. 1974. A study of proteolytic activities in the mosquito parasite Lagenidium giganteum. Honors thesis to Biology Curriculum, Univ. of N. Carolina.

Giebel, P. E. and Domnas, A.J. 1976. Soluble trehalases from larvae of the mosquito, Culex pipiens and the fungal parasite Lagenidium giganteum. Insect Biochem., 6, 303-311.

Giebel, P. E. and Domnas, A. J. 1978. Carbohydrate and amino acid metabolism in Culex pipiens quinquefasciatus infected with Lagenidium giganteum. Proc. Int'l. Congr. Invertebr. Pathol., Prague, p. 43.

Gleason, J. H. 1968. Nutritional comparison in Leptomitales. Amer. J. Bot., 55, 1003-1010.

Hendrix, J. W. 1970. Sterols in growth and reproduction in fungi. Ann. Rev. Phytopathol., 8, 111-130.

Lightner, D. V. and Fontaine, C. T. 1973. A new disease of the white shrimp Penaeus setiferus. J. Invertebr. Pathol., 22, 94-99.

McInnis, T. M., Jr. and Domnas, A. 1973. The properties of trehalase from the mosquito-parasitizing water mold Lagenidium sp. J. Invertebr. Pathol., 22, 313-320.

Moosa, H. 1977. Ammonia utilization by the aquatic Phycomycete, Lagenidium giganteum. Honors thesis to the Biology Curriculum, Univ. of N. Carolina.

Sparrow, F. K. 1970. "Aquatic" Phycomycetes, 2nd Edition. University of Michigan Press, Ann Arbor, Michigan.

Umphlett, C. J. 1973. A note to identify of a certain isolate of *Lagenidium* which kills mosquito larvae. Mycologia, *65*, 970-972.

Chapter 16

FUNGAL DISEASES OF MARINE CRUSTACEA

by Donald V. Lightner*

INTRODUCTION

Fungi are important decomposers in both terrestrial and aquatic ecosystems. The vast majority of species of fungi pose no threat to living animals and plants. Yet, other species have adapted to become true parasitic forms, and still others are capable of becoming opportunistic pathogens under certain conditions in the right host. It is the purpose of this review to discuss the fungus-caused diseases of marine invertebrates. Many of the so-called fungus-caused diseases in marine invertebrates are rather poorly understood and often the causative microorganism was not isolated, cultured and characterized. In many of these, implication of a fungal etiology was purely speculative. Such was the case in a much studied disease of oysters of the Gulf of Mexico and Atlantic Coasts in which the causative microorganism was thought to be a fungus for nearly three decades. Originally called Dermocystidium marinum (Mackin et al., 1950) and later Labyrinthomyxa marina (Mackin and Ray, 1966; Quick and Mackin, 1971), the organism was finally discovered by electron microscopic study of its spores to be a protozoan rather than a fungus. Now called Perkinsus marinus, the organism is classified as a member of the sporozoa (Levine, 1978).

This review of the pathogenesis of fungal diseases in invertebrates will, by necessity, be limited almost exclusively to the few diseases of crustaceans that are of known fungal etiology.

*Environmental Research Laboratory, University of Arizona, Tucson, Arizona 85706

Several excellent reviews have been published on fungus diseases of crustaceans and other invertebrates. The reader is referred to these works for further information on the taxonomy, host range, ecology, etc. of the fungi reviewed in the present paper (Johnson and Sparrow, 1961; Unestam, 1973; Bland, 1975; Sindermann, 1977).

ASCOMYCETES

Black Mat Syndrome

An ascomycete classified recently as *Trichomaris invadens* (Hibbits et al., 1981) is responsible for a disease called black mat syndrome (Sparks and Hibbits, 1979) in the tanner or snow crab, *Chionecetus bairdi* (Crustacea: Decapoda). Earlier workers (Van Hyning and Scarborough, 1973) called the condition "shell syndrome" in *C. bairdi* and isolated the microfungus *Phoma fimeti* from scrapings of the carapaces of four heavily infected tanner crabs. Johnson and Sparrow (1961) listed members of the genus *Phoma* as terrestrial and as marine or brackish water saprophytes occurring on substrates such as dead marsh grasses and among the fouling organisms on the hulls of ships. Because of distinct morphological differences between *Phoma* and the fungus present on tanner crabs with the disease, Sparks and Hibbits (1979) and Hibbits et al. (1981) concluded that *T. invadens* was the primary cause of the disease.

The syndrome is easily recognized in tanner crabs by the black mass of encrusting fungus (Fig. 16-1) on the external surface of the carapace. The encrustation begins as small discrete black spots on the dorsal cephalothorax and then enlarges and coalesces to cover much of the dorsal aspect of the cephalothorax, eventually spreading to the appendages and the ventral side of the crab. Microscopic examination of the sectioned carapace revealed that the mats consist of a mass of fungal hyphae and fruiting bodies (Fig. 16-2). Surface encrusting hyphae and fruiting bodies possess a black pigment which is apparently derived from the fungus and not from the host crab. Using Grocott's method for staining fungi (GMS), Sparks and Hibbits (1979) demonstrated that fungal hyphae readily perforated the thick chitinous carapace, invaded the underlying hypodermis and proliferated laterally, replacing virtually all hypodermal tissues (Fig. 16-3). The chitinous covering of the eyestalk was also penetrated by hyphae which proliferated by elongation and branching to essentially destroy the entire eyestalk hypodermis. As growth of the fungus continued, eyestalk muscles were invaded with resultant swelling, disorganization and destruction. Hemal spaces and nervous elements of the eye were also invaded. Study of GMS preparations of other tissues of infected tanner

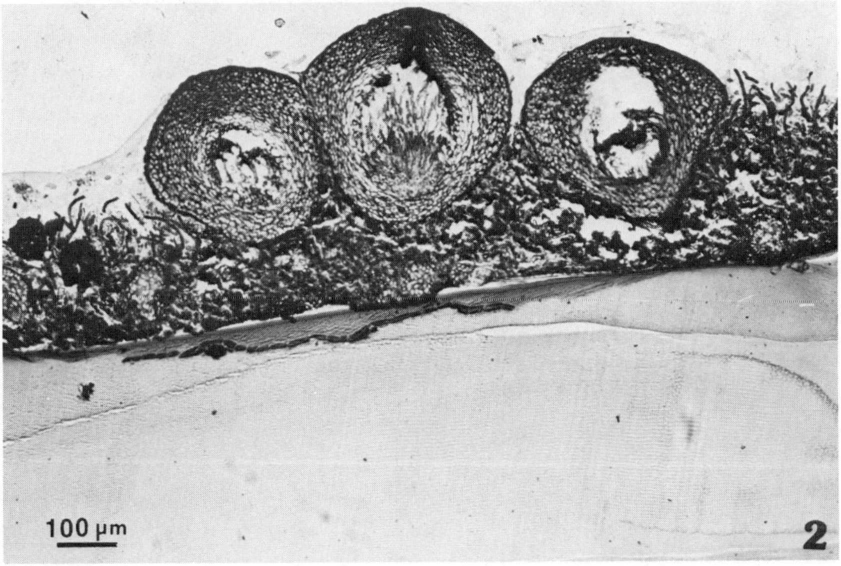

Figure 16-1. Tanner crab (Chionoecetes bairdi) with black mat syndrome. (From Sparks and Hibbits, 1979).

Figure 16-2. Histological section through the carapace of a tanner crab with black mat syndrome. The "black mat" is shown to be composed of a black pigment matrix in which are embedded fungal hyphae and fruiting bodies. GMS staining. (From Sparks and Hibbits, 1979).

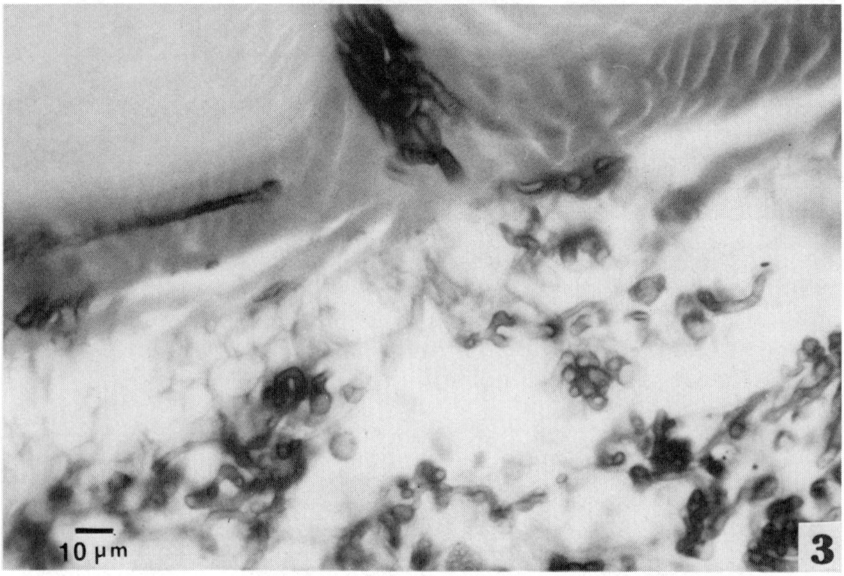

Figure 16-3. Histological section through the carapace of a tanner crab with black mat syndrome. Hyphae are shown penetrating the endocuticle layer of the carapace and proliferating laterally within the hypodermis. GMS staining. (From Sparks and Hibbits, 1979).

crabs revealed fungal hyphae in the connective tissue sheaths of the esophagus, stomach, heart, thoracic ganglion, hemopoetic tissue, ovary, hepatopancreas and antennal gland. Curiously, no pathological effects (other than tissue invasion) nor host inflammatory response were reported in any of the affected tissues.

In both surface encrustations and internal tissues, the hyphae were sparingly septate and sparingly branched. Hyphae grew tortuously and frequently formed tight bundles, especially in internal tissues. In addition to being nonpigmented, hyphae in internal infections were about half the diameter of those on the surface of the exoskeleton (approximately 2.3 to 2.6 μm). Fruiting bodies (asci) were confined to the surface mats and no evidence of reproduction, other than proliferative growth and branching of the hyphae, was observed in internal tissues (Sparks and Hibbits, 1979).

Because of the apparent ease with which *T. invadens* penetrates the tanner crab exoskeleton and its extensive proliferation and invasion of many internal tissues, Sparks and Hibbits (1979) assumed black mat syndrome to be a serious and perhaps virulent disease. It is unlikely that crabs with all or much of the carapace and eyestalk hypodermis destroyed can successfully molt. However, moribund tanner crabs with the disease have not yet been observed, despite up to 75% incidence of the disease in some populations of tanner crabs in Alaskan waters.

PHYCOMYCETES

Lagenidium Disease

In 1941, Dr. C. L. Newcombe (then Director of the Virginia Fisheries Laboratory) discovered a fungus in the ova of the blue crab, *Callinectes sapidus*. The following year the fungus was described and named *Lagenidium callinectes* by Couch (1942). Since then, *L. callinectes*, or similar *Lagenidium* species have been found to cause disease in the eggs or larvae of related crabs and in the barnacles *Chthamalus fragailis* and *Chelonibia patula* (Johnson, 1958; Johnson and Bonner, 1960), in numerous species of penaeid shrimp (Lightner and Fontaine, 1973; Lightner, 1977), in the dungeness crab, *Cancer magister* (Armstrong et al., 1976), and in the lobster, *Homarus americanus* (Nilson et al., 1976; Fisher et al., 1976). *Artemia salina* eggs and larvae have also been experimentally infected (Bland and Lightner, unpublished).

Lagenidium was found to only infect the peripheral portion of the egg mass ("sponge") of even heavily infected sponges of brooding female blue crabs (*C. sapidus*) in a survey of Chesapeake Bay (Rogers-Talbert, 1948). Rogers-Talbert concluded that *Lagenidium* could only be responsible for a maximum reduction of 25% in viability and hatching of the approximately two million eggs carried in the sponge, and therefore Lagenidium could not be regarded as a factor in fluctuations of crab populations in Chesapeake Bay. In laboratory studies, blue crab zoea which had hatched normally were infected and killed by Lagenidium (Couch, 1952; Rogers-Talbert, 1948).

Lagenidium has been particularly troublesome in aquaculture of marine decapod crustaceans, reportedly being responsible for major epizootics in the larvae of most of the penaeid shrimp reared throughout the world (Lightner, 1977) and in the lobster, H. americanus, and the dungeness crab, C. magister, reared on the Pacific Coast of the United States (Nilson et al., 1976; Armstrong et al., 1976). The pathology of Lagenidium in all host species is similar, but in wild crustacean populations, it is known as an egg parasite, while in aquaculture it affects both eggs and larvae.

Rogers-Talbert (1948) described the pathogenesis of Lagenidium in crab eggs in experimental infections. Egg-to-egg transmission was extremely rapid, with the entire exposed group of crab eggs being destroyed by the fungus in three or four days, even when only a few eggs were infected on the first day. In aquaria, the eggs of healthy egg-bearing crabs quickly became infected when crabs with diseased eggs were introduced. Infection is brought about by zoospores which settle on the eggs and encyst. The spore germinates by sending a delicate tube through the outer egg covering and the egg membrane. Once the tube has entered the protoplasm of the egg, it enlarges to the full diameter of the hypha, and as the hypha grows, the cyst becomes entirely empty. The empty cyst of the spores has a thick and persistent wall and can be recognized for several hours after infection. Several spores may infect one egg and the mycelium from one spore may produce several sporangia. Eggs thus infected are totally destroyed by the fungus within 48 hours, and are filled with a mycelium of branched, sparingly septate, thin-walled hyphae. Once the nutrient material of the egg has been totally exhausted, sporulation begins with emergence of the special extramatrical hyphae from the egg. Into the tips of these special hyphae flows sporoplasm from sporangia (located in the mycelium within the eggs) swelling the hyphal tip into a vesicle. Zoospores mature in the vesicle and are formed and released within about an hour (Couch, 1942; Rogers-Talbert, 1948).

Lagenidium is also known to infect the ova of brooding American lobsters, H. americanus, causing egg mortality rates similar to those previously described for the blue crab. As in the blue crab, the fungus occurs primarily on the surface of the lobster's egg mass, and the rate of disease proceeds much more slowly in egg infections than in larval infections

(Fisher et al., 1978). In the lobster, Lagenidium infections may be severe with mortality rates in laboratory-reared lobster larval populations exceeding 90% (Nilson et al., 1976).

Lagenidium is an important disease in the culture of other decapods (Bland, 1975). Armstrong et al. (1976) found the fungus caused about 40% mortality of the second stage zoea of the dungeness crab, C. magister. In the penaeid shrimp, mortality rates of up to 100% have been reported for larvae of many of the culture penaeid species (Lightner, 1977). Lagenidium infections of penaeid larvae may serve as a model for the pathogenesis of this disease in the larvae of decapods. Larvae of the brown shrimp, Penaeus aztecus, were experimentally infected with Lagenidium by exposing a population of larvae to zoospores of the fungus. Infection occurs when a zoospore settles upon a susceptible larvae, encysts, and germinates. If the germinal hypha succeeds in penetrating the cuticle, an infection is established. The earliest sign of infection in natural or artificially infected larval penaeid shrimp is the presence of hyphae within the appendages. Often, an empty cyst case may remain on the end of a hypha on the larval shrimp's surface (Fig. 16-4).

Once Lagenidium has successfully established infection, the fungal mycelium gradually invades and replaces nearly all of the striated muscle tissue of the larval shrimp. The thorax, abdomen, swimming appendages and even the eyestalks become filled with hyphae (Fig. 16-5). Massive tissue destruction, particularly of the striated muscle, results in immoblization of affected shrimp as much as one hour before death. An occasional feeble movement of an appendage or contraction of the gut musculature is the only sign of life in these shrimp.

Soon after the death of an infected larva, the process of sporulation begins with the emergence from the larva of extramatrical hyphae or discharge tubes. The apical end of these discharge tubes swell, forming a vesicle as it fills with units of sporoplasm which flow from a sporangium located on an intramatrical hypha which is within the body of the larva. Zoospores develop from the amorphous mass of sporoplasm in the vesicle and are released when their movements rupture the vesicle. Zoospores are reniform, motile by two flagella that originate from the lateral groove, and are 8.7 by 12.0 μm in size.

Siropidium

Another phycomycetous fungus, Siropidium sp., causes a disease similar to Lagenidium in cultured larvae penaeid shrimp (Lightner, 1977) and a related fungus, S. zoophthorum, causes a severe disease in cultured larval oysters, Crassostrea

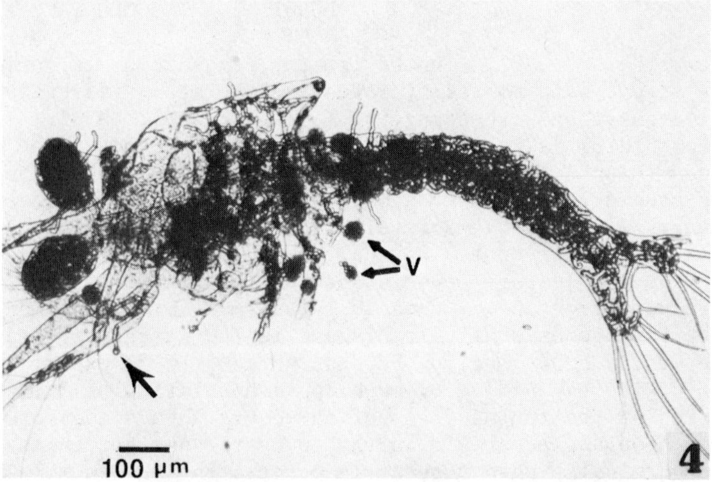

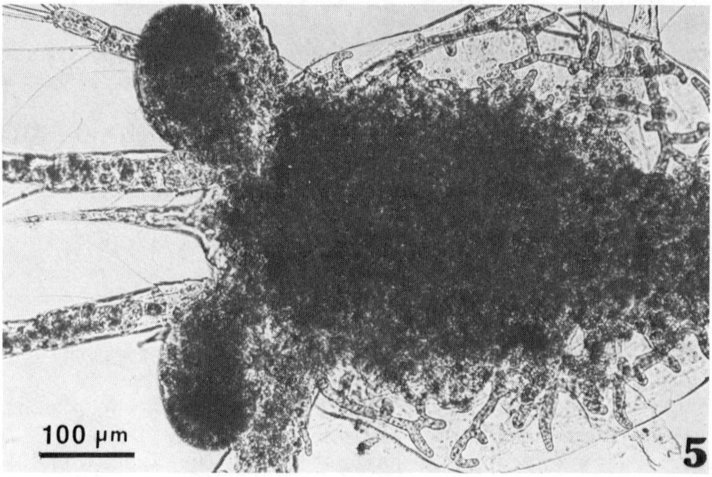

Figure 16-4. Second protozeal stage larval white shrimp (*Penaeus setiferus*) that is heavily infected with *Lagenidium callinectes*. Extramatrical hyphae, some with terminal vesicles (V), are shown protruding from the shrimp. Also shown is a spore case (arrow) from which the infection may have been initiated. No stain. (From Lightner and Fontaine, 1973).

Figure 16-5. Larval white shrimp with hyphae of *L. callinectes* occupying much of the space in the cephalothorax, eyestalks and appendages. No stain. (From Lightner and Fontaine, 1973 ; Figures 16-1 through 16-5 reprinted by permission of Academic Press.)

virginica (Sindermann, 1977). In both the larval shrimp and larval lobsters, the disease follows a pathogenesis similar, if not identical, to that of Lagenidium in larval shrimp. Siropidium, however, is easily distinguished from Lagenidium by its method of zoospore formation and discharge.

Haliphthoros

A phycomycetous marine fungus, Haliphthoros milfordensis, is known to be a serious pathogen of cultured European and American lobster, and to infect the ova and developing eggs of the oyster drill, a gastropod Urosalpinx cinerea, and the eggs of the pea crab, Pinnotheres pisum. The fungus has been found experimentally to be capable of infecting the eggs of the blue crab, Callinectes sapidus, the egg, embryo and adult stages of the brine shrimp, Artemia salina, and adults of the pink shrimp, Penaeus duorarum, and to cause 100% mortality for each of these species within five days (Tharp and Bland, 1977). In cultured postlarval lobsters, the fungus has caused 46% mortality in 22 days (Fisher et al., 1975).

In crustaceans, H. milfordensis may penetrate the exoskeleton through small wounds, or the zoospores may have chitinase and be capable of creating portals of entry. Germination of encysted spores on the crustacean exoskeleton can occur in 2-3 hours (Tharp and Bland, 1977). In the lobster, the initial infection normally occurs at the soft flexible joints of the appendages, on the ventral abdomen and, most frequently, on the inner wall of the branchial chamber (Fisher et al., 1975). The greatest destruction of tissue occurs at the latter site (Figs. 16-6, 16-7, 16-8). A strong cellular inflammatory response accompanies growth of the fungus in the lobsters' tissues, with invading hyphae becoming encapsulated and melanized by hemocytes. As a result, the site of infection becomes reddish-brown and is grossly visible. Death is thought to result from tissue destruction and from impairment of molting by adhesions formed by inflammatory cells between the exoskeleton and underlying tissues (Fisher et al., 1978). Young postlarval lobsters are not always successful in their attempts to encapsulate the fungus and frequently succumb to invasion. Larger lobsters are less susceptible, probably due either to a thicker exoskeleton or to a more successful hemocytic response. Although larval lobsters should be the most susceptible, only a few cases have been observed, and no serious epizootics have been reported (Fisher et al., 1978).

Haliphthoros sp. is a potentially serious pathogen of penaeid shrimp. Bland (1975) found Haliphthoros sp. responsible for an epizootic of aquarium-held adult pink shrimp, P. duorarum, in which all 24 shrimp held in the tank died within a two-week

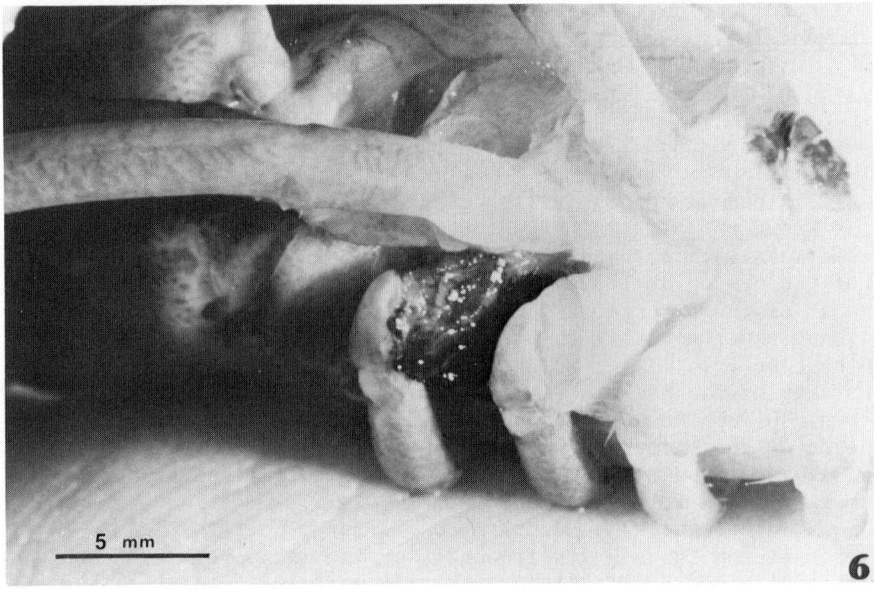

Figure 16-6. A: External melanization on the ventral abdomen of a juvenile lobster (Homarus americanus) caused by infection with the fungus, Haliphtoros milfordensis. (From Fisher et al., 1978).

Figure 16-7. The lobster's inflammatory response to the invading hyphae results in hemocyte encapsulation and melanized hyphae within the tissues. No stain. (From Fisher et al., 1978).

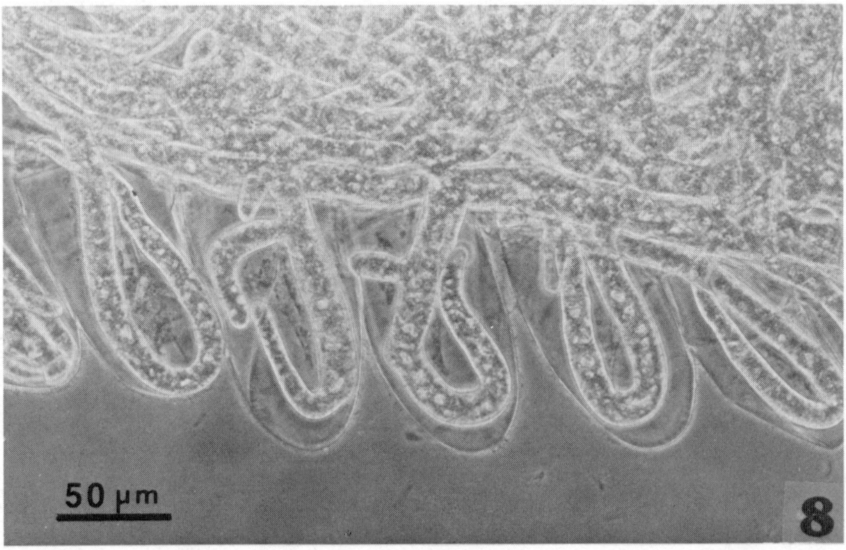

Figure 16-8. Hyphae of H. milfordensis, shown within the cuticle of several gill processes, has completely consumed the soft gill tissues. (From Fisher et al., 1978; Figures 16-6 through 16-8 reprinted by permission of Elsevier Scientific Publishing Co.)

period. The infection was confined mostly to the gills, but in heavily infected individuals, the eyes were also infected. There was no apparent invasion of the fungus into the muscle tissue.

Ichthyosporidium

Ichthyosporidium, an important fungus in marine and freshwater fishes, has been found on rare occasions to infect the marine copepods Calanus finmarchicus, Paracalanus parvus, Calusocalanus arcuicornis and Acartia clausi (Jepps, 1937, in: Johnson and Sparrow, 1961; Reichenbach-Klinke and Landolt, 1973). While the significance of finding a few copepods infected with the fungus may only be speculative, these copepods may be important in the transmission cycle of the fungus in its principal hosts, which are fish.

Atkinsiella

An unusual phycomycete, Atkinsiella dubia, has been described from the egg mass of several crab species of the genera Hyas, Oregonea, Pugettia, Chorilia, Skyra, Chianectes and Cancer from the coastal waters of the Pacific Northwest. In coastal waters of New England, the fungus (originally known there as Plectospira dubia (Atkins, 1954) until redescribed as Atkinsiella dubia by Vishniac (1958), was found infecting the eggs of the crab Pinnotheres pisum. Under experimental conditions, A. dubia was found by Atkins (1954) to infect eggs of the crustaceans Typton spongicola, Crangon vulgaris, Leander serratus, Macropodia sp. and Portunus depurator. All of the cases of egg parasitism by A. dubia were from crustacean eggs that were experimentally exposed to the fungus or from egg sponges stored at refrigerator temperatures for several weeks (Sparrow, 1973). No signs of infection by A. dubia have been observed in the adult brooding crabs from which the eggs were taken and later found to harbor the fungus (Atkins, 1954). Therefore, while this most interesting fungus is a potentially serious pathogen of crustacean eggs, its importance has yet to be demonstrated.

Atkins (1954) described the pathology of A. dubia in the eggs of the crab species listed above. In every case, the first sign of infection in the egg (even in eggs containing embryos with the heart beating) was the development of a space between the inner and outer egg membranes, and in that space the young infecting hyphae first appeared. The space increased with the growth of infection and this, together with some bulging of the sporangia beyond the egg, increased the size of infected eggs. The hyphae were generally stout with widths of about 27 to 50 µm, irregularly branched, and contained groups of refractile granules in an otherwise hyaline cytoplasm. Within the host the growth soon became irregular, with the branches from the mycelium becoming inflated and bulbous, forming sprangia that were up to 100 µm wide. The inflated sporangia were, for the most part, within the host tissue, but characteristically bulged beyond the egg making much of the sporangium extramatrical. Efferent hyphae formed on the sporangia, and from the sporangia and in the proximal portions of these hyphae, zoospores formed. Mature zoospores are eventually ejected from pores at the tips of the efferent hyphae, thus completing the cycle once a suitable new host is infected (Atkins, 1954). *Atkinsiella dubia* has not been described from cultured crustaceans, although Lightner (unpublished) found a fungus tentatively identified by Bland (personal communication) as *A. dubia* in captive wild adult *Penaeus aztecus*, where it caused yellowish-white friable lesions in the cuticle (Figs. 16-9 and 16-10).

FUNGI IMPERFECTI

Fusarium

Only one member of this group of common fungi has been recognized as a serious pathogen of crustaceans. That fungus, *Fusarium* sp., has been described from numerous crustacean species from many locations in the world. Its distribution appears to be ubiquitous in both marine and freshwater environments. Only one species, *F. solani*, has so far been identified from crustacea by taxonomists; those isolates were from a shrimp in Japan (Egusa and Ueda, 1972) and shrimp and lobsters in North America (Lightner et al., 1979; Claydon et al., 1977a).

A disease resulting from *Fusarium* infection in crustacea was first described as "burn spot disease" in the European crayfish (*Astacus astacus*) (see Chapter 17) and in a crab (Mann and Pieplow, 1938, *in*: Unestam, 1973). Mann and Pieplow identified the causative fungus in these crustaceans to be *Ramularia astaci*. Sordi (1958) reported a disease of the gills of the European lobsters *Palinurus vulgaris* and *Homarus vulgaris* caused by the fungi *R. branchialis* and the closely related fungus, *Didmyaria palinuri*. Since then, the genus *Ramularia* has been listed as a synonym for the genus *Fusarium* (Booth, 1971).

The advent of experimental aquaculture of various marine and freshwater decapod crustaceans has been accompanied by reports of mycoses caused by *Fusarium* spp. in many of the decapods being cultured. Reports of infections caused by *Fusarium* spp. are listed in Table 16-1.

TABLE 16-1. KNOWN HOSTS AND GEOGRAPHIC LOCATIONS OF REPORTED *FUSARIUM* SP.-CAUSED DISEASE IN DECAPOD CRUSTACEANS

Host Species	Lesion Type	Geographic Location	Reference
European crayfish (*Astacus astacus*)	Brown or black spots on exoskeleton, penetrating into underlying masculature	Germany	Mann and Pieplow, 1958 (*In*: Unestam, 1973)
European lobster (*Homarus vulgaris*) and Spiny lobster (*Palinurus vulgaris*)	Gill infections	Italy	Sordi, 1958
American lobster (*H. americanus*)	Orange to black expanding lesions beginning on cuticle, appendages and gills and penetrating into underlying tissues and musculature	New York, California	Lightner and Fontaine, 1975 and Fisher et al., 1978
Freshwater prawn (*Macrobrachium rosenbergii*)	Blackened cuticular lesions	Florida	Burns et al., 1979
Kurma prawn (*Penaeus japonicus*)	Blackened gill lesions	Tahiti, Japan	Egusa and Ueda, 1972, Guary et al., 1974 and Aquacop, 1977
Pink shrimp (*P. duorarum*)	Blackened expanding lesions in gills, on cuticle and appendages, expanding into underlying tissues and musculature	Florida	Nimmo et al., 1977
	Unblackened lesions in gills and on antennal scales	Texas	Johnson, 1974
White and pink shrimps (*P. setiferus* and *P. duorarum*)	Blackened lesions in gills	North Carolina	Bland, *In*: Lightner, 1977
Mexican white shrimp (*P. vannamei*)	Expanding white plaque on eyes	Florida	Laramore et al., 1977
California brown (*P. californiensis*), blue (*P. stylirostris*), and Mexican white shrimp (*P. vannamei*)	Expanding blackened lesions on cuticle, appendages or in the gills, penetrating into underlying tissues and musculature	Mexico	Lightner, 1975 and Lightner et al., 1979

Figure 16-9. An adult brown shrimp (*P. aztecus*) with probable *Atkinsiella dubia* lesions on the cuticle of the abdomen.

Figure 16-10. A wet mount of a probable *A. dubia* lesion in the carapace of the brown shrimp in Fig. 16-10. Note the peculiar sporangium and discharge tubes which are characteristic of *A. dubia*. No stain.

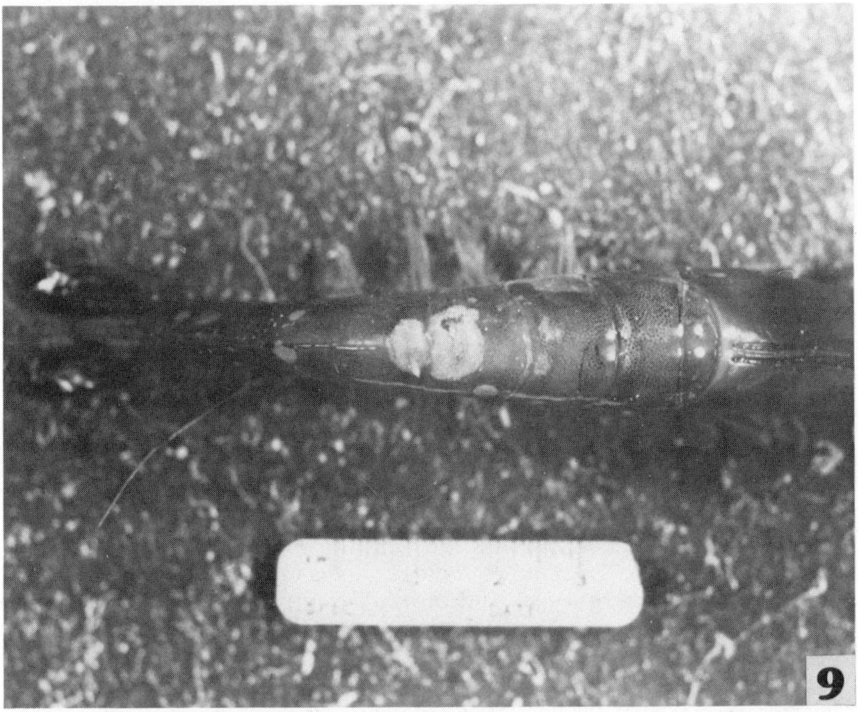
9

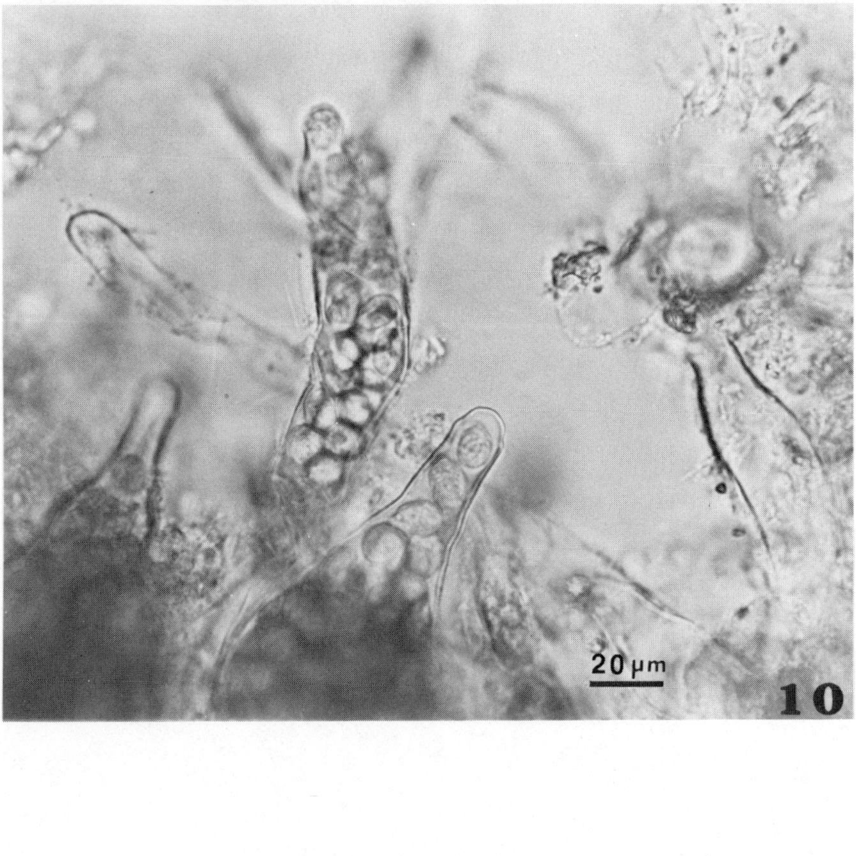
20 μm
10

In decapod crustaceans, Fusarium typically is an opportunistic pathogen, invading surface tissues damaged by wounds, abrasions, chemicals or other microorganisms. Hence. Fusarium lesions typically begin as inconspicuous focal lesions in the gills, on the appendages or on the exoskeleton proper. As the lesions expand, they become increasingly inflamed and become grossly visible as they become melanized. These melanized lesions have given rise to the names "black gill disease" (Egusa and Ueda, 1972), "burn spot disease" and "black spot disease" (Mann and Pieplow, 1938, in: Unestam, 1973) which have been given to Fusarium-caused disease in various decapod crustaceans.

The pathology of Fusarium is similar in all decapods. Therefore, a description of the pathogenesis in the penaeid shrimp will be given here as a representative model of that occurring in all the decapods.

In the penaeid shrimp, F. solani infects dead or damaged tissue, such as wounds and abrasions resulting from crowding, gills damaged from chemicals or lesions from other disease processes such as bacterial "shell disease." Fusarium solani lesions are visible as expanding cuticular or subcuticular lesions that are usually blackened with melanin, occasionally with reddened necrotic centers, that occur on the general body surface, in the gills or on the tips of the appendages (Figs. 16-11 and 16-12). Once established, the infection is subacute to chronic, usually progressive, and eventually leads to death of the infected host. Cause of death can be attributed to tissue destruction, due to toxins (Claydon et al., 1977a, b) produced by the fungus or to secondary bacterial infections.

Fresh mounts of F. solani-infected tissue typically contain hyphae and conidia of the fungus, and demonstration of the characteristic macroconidia provides for a positive diagnosis of a Fusarium-caused mycosis (Fig. 16-13). Histopathology shows F. solani-caused lesions to be composed of very large numbers of infiltrating hemocytes, necrotic tissue debris, and fungal hyphae and conidia (Figs. 16-14 and 16-15). Typically, hyphae are encapsulated by multiple layers of hemocytes, which frequently are melanized (Fig. 16-16) (Lightner et al., 1979; Lightner et al., 1981).

The epizootiology of Fusarium disease in the penaeids was found to be dependent on host species and host age. Of the penaeid species reared at the University of Arizona shrimp culture facility at Puerto Peñasco, Sonora, Mexico, the California brown shrimp (Penaeus californiensis) was found to have the highest susceptibility to infection, with the blue (P. stylirostris) and the Mexican white shrimp (P. vannamei) being considerably less susceptible (Table 16-2). Populations of P. californiensis were found to be susceptible at a younger age to F. solani, to have a higher incidence of infection, and to have a higher mortality rate per unit of time than was

TABLE 16-2. SUSCEPTIBILITY AND INCIDENCE OF *FUSARIUM SOLANI* INFECTIONS IN THREE SPECIES OF *PENAEUS* AT PUERTO PEÑASCO, MEXICO

Species	Relative Susceptibility	Susceptible Age	Incidence/1000 Population
Penaeus californiensis[a] (Calif. Brown)	High	30-40 weeks	55[a]
P. stylirostris[b] (Blue)	Moderate	40-50 weeks	6.5[b]
P. vannamei[c] (Mexican White)	Low	68 weeks	3.5[c]

[a]Born 4/2/76; for January to March 1977.

[b]Born 5/1/76; for January to March 1977.

[c]Born 9/4/75; for January to March 1977.

(from Lightner et al., 1979)

Figure 16-11. California brown shrimp (*P. californiensis*) infected with *Fusarium solani* in the gill region especially above the 4th and 5th pereiopods.

Figure 16-12. The abdomen of a California brown shrimp with a large well developed *F. solani*-caused lesion. (From Lightner et al., 1979).

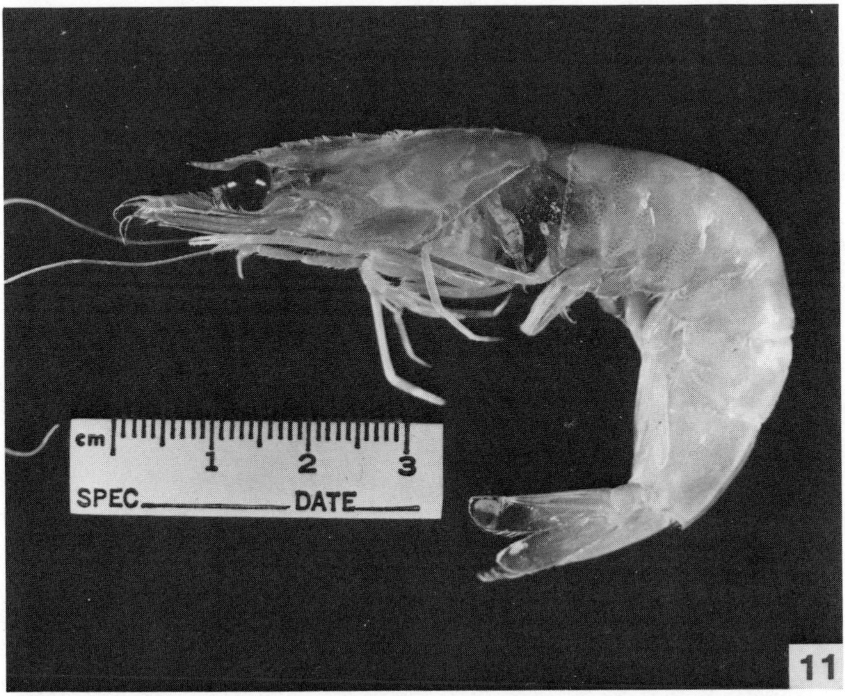

11

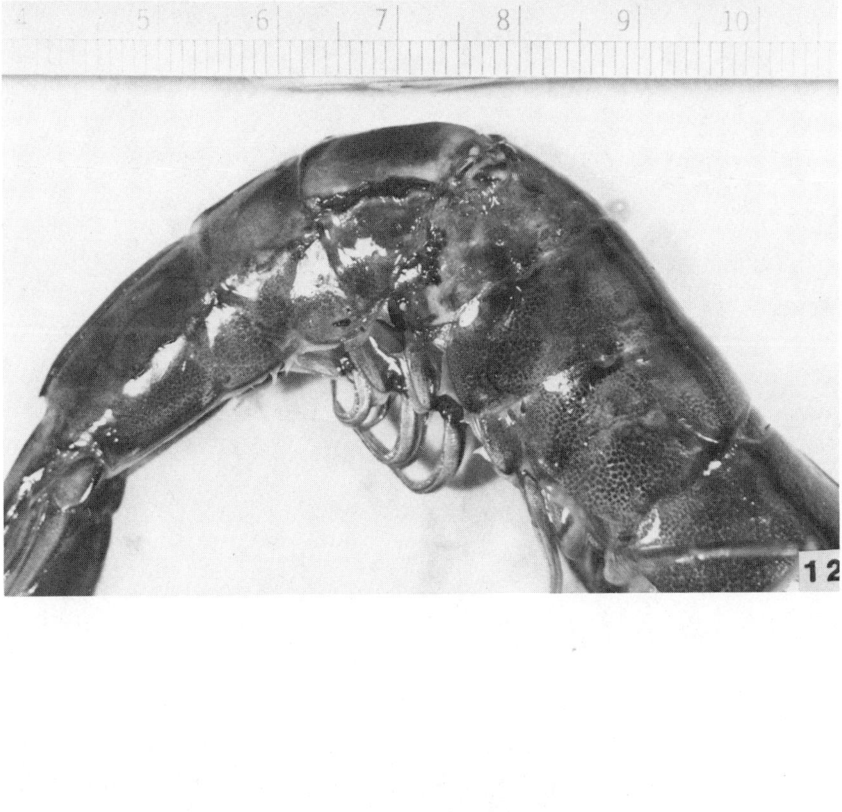

12

Figure 16-13. Wet mount of macro- and microconidia from the surface of a *F. solani*-caused lesion in the gills of a California brown shrimp. No stain. (From Lightner et al., 1979).

Figure 16-14. Histological section through a cuticular lesion due to *F. solani* in the brown shrimp *P. californiensis*. Note the intense cellular inflammatory response and the encapsulated and melanized hyphae (arrows) present within the lesion. Masson trichrome.

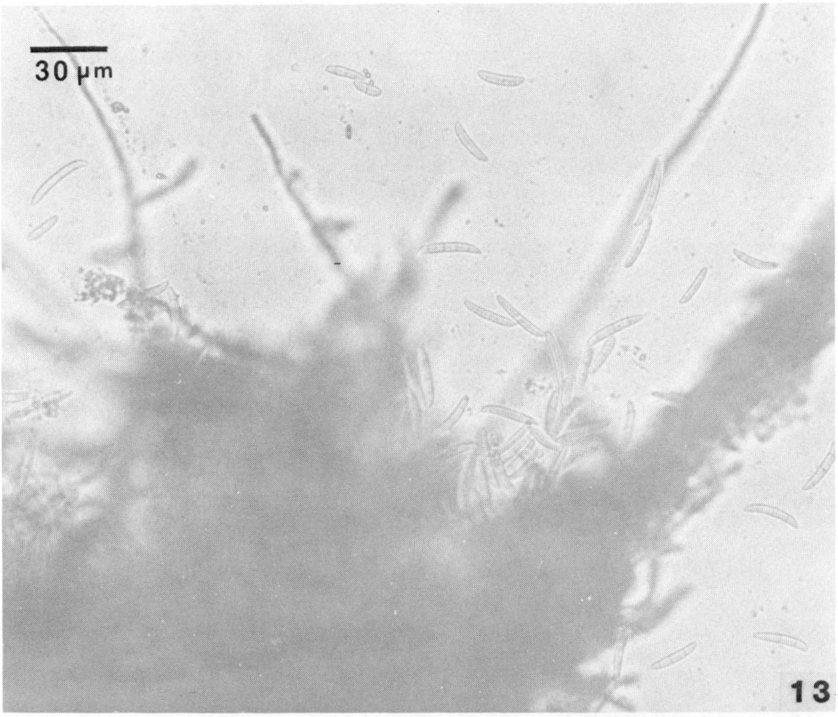
30 µm
13

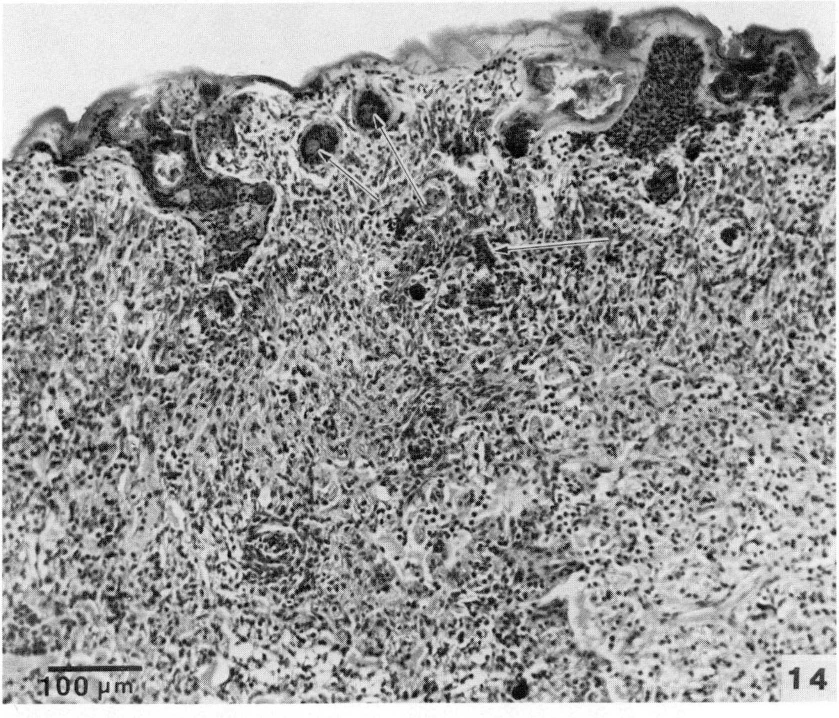
100 µm
14

Figure 16-15. Light photomicrograph of hemocyte encapsulated hyphae (arrows) of F. solani in a cuticular lesion on P. californiensis. Granular and hyaline hemocytes are shown involved in capsule formation and free in the tissue spaces. Orthotoluidine blue staining.

Figure 16-16. Electron micrograph of a melanized, hemocyte encapsulated F. solani hypha (H) within a cuticular lesion of a California brown shrimp. Lead and uranylacetate staining.

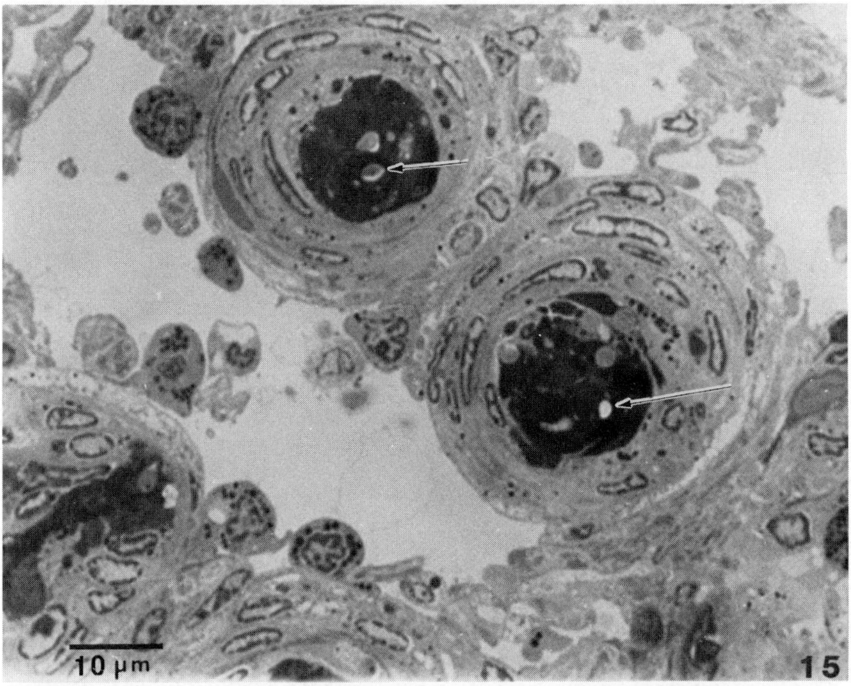
10 μm
15

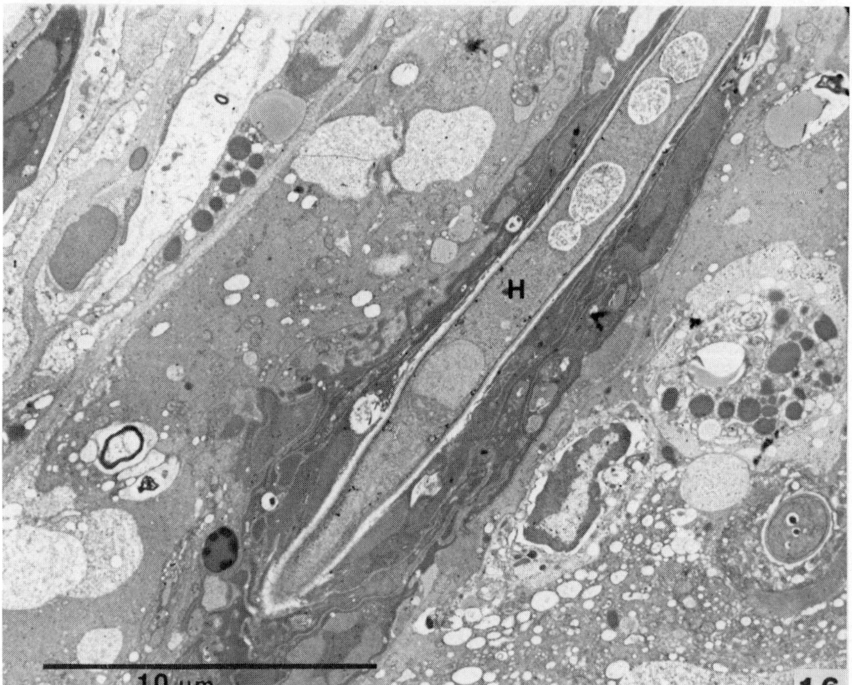
H
10 μm
16

observed in the other two species (Fig. 16-17). Penaeus stylirostris was found to be intermediate in susceptibility and mortality rate between P. californiensis and P. vannamei (Lightner et al., 1979).

The pathogenesis of F. solani was studied in artificially infected penaeid shrimp (Lightner et al., 1981). Shrimp in these studies were artificially infected with F. solani by wounding the experimental shrimp by clipping off the edge of the cuticle of the carapace, a pleural plate, a small portion of an appendage, or a gill process, and then swabbing the wound immediately with a cotton swab that had been dipped in a dense saline suspension of F. solani conidia (> 10^4 conidia/ml). Within about 1 minute the shrimp's blood issuing from the wound clotted, trapping the introduced conidia on the wound's surface. Using this technique, patent Fusarium infectionswere produced with success rates of 100% within 14 days in P. californiensis weighing 15 g or more. About 30% successful infection rates were produced in 90 days in 20 g P. stylirostris. Developing Fusarium lesions (as determined by the presence of hyphae and hemocytic inflammation to those hyphae) were detected in the wound vicinity in artificially infected P. californiensis by histological examination as early as 5 hours post-infection. Death of greater than half of the experimental P. californiensis in this study had occurred by 24 days post-infection (Lightner et al., 1981).

Before the successful "wound and swab" method of artificially establishing Fusarium infections was developed, attempts were made to induce the disease by injection of suspensions of conidia of F. solani into healthy P. aztecus and P. setiferus (Solangi and Lightner, 1976). None of the experimental shrimp so exposed developed patent Fusarium lesions. Instead, the challenged shrimp of both species were able to completely eliminate the introduced conidia by hemocytic activity. The conidia were encapsulated and then lysed when the challenge dose was about 10^3 conidia or less per shrimp (Fig. 16-18; Table 16-3). However, when the challenge dose of conidia exceeded about 10^4 conidia per shrimp, shrimp of both species died within 24 hours post-infection. These deaths were probably due to mechanical blockage of hemolymph circulation through the gill lamellar tips where large numbers of conidia had lodged because they were too large to make the turn from the afferent hemolymph channel to the efferent hemolymph channel. Shrimp that received sufficient Fusarium inoculum but did not die within 24 hours due to mechanical blockage of the gills, died of massive Fusarium invasion of the tissues within 1 to 6 days post-infection. Typically, these infections were accompanied by little or no hemocytic response, probably because most of the circulating hemocytes had been removed from the circulation to encapsulate the large number of conidia that were introduced

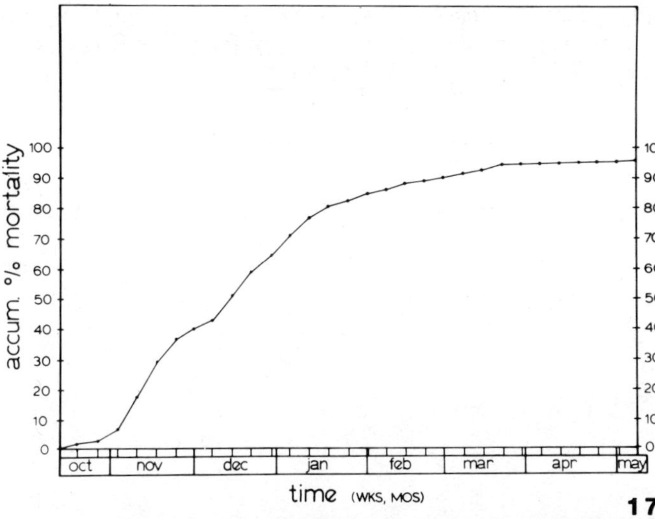

17

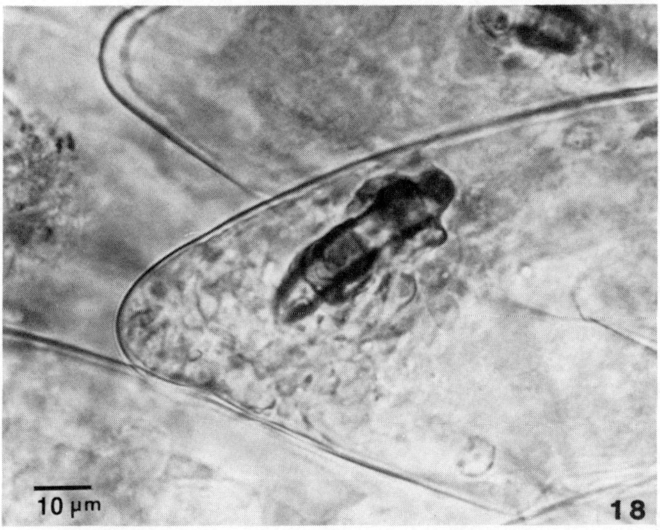

Figure 16-17. Accumulated percent mortality by week in a population of raceway-reared California brown shrimp with a severe epizootic due to F. solani. (From Lightner et al., 1979).

Figure 16-18. Photomicrograph of a melanized, hemocyte encapsulated F. solani macroconidium in the hemocoel of a gill lamellae. No stain. (From Solangi and Lightner, 1976; reprinted by permission of Academic Press, Inc.)

TABLE 16-3. INFECTION OF BROWN (*P. AZTECUS*) AND WHITE (*P. SETIFERUS*) SHRIMP WITH VARIOUS CONCENTRATIONS OF *FUSARIUM SOLANI*

No. spores/ 0.1 ml	No. shrimp	Hr. to first mortality	No. Survivors after 5 days
		Brown, *P. aztecus*	
24	5	24	4
240	5	24	4
240,000	5	12	2[a]
600,000	5	8	2
900,000	5	6	0[b]
		White, *P. setiferus*	
24	5	--	5
240	5	24	4
240,000	5	36	3
600,000	5	12	2
900,000	5	12	1[c]

[a]One apparent death (trapped in the air-lift pipe).

[b]After 72 h.

[c]Died on the sixth day

(from Solangi and Lightner, 1975; reprinted by permission of Academic Press, Inc.)

(Table 16-2). Hence, patent, slow developing *Fusarium* lesions, characterized by massive hemocytic inflammation, could not be induced by the injection of conidia of *F. solani* (Solangi and Lightner, 1976).

In another study, blood chemistries, hemocyte cell count and hematopoietic mitotic index were investigated in normal and in naturally and artificially *F. solani*-infected blue shrimp and California brown shrimp (Lightner et al., 1981). Significant differences in hemolymph alkaline phosphatase, Serum Glutamic-Oxaloacetic Transaminase (SGOT), glucose, total protein and hemocyte count were observed between uninfected control and *Fusarium*-infected shrimp of both species, and among infected shrimp, depending on the degree of severity of the *Fusarium* infection (Table 16-4). Hematopoietic mitotic indices were greatly increased, as would be expected, in *Fusarium*-infected shrimp (Table 16-5). Similar results were found in adult blue shrimp with natural *Fusarium* infections (Lightner et al., 1981).

Fusarium solani was shown to produce several mycotoxins that are highly toxic to insects (Claydon et al., 1977a, b). The naphthazarin pigments anhydrofusarubin, javanicin and fusarubin as well as fusaric acid were toxic when injected into adult blowflies (*Calliphora erythrocephala*). Various strains of *F. solani* differ in the amount of naphthazarin pigments they produce when cultured under identical conditions on Sabourand dextrose agar. Limited pathogenetic studies indicate that those isolates producing more of the naphthazarin pigments are the most pathogenic to penaeid shrimp, while those that produce little or none of the pigments are unable to produce advanced infections and, hence, are little more than epicommensal organisms (Lightner, unpublished). Perhaps these pigmented mycotoxins are most responsible for the striking histopathology and clinical changes in blood chemistries observed in penaeid shrimp with patent *Fusarium* disease.

TABLE 16-4. HEMOLYMPH CHEMISTRY AND CELL COUNT VALUES FOR *FUSARIUM SOLANI*-INFECTED AND UNINFECTED CALIFORNIA BROWN SHRIMP (*P. CALIFORNIENSIS*)

Hemolymph Test	Control	Infected
Alkaline Phosphatase	0 - 46 U/l	14 - 95 U/l*
SGOT	0 - 113 U/l	55 - 300 U/l*
Glucose	7 - 363 mg/dl	4 - 154 mg/dl*
Ammonia (BUN)	3 - 19 mg/dl	2 - 22 mg/dl
Total Hemocyte Count (Cells/mm)		
Clottable Blood	5,800-40,800	4,800-53,600
Unclottable Blood	---	883- 8,350*
Total Blood Protein		
Clottable Blood	5-15 g/dl	9-15 g/dl
Unclottable Blood	---	4- 8 g/dl*

*Statistically significant difference by t-test at 0.005 level.

(from Lightner et al., 1981)

TABLE 16-5. MITOTIC INDEX IN STAINED SECTIONS OF EPIGASTRIC AND MAXILLIPED HEMATOPOIETIC TISSUES OF *FUSARIUM SOLANI*-INFECTED AND UNINFECTED *P. CALIFORNIENSIS*

Hematopoietic Tissue	Mitotic Index[1]			
	Control		Infected	
	Range	Average	Range	Average
Epigastric	0	0	14	6.2*
Maxilliped	0-2	0.6	0-36	11.9*

[1]No. mitotic figures/450x field.

*Statistically significant difference by t-test at 0.005 level.

(from Lightner et al., 1981)

REFERENCES

Aquacop, 1977. Observations on diseases of crustacean cultures in Polynesia. Proc. 8th Ann. Mtg. World Mariculture Soc., 8, 685-703.

Armstrong, D. A., Buchanan, D. V. and Caldwell, R. S. 1976. A mycosis caused by Lagenidium sp. in laboratory-reared larvae of the dungeness crab, Cancer magister, and possible chemical treatments. J. Invertebr. Pathol., 28, 329-336.

Atkins, D. 1954. A marine fungus Plectospira dubia n. sp. (Saprolegniaceae), infecting crustacean eggs and small crustacea. J. Marine Biol. Assoc. U.K., 33, 721-732.

Bland, C. E. 1975. Fungal diseases of marine crustacea. Proc. U.S.-Japan Natural Resources Prog., Symposoim on Aquaculture Disease, Tokyo. pp. 41-48.

Booth, C. 1971. The Genus Fusarium. Commonwealth Agricult. Bureau. 237 pp.

Burns, C. D., Berrigan, M. E. and Henderson, G. E. 1979. Fusarium sp. infections in the freshwater prawn Macrobrachium rosenbergii (DeMan). Aquaculture, 16, 193-198.

Claydon, N., Grove, J. F. and Pople, M. 1977a. Insecticidal secondary metabolic products from the entomogenous fungus Fusarium solani. J. Invertebr. Pathol., 30, 216-223.

Claydon, N., Grove, J. F. and Pople, M. 1977b. Fusaric acid from Fusarium solani. Phytochem., 16, 603.

Couch, J. N. 1942. A new fungus on crab eggs. J. Elisha Mitchell Sci. Soc., 58, 158-162.

Egusa, S. and Ueda, T. 1972. A Fusarium sp. associated with black gill disease of the Kuruma prawn, Penaeus japonicus Bate. Bull. Japan. Soc. Sci. Fisheries, 38, 1253-1260.

Fisher, W. S., Nilson, E. H. andShelser, R. A. 1975. Effect of the fungus Halipthoros milfordensis on the juvenile stages of the American lobster (Homarus americanus). J. Invertebr. Pathol., 26, 41-45.

Fisher, W. S., Nilson, E. H., Follett, L. F. and Shelser, R. A. 1976. Hatching and rearing lobster larvae (Homarus americanus) in a disease situation. Aquacult., 7, 75-80.

Ganaros, A. E. 1957. Marine fungus infecting eggs and embryos of Urosalpinx cinerea. Science, 125, 1194.

Guary, J. C., Guary, M. and Egusa, S. 1974. Infections bacteriennes et fongiques de crustaces peneides (Penaeus japonicus Bate) en elevage. Colloque sur L'Aquaculture. Actes de Colloques. No. 1, 1974 CNEXO ed. pp. 125-135.

Hibbits, J., Hughes, S. E. and Sparks, A. K. 1981. Trichomaris invadens gen. et sp. nov., an ascomycete parasite of the tanner crab Chinoecetus bairdi (Crustacea: Brachyura). Can. J. Bot., submitted.

Johnson, S. K. 1974. Fusarium sp. in laboratory-held pink shrimp. Texas A&M Univ. Agr. Ext. Serv. Fish Disease Diagnostic Laboratory Leaflet No. 51. 1 p.

Johnson, T. W. 1958. A fungus parasite in ova of the barnacle Chthamalus fragilis denticulata. Biol. Bull., 114, 205-214.

Johnson, T. W. Jr. and Sparrow, F. K. Jr. 1961. Fungi in Oceans and Estuaries. J. Cramer, Hafner Publ. Co., N.Y. 668 pp.

Johnson, T. W. Jr. and Bonner, R. R. Jr. 1960. Lagenidium callinectes Couch in barnacle ova. J. Elisha Mitchell Sci. Soc., 76, 147-149.

Laramore, C. R., Barkate, J. A. and Persyn, H. O. 1977. Fusarium infection in eyes of mature shrimp (Penaeus vannamei). Texas A&M Univ. Agr. Ext. Serv. Fish Disease Diagnostic Laboratory Leaflet No. 59. 1 p.

Levine, N. D. 1978. Perkinsus gen. n. and other new taxa in the protozoan phylum Apicomplexa. J. Parasitol., 64, 549.

Lightner, D. V. 1975. Some potentially serious disease problems in the culture of penaeid shrimp in North America. Proc. U.S.-Japan Natural Resources Program, Symposium on Aquaculture Diseases, Tokyo. pp. 75-97.

Lightner, D. V. 1977. Larval mycosis of shrimps. In Disease Diagnosis and Control in North American Marine Aquaculture, C. J. Sindermann, ed. Devel. in Aquacult. and Fisheries Sci. Elsevier Sci. Publ. Co., N.Y. 6, 36-41.

Lightner, D. V. and Fontaine, C. T. 1973. A new fungus disease of the white shrimp Penaeus setiferus. J. Invertebr. Pathol., 22, 94-99.

Lightner, D. V. and Fontaine, C. T. 1975. A mycosis of the American lobster, Homarus americanus, caused by a Fusarium sp. J. Invertebr. Pathol., 25, 239-235.

Lightner, D. V., Moore, D. and Danald, D. A. 1979. A mycotic disease of cultured penaeid shrimp caused by the fungus Fusarium solani. In Proc. 2nd Biennial Crustacean Health Workshop, D. H. Lewis and J. K. Leong, eds. Texas A&M Univ. Sea Grant Publ. No. 79-114. pp. 137-158.

Lightner, D. V., Danald, D. A., Redman, R. M. and Hose, J. E. 1981. Pathogenesis of Fusarium solani in natural and artificial infections in two penaeid shrimp. Proc. World Maricult. Soc., in press.

Mackin, J. G. and Ray, S. M. 1966. The taxonomic relationship of Dermocystidium marinum Mackin, Owen and Collier. J. Invertebr. Pathol., 8, 544-545.

Mackin, J. G., Owen, H. M. and Collier, A. 1950. Preliminary note on the occurrence of a new protistan parasite, Dermocystidium marinum n. sp. in Crassostrea virginica (Gmelin). Science, 111, 328-329.

Nilson, E. H., Fisher, W. S. and Shleser, R. A. 1976. A new mycosis of larval lobster (Homarus americanus). J. Invertebr. Pathol., 27, 177-183.

Nimmo, D. R., Lightner, D. V. and Bahner, L. H. 1977. Effects of cadmium on the shrimps, Penaeus duorarum, Palaemonetes pugio and Palaemonetes vulgaris. In Physiological Responses of Marine Biota to Pollutants, F. J. Vernberg, A. Calabrese, F. P. Thurberg, and W. B. Vernberg, eds. pp. 131-183.

Quick, J. A. Jr. 1971. Oyster parasitism by Labyrinothomyxa marina in Florida. Fla. Dept. Nat. Resources, Prof. Pap. Ser. No. 13, 55 pp.

Reichenbach-Klinke, H. H. and Landolt, M. 1973. Fish Pathology. A Guide to the Recognition and Treatment of Diseases and Injuries of Fishes, with Emphasis on Environmental and Pollution Problems. T.F.H. Publications, N.J. 512 pp.

Rogers-Talbert, R. 1948. The fungus Lagendium callinectes Couch (1952) on eggs of the blue crab in Chesepeake Bay. Biol. Bull., 95, 214-288.

Sindermann, C. J.,ed. 1977. Larval mycosis of oysters. Disease Diagnosis and Control in North American Marine Aquaculture. Devel. in Aquaculture and Fisheries Science, Elsevier Sci. Publ. Co., N.Y. 6, 195-197.

Sindermann, C. J. 1979. Epizootics in crustacean populations. In Proc. of the Second Biennial Crustacean Health Workshop, D. H. Lewis and J. K. Leong, eds. Texas A&M Univ. Sea Grant Publ. No. 79-114, pp. 1-37.

Solangi, M. A. and Lightner, D. V. 1976. Cellular inflammatory response of Penaeus aztecus and P. setiferus to the pathogenic fungus, Fusarium sp., isolated from the California brown shrimp. P. californiensis. J. Invertebr. Pathol., 27, 77-86.

Sordi, M. 1958. Micosi dei crostacei decapodi marini. Rivista di Parassitologia, 19, 131-137.

Sparks, A. K. and Hibbits, J. 1979. Black mat syndrome, and invasive mycotic disease of the tanner crab, Chionoecetes bairdi. J. Invertebr. Pathol., 34, 184-191.

Sparrow, F. K. 1973. The peculiar marine phycomycete Atkinsiella dubia from crab eggs. Arch. Mikrobiol., 93, 137-144.

Tharp, T. P. and Bland, C. E. 1977. Biology and host range of Haliphthoros milfordensis Vishniac. Can. J. Bot., 55, 2936-2944.

Unestam, T. 1973. Fungal diseases of crustacea. Rev. Med. Vet. Mycol., 8, 1-20.

Van Hyning, J. M. and Scarborough, A. M. 1973. Identification of fungal encrustation of the shell of the snow crab (Chionoecetes bairdi). J. Fisheries Res. Board Can., 30, 1738-1739.

Vishniac, H. 1958. A new marine Phycomycete. Mycologia, 50, 66-79.

Chapter 17

FUNGAL DISEASES OF FRESHWATER AND TERRESTRIAL CRUSTACEA

by T. Unestam*

INTRODUCTION

In a previous review on fungal diseases of Crustacea (Unestam, 1973) it was suggested that fungal parasites play a greater role in the invertebrate groups than among vertebrates. Recent information has not contradicted this possibility. Certain properties of hyphal fungi, such as cell wall chemistry and the abilities to penetrate the host mechanically, to react chemotropically, and to translocate intracellular material over long distances, make the fungal parasites distinct from bacteria and most other pathogens. These fungal properties may favor parasitic success more in invertebrates than in vertebrates.

In this Chapter, recent findings on fungal diseases of Crustacea, most published after 1972, are reviewed (see also Table 17-1). Fungal diseases of marine Crustacea are discussed by Lightner in Chapter 16. Recently Vey (1976) and Manier (1978) have also published reviews on these topics.

BIOLOGY OF PATHOGENS

Ascomycetes and Fungi Imperfecti

Only limited experimental work has been done with these pathogens. The yeast Metschnikowia (Monospora) bicuspidata was the first fungus ever studied experimentally in crustacean pathology (Metschnikoff, 1884). This fungus was studied

*Department of Forest Mycology and Pathology, Swedish University of Agricultural Sciences, Uppsala, Sweden

TABLE 17-1. FUNGAL PATHOGENS IN NATURAL DISEASES OF CRUSTACEA

Ecological and physiological research published between 1972 and 1979.

Pathogen	Crustacean host	Research Field	Reference
(Ascomycetes and Fungi Imperfecti):			
Metschnikowia bicuspidata (yeast)	Artemia salina (brine shrimp)	Spore discharge, cytochemistry	Lachance et al., 1976
Fusarium solani	Austropotamobius pallipes (freshwater crayfish) in culture	Life cycle	Vey & Vago, 1973
Candida (?) sp.	Armadillidium vulgare (terrestrial isopod)	Life cycle	Yousfi, 1976
cryptococcaceous yeast	Proasellus coxalis (freshwater isopod)	Host cytology	Manier & Loubes, 1978b
(Hyphochytriales):			
Hyphochytrium peniliae	Penilia avirostris, Podon sp., Evadne sp. (plankton caldocerans)	Life cycle, host range, cyclic appearance	Zelezinskaya, 1972
(Blastocladiales):			
Catenaria anguillulae	Macrocyclops albidus and Eucyclops servulatus (copepod)	Life cycle, ultrastructure, taxonomy	Manier, 1977
Coelomomyces psorophorae	Cyclops vernalis (copepod)	Host alternation, life cycle	Whisler et al., 1974, 1975
C. punctatus + C. dodgei	Cyclops vernalis	Host alternation, taxonomy, genetics	Federici & Roberts, 1976; Federici & Chapman, 1977; Federici, 1979
C. opifexi	Tigriopus sp. (copepod)	Host alternation	Pillai et al., 1976
C. chironomi	Heterocypris incongruens	Host alternation	Weiser, 1976
Callimastix cyclopis (Coelomomyces psorophorae)	Microcyclops sp. (copepod)	Taxonomy, ultrastructure	Mainer & Loubes, 1978a

(Lagenidiales):			
Blastulidium poedophthorum	*Simocephalus vetulus* (Cladoceran)	Ultrastructure, taxonomy	Manier, 1976
Lagenidium callinectes	*Callinectes sapidus* (blue crab ova)	Field ecology, fungal ultrastructure	Bland & Amerson, 1973, 1974
Haliphthorus milfordensis	*Penaeus duoranum* (pink shrimp)	Taxonomy, host range	Tharp & Bland, 1977
(Saprolegniales):			
Aphanomyces astaci	*Astacus astacus* (crayfish)	Physiology of zoospore-host interaction	Svensson & Unestam, 1975; Svensson, 1978
A. astaci	*Astacus astacus*, *Pacifastacus leniusculus* (crayfish)	Ultrastructure of infection	Nyhlén & Unestam, 1980
A. astaci	Crayfish	Ecological range, vector	Unestam et al., 1977
A. astaci	Crayfish	Defense mechanism	Söderhäll et al., 1979; Söderhäll & Ajaxon, 1981; Söderhäll & Unestam, 1979; Unestam & Ajaxon, 1976
Dictyuchus sp.	*Pacifastacus leniusculus* (crayfish egg mass)	Life cycle under reared conditions	Vey, 1977
(Trichomycetes)	Various species	Review, taxonomy, biology, symbiosis	Moss, 1979
(Various fungi):			
Yeasts, chytrids, water molds	Cladocerans	Field ecology cyclic appearance	Green, 1974
Ellobiopsidae	Various Crustacea	Bibliobraphy	Vader, 1973
Septate fungus	*Palinurus elephas*	Histology	Alderman, 1973

cytochemically by Lachance et al. (1976). The parasite is found in *Daphnia magna* (Cladocera), as well as in the brine shrimp, *Artemia salina* (Phyllopoda). The authors discovered that the needle-shaped ascospores of this species are forcibly and violently expelled from the asci by means of a swelling ectoplasmic mucilage (Fig. 17-1). This mechanism may help the fungus to attach to and mechanically penetrate the host. Penetration may therefore not be entirely passive, as was believed by Metschnikoff (1884) (Fig. 17-2).

Metschnikowa includes terrestrial species infecting marine copepods as well as marine species experimentally parasitizing freshwater crustaceans and the species may in fact include aggressive, facultative parasites with wide host ranges (see review by Unestam, 1973). Thus, the yeasts could infect the genera *Daphnia*, *Dasyhelea*, *Artemia*, *Calanus*, and *Eurytemora*. They may also be classified as opportunistic parasites, many of which take advantage of host weaknesses caused by environmental extremes, over-population or other temporary stresses to the individual or group. Such temporary disease-yielding conditions may occur during endogenic, periodic fluctuations in density of the animal population or during normal variation in the environment and may be important factors in many fungal diseases in Crustacea (see below, Resistance to Invasion).

Hyphae of *Fusarium solani* infect natural and experimental wounds of the freshwater crayfish, *Austropotamobius pallipes*, and kill the animal after penetrating into the body cavity (Vey and Vago, 1973). After invading the dead animal the external hyphae form clamydospores and conidia. Nonspecific wound parasites of this kind are normally not serious to natural populations but are likely to be so in dense artificial populations where many fungi are a nuisanse (see Chapter 16). *Fusarium solani* is a root pathogen of soybean and may also infect man and other vertebrates (O'Day et al., 1979); it must consequently be classified as a nonspecific opportunist. *Fusarium* is a genus found both on marine and freshwater crustaceans and close relatives of *Fusarium*, *Ramularia* spp., have earlier been found parasitizing crayfish and lobsters (Unestam, 1973; see Chapter 16).

Aquatic Phycomycetes

Blastocladiales: One of the most important events in arthropod myco-pathology in recent years was the discovery of the complete life cycle of the obligate parasite of mosquito larvae, *Coelomomyces*. For decades, attempts to culture this promising mosquito control agent in the laboratory had met with little success. Finally, Whistler et al. (1974, 1975) reported that *C. psorophorae* parasitizing the larva of the midge *Culiseta vernalis* has an obligate, alternate host, the copepod *Cyclops vernalis*. Isogametes develop in wall-less gametangia inside the copepod. After being released, the gametes fuse and produce a diploid zygote that subsequently infects the mosquito host. Zoospores from resistant sporangia in the hemocoel of the latter host then infect the copepod.

Almost immediately, biologists in different parts of the world discovered intermediate hosts for species of *Coelomomyces* among copepods and ostracods (where it had been called *Callimastix* in the crustacean hosts), thereby completing the life cycles of previously well-known species. Thus, both *C. punctatus* and *C. dodgei* require *Cyclops vernalis* as well as the larva of the mosquito, *Anopheles quadrimaculatus*, to complete a 2 week life cycle in the laboratory. The mosquito is susceptible in all instars (Federici and Roberts, 1976; Federici and Chapman, 1977). Similarly, the copepod *Tigriopus* sp. alternates obligately with two brackish water mosquitoes, *Opifex fuscus* and *Aedes australis*, as hosts of *C. opifexi* in New Zealand (Pillai et al., 1976). Weiser (1976) reported that *Chironomus plumosus* larvae and the ostracod, *Heterocypris incongruens*, are alternate hosts of *C. chironomi* in Europe. The ostracod feeds on the larva, probably supporting infection of the latter.

Federici (1979) hybridized *in vitro* two species of *Coelomomyces*, *C. dodgei* and *C. punctatus*. The hybrid sporangia in the mosquito host, *Anopheles quadrimaculatus*, were often morphologically similar to those of another species, *C. lattivitatus*, indicating this species may be a naturally occurring hybrid between the former two species.

Manier (1978) and Manier and Loubes (1978a) studied ultrastructurally and taxonomically *Callimastix cyclopis* (*Coelomomyces psorophorae*) in a copepod, *Microcyclops* sp. Found in the malaria region of Tchad, it may be of special interest in biological control of vectoring mosquitoes.

Oomycetes: An oomycete destroying the whole external egg mass of the freshwater crayfish, *Pacifastacus leniusculus*, without harming the adult, was found under artificial culture conditions (Vey, 1977). The fungus, *Dectyuchies* sp., penetrated the egg chorion and grew progressively inside and, later, outside the egg, where it produced zoospores. It is not presently known whether this fungus is an adapted parasite in nature (as are *Lagenidium* spp., see below), able to penetrate a healthy egg, or only an aquaculture parasite (cf. Unestam, 1973), or even a saprophytic mold, temporarily surrounding and feeding on eggs under stressed culture conditions.

The crayfish plague in Europe, caused by *Aphanomyces astaci*, has spread further in Europe since 1972 (Unestam, 1973) and no resistance has yet appeared in the European crayfish, *Astacus astacus*. The pathogen was probably introduced from

North America where all crayfish are highly and fully resistant. In highly resistant crayfish, soft parts of the cuticle are invaded and become dark due to melanin formation in the limited, infected spots. This is true for Pacifastacus leniusculus, which has recently been introduced in Europe and is a reservoir of the disease (Unestam et al., 1977). Since all crayfish of the world, except the American species, seem to be lethally susceptible, such introductions may be very hazardous.

In 1977 we discovered (Unestam et al., unpublished) that, in Sweden, even previously resistant P. leniusculus may suddenly become completely and lethally susceptible to the pathogen it carries, at least under rearing conditions. The conditions causing the complete breakdown of resistance are not yet known. This breakdown indicates a delicate balance between virulence of this very specialized pathogen and host resistance.

ULTRASTRUCTURE AND BIOCHEMISTRY OF INVASION

The yeast, Metschnikowia bicuspidata, penetrates the intestinal wall of Daphnia passively with help of the movements of the intestine (Fig. 17-2) (Metschnikoff, 1884), perhaps also with the aid of the mechanical force of the expelled spore (Fig. 17-1) (Lachance et al., 1976). Metschnikowia sp., found in the intestine of the marine copepod Calanus plumcrus, had no chitinolytic capacity but dissolved mucopolysaccharides enzymatically (Seki and Fulton, 1969), perhaps a parasitic specialization to mucous tissues.

Invasion through the exoskeleton must always take place through wounds or by means of extracellular enzymes and/or mechanical force of the penetration tube or hyphal apex through intact cuticle layers. The mechanisms of zoospore encystment, adhesion, and subsequent germination of the crayfish plague fungus, A. astaci, was studied experimentally in vivo and in vitro (Svensson and Unestam, 1975; Svensson, 1978). A steep gradient of leaking sodium chloride and other soluble compounds from superficial wounds, and even leaking from the intact cuticle, seem to osmotically trigger encystment (with cell wall formation within 15 seconds) and subsequent germination of the spore (Fig. 17-3). This mechanism may represent part of the specific interaction between host and parasite. The spore adheres firmly to the surface of the intact epicuticle, probably by means of its own sticky, lipophilic surface, but no "cementing" substance (cf. Zacharuk, 1970a; Travland, 1979, and Chapters 13 and 14) between the cuticle surface and the spore cyst wall was seen in the transmission electron microscope (Nyhlén and Unestam, 1975). The firmly fixed cyst erodes the surface, "folding"

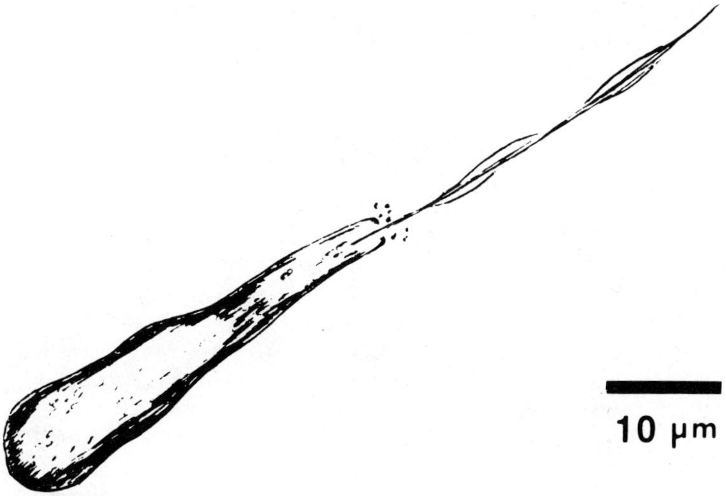

Figure 17-1. A needle-shaped spore of the yeast, Metschnikowia biscuspidata, being forcibly expelled from the ascus. Such a violently released, sharp spore may penetrate the host surface and be part of a parasitic adaptation (drawing based on Lachanche et al., 1976; reprinted by permission of the National Research Council of Canada).

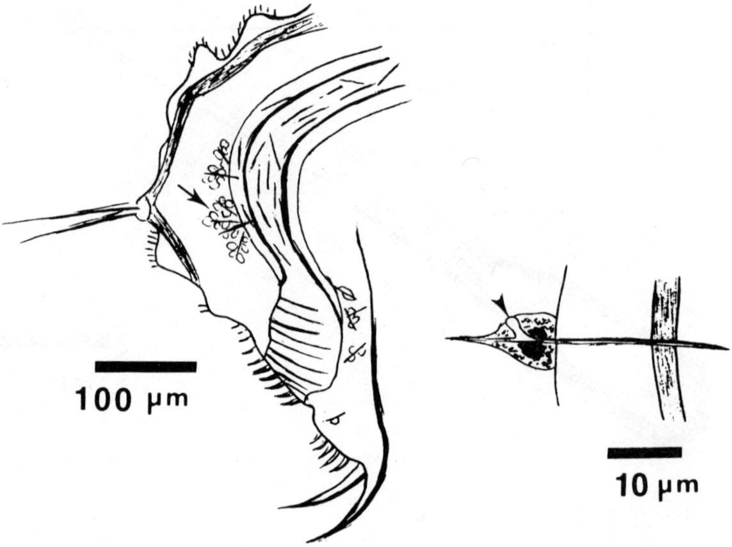

Figure 17-2. The needle-shaped spores of the yeast, Metschnikowia bicuspidata, penetrating the intestinal wall of Daphnia magna. Left: Cells reaching inside the hemocoel are encapsulated (arrow) or phagocytized by hemocytes. Right: a penetrating and subsequently germinating (arrow) spore being phagocytized. (Redrawn from Metschnikoff, 1884.) These observations on induced cellular resistance mechanisms were the first described in animals.

away the outermost lipid layer, and forms a penetration tube which penetrates through the protein epicuticle, now exposed (Fig. 17-3). The spore content is injected into the chitino-proteinaceous endocuticle, where it forms a hypha. In susceptible crayfish species, this cuticle parasite then develops its dense mycelium within the endocuticle and finally sporulates into the water.

Aphanomyces astaci appears to use extracellular lipolytic and proteolytic activities, in addition to mechanical force, during penetration of the two epicuticle layers. Protease activity as well as subsequently induced, weak, chitinolytic activity are evident when the hypha is formed and grows in the endocuticle (Nyhlén and Unestam, 1980; Söderhäll et al., 1978). The different phases of penetration are shown in Fig. 17-3. The hyphae prefer penetrating the endocuticle in its cleavage planes, i.e. parallel to the surface, between and parallel to the chitin fibrils; the weak, extracellular chitinase activity has probably only a supplementary function.

In insects, histolysis of the cuticle is also normally combined with mechanical penetration, e.g. in the cleavage planes, by fungal parasites (Zacharuk, 1970b; Mohamed et al., 1978; Lambiase and Yendol, 1977; see also Chapters 13 and 14). Thus, Samsinakova and Misikova (1973), working with a number of Deuteromycetes, found that some lipolytic and proteolytic activity may be necessary for an effective chitinase activity and for penetration of the cuticle of the larva of *Galleria mellonella*. Samsinakova et al. (1977) stressed that protease and chitinase activity of each pathogenic strain of *Beauveria bassiana* was related to the degree of virulence against the Colorado beetle larva.

The soft cuticle of arthropods immediately after ecdysis is sometimes particularly vulnerable to fungal penetration (Vey and Fargues, 1977; Unestam and Ajaxon, 1976). In crayfish, we found that the cuticle at this stage lacks the outer "lipid" layer of the epicuticle (unpublished), and has a very low phenoloxidase activity (Unestam and Ajaxon, 1976). Both the lipid layer and the phenoloxidase may normally be major obstacles for fungal invasion (see below).

RESISTANCE TO INVASION

The virtual elimination of the cladoceran *Peniliae avirostris* in part of the Black Sea during 1963-1964 and again 10 years later, was caused by *Hyphochytrium peniliae* (*Hyphochytriales*) (Zelezinskaya, 1972). Sudden, periodic outbreaks of phycomycetous and other diseases of Cladocera were also observed by Green (1974).

Periodic fluctuations in the occurrence of phycomycetous diseases among planktonic crustaceans is a well-known phenomenon (Unestam, 1973), but its reasons are not clear.

Figure 17-3. Electron micrographs of a specifically adapted parasite, the crayfish plague fungus, Aphanomyces astaci, penetrating the soft cuticle of a susceptible crayfish host, Astacus astacus.

A. The spore has attached firmly to the lipid surface, eroded this surface, and, by expanding its cell wall, torn the surface lipid layer (1) apart. A penetration peg (pp) has been initiated and has induced an electron dense host reaction (r) in the epicuticle (e).

B. The spore content (protoplasm) is vertically injected through the penetration tube (pt) in the proteinaceous epicuticle (e), and into the chitinous cuticle (cut) and the spore cell wall is left behind on the cuticle surface. Lysis and mechanical pressure is apparent around the penetration tube. (The lipid layer of the epicuticle had detached during preparation for electron microscopy.)

C. The injected protoplasm has formed a hypha (h) penetrating inside and feeding on the cuticle (cut). Before weakening the animal the fungus is not able to reach further into the hemolymph, where resistance is too strong. (A and C from Nyhlén and Unestam, 1975; reprinted by permission of Academic Press, Inc.)

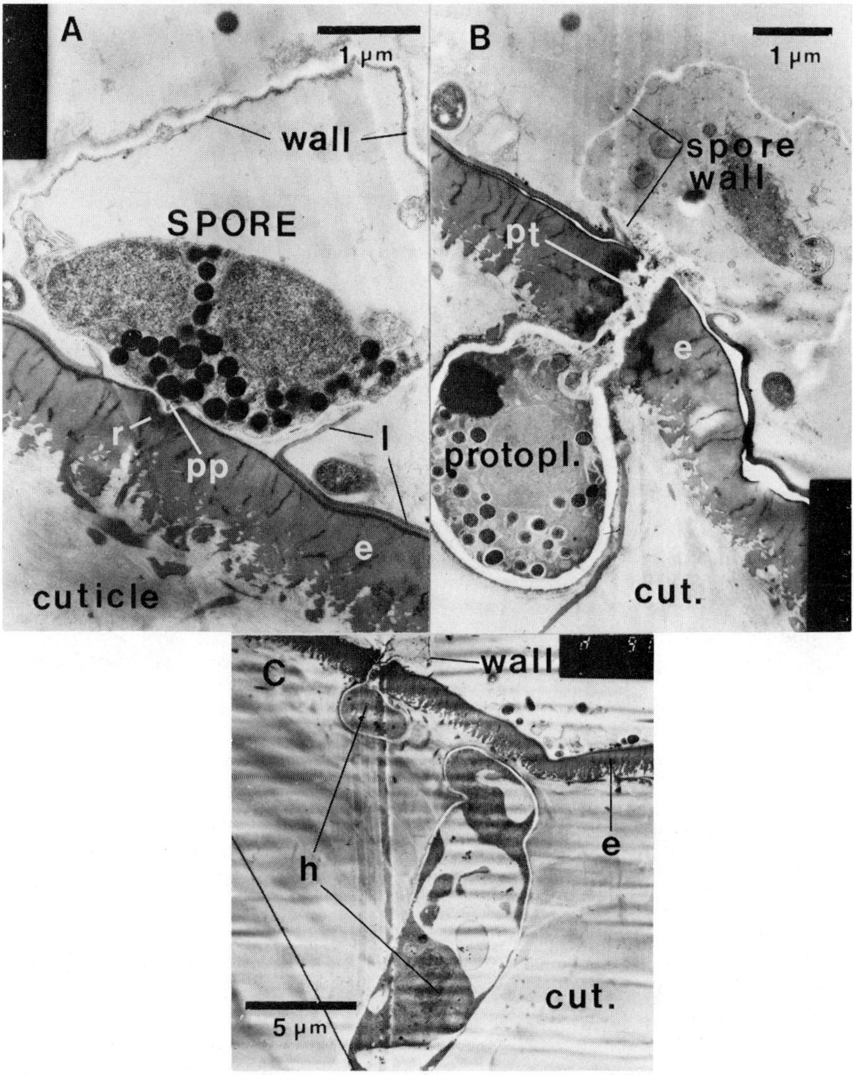
A
1 µm
wall
SPORE
r
pp
l
e
cuticle
B
1 µm
spore wall
pt
e
protopl.
cut.
C
wall
h
e
cut.
5 µm

From Green's (1974) and earlier reports it appears that such variation between years, despite continuous abundance of hosts, may be due to an ecological feedback mechanism, "cyclic selection". Thus, genetic development of resistance through increasing selection pressure on the short-lived plankton animals may, after a few generations, lead to a drastic reduction in infection potential of the pathogen. This would, as a consequence, result in a decreasing selection pressure and, therefore, a subsequent loss in resistance. This, in turn, may finally lead to rapid development of the parasite and a sudden outbreak of disease in the "defenseless" population. The nature of such resistance is still not known in these planktonic species.

In the larger, longer-lived crustaceans, such genetic feedback systems are probably of lesser importance. Instead the animal must have a number of consistant barriers as well as pre-existing, inducible defense mechanisms, triggered by invasion of a parasite. Periodic occurrence of disease, as for example *Lagenidium callinectes* on crab ova, must be a function of variation in the environment (Bland and Amerson, 1974). Dependence on salinity and temperature, in combination, was earlier found in the pathogenic relation between another ova parasite, *L. chthamalophilum*, and two barnacle hosts (Johnson, 1960).

A number of barriers and induced defense processes are found in the decapods and many of them seem to be of importance to fungal penetration in eggs, cuticle or hemolymph (Fig. 17-5) as shown in the following:

The egg

Lagenidium species specifically parasitizing crustacean eggs have a wide host range in contrast to other parasites in the same genus. *Lagenidium callinectes* is found on several crabs as well as on barnacles (Bland and Amerson, 1974) and *L. chthamalophilum* is found on different barnacle species (Johnson, 1960). *Blastulidium peodophthorum* (Manier, 1976) and possibly *Dictyuchus* sp., (Vey, 1977) are specialized parasites among the Oomycetes, which have a high affinity for eggs of Crustacea. The adult is normally not penetrated by such egg pathogens. So, the defense properties must be different in eggs and cuticle.

The crayfish plague fungus, on the other hand, is specialized to penetrate into and live in the crayfish cuticle. It is never found attacking the eggs under the abdomen of the female, even when heavily infected cuticle of the same female is in close contact with the eggs (Unestam et al., 1977; and unpublished results). Intact eggs are also not infected by zoospores *in vitro*. Empty egg shells are penetrated by the fungus inside, however. The surface of the intact egg

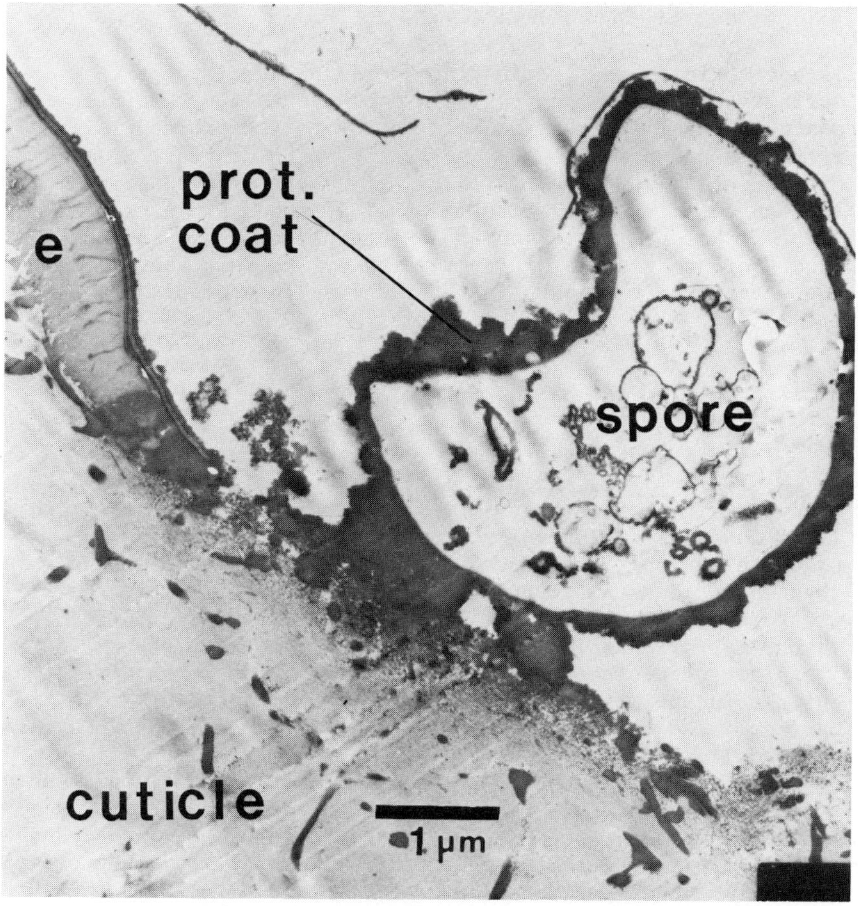

Figure 17-4. A spore attached to a superfically wounded soft cuticle (cut) of the crayfish, Astacus astacus. The cell wall is glued onto the naked, chitinous cuticle (epicuticle (e) stripped off) by means of a clotting protein (prot) released from the cuticle. Such clotted protein has also attached to the entire cell wall and contains phenoloxidase which has been activated by the β-1,3-glucans of the cell wall. This defense enzyme has produced electron dense (dark) melanin and the spore has finally died.

is not toxic to the germinating spores. Therefore, the surface of the crayfish egg seems to have an inert barrier against fungal spore attachment and/or to the extracellular enzymes of A. astaci. This surface must be distinct from that of the intact cuticular surface. If such a difference between egg and cuticle exists also in other Decapoda, this can explain part of the specialization of the egg parasites. Since eggs of different decapods are penetrated by the same specialized egg parasite, it appears that the egg surface of different species may have similar properties.

Barriers of the cuticle and the hemolymph

The different layers of the crustacean cuticle, such as the lipid layer, the proteinaceous epicuticle and the heavy chitin-protein layer, represent barriers (Fig. 17-3 and 17-4) and will exclude most fungi. Only fungi having protease and chitinase, such as the crayfish plague fungus (Nyhlén and Unestam, 1980), can invade through or into the exoskeleton. Lipolytic activity, eroding the surface lipid layer, may also assist penetration. Consequently, the inhibitor of proteases which we recently found in the crayfish cuticle (unpublished) may also be part of the cuticular barrier. The same type of inhibitor was found also in crayfish blood cells.

When inside of the cuticle a parasite may meet non-induced, toxic barriers of the hemolymph (Fig. 17-4). No such toxin against fungi have been described in Crustacea, but non-induced as well as induced bactericidal activity with varying specificity exists in several decapods (see review by Shapiro, 1975).

Induced response in cuticle and hemolymph

Mechanisms of acquired, specifically hightened resistance against microbes after vaccination have been found in crayfish and lobsters (reviewed by Shapiro, 1975). In crayfish vaccination probably results in increased numbers of phagocytes and is not entirely specific. Vaccination with fungal pathogens has rarely been attempted (Unestam, 1973).

As reviewed by Shapiro (1975), agglutinins, often lectin-like and sometimes acting as "opsonins" for phagocytosis of microorganisms in general, seem to be common in the blood of decapods. Precipitin-like factors also exist. Immune responses are usually mixtures of humoral and cellular elements. The humoral elements lack the specificity of vertebrate immunoglobulins. Clearing of foreign proteins in crayfish shows at least some specificity and is distinct from the metabolism of "self" proteins (Sloan et al., 1975).

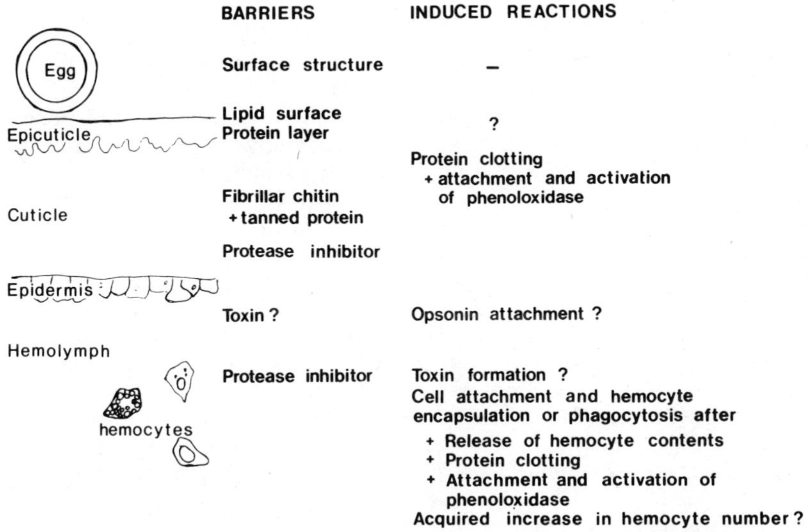

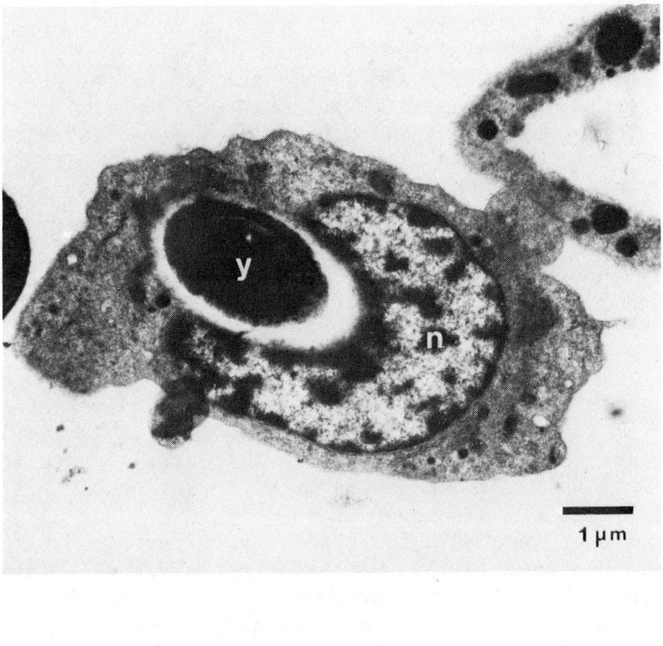

Figure 17-5. Defense barriers and reactions against fungal penetration of the egg, the cuticle, and the hemolymph in decapod crustaceans.

Figure 17-6. Electron micrograph of a parasitic yeast cell (Cryptococcaceae) (y) phagocytized by a hemocyte of the isopod Proasellus coxalis (n, hemocyte nucleus). (Photo kindly supplied by Dr. J.-F. Manier).

Induced response in the Decapoda against fungi has been found at all levels, from the epicuticle into the hemolymph (Fig. 17-5). Most of the mechanisms of active, induced response to invasion described in the literature have not been tested against fungi, however, partly because hyphal growth makes quantification very difficult. The importance of such response in the blood system is obvious since certain fungi may successfully overcome the cuticular resistance without being allowed to enter the adjacent blood system (Unestam, 1973; Alderman, 1973).

The crustacean cuticle is therefore not only a mechanical barrier to invasion but also contains a system of inducible defense mechanisms. The cuticle of crayfish contains a fibronogen-like, clottable protein (Durliat and Vranckx, 1978) and proteins, including phenoloxidase, which are able to coat fungal spores (Söderhäll et al., 1979) (see below and Fig. 17-5).

Three different types or stages of blood cells are found in Crustacea: granular, semigranular, and hyaline. They are involved in phagocytosis of small foreign particles and blood cell encapsulation of large objects, both induced processes. Phagocytosis (Figs. 17-2 and 17-6) seems to be the defense line beyond which parasitic yeasts cannot reach in surviving animals (Metschnikoff, 1884; Manier and Loubes, 1978b). In addition, some of these blood cells explode and release their granular material when in contact with a foreign surface. The blood cells also release clottable proteins as well as phenoloxidase (Fig. 17-5). The extreme fragility of many of these cells often makes it difficult to separate humoral from cellular involvement (Shapiro, 1975; Bauchau and Mengeot, 1978; Ratcliffe and Rowley, 1979).

The crayfish model and the involvement of fungal glucans

Encapsulation, release of granular material onto the fungal cell wall surface, and slow melanization of this material (due to its content of phenoloxidase) are all parts of the active, induced defense process in the blood of crayfish (Unestam, 1973). Melanization *in vivo* is closely correlated with inhibition of fungal penetration, both in cuticle and hemolymph. Phenol oxidation, which preceeds melanization and probably produces local, high quantities of short-lived, toxic quinones and radicals, is specifically induced by the β-1,3-glucans of fungal cell walls, both in cuticle and blood. Thus phenoloxidase, located in the outer cuticle and in the granules of explosive blood cells, is activated by the same glucan of most fungi, both *in vivo* and *in vitro* (Unestam and Ajaxon, 1976; Söderhäll and Unestam, 1979). The eliciting effect of the β-1,3 glucose unit of the glucan is very specific and is not mimiced by any other sugar.

Only five such units in the glucan, i.e. a pentose, are necessary and extremely small concentrations are required.

Further biochemical work in our laboratory on the hemolymph phenoloxidase system showed that, in contrast to activation, attachment of the enzyme to fungal walls is a non-specific process, mediated by a clottable protein, and that attachment and enzyme activation are closely associated (Söderhäll and Unestam, 1979; Söderhäll et al., 1979). The same also seems to be involved in both clotting and phenoloxidase activation and the presence of calcium ions is necessary in both (Söderhäll, in press). Clotting resulting in attachment is also induced by β-1,3-glucans. In spite of this, clotting and activation of the phenoloxidase are apparently not linked but can be separated.

The non-specific coating of spores and other "not-self" surfaces with protein from crayfish hemocytes resembles closely the opsonization of bacteria with hemolymph protein in crayfish (Tyson and Jenkin, 1973). The protease- and calcium-dependent clotting in the horseshoe crab, _Limulus_, induced by bacterial endotoxin (Tai and Liu, 1977), shows remarkable biochemical similarities with the crayfish coating system. The fibrinogen of the spiny lobster, _Panilurus interruptus_, with biochemical properties comparable to mammalian immune proteins (Fuller, 1978), may be a related protein. All these arthropod systems may therefore have similar, if not the same functions.

We have also shown that synthetic and native crayfish melanin as well as some of its precursors are indeed toxic to fungal growth _in vitro_ (Söderhäll and Ajaxon, 1981).

On the basis of our results a simplified model for defense against fungal penetration in crayfish can be described as follows. In susceptible as well as in highly resistant animals the crayfish plague fungus has to enzymatically and mechanically penetrate the lipid cuticle surface. Further penetration is facilitated by fungal protease and chitinase and is inhibited by a protease inhibitor. Meanwhile the cuticle phenoloxidase is activated by the fungal cell wall surface glucans and the glucan parts of the released, extracellular fungal enzymes. The penetration also probably triggers a signal for the epidermis (or blood cells) to produce phenols which are then translocated to the outer cuticle, where the activated phenoloxidase probably converts them into toxic products on the cell wall surface. When, as in resistant animals, this occurs on the hyphal tip of the fungus, fungal growth ceases, or is drastically retarded. The "missing factor" in the cuticle of a susceptible crayfish is not yet known but is probably not phenoloxidase activity.
If a hypha happens to reach inside the epidermis of resistant or susceptible crayfish it is immediately encapsulated by sticky blood cells which release their phenoloxidase and deposit it by means of sticky proteins on the fungal cell wall.

The immobilized phenoloxidase, now in contact with the cell wall glucans, becomes activated and oxidizes phenols released from the blood cells; fungal invasion into the blood is always stopped. The high efficiency of the hemocytes in preventing hyphae from invading the hemolymph is apparently the same in both susceptible and resistant species. Death in the susceptible species is apparently preceded by extensive fungal growth in the cuticle which finally weakens the animal so that hyphae can finally enter the hemolymph.

Phenol oxidation in mycopathology of arthropods

In the Decapoda local melanization, often seen as a "sheath" of dark chitin around penetrating hyphae or dark staining of the whole infected area, is very common (see Chapter 16; Alderman, 1973; Unestam, 1973, Unestam and Ajaxon, 1976). Thus, in the burn spot and similar disease of lobsters, crabs, and freshwater crayfish dark necrotic spots of discoloration in the calcified exoskeleton are seen where the parasitic fungus is present. The same is seen in the crayfish plague, both around hyphae in the soft cuticle of highly resistant species and around hyphae of low-virulent fungal strains in the susceptible species.

Also in insects melanin deposition on parasites is a common phenomenon (Whitcomb et al., 1974) and is often considered a major defense mechanism. The process of melanin deposition on the parasite in the hemolymph (Sroka and Vinson, 1978; Ratcliffe, and Rowley, 1979) seems often to be essentially the same as the one in crayfish (Unestam and Ajaxon, 1976). Phenol oxidation in the insect cuticle is also often associated with parasite invasion, but has not yet been thoroughly investigated (Aoki and Yanase, 1970) and was not even mentioned in the review by Whitcomb et al. (1974).

Whether the fungal β-1,3-glucans have the same key role in triggering these processes in arthropods in general as they have in crayfish, is not yet known. Interestingly, fungal β-1,3-glucans are involved in eliciting defense mechanisms also in vertebrates (Whistler et al., 1976; Kokoshis et al., 1978) and in plants (Albersheim and Valent, 1978). Possibly this glucan, so typical to the fungal cell wall, is used by a great variety of organisms as a general warning signal for hyphal invasion.

Research on active (cellular and non-cellular, acquired and pre-existing) defense mechanisms in the Crustacea, particularly the Decapoda, has enjoyed success in recent years and the results may not only provide further understanding of the pathology of this group but will also provide for comparison with vertebrate immunosystems in the search for common, probably "primitive" animal defense systems.

ACKNOWLEDGEMENT

I thank Dr. K. Söderhäll for his critical preview of this manuscript.

REFERENCES

Albersheim, P. and Valent, B. S. 1978. Host-pathogen interaction in plants. Plants, when exposed to oligosaccharides of fungal origin, defend themselves by accumulating antibiotics. J. Cell Biol., 78, 627-643.

Alderman, D. J. 1973. Fungal infection of crawfish (*Palinurus elephas*) exoskeleton. Trans. Br. Mycol. Soc., 61, 595-597.

Aoki, J. and Yanase, K. 1970. Phenol oxidase activity in the integument of the silkworm *Bombyx mori* infected with *Beauveria bassiana*. J. Invertebr. Pathol., 16, 459-464.

Bauchau, A. G. and Mengeot, J. C. 1978. Structure et fonction des hémocytes chez les Crustaceas. Arch. Zool. Experim. Generale, 119, 227-248.

Bland, C. E. and Amerson, H. V. 1973. Electron microscopy of zoosporogenesis in the marine phycomycete, *Lagenidium callinectes* Couch. Arch. Mikrobiol., 94, 47-64.

Bland, C. E. and Amerson, H. V. 1974. Occurrence and distribution in North Dakota waters of *Lagenidium callinectes* Couch, a fungal parasite of blue crab ova. Chesapeake Sci., 15, 232-235.

Durliat, M. and Vranckx, R. 1978. Changes in water-soluble proteins from integument of *Astacus leptodactylus* during the molt cycle. Comp. Biochem. Physiol., 59B, 123-128.

Federici, B. A. 1979. Experimental hybridization of *Coelomomyces dodgei* and *Coelomomyces punctatus*. Proc. Natl. Acad. Sci. USA, 75, 4425-4428.

Federici, B. A. and Chapman, H. C. 1977. *Coelomomyces dodgei*: Establishment of an *in vitro* laboratory culture. J. Invertebr. Pathol., 30, 288-297.

Federici, B. A. and Roberts, D. W. 1976. Experimental laboratory infection of mosquito larvae with fungi of the genus *Coelomomyces*: II. Experiments with *Coelomomyces punctatus* in *Anopheles quadrimaculatus*. J. Invertebr. Pathol., 27, 333-341.

Fuller, G. M. 1978. Are lobster fibrogen and cold-insoluble globulin related molecules? Ann. N.Y. Acad. Sci., 312, 31-37.

Green, J. 1974. Parasites and epibionts of Cladocera. Trans. Zool. Soc. Lond., 32, 417-516.

Johnson, T. W., Jr. 1960. Infection potential and growth of *Lagenidium chthamalophilum*. Am. J. Bot., 47, 383-385.

Kokoshis, P. L., Williams, D. L., Cook, J. A., and DiLuzio, N. R. 1978. Increased resistance to Staphylococcus aureus infection and enhancement in serum lysozyme activity by glucan. Science, 199, 1340-1342.

Lachance, M. A., Miranda, M., Miller, M. W., and Phaff, H. J. 1976. Dehiscence and active spore release in pathogenic strains of the yeast Metschnikowia bicuspidata var. australis: Possible predatory implication. Can. J. Microbiol., 22, 1756-1761.

Lambiase, J. T. and Yendol, W. G. 1977. The fine structure of Entomophthora apiculata and its penetration of Trichoplusia ni. Canad. J. Microbiol., 23, 452-464.

Manier, J.-F. 1976. Cycle and ultrastructure of Blastulidium poedophthorum Perez, 1903 (Phycomycetes, Lagenidiales), parasite of eggs in Simocephalus vetulus (Mull.) Schoedler (Crustacea, Cladocera). Protistologica, 12, 225-238.

Manier, J.-F. 1977. Cycle and ultrastructure of a Catenaria (Phycomycetes, Blastocladiales) parasite of Crustacea, Cyclopodia. Ann. Parasitol. Hum. Comp., 52, 363-376.

Manier, J.-F. 1978. Mycoes des Crustacés. Arch. Inst. Pasteur Tunis, 55, 401-418.

Manier, J.-F. and Loubes, C. 1978a. Callimastix cyclopis Weissenberg, 1912 (Phycomycète, Blastocladiale) parasite d'un Microcyclops Claus, 1893 (Copepode, Cyclopoide) récolté au Tchad: Particularités ultrastructurales. Protistologica, 14, 493-501.

Manier, J.-F. and Loubes C. 1978b. Leuvres Cryptococcaceae, pathogènes pour Proasellus coxalis Dolfus (Crustacé, Isopode). Bull. Soc. Zool. France, 103, 483-490.

Mann, H. and Pieplow, U. 1938. Die Brandfleckenkrankheit und ihre erreger. Z. Fisch. u. d. Hilfswiss., 36, 225-240.

Metschnikoff, E. 1884. Uber eine Sprosspilzkrankheit der Daphnien. Beitrag zur Lehre uber den Kampf der Phagocyten gegen Krankheitserreger. Virch. Arch. Pathol. Anat. Physiol., 96, 177-195.

Mohamed, A. K. A., Sikorowski, P. P., and Bell, J. V. 1978. Histopathology of Nomuraea rileyi in larvae of Heliothis zea and in vitro enzymatic activity. J. Invertebr. Pathol., 31, 345-352.

Moss, S. T. 1979. Commensalism of the Trichomycetes. In Insect-Fungus Symbiosis. S. Batra, ed. Allanheld. Osmum. Montclair, N.Y. USA. pp. 175-227.

Nyhlén, L. and Unestam, T. 1975. Ultrastructure of penetration of the crayfish integument by a fungal parasite, Aphanomyces astaci, Oomyctes. J. Invertebr. Pathol., 26, 353-366.

Nyhlén, L. and Unestam, T. 1980. Wound reactions and Aphanomyces astaci growth in crayfish cuticle. J. Invertebr. Pathol., 36, 187-197.

O'Day, D. M., Akrabawi, P. L., Richmond, L. D., Jones, B. R. and Clayton, Y. 1979. An animal model of Fusarium solani endophthalamitis. Br. J. Ophthalmol., 63, 277-280.

Pillai, J. S., Wong, T. L. and Dodgshun, T. J. 1976. Copepods as essential hosts for the development of a Coelomomyces parasitizing mosquito larvae. J. Med. Entomol., 13, 45-50.

Ratcliffe, N. A. and Rowley, A. F. 1979. A comparative synopsis of the structure and function of the blood cells of insects and other invertebrates. Devel. Comp. Immunol., 3, 189-243.

Samsinakova, A., Bajan, C., Kalalova, S., Kmitowa, K. and Wojciechowska, M. 1977. The effect of some entomophagous fungi on the Colorado beetle and their enzyme activity. Bull. Acad. Polon. Sci. Biol., 25, 521-526.

Samsinakova, A. and Misikova, S. 1973. Enzyme activities in certain entomophagous representatives of Deuteromycetes (Moniliales) in relationship to their virulence. Cesk. Mykol., 27, 55-60.

Shapiro, H. 1975. Immunity in decapod crustaceans. Am. Zool., 15, 13-19.

Sloan, B., Yocum, C. and Clem, L. W. 1975. Recognition of self from non-self in curstaceans. Nature, 258, 521-523.

Söderhäll, K. and Ajaxon, R. 1981. Effect of quinones and melanin on mycelial growth of Aphanomyces spp. and extracellular protease of Aphanomyces astaci, a parasite on crayfish. J. Invertebr. Pathol. (in press).

Söderhäll, K., Häll, L., Unestam, T. and Nyhlén, L. 1979. Attachment of phenoloxidase to fungal cell walls in arthropods immunity. J. Invertebr. Pathol., 34, 285-294.

Söderhäll, K., Svensson, E. and Unestam, T. 1978. Chitinase and protease activities in germinating zoospore cysts of a parasitic fungus, Aphanomyces astaci, Oomycetes. Mycopathologia, 64, 9-11.

Söderhäll, K. and Unestam, T. 1979. Activation of prophenoloxidase. The specificity of cell wall glucan activation and activation by purified fungal glycoproteins. Can. J. Microbiol., 25, 404-416.

Sroka, P. and Vinson, S. B. 1978. Phenoloxidase activity in the hemolymph of parasitized and unparasitized Heliothis virescens. Insect Biochem., 8, 399-402.

Svensson, E. 1978. Interactions between a parasitic fungus, Aphanomyces astaci, Oomycetes, and its crayfish host. 1. Motility, encystment attachment, and germination of the zoospore. Acta Universitatis Upsalliensis, 457, 18 pp.

Svensson, E. and Unestam, T. 1975. Differential induction of zoospore encystment and germination in Aphanomyces astaci, Oomycetes. Physiol. Plantar., 35, 210-216.

Tai, J. Y. and Liu, T. Y. 1977. Studies on Limulus amoebocyte lysate. Isolation of pro-clotting enzyme.

J. Biol. Chem., 252, 2178-2181.

Tharp, T. P. and Bland, C. E. 1977. Biology and host range of Haliphthorus milfordensis. Can. J. Bot., 55, 2936-2944.

Travland, L. B. 1979. Initiation of infection of mosquito larvae (Culiseta inornata) by Coelomomyces psorophorae. J. Invertebr. Pathol., 33, 95-105.

Tyson, C. J. and Jenkin, C. R. 1973. The importance of opsonic factors in the removal of bacteria from the circulation of the crayfish (Parachaeraps bicarinatus). Austral. J. Exp. Biol. Med. Sci., 51, 609-615.

Unestam, T. 1973. Fungal diseases of Crustacea. Rev. Med. Vetr. Mycol., 8, 1-20.

Unestam, T. and Ajaxon, R. 1976. Phenol oxidation in soft cuticle and blood of crayfish compared with that in other arthropods and activation of the phenol oxidase by fungal and other cell walls. J. Invertebr. Pathol., 27, 287-295.

Unestam, T., Söderhäll, K., Nyhlén, L., Svensson, E. and Ajaxon, R. 1977. Specialization in crayfish defence and fungal aggressiveness upon crayfish plague infection. In Proceedings, 3rd International Symposium on Freshwater Crayfish, Kuopio, Finland, 1976, pp. 321-331.

Vader, W. 1973. A bibliography of the Ellobiopsidae, 1959-1971, with a list of Thallassomyces species and their hosts. Sarsia, 52, 175-180.

Vey, A. 1976. Mycoses des Crustacés. Oceanis, 3, 209-218.

Vey, A. 1977. Studies on the pathology of crayfish under rearing conditions. In Proceedings, 3rd International Symposium on Freshwater Crayfish, Kuopio, Finland, 1976, pp. 311-319.

Vey, A. and Fargues, J. 1977. Histological and ultrastructural studies of Beauveria bassiana infection in Leptinotarsa decemlineta larvae during ecdysis. J. Invertebr. Pathol., 30, 207-215.

Vey, A. and Vago, C. 1973. Protozoan and fungal diseases of Austropotamobius pallipes Lereboullet in France. In Proceedings, 1st International Symposium on Freshwater Crayfish, Austria, 1972, pp. 165-179.

Weiser, J. 1976. The intermediary host for the fungus Coelomomyces chironomi. J. Invertebr. Pathol., 28, 273-274.

Whitcomb, R. F., Shapiro, M. and Granados, R. R. 1974. Insect defense mechanisms against microorganisms and parasitoids. In The Physiology of Insecta, M. Rockstein, ed. 2nd ed., Vol. 5, Academic Press, N.Y. pp. 447-535.

Whisler, H. C., Zebold, S. L. and Shemanchuk, J. A. 1974. Alternate host for mosquito parasite Coelomomyces. Nature, 251, 715-716.

Whisler, H. C., Zebold, S. L. and Shemanchuk, J. A. 1975. Life history of Coelomomyces psorophorae. Proc. Nat'l. Acad. Sci. U.S.A., 72, 693-696.

Whistler, R. L., Bushway, A. A. and Singh, P. P. 1976. Noncytolytic, antitumor polysaccharides. Carboh. Chem. Biochem., 32, 235-275.

Yousfi, A. 1976. Recherches sur la pathologie des crustacés isopods oniscoides. Doctoral thesis, Technical Science University of Languedoc, Montpellier, 159 pp.

Zacharuk, R. Y. 1970a. Fine structure of the fungus Metarrhizium anisopliae infecting three species of larval Elateridae (Coleoptera). 2. Conidial germ tubes and appressoria. J. Invertebr. Pathol., 15, 81-91.

Zacharuk, R. Y. 1970b. Fine structure of the fungus Metarrhizium anisopliae infecting three species of larval Elateridae (Coleoptera). 3. Penetration of the host integument. J. Invertebr. Pathol., 15, 372-396.

Zelezinskaya, L. M. 1972. Investigation of the fungus infections of marine Cladocera. Gidrobiol Zh., 8, 86-87.

SECTION IV

RESISTANCE

Chapter 18

RESISTANCE OF INSECT SPECIES TO MICROBIAL PATHOGENS

by D. T. Briese*

INTRODUCTION

This chapter discusses the problem of intraspecific differences in the levels of innate resistance shown by insects to microbial pathogens. Several other aspects that bear on the resistance of insects to these pathogens are the degree of specificity shown by a pathogen for a particular host insect, the increase in resistance during development of immature insects, and the whole spectrum of interactions between the environment, the insect and the pathogen. However, the existence of genetically determined differences in resistance within a particular species, and the possibility that it may be deliberately or unintentionally selected for, has important implications for insect culture and insect control. Despite the lessons that can be learned from resistance to chemical insecticides, and the obvious economic importance of acquired resistance to pathogens, the subject has received relatively little attention. Burges (1971) found that, while the existence of resistant strains of insects had been demonstrated, there was insufficient evidence to say whether or not it would be of widespread importance in practice.

As indicated by the low slopes of dosage-mortality regressions (Burges and Thompson, 1971), there is typically a large degree of variability in response to a pathogen within populations of an insect species. However, in order to establish the genetic basis for these differences, evidence should be obtained for differences between populations. This evidence is usually presented in one of three forms: i. as

*CSIRO, Division of Entomology, Canberra City, A.C.T. Australia.

differences in response to a single dosage; ii. as differences between LD_{50} values (mean lethal dosage, derived from several measurements of dosage-mortality); and iii. as differences in LD_{50} values given together with values for the slopes of dosage-mortality regressions. These represent increasingly informative steps, for, although single-dosage differentiation of populations may infer some degree of resistance, it is only when slopes are known that the cases where more susceptible individuals have been eliminated from a population can be distinguished from those in which "true" resistance (*sensu* Burges, 1971) exists.

Having established the existence of differential susceptibility between races or strains of the same species, three further steps should be undertaken to clarify the nature of intraspecific resistance: ascertaining the genetic mechanisms that control such differences; examining the possibility of selection for them; and elucidating differences in the physiological or biochemical defense mechanisms of the insect that may account for them.

In the following sections, evidence arising from these four stages of work will be examined for each group of pathogens.

VIRUSES

Most work on insect-virus interactions has concerned the Lepidoptera, and examples presented in Table 18-1 indicate that substantial differences in response to several groups of virus have been found within lepidopteran species. The silkworm, *Bombyx mori*, which has been studied more intensively than other species, has some populations which are several hundred times more resistant than others to cytoplasmic polyhedrosis virus (CPV) and nuclear polyhedrosis virus (NPV), whereas others show virtual immunity to the non-occluded viruses, flacherie virus (FV), small flacherie virus (SFV) and densonucleosis virus (DNV), which may cause high mortality to some populations. Differential susceptibility of pest Lepidoptera to both NPV and CPV, as well as to granulosis viruses (GV) has been reported, but generally the relative resistance levels are considerably lower than those found in the silkworm (Table 18-1). This may reflect the fact that there are about 700 different races of silkworm, some of which have been artificially selected for resistance to disease (Yokoyama, 1979), whereas studies on pest insects have rarely considered more than a few populations, none of which have been so selected.

In the case of the codling moth *Laspeyresia pomonella* and its GV, Sheppard and Stairs (1977) found a distinct bimodality in the time-mortality response of a laboratory strain tested with GV. Their evidence suggested that 20% of the population was very resistant to the virus, indicating that with

TABLE 18-1. DIFFERENCES IN SUSCEPTIBILITY TO A VIRUS BETWEEN POPULATIONS OF LEPIDOPTERAN SPECIES

A. Cases in which differences are based on the response to a single dose of the virus

Insect Species	No. of Populations	Origin of Populations	Instar Tested	Virus	Mortality Range	Reference
Pieris brassicae	3	1 laboratory 2 field	1st	GV	61-100%	Sidor, 1959
Pieris brassicae	3	1 laboratory 2 field	1st	CPV	0-41.5%	Sidor, 1959
Mamestra brassicae	3	1 laboratory 2 field	3rd	CPV	0-40%	Maleki-Milani, 1970
Heliothis zea	5	field	1st	NPV	38-56%[a]	Ignoffo and Allen, 1972
Epiphyas postvittana	12	field	3rd	NPV	4-84%	Geier and Briese, 1979

B. Cases in which differences are based on LD_{50} values, but slopes of regression lines have not been reported

Insect Species	No. of Populations	Origin of Populations	Instar Tested	Virus	Range of Resistance factors	Reference
Bombyx mori	5	Culture	4th	CPV	6-166	Aruga and Watanabe, 1964
	11	Culture	4th	CPV	1-250	Watanabe, 1966a
	15	Culture	3rd	CPV	6-1380	Watanabe et al., 1974
	11	Culture	4th	NPV	2-630	Watanabe, 1966a
	15	Culture	4th	FV	$1\text{-}10^4$ [b]	Watanabe et al., 1974
	20	Culture	3rd	FV	1-344	Uzigawa and Aruga, 1965
	2	Culture	2nd	SFV	$> 10^5$ [b]	Furuta, 1978
	2	Culture	3rd	DNV	$> 10^5$ [b]	Watanabe and Maeda, 1978
Spodoptera frugiperda	2	Laboratory	1st	NPV_1 [c]	3	Reichelderfer and Benton, 1974
	2	Laboratory	1st	NPV_2 [c]	4	Reichelderfer and Benton, 1974
	2	Laboratory	1st	NPV_3 [c]	5	Reichelderfer and Benton, 1974

TABLE 18-1 (continued)

C. Cases in which differences are based on LD_{50} values, and the slopes of the regression lines were determined

Insect Species	No. of Populations	Origin of Populations	Instar Tested	Virus	Range of Resistance Factors	Range of Slopes	Reference
Bombyx mori	2	Culture	4th	CPV	83	1.71-1.88	Watanabe, 1966b
Bombyx mori	2	Culture	4th	CPV	11	0.84-1.04	Watanabe, 1966c
Pieris rapae	2	Field	4th	GV	No difference		Martignoni and Schmid, 1961
Phyrgandia californica	2	Field	5th	NPV	<2	0.77-1.21	Martignoni and Schmid, 1961
Plodia interpunctella	2	Laboratory	1st	GV	7	1-1.2	Hunter and Hoffman, 1973
Spodoptera littoralis	2	1 laboratory 1 field	5th	NPV	12	0.84-1.09	Klein and Podoler, 1978
Epiphyas postvittana	3	2 laboratory 1 field	3rd	NPV	50-160	1.77-2.56	Briese et al., 1980
Phthorimaea operculella	6	Field	3rd	GV	1-10	0.60-0.92	Briese and Mende, 1981
Phthorimaea operculella	6	Field	3rd	GV	1-11	0.69-1.03	Briese and Mende, 1981

[a]Differences not significant; [b]the most resistant strains showed virtual immunity to the virus; [c]NPV obtained from three different insect hosts

appropriate methods it might be possible to demonstrate the presence of insects with relatively high resistance within a population. In such cases it should be possible to select these individuals and confirm the suggestion by comparing their progeny with the original population.

Several studies have examined the mode of inheritance of resistance (Table 18-2). These show that genetic control can occur in several different ways, as a dominant or recessive major gene, as genes lacking dominance or as a polygenic complex. The studies mentioned in Table 18-2 have only considered the mode of inheritance for large shifts in the response of single species. Most studies that consider more than just a few populations have shown LD_{50} values to be widely distributed rather than clumped at a few extremes, which indicates that other genetic factors might be involved. Aratake (1973a, b) also found variability within as well as between strains of silkworm treated with CPV or NPV, and observed marked F_1 heterosis, which supports Watanabe's (1971) contention that several other genes probably modify the levels of resistance shown in any one strain. Much of the variability in susceptibility between populations may be subject to polygenic control.

Despite several reported failures, it is possible to select for resistance after only a few generations (Table 18-3). In the case of the CPV of B. mori, Watanabe (1967) found that a low selection pressure produced no difference in resistance over the first four generations (Table 18-3), but when he raised the selection pressure to over 60% he rapidly obtained an increase in resistance in the remaining four generations. Briese (unpublished data) showed a significant increase in resistance of the potato moth, Phthorimaea operculella to GV after only five generations. However, the slope of the dosage-mortality regression for the selected line was also significantly higher, suggesting that the greater resistance was due mainly to the elimination of more susceptible individuals from the population. Watanabe (1967) also found that regression slopes were slightly steeper for the selected strains than for controls, and suggested that this might be due to a reduction in genetic variability during the course of selection. The failure to select for resistance in certain cases does not necessarily mean that such selection is impossible for that particular species. It may merely be the result of the absence of a resistant gene or genes in a small laboratory population with limited genetic variability. Finally, we should mention here the related phenomenon of virus induction. Aizawa et al. (1961), in 13 generations, selected for resistance to the induction by stress of latent virus within a silkworm strain. Dosage-mortality data indicated that the mechanism of resistance to virus induction is independent of that to virus infection.

TABLE 18-2. GENETIC MECHANISMS OF INFRASPECIFIC RESISTANCE TO A VIRUS

Insect Species	Virus	Inferred Genetic Mechanism	Author
Bombyx mori	CPV	Dominant major gene	Watanabe, 1965
	FV	Recessive major gene	Funada, 1968
	SPV	Recessive major gene	Furuta, 1978
	DNV	Recessive major gene	Watanabe and Maeda, 1978
Spodoptera frugiperda	NPV	One or more genes lacking dominance	Reichelderfer and Benton, 1974
Epiphyas postvittana	NPV	Polygenic	Briese et al., 1980
Phthorimaea operculella	GV	Dominant major gene	Briese and Mende (unpublished data)

TABLE 18-3. ATTEMPTS TO ARTIFICIALLY SELECT FOR INCREASED RESISTANCE TO A VIRUS IN AN INSECT POPULATION (See Text for Details)

Insect Species	Origin of Population	Virus	No. of Generations Selected	Selection Pressure (% Mortality)	Resistance Factor	Author(s)
Lepidoptera						
Bombyx mori	Culture	FV	5	Not given	11-28	Uzigawa and Aruga, 1965
Bomybx mori	Culture	CPV	8	45-83%	10-14[a]	Watanabe, 1967
Bombyx mori	Culture	CPV	8	53-92%	14-20[a]	Watanabe, 1967
Heliothis zea	Laboratory	NPV	25	5-70%	No difference	Ignoffo and Allen, 1972
Heliothis armigera	Laboratory	NPV	22	40-80%	No difference	Whitlock, 1977
Heliothis armigera	Laboratory	GV	22	15-90%	No difference	Whitlock, 1977
Phthorimaea operculella	Laboratory	GV	10	40-70%	6-10	Briese (unpublished data)
Hymenoptera						
Apis mellifera	Laboratory	CBPV	4	< 25%	ca. 10^4 [b]	Kulincevic and Rothenbuhler, 1975; Rinderer et al., 1975

[a]Values for selection on two different strains of virus

[b]Estimated from raw data of Rinderer et al., 1975

In addition to these deliberate attempts, there is strong circumstantial evidence that natural selection for increased resistance has occurred in inbreeding populations (Table 18-4). These examples all have the following features in common: i. a laboratory population of an insect has been found to have a significantly higher level of resistance to a virus than either field populations or other laboratory populations; ii. these populations have known histories of unintentional exposure to that virus and consist of descendants of the survivors of those outbreaks; iii. where measured, the slopes of dosage-mortality regression for resistant lines do not differ significantly from those for susceptible lines, indicating a true shift in the resistance level; iv. the level of resistance appears to be fairly stable.

Natural selection in outbreeding field populations has been documented in only one case (Table 18-4) where actual dosage responses were measured. Martignoni (1957) found an increase in LD_{50} for GV of the larch bud moth *Eucosma griseana* of about 38 times following a natural epizootic in the field. This, however, was accompanied by a considerably steeper slope of regression, which indicates that the GV epizootic merely removed most of the more susceptible individuals from the population. It seems probable that the mean level of resistance could have reverted once again, as genes from the susceptible survivors or immigrants spread through the population, though this was not investigated. Two other cases, sometimes cited in support of field-developed resistance, are not included in Table 18-4 because of their speculative nature. Bergold (1951) considered it possible that the spruce budworm may have high resistance to a polyhedrosis virus because LD_{50} values were several orders of magnitude greater than for other insect-virus relationships, while Bird and Elgee (1957) thought that the sawfly (*Diprion hercyniae*) may have developed resistance to its virus, because the virus failed to reach epizootic proportions under conditions in which it has previously done so. Considering that viruses have been used as a control agents for a relatively short period and that few detailed observations of natural epizootics have been made, the situation remains as Burges (1971) stated, i.e. that there is no substantial evidence for the development of true resistance to viruses in the field.

The final stage in any study of resistance is to determine physiological or biochemical processes by which it occurs. Most data on this aspect have come from Japanese studies of the silkworm. Watanabe (1966d) found 1000 to 10,000 fold lower LD_{50} values when silkworm larvae were given subcutaneous injections of CPV than when given it perorally, suggesting that much of the ingested virus did not enter the susceptible epithelial cells of the midgut. Therefore, resistance might depend more on inhibitory mechanisms in the gut and less on suppression of virus multiplication within cells. This may

TABLE 18-4

A. Inbred laboratory populations in which virus outbreaks are known to have previously occurred. The reported levels of resistance appear to be stable.

Insect Species	Virus	Resistance Factor[a]	Author
Pieris brassicae	GV	10	David and Gardiner, 1960, 1965
Spodoptera frugiperda	NPV	12	Klein and Podoler, 1978
Epiphyas postvittana	NPV	160	Briese et al., 1980
Phthorimaea operculella	GV	30	Briese and Mende, 1981

B. Outbreeding field populations in which virus outbreaks are known to have previously occurred. The stability of the increased level of resistance is unknown.

Eucosma griseana	GV	38[c]	Martignoni, 1957

[a]The level of resistance relative to a known susceptible strain

[b]Estimated from the original data by Burges (1971)

[c]Level of resistance relative to that measured for the same population before an outbreak of virus

apply to most Lepidoptera, as Stairs (1965) also found a large difference in response to NPV administered perorally or subcutaneously to the wax moth, *Galleria mellonella*.

Antiviral substances in silkworm gut juice, such as the red fluorescent protein which agglutinates virus particles (Hayashiya et al., 1978), have been found to inactivate both NPV (Aratake and Ueno, 1973) and CPV (Aratake et al., 1974). While the activity of these substances varied between strains it was not correlated with interstrain differences in susceptibility to peroral infection (Aratake et al., 1974; Watanabe, 1966d), indicating that they only played a minor role in the phenomenon of intraspecific resistance. Gut juices from older larvae showed a higher antiviral activity (Aratake and Ueno, 1973), suggesting that this might contribute to the development of maturation immunity.

Watanabe and Takamiya (1976) found that silkworm larvae reared in constant light were more resistant to CPV than those reared in constant darkness. This was associated with an increase in the ratio of columnar to goblet cells in the midgut of those larvae. There is also a positive correlation between resistance to CPV, which directly affects the midgut cells, and to NPV which must pass through the midgut cells to attack susceptible tissues beyond (Watanabe, 1966a), indicating that a common mechanism involving these cells might be important in both cases. This barrier to infection also seems to be important in maturation immunity, for Kobayashi et al. (1969) found that the reduction in susceptibility to NPV with age was related to the inhibitory properties of the intestine. The decrease in susceptibility with age was more rapid in resistant than in susceptible larvae. In the case of CPV, old columnar cells infected with the virus are discharged into the midgut lumen at larval molt and new cells develop to take their place (Inoue and Miyagawa, 1978). The degree of this epithelial regeneration varied with the strain of silkworm. In the silkworm, new columnar cells became reinfected with CPV. However if these new cells should be immune, this could result in resistance to the virus. This does in fact occur in the fall webworm *Hyphantria cunea*, which regenerates columnar cells that are subsequently immune, following infection with *Bombyx* CPV (Yamaguchi, 1979).

Similar mechanisms seem to be involved with resistance of *Bombyx* to the non-occluded flacherie virus (FV). Following peroral entry, this virus multiplied to the same extent in both susceptible and resistant larvae, but in the resistant strain infected goblet cells were discharged into the gut lumen at each molt and regenerative cells developed into new goblet cells to replace them (Inoue, 1974). Although this mechanism is similar in action to that for CPV, it seems to have a different basis, for an analysis of data given by Watanabe et al. (1974) shows that there is not significant correlation between susceptiblity to FV and to CPV.

There is a degree of differential susceptibility between strains of silkworm given subcutaneous injections of CPV (Watanabe, 1966d), suggesting that mechanisms for resistance might also operate here. These could include the presence of viral inhibitory factors in the hemolymph (Horie and Watanabe 1980) or the phagocytosis of virus particles. In support of the latter possibility, Stairs (1964) found that blockage of phagocytes with India ink increased the susceptibility of G. mellonella to its NPV by 13 times.

The available evidence suggests that the major barrier to infection by virus particles is the midgut epithelium, and that most of the intraspecific differences in susceptibility are due to differences in midgut cellular defense mechanisms. These mechanisms may be further augmented by differences occurring prior to this barrier (antiviral substances in the gut juice) and subsequent to it (phagocytosis, inhibitory factors). For example, David (1978) reported that strains of P. brassicae, susceptible to and resistant to peroral infection were similarly susceptible or resistant when free CPV virions were injected into the hemolymph, which suggests that the major barrier in this species occurs subsequent to infection of the midgut. The existence of several defenses against viral infection is consistent with the polygenic nature of resistance. Major shifts in resistance levels are however likely to occur through a change in the primary defense mechanism, i.e. the midgut barrier.

Apart from the Lepidoptera, the only other insect in which intraspecific differences in susceptibility to viruses have been demonstrated is the honey bee, Apis mellifera. Bailey (1967) suggested that there were differences in susceptibility of honey bees to sacbrood virus (SBV), based on post-mortems of dead larvae from different colonies. Differences in the response of colonies to chronic bee paralysis virus (CBPV) have also been reported (Bailey, 1965; Kulincevic and Rothenbuhler, 1975). CBPV particles apparently occur in many normal bees, and most are able to resist viral spread in their tissues. It has been suggested that bees only become paralyzed if they have inherited susceptibility from the queen (Bailey, 1965). Rinderer et al. (1975) were able to demonstrate differences between three genetically different stocks of honey bee to "hairless-black syndrome", caused by a virus serologically indistinguishable from CBPV (Rinderer and Green, 1976). Their data indicate a difference of about 10^4 in LD_{50} between a strain selected for resistance and one selected for susceptibility, though there appeared to be large within-strain variability. These two strains had been selected from a common parent over a period of four generations (Kulincevic and Rothenbuhler, 1975) (Table 18-4), providing further evidence for the genetic nature of resistance to CBPV. However, neither the mode of inheritance nor the mechanism of resistance have yet been studied.

BACTERIA

It has long been thought that races of the honey bee vary in their susceptibility to bacteria responsible for European foulbrood, and the replacement of German bees by supposedly more resistant Italian bees was thought to have been responsible for the virtual elimination of this disease from North America (Rothenbuhler, 1979). Recent studies by Cherepov (1970) confirmed that there are some differences in the survival rates of infected worker and drone larvae from five different Russian races. No further studies on this resistance have been published.

The best documented case of resistance to bacteria by an insect also involves honey bee strains with differential susceptibility to the causative agent of American foulbrood disease, *Bacillus larvae*. Rothenbuhler and Thompson (1956) treated bee larvae with a single dose of *B. larvae* spores and found significantly lower mortality of larvae from colonies that had been bred for resistance than of those bred for susceptibility. Hoage and Rothenbuhler (1966) confirmed this difference between strains when they found a significant, albeit small, difference (resistance factor of 1.9) in LD_{50} values. The slope of the dosage-mortality regression was found to be lower in the more resistant line. This indicates increased variability, which might be due to the presence of a mixture of resistant and susceptible insects in the population or to heterozygosity of a semidominant resistance gene (Burges, 1971). In either case there is the potential for selection of a more resistant strain. Hoage and Peters (1969) demonstrated that resistance could be rapidly selected for; using a hybrid of the resistant and susceptible parent lines, they increased the level of resistance by 45% within three generations. Although the genetics of this physiological resistance has not been fully determined, it is known that matings between resistant queens and susceptible drones results in progeny having lesser resistance than the resistant parental line (Lewis and Rothenbuhler, 1961).

The mechanism of resistance appears to be related to maturation immunity, i.e. the increase in resistance of larvae with age. Bamrick and Rothenbuhler (1961) found that susceptible larvae were immune at 48 hours of age, whereas resistant larvae became immune by 36 hours. The maximum difference in susceptibility occurred at age 21 hours when there was greater spore germination and multiplication of bacteria in the susceptible strain (Bamrick, 1967). Since Sutter et al. (1968) found that resistant larvae developed more rapidly than susceptible larvae of the same age, it appears that resistance is attributable to a genetically determined increase in the rate of early larval development, leading to more rapid acquisition of maturation immunity. The precise mechanism of this developing immunity remains to be determined

TABLE 18-5. INTRASPECIFIC DIFFERENCES IN THE SUSCEPTIBILITY OF INSECT SPECIES TO *BACILLUS THURINGIENSIS*

Insect Species	No. of Populations Tested	Origin of Populations	Instar Tested	Preparation of *B. thuringiensis* used (Serotype in brackets)	Resistance factor (rounded to whole number)	Author
LEPIDOPTERA (Susceptible primarily to δ-endotoxin (Type I and II of Heimpel and Angus, 1959))						
Bombyx mori	3	Culture	3rd & 4th	var. *thuringiensis* (I) "Bactospeine"	6-50[a]	Aizawa et al., 1962
	3	Culture	3rd & 4th	var. *thuringiensis* (I) "Thuricide"	6-15[a]	Aizawa et al., 1962
	6	Culture	4th	strain A. (?)	2-6 [a]	Aizawa, 1971
	6	Culture	4th	strain B. (?)	7-41[a]	Aizawa, 1971
*Plodia interpunctella**	13	4 field 9 laboratory	1st	var. *kurstaki* HD-1 (III) "Dipel"	6	Kinsinger and McGaughey, 1979
Ephestia cautella	7	2 field 5 laboratory	1st	var. *kurstaki* HD-1 (III) "Dipel"	10	Kinsinger and McGaughey, 1979
DIPTERA (Susceptible primarily to β-exotoxin)						
Musca domestica	4	laboratory[b]	24 h old	var. *thuringiensis* (I) autoclaved exotoxin filtrate	No difference	Galichet, 1967
	5	laboratory[b]	48 h old	var. *thuringiensis* (I)	24-93%[c]	Hall and Arakawa, 1959

[a]Range over several series of treatments; [b]Includes strains resistant to chemical insecticides;

[c]Percentage mortality at a single dose;

*Although *P. interpunctella* is susceptible to crystals alone, a reduction in LD_{50} does occur when a 50:50 mixture of spores and crystals is used (McGaughey, 1978). Therefore this species seems to be intermediary between types I and II and type III of Heimpel and Angus' classification.

though it appears to be related to the ability of the bacteria to build up a population sufficient to penetrate the peritrophic membrane and midgut epithelium, and perhaps to the thickness of the peritrophic membrane (Bamrick, 1964; Davidson, 1973).

In social insects, relative susceptibility to infection can also be mediated by behavioral factors, such as the speed with which diseased larvae are detected and removed by honey bee workers (Rothenbuhler, 1964a). These factors are also genetically controlled (Rothenbuhler, 1964b), and, in combination with physiological factors determine the overall level of resistance shown by a colony.

Bacillus popilliae, the causative agent of milky disease in larvae of scarabeid beetles, including the Japanese beetle *Popillia japonica* and the Oriental beetle *Anomala orientalis*, has been successfully used to control these insects in the USA since 1939. In Connecticut, control was still effective in 1962, but by 1974 the insect populations were increasing despite an extensive build up of disease inoculum in the soil (Dunbar and Beard, 1975). These authors found that the infectivity of the *B. popilliae* spores had diminished significantly, and also presented evidence that an increase in resistance to infection may have developed in the larvae. Splittstoesser et al. (1978) have observed defense mechanisms in another scarabeid species, *Amphimallon majalis*, against *B. popilliae*. These involved encapsulation by circulating hemocytes and the extrusion of infected columnar epithelium followed by regeneration (see Chapter 7). This latter process is similar to that involved in the resistance of Lepidoptera to viruses, and, if it is subject to genetic variability in scarab larvae, could provide a mechanism by which increased resistance to *B. popilliae* can occur.

The most intensely studied entomopathogenic bacterium is the sporeformer *Bacillus thuringiensis*, which differs from *B. larvae* and *B. popilliae* in that it produces metabolites toxic to many insect hosts. The remainder of reports on intraspecific resistance to bacteria concern this species. Only a few studies report differences between field or laboratory populations of a species that had not been previously subject to artificial selection (Table 18-5). They indicate only low to moderate degrees of resistance. For both the almond moth, *Ephestia cautella*, and the Indian meal moth, *Plodia interpunctella*, treated with a spore/crystal complex, the slopes of dosage-mortality lines did not differ significantly (Kinsinger and McGaughey, 1979), indicating real differences in resistance of the populations, not merely a reduction in variability due to elimination of more susceptible forms. These two species show different responses to *B. thuringiensis* (McGaughey, 1978); *E. cautella* is primarily susceptible to the crystal toxin, while *P. interpunctella*, which was also susceptible to the crystal toxin alone, showed greatest mortality when treated with a 50:50 spore/crystal mixture. Consequently, mechanisms of resistance shown by these species are possibly different. Of the other species cited in Table 18-5, the silkworm, B. mori is known to be susceptible to the δ-endotoxin of the crystal (Burgerjon and Martouret, 1971). The house fly, M. domestica, although affected in some instances by δ-endotoxin, is primarily susceptible to the β-exotoxin (Rogoff et al., 1969). Although there were some differences in LD_{50} values among strains of these species, the slopes of regression lines were not given.

For several pest insects, attempts have been made to select for increased resistance to several different preparations of B. thuringiensis in the laboratory to determine whether this might occur in the field should it be used as an insecticide. Four of the studies indicated some increase in resistance and three showed no change (Table 18-6). The studies in which a shift in resistance was obtained all concerned Diptera, in which the resistance is considered to be manifest against the β-exotoxin produced by the strains of B. thuringiensis serotype I (Devriendt and Martouret, 1976). It may be more difficult to select for resistance in Lepidoptera which may be affected by either the crystal protein, spore germination and multiplication, β-exotoxin or combinations of these components (Burgerjon and Martouret, 1971); e.g. in Table 18-6, A. kuehniella is only affected by a combination of spores and crystals (Heimpel and Angus, 1959). Bacillus thuringiensis has been used to control the cabbage looper Trichoplusia ni, a species susceptible to δ-endotoxin, under natural conditions since the 1950's, and to date no evidence of selection for resistance has been presented (D.E. Pinnock, pers. comm.). However, the two studies of Lepidoptera which produced no change in susceptibility involved high selection pressures for only a few generations (Yamvrias, 1962; Devriendt and Martouret, 1976), and the study on a Dipteran which failed was maintained for a large number of generations at a low selection pressure (Feigin, 1963). The greatest increase in resistance occurred in the study by Harvey and Howell (1965) on M. domestica in which high selection pressures were maintained over a large number of generations (Table 18-5), which suggests that both these factors may be necessary to produce a shift in response.

Harvey and Howell (1965) used "Bakthane", a commercial preparation of B. thuringiensis, as the selection agent, and demonstrated by using autoclaved Bakthane that the resistance shown in M. domestica was against the exotoxin. They obtained differences in susceptibility of a similar order of magnitude to those found between some unselected populations of insects (Table 18-5). Although the slopes of the dosage - mortality regressions were somewhat steeper for the selected line, they were not significantly different, which suggests there had been

TABLE 18-6. ATTEMPTS TO ARTIFICIALLY SELECT FOR INCREASED RESISTANCE TO BACILLUS THURINGIENSIS IN AN INSECT POPULATION (See Text for details)

Insect Species	Origin of Population	Preparation of B. thuringiensis used	No. of Generations used	Selection Pressure (% mortality)	Resistance factor (rounded to whole number)	Author
DIPTERA (Susceptible primarily to β-exotoxin)						
Musca domestica	Field	var. thuringiensis (I)	27	ca. 40%	No difference	Feigin, 1963
	Laboratory	var. thuringiensis (I) "Bakthane"	50	50-90%	8-14[a]	Harvey and Howell, 1965
	Field	var. thuringiensis (I) "Bakthane"	10	Not stated	2[b]	Harvey and Howell, 1965
	Laboratory	var. thuringiensis (I)	25	Not stated	6	Wilson and Burns, 1968
Drosophila melanogaster	Laboratory	AL 174 AS exotoxin standard	30	Not stated	3	van Herrewege, 1969 (cited in Devriendt and Martouret, 1976)
LEPIDOPTERA (Susceptible only to spore/crystal combination (Type III of Heimpel and Angus, 1959))						
Anagasta kuehniella	Laboratory	var. alesti E58 (III)		60-90%	No difference	Yamvrias, 1962
LEPIDOPTERA (Susceptible primarily to δ-endotoxin (Types I and II of Hemipel and Angus, 1959))						
Plutella maculipennis	Laboratory	var. thuringiensis E61 (I) crystal/spore complex	10	90%	No difference	Devriendt and Martouret, 1976

[a]Range of several series of comparisons from generations 27 to 50; [b]Not significantly different

a true shift in the level of resistance. Harvey and Howell (1965) observed that when the selection pressure was removed, there was a significant decline in the resistance of the selected strain over 20 generations. As the population continued to inbreed, this suggests that increased resistance may have been accompanied by a reduction in fitness under "normal" conditions.

To date there have been no attempts to understand either the genetics of intraspecific resistance to *B. thuringiensis* or differences in the physiological and biochemical processes between strains of varying susceptibility. Because of the complex of activities of *B. thuringiensis*, these processes are likely to be more complex and more characteristic of individual species than appears to be the case with insect defense mechanisms against viruses. Kinsinger and McGaughey (1979) found no correlation between the response of *E. cautella* and *P. interpunctella* populations to *B. thuringiensis* and midgut pH, a factor considered to contribute to differences in susceptibility between species (Burgerjon and Martouret, 1971). The findings and Hall and Arakawa (1959) and Galichet (1967) that strains of the house fly, resistant to a broad spectrum of chemical insecticides, do not manifest increased resistance to the thermostable β-exotoxin of *B. thuringiensis* are also pertinent, for they may indicate that the mechanism conferring resistance to chemical insecticides and that conferring resistance to the bacterial toxins are not related.

FUNGI

Most evidence for intraspecific differences in the susceptibility of insect species to fungal attack comes from studies of two domesticated insects, the silkworm *B. mori* and honey bee *A. mellifera*. Chalkbrood, a disease of larval honey bees caused by the fungus *Ascosphaera apis*, is becoming an increasing problem for bee-keepers. However, variability in susceptiblity of bee colonies to the fungus has been reported in several instances (Moeller and Williams, 1976; Menapace, 1978). DeJong (1976) administered standard doses of *Ascosphaera* spores to 3-day old larvae reared under constant conditions and found infection rates that varied between colonies from 0.5% to 78%. DeJong (1976) and others have suggested selection for resistance as a control for chalkbrood, but selection of resistant lines has not yet been reported. DeJong found that the rate at which infected larvae were removed by worker bees influenced the apparent susceptibility of a colony. Thus there appears to be a behavioral factor in resistance to chalkbrood similar to that described for American foulbrood resistance.

Differences in the susceptiblity of silkworm larvae to three species of fungus have been reported (Table 18-7), and,

TABLE 18-7. DIFFERENCES IN SUSCEPTIBILITY OF SILKWORM POPULATIONS TO INFECTION BY SEVERAL SPECIES OF FUNGUS

Fungus	No. of Populations Tested	Range of Susceptibility	Author
Beauveria bassiana	18	19 - 100 %[a]	Aratake, 1961
Beauveria bassiana	12	47 - 100 %[a]	Aratake, 1961
Isaria farinosa	8	0 - 50.5 %[b]	Aratake et al., 1961
Isaria farinosa	6	2.2 - 86.1 %[b]	Aratake et al., 1961
Aspergillus flavus isolate S85	4	60 -1260 [c]	Kawakami, 1975
A. flavus isolate K1	4	60 - 540	Kawakami, 1975

[a]Mortality at a single dose

[b]Mortality due to natural infection

[c]LC_{50} (Spores/mm^3)

in the case of resistance to white (Beauveria bassiana) and yellow muscardines (Isaria farinosa), a thorough analysis of the genetic mechanisms has been made (Aratake, 1961; Aratake et al., 1961). These workers found that essentially the same mechanism controlled resistance in both cases. The polygenic mechanisms involved paired and multiple alleles on two chromosomes. No genetic mechanism was determined for resistance to Aspergillus flavus (see Table 18-7). The populations used originated from quite distinct Chinese and Japanese geographic races (Yokoyama, 1979), and therefore it is reasonable to assume that the basis for the differences was genetic rather than environmental.

Aratake (1961) found that the growth of germ tubes of Beauveria bassiana was greater in the hemolymph of susceptible larvae than in resistant ones, indicating that the physiological mechanism might incorporate properties of the hemolymph. In the case of resistance to Aspergillus, increased resistance was neither related to percentage spore germination nor to length of germ tubes (Kawakami, 1975). However, Koidsumi (1957) found that removal of cuticular lipids from silkworm larvae greatly increased their susceptibility to attack by Aspergillus, and further found that antifungal activity was associated with the amount of free medium-chain fatty acids. An extract of the living integument and exuviae of resistant Chinese race silkworms showed a greater antifungal action than that from the susceptible Japanese race (Koidsumi and Wada, 1955), and this was attributed to a greater concentration of the free fatty acids in the resistant silkworms. This is a defense mechanism against fungal penetration of the cuticle which shows variation between populations of an insect species.

Many fungi apparently owe their insect pathogenicity to toxic metabolites. Aspergillus produces an insecticidal toxin Aflatoxin B_1 (AFB_1). Llewellyn and Chinnici (1978) found variability in the response to five wild-type laboratory strains of the fruit fly, Drosophila melanogaster, to AFB_1 administered perorally. Studies of hybrid crosses indicated that the variable resistance to AFB_1 in Drosophila is due to a complex polygenic system, in which each of the two major autosomes carries genes exerting some influence on resistance (Chinnici and Llewellyn, 1979). The mechanism by which resistant Drosophila avoid the toxic affects of AFB_1 has not yet been elucidated. These are therefore potentially at least two mechanisms of defense against fungal attack which are subject to genetic variation; the cuticular defense against fungal penetration, as exhibited by silkworm, and the resistance to toxic metabolites as shown by Drosophila.

Only a few cases of differential susceptibility to fungi in pest species have been reported. Bell and Hamalle (1971) noted briefly that a laboratory population of the wireworm, Conoderus falli, treated with the fungus, Metarrhizium anisopliae, showed significantly longer LT_{50} values than a population originating in the field; slopes of the responses were not significantly different. The authors implied without direct evidence that the differences may be due to differences in environmental factors between laboratory and field, however genetic differences in the defense mechanism against Metarrhizium are also possible. In contrast, Ignoffo et al. (1976) found no significant differences among the LT_{50} values of three strains of velvet bean caterpillar, Anticarsia gemmatalis, treated with the fungus Nomuraea rileyi.

In two studies, large intraspecific differences were demonstrated in the susceptiblity of the pea aphid, Acyrthosiphon pisum, to fungi of the genus Entomophthora. Papierok and Wilding (1979) found that aphids from one laboratory population were not infected by either of two strains of E. obscura at dosages which produced 100% infection in another aphid population. In another study, using progeny from ten specific females, R. J. Milner (unpublished data) found that offspring of four females almost all succumbed to E. exitialis spores, while offspring of the remaining six females were unaffected by the same spore dosage. The resistant and susceptible females were derived from the same population, which had been reared in the laboratory for only eight generations. Bioassays of this population with E. exitalis produced infection rates which quickly rose to ca. 24%, with no further response above this infection level. This population thus appears to contain a mixture of resistant and susceptible individuals in a ration of approximately 3:1. Being chiefly parthenogenetic, aphids pose special problems for studies on the mode of inheritance and the possible selection of resistance. The large intraspecific differences in susceptiblity to Entomophthora sp. suggest that the pea aphid might be a good experimental insect for studying the biochemical or physiological processes involved in defense against fungal attack.

There is only one report of deliberate selection for resistance to fungi. Vansulin (1974) obtained a reduction in mortality of 94% to 48% over eight generations in a strain of the mosquito Culex pipiens molestus treated with a constant dose of "Boverin" (a commercial preparation of Beauveria bassiana). However, as slopes of dosage-mortality regressions were not determined, it is unclear whether more susceptible individuals were simply eliminated or whether a resistance gene (or genes) had been selected.

PROTOZOA

Probably the first example of intraspecific differences in the resistance of an insect species to a microbial pathogen involved a protozoan. According to Steinhaus (1949), in the late 1860's Pasteur was able to select a stock of silkworm, B. mori, resistant to disease caused by the microsprodan Nosema bombycis. Very little further work has been carried out on the possibility of resistance of protozoans, although some, such as Nosema locustae and Vairimorpha necatrix, are candidate biological control agents.

According to Weiser (1969), certain races of the moths Antherea pernyi and Platysamia cecropia are more resistant to microsporidans than others, and such races also occur in the silkworm and honey bee. Several workers in the U.S.S.R. have also reported variability in susceptibility to the microsporidan Nosema apis among races of the honey bee (Table 18-8). It is not known to what extent these differences may reflect the "normal" variability in susceptibility to infection, but in some cases (e.g. Sidorov et al., 1975; Gapanova and Mel'nik, 1975) the differences are very large indicating the possibility of discrete differences in response. Moreover, Sidorov and Mikhailenko (1969) found a significant interstrain differences in the number of spores in the mid- and hindguts of experimentally infected bees, and observed a 9-fold difference in the multiplication rate of spores after the infection of different populations.

Sidorov et al. (1975) found that the F_1 hybrid bee lines were relatively more resistant. This may be due to a heterosis effect, and indicates that the differences in susceptibility are likely to be genetic.

Studies on the related topic of the susceptibility of insect vectors to their vectored protozoan parasites give an indication of the type of genetic mechanism that might control differences in susceptibility of insect-pathogenic protozoa. For example Kilama and Craig (1969) found that there was a wide range in the susceptibility of the mosquito Aedes aegypti to infection with Plasmodium gallinaceum. Resistance could be rapidly selected for and was found to be controlled by a single autosomal recessive gene.

No studies have been reported which elucidate differences in defense mechanisms between protozoa-susceptible and resistant strains of an insect species. Weiser (1969) reported that individuals of susceptible strains are larger, eat more and regenerate their tissues more slowly than more resistant larvae. This suggests that rapid regeneration of infected tissues may confer increased resistance to protozoa, similar to defense mechanisms observed in insects resistant to some viruses and bacteria.

TABLE 18-8. VARIABILITY IN SUSCEPTIBILITY OF POPULATIONS OF THE HONEY BEE *APIS MELLIFERA* TO INFECTION BY THE MICROSPORIDAN *NOSEMA APIS*

Number of Populations of *Apis mellifera* tested	Range of Infection Rates Obtained	Author
4	18.8-51.9%	Cherepov and Kuznetsova, 1969
5	77.5-95.1%	Sidorov and Mikhailenko, 1969
Several	2-5 fold difference	Andreev, 1973
7 (low dose)	14.3-95.0%	Sidorov et al., 1975
7 (high dose)	51.6-93.3%	Sidorov et al., 1975
4	12-86%	Gapanova and Mel'nik, 1975

NEMATODES

Only two reports have been published on resistance to insect pathogenic nematodes, both concerning mosquitoes (Table 18-9). In these cases, resistance was unintentionally produced over a large number of generations through the practice of using pupae which had been exposed to nematodes but escaped infection to maintain the laboratory colony. The susceptibility to nematode infection was reduced by 50% during the course of selection in both mosquito species. Petersen (1978) found no evidence of humoral defense mechanisms, and could not determine the mode of resistance. However, Woodard and Fukuda (1977) found that larvae of their selected strain were much more active than those of the unselected strain, and defended themselves against attacking nematodes. Thus, there appears to be a behavioral component to resistance, although as Petersen (1978) points out there is a possibility that differences in the nature and thickness of the cuticle could retard penetration by the nematode and contribute to the reduction in susceptibility.

Other nematodes such as Neoaplectana spp. which have been used in biological control, may invade their insect host perorally. Although there is presently no evidence of resistance to these pathogens, resistance has been documented to filarial nematodes that enter their insect vectors perorally. A sex-linked recessive gene controls the susceptibility of the mosquito A. aegypti to several nematode species, and variation in the susceptibility shown by field populations can be correlated with differences in the frequency of this gene (MacDonald, 1967). Kartman (1953) found differences in susceptibility to infection by the nematode Dirofilaria immitis between populations of the mosquito A. aegypti from different geographic zones, and was able to select for increased resistance in eight generations. The differences in susceptibility were inversely correlated to the percentage encapsulation of microfiliaria. Such encapsulation of invading Neoaplectana carpocapsae has been observed in several mosquito species (Bronskill, 1962). This is a mechanism of defense against those nematodes which enter perorally, and is one which appears to be subject to genetic variability.

DISCUSSION

There is now evidence for intraspecific differences in the innate resistance shown by insects to microbial pathogens. There is a disparity in the numbers of studies reported for the major groups of pathogens; the most frequent reports concern viruses, followed by bacteria, fungi, protozoa and nematodes respectively. This may reflect more the amounts of research in these areas than real differences between groups.

TABLE 18-9. RELATIVE SUSCEPTIBILITY TO ATTACK BY MERMITHID NEMATODES OF SELECTED AND UNSELECTED STRAINS OF TWO MOSQUITO SPECIES

Nematode Species	Mosquito Species	Strain	No. of Generations selected	Parasite to Host Ratio	Percentage Insects Infected	Percentage Nematodes Penetrating	Author
Diximermis peterseni	*Anopheles quadrimaculatus*	Selected	< 104	5 : 1	42	15	Woodard and Fukuda, 1977
		Unselected		5 : 1	90	56	
Romanomermis culicivorax	*Culex pipiens quinquefasciatus*	Selected	≃ 400	5 : 1	26.7 ± 3.5	-	Petersen, 1978
		Unselected		5 : 1	57.1 ± 8.2	-	

There is also an imbalance in the distribution of studies among groups of insects. The majority of cases concern the principal species cultured by man, the silkworm, *B. mori*, and the honey bee, *A. mellifera*. Both have been subject to intensive selection for improved performance and production characteristics, including resistance to disease. For *B. mori*, which has about 700 genetically distinct races (Yokoyama, 1979), there exist populations relatively resistant to five viruses, a bacterium, three fungi and a protozoan, while *A. mellifera*, for which there are also many different races (Rothenbuhler, 1979), has populations that show differential susceptibility to two viruses, one and possibly a second bacterium, a fungus and a protozoan.

There have been no concerted efforts to investigate resistance in pest insects and the literature consists mainly of *ad hoc* examples. This may appear to be surprising in view of the potential economic costs associated with the development of resistance to a microbial pathogen in field use, but can be accounted for partly by two factors. The first is that, in any particular laboratory, work with pest insects usually involves only one or a few strains. A deliberate effort should be made to test the responses of substantial numbers of host populations (preferably field-sampled) to candidate microbial control agents. The second factor is the attitude that because there is yet no clear-cut evidence for resistance to pathogens under natural conditions, insects are highly unlikely to develop it (e.g. McBain-Cameron, 1967; Jaques, 1973). The information presented in this chapter should, to some extent, dispel this.

While it is possible to select for resistance, this has so far been done only by restricting the normal gene flow. In the case of domesticated insects, in which breeding is controlled, selection provides an important tool for minimizing the impact of disease. However, if resistance arises in pest insects in the field, an inflow of susceptible genes from either individuals in the population that have avoided the pathogen or from immigrants can slow or suppress the development of resistance, as has been shown in models of resistance to chemical insecticides (Georghiou and Taylor, 1977). The critical evidence of a clearly demonstrated and stable increase in the frequency of a gene or genes conferring resistance to a microbial pathogen in a field population of insects is still lacking. It is fortunate perhaps that there seems to have been no such selection in any of the situations where microbial pathogens have been used in field control programs. However, the existence of significant natural variability between field populations, such as found for *P. operculella* (Briese and Mende, 1981) and *E. postvittana* (Geier and Briese, 1979; Briese et al., 1980), and the observations that these are heritable (Briese et al., 1980, and unpublished data), demonstrate that differential resistance, and the possibility of selecting for it, must be considered when using pathogens as biological control agents.

Studies of the genetics and the function of resistance mechanisms would help to refine decision-making aimed at controlling the spread of resistance. The possibility of cross-resistance to pathogens is another important aspect. For example, there appears to be a relationship between resistance to NPV and CPV in *B. mori*; one strain of this species (C124) is highly resistant to three non-occluded viruses while another (N124) is susceptible to all, suggesting that cross-resistance might occur.

Finally, microbial pathogens are living organisms and just as amenable to selection for changes in virulence as is the insect host for resistance. This characteristic can be used to advantage by selecting for more virulent forms of a pathogen to combat developing resistance.

REFERENCES

Aizawa, K. 1971. Strain improvement and preservation of virulence of pathogens. *In* Microbial Control of Insects and Mites, H. D. Burges and N. W. Hussey, eds., Academic Press, London and N.Y. pp. 655-672.

Aizawa, K., Foruta, Y. and Nakamura, K. 1961. Selection of a resistant strain to virus induction in the silkworm *Bombyx mori*. J. Sericult. Sci. Japan, *30*, 405-412. (In Japanese).

Aizawa, K., Kawarabata, T. and Sato, F. 1962. Response of the silkworm *Bombyx mori*, to *Bacillus thuringiensis* Berliner. J. Sericult. Sci. Japan, *31*, 253-257. (In Japanese).

Andreev, P. G. 1973. Susceptibility of various races of honeybee to Nosema disease in the conditions of Kuibyshev oblast. (In Russian) Cited in Apicult. Abst. *28*, 120.

Aratake, Y. 1961. Genetical analyses of the infection with muscardines of *Bombyx mori* L. I. Infection with the white muscardine. Bull. Sericult. Exptl. Sta. (Tokyo), *17*, 136-165. (In Japanese)

Aratake, Y. 1973a. Difference of the resistance to infectious Flacherie virus between the strains of the silkworm, *Bombyx mori* L. Sansi-Kenkyu, *86*, 48-57. (In Japanese).

Aratake, Y. 1973b. Strain differences of the silkworm *Bombyx mori* L. in the resistance to a nuclear polyhedrosis virus. J. Sericult. Sci. Japan, *42*, 230-238.

Aratake, Y. and Ueno, H. 1973. Inactivation of a nuclear-polyhedrosis virus by the gut-juice of the silkworm, *Bombyx mori* L. J. Sericult. Sci. Japan, *42*, 279-284. (In Japanese)

Aratake, Y., Kayamura, T. and Watanabe, H. 1974. Inactivation of a cytoplasmic-polyhedrosis virus by gut-juice of the silkworm, Bombyx mori L. J. Sericult. Sci. Japan, 43, 41-44. (In Japanese)

Aratake, Y., Kikukawa, S. and Kayamura, T. 1961. Genetical analyses of the infection with muscardines of Bombyx mori L. II. Infection with the yellow muscardine. Bull. Sericult. Exptl. Sta. (Tokyo), 17, 166-171. (In Japanese).

Aruga, H. and Watanabe, H. 1964. Resistance to per os infection with cytoplasmic-polyhedrosis virus in the silkworm, Bombyx mori (Linnaeus). J. Insect Pathol., 6, 387-394.

Bailey, L. 1965. Paralysis of the honeybee, Apis mellifera Linnaeus. J. Invertebr. Pathol., 7, 132-140.

Bailey, L. 1967. The incidence of virus diseases in the honeybee. Ann. Appl. Biol., 60, 43-48.

Bamrick, J. F. 1964. Resistance to American foulbrood in honeybees. V. Comparative pathogenesis in resistant and susceptible larvae. J. Insect Pathol., 6, 284-304.

Bamrick, J. F. 1967. Resistance to American foulbrood in honeybees. VI. Spore germination in larvae of different ages. J. Invertebr. Pathol., 9, 30-34.

Bamrick, J. F. and Rothenbuhler, W. C. 1961. Resistance to American foulbrood in honeybees. IV. The relationship between larval age at inoculation and mortality in a resistant and in a susceptible line. J. Insect Pathol., 3, 381-390.

Bell, J. V. and Hamalle, R. J. 1971. Comparative mortalities between field-collected and laboratory reared wireworm larvae. J. Invertebr. Pathol., 18, 150-151.

Bergold, G. H. 1951. The polyhedral disease of the spruce budworm, Choristoneura fumiferana (Clem.) (Lepidoptera: Tortricidae). Can. J. Zool., 29, 17-23.

Bird, F. T. and Elgee, D. E. 1957. A virus disease and introduced parasites as factors controlling the European spruce sawfly, Diprion hercyniae (Htg.), in central New Brunswick. Can. Entomol., 89, 371-378.

Briese, D. T. and Mende, H. A. 1981. Differences in susceptibility to a granulosis virus between populations of the potato moth Phthorimaea operculella (Zeller). Bull. Ent. Res., in press.

Briese, D. T., Mende, H. A., Grace, T. D. C. and Geier, P. W. 1980. Resistance to a nuclear polyhedrosis virus in the light-brown apple moth Epiphyas postvittana (Lepidoptera: Tortricidae). J. Invertebr. Pathol., 36, 211-215.

Bronskill, J. F. 1962. Encapsulation of rhabditoid nematodes in mosquitoes. Can. J. Zool., 40, 1269-1275.

Burgerjon, A. and Martouret, D. 1971. Determination and significance of the host spectrum of Bacillus thuringiensis. In Microbial Control of Insects and Mites, H. D. Burges and N. W. Hussey, eds., Academic Press, London and N.Y. pp. 305-326.

Burges, H. D. 1971. Possibilities of pest resistance to microbial control agents. In Microbial Control of Insects and Mites, H. D. Burges and N. W. Hussey, eds., Academic Press, London and N.Y. pp. 445-457.

Burges, H. D. and Thomson, E. M. 1971. Standardization and assay of microbial insecticides. In Microbial Control of Insects and Mites, H. D. Burges and N. W. Hussey, eds. Academic Press, London and N.Y. pp. 591-622.

Cherepov, V. T. 1970. Resistance to European foulbrood disease among the main honeybee races. Veterinariya (Moscow), 6, 54-56. (In Russian)

Cherepov, V. T. and Kuznetsova, N. F. 1969. Race variability of bee resistance to Nosema. XXII Int'l. Beekeep. Congr. Summ., 180-181. (In Russian).

Chinnici, J. P. and Llewellyn, C. 1979. Reduced Aflatoxin toxicity in hybrid crosses of Aflatoxin B_1 sensitive and resistant strains of Drosophila melanogaster (Diptera). J. Invertebr. Pathol., 33, 81-85.

David, W. A. L. 1978. The granulosis virus of Pieris brassicae (L.) and its relationship with its host. Adv. Virus. Res., 22, 112-161.

David, W. A. L. and Gardiner, B. O. C. 1960. A Pieris brassicae (Linnaeus) culture resistant to a granulosis J. Insect Pathol., 2, 106-114.

David, W. A. L. and Gardiner, B. O. C. 1965. Resistance of Pieris brassicae (Linnaeus) to granulosis virus and the virulence of the virus from different host races. J. Invertebr. Pathol., 7, 285-290.

Davidson, E. W. 1973. Ultrastructure of American foulbrood disease pathogenesis in larvae of the worker honey bee, Apis mellifera. J. Invertebr. Pathol., 21, 53-61.

De Jong, D. 1976. Experimental enhancement of chalkbrood infections. Bee World, 57, 114-115.

Devriendt, M. and Martouret, D. 1976. Absence de resistance a Bacillus thuringiensis chez la teigne des cruciferes, Plutella maculipennis (Lep: Hyponomeutidae). Entomophaga, 21, 189-199.

Dunbar, D. M. and Beard, R. L. 1975. Present status of milky disease of Japanese and Oriental beetles in Connecticut. J. Econ. Entomol., 68, 453-457.

Feigin, J. M. 1963. Exposure of the housefly to selection by Bacillus thuringiensis. Ann. Ent. Soc. Am., 56, 878-879.

Funada, T. 1968. Genetic resistance of the silkworm Bombyx mori L. to an infection of a flacherie virus. J. Sericult Sci. Japan, 37, 281-287. (In Japanese)

Furata, Y. 1978. On the heredity of the resistance to small flacherie virus in the silkworm Bombyx mori. J. Sericult Sci. Japan, 47, 241-242. (In Japanese)

Galichet, P. F. 1967. Sensitivity to the soluble heat-stable toxin of Bacillus thuringiensis of strains of Musca domestica tolerant to chemical insecticides. J. Invertebr. Pathol., 9, 261-262.

Gapanova, V. S. and Mel'nik, V. N. 1975. Prophylaxis of Nosema disease in honeybees. Veterinariya (Moscow), 9, 80-82. (In Russian)

Geier, P. W. and Briese, D. T. 1979. The light-brown moth Epiphyas postvittana (Walker). 3. Differences in susceptibility to a nuclear polyhedrosis virus. Aust. J. Ecol., 4, 187-194.

Georghiou, G. P. and Taylor, C. E. 1977. Genetic and biological influences in the evolution of insecticide resistance. J. Econ. Entomol., 70, 319-323.

Hall, I. M. and Arakawa, K. Y. 1959. The susceptibility of the housefly, Musca domestica Linnaeus, to Bacillus thuringiensis var. thuringiensis Berliner. J. Insect Pathol., 1, 351-355.

Harvey, T. L. and Howell, D. E. 1965. Resistance of the housefly to Bacillus thuringiensis. J. Invertebr. Pathol., 7, 92-100.

Hayashiya, K., Uchida, Y. and Himeno, M. 1978. Mechanism of antiviral action of red fluorescent protein (RFP) on nuclear-polyhedrosis virus (NPV) in silkworm larvae. Jap. J. Appl. Ent. Zool., 22, 238-242. (In Japanese).

Heimpel, A. M. and Angus, T. A. 1959. The site of action of crystalliferous bacteria in Lepidoptera larvae. J. Insect Pathol., 1, 152-170.

Hoage, T. R. and Peters, D. C. 1969. Selection for American foulbrood resistance in larval honeybees. J. Econ. Entomol., 62, 896-900.

Hoage, T. R. and Rothenbuhler, W. C. 1966. Larval honeybee response to various doses of Bacillus larvae spores. J. Econ. Entomol., 59, 42-45.

Hunter, D. K. and Hoffmann, D. F. 1973. Susceptibility of two strains of Indian meal moth to a granulosis virus. J. Invertebr. Pathol., 21, 114-115.

Ignoffo, C. M. and Allen, G. E. 1972. Selection for resistance to a nuclear polyhedrosis virus in laboratory populations of the cotton bollworm, Heliothis zea. J. Invertebr. Pathol., 20, 187-192.

Ignoffo, C. M., Puttler, B., Hostetter, D. L. and Dickerson, W. A. 1976. Susceptibility of the cabbage looper Trichoplusia ni and the velvetbean caterpillar Anticarsia gemmatalis to several isolates of the entomopathogenic fungus Nomuraea rileyi. J. Invertebr. Pathol., 28, 259-262.

Inoue, H. 1974. Multiplication of an infectious-flacherie virus in the resistant and susceptible strains of the silkworm Bombyx mori. J. Sericult. Sci. Japan, 43, 318-324. (In Japanese).

Inoue, H. and Miyagawa, M. 1978. Regeneration of midgut epithelial cells in the silkworm, Bombyx mori, infected with viruses. J. Invertebr. Pathol., 32, 373-380.

Jaques, R. P. 1973. Methods and effectiveness of distribution of microbial insecticides. Ann. N.Y. Acad. Sci., 217, 109-120.

Kartman, L. 1953. Factors influencing infection of the mosquito with Dirofilaria immitis (Leidy, 1956). Expt'l. Parasitol., 2, 27-78.

Kawakami, K. 1975. Susceptibility of several varieties of the silkworm Bombyx mori L. to Aspergillus disease and germination of fungus spores in larval hemolymph. J. Sericult. Sci. Japan., 44, 39-44. (In Japanese).

Kilama, W. L. and Craig, G. B. Jr. 1969. Monofactorial inheritance of susceptibility to Plasmodium gallinaceum in Aedes aegypti. Ann. Trop. Med. Parasitol., 63, 419-432.

Kinsinger, R. A. and McGaughey, W. H. 1979. Susceptibility of populations of Indian meal moth and Almond moth to Bacillus thuringiensis. J. Econ. Entomol., 72, 346-349.

Klein, M. and Podoler, H. 1978. Studies on the application of a nuclear polyhedrosis virus to control populations of the Egyptian cotton worm Spodoptera littoralis. J. Invertebr. Pathol., 32, 244-248.

Kobayashi, M., Yamaguchi, S. and Yokoyama, Y. 1969. Influences of the larval development on the susceptibility of the silkworm Bombyx mori L. to the nuclear-polyhedrosis virus. J. Sericult. Sci. Japan, 38, 481-487. (In Japanese).

Koidsumi, K. 1957. Antifungal action of cuticular lipids in insects. J. Insect Physiol., 1, 40-51.

Koidsumi, K. and Wada, Y. 1955. Studies on the antimicrobial function of insect lipids. IV. Racial difference in the antifungal activity in the silkworm integument. Jap. J. Appl. Zool., 20, 184-190.

Kulincevic, J. M. and Rothenbuhler, W. C. 1975. Selection for resistance and susceptibility to hairless-black syndrome in the honeybee. J. Invertebr. Pathol., 25, 289-295.

Lewis, L. F. and Rothenbuhler, W. C. 1961. Resistance to American foulbrood in honeybees: III. Differential survival of the two kinds of larvae from two-drone matings. J. Insect Pathol., 3, 197-215.

Llewellyn, G. C. and Chinnici, J. P. 1978. Variation in sensitivity to Aflatoxin B_1 among several strains of Drosophila melanogaster (Diptera). J. Invertebr. Pathol., 31, 37-40.

Longworth, J. F. and Kalmakoff, J. 1978. An ecological approach to the use of insect pathogens for pest control. In Microbial Ecology, M. W. Loutit and J. A. R. Miles, eds., Springer-Verlag, Berlin. pp. 269-271.

MacBain-Cameron, J. W. 1967. Suitability of pathogens for biological control. In Insect Pathology and Microbial Control, P. A. Van der Laan, ed., North Holland Publ. Co., Amsterdam. pp. 182-196.

MacDonald, W. W. 1967. The influence of genetic and other factors on vector susceptibility to parasites. In Genetics of Insect Vectors of Disease, J. W. Wright and R. Pal, eds. Elsevier, Amsterdam. pp. 567-584.

McGaughey, W. H. 1978. Response of Plodia interpunctella and Ephestia cautella larvae to spores and parasporal crystals of Bacillus thuringiensis. J. Econ. Entomol., 71, 687-688.

Maleki-Malani, H. 1970. Sur quelques caractéres de virose cytoplasmique de la nocteulle du chou Mamestra (Barathra) brassicae (Lep. Noctuidae). Entomophaga, 15, 315-325.

Martignoni, M. E. 1957. Contibuto alla conoscenza di una granulosi di Eucosoma griseana (Hübner) Tortricidae, Lepidoptera) quale Fattore linitaute il pullulanemts dell'insetts nella Engadina alta. Mitt. Schweiz. Anst. Forstl. Versuchw., 32, 371-418.

Martignoni, M. E. and Schmid, P. 1961. Studies on the resistance to virus infection in natural populations of Lepidoptera. J. Insect Pathol., 3, 62-74.

Menapace, D. M. 1978. Chalkbrood infection and detection in colonies of honeybees Apis mellifera. Am. Bee J., 118, 158-159.

Moeller, F. E. and Williams, P. H. 1976. Chalkbrood research at Madison, Wisconsin. Am. Bee J., 116, 484.

Papierok, B. and Wilding, N. 1979. Mise an évidence d'une différence de sensibilité entre 2 clones du Puceron du Pois, Acyrthosiphon pisum Harr. (Homopteres: Aphididae), exposés á 2 souches du Champignon Phycomycéte: Entomophthera obscura Hall & Dunn. C. R. Acad. Sc. Paris, Ser. D, 288, 93-95.

Petersen, J. J. 1978. Development of resistance by the southern house mosquito to the parasitic nematode Romanomermis culicivorax. Environ. Entomol., 7, 518-520.

Reichelderfer, D. F. and Benton, C. V. 1974. Some genetic aspects of the resistance of Spodoptera frugiperda to a nuclear polyhedrosis virus. J. Invertebr. Pathol., 23, 378-382.

Rinderer, T. E. and Green, T. J. 1976. Serological relationship between chronic bee paralysis virus and the virus causing hairless-black syndrome in the honeybee. J. Invertebr. Pathol., 27, 403-405.

Rinderer, T. E., Rothenbuhler, W. C. and Kulincevic, J. M. 1975. Responses of three genetically different stocks of the honeybee to a virus from bees with hairless-black syndrome. J. Invertebr. Pathol., 25, 297-300.

Rogoff, M. H., Ignoffo, C. M., Singer, S., Gard, I. and Prieto, A. P. 1969. Insecticidal activity of thirty-one strains of Bacillus against five insect species. J. Invertebr. Pathol., 14, 122-129.

Rothenbuhler, W. C. 1964a. Behaviour genetics of nest-cleaning in honeybees. I. Responses of four inbred lines to disease-killed larvae. Animal Behaviour, 12, 578-583.

Rothenbuhler, W. C. 1964b. Behaviour genetics of nest-cleaning in honeybees. IV. Responses of F_1 and backcross generations to disease killed brood. Am. Zool., 4, 111-123.

Rothenbuhler, W. C. 1979. Semi-domesticated insects: Honeybee breeding. In Genetics in Relation to Insect Management, M. A. Hoy and J. J. McKelvey Jr., eds., Rockefeller Found. pp. 84-92.

Rothenbuhler, W. C. and Thompson, V. C. 1956. Resistance to American foulbrood in honeybees. I. Differential survival of larval of different genetic lines. J. Econ. Entomol., 49, 470-475.

Sheppard, R. F. and Stairs, G. R. 1977. Dosage-mortality and time-mortality studies of a granulosis virus in a laboratory strain of the codling moth, Laspeyresia pomonella. J. Invertebr. Pathol., 29, 216-221.

Sidor, C. 1959. Susceptibility of larvae of the large white butterfly (Pieris brassicae L.) to two virus diseases. Ann. Appl. Biol., 47, 109-113.

Sidorov, N. G. and Mikhailenko, G. P. 1969. Resistance of different races of bees to experimental infection with Nosema disease. Pchelovodstvo, 89, 14-15. (In Russian)

Sidorov, N. G., Koptev, V. S. and Mugalimov, I. A. M. 1975. Resistance of honeybees to Nosema disease and a genetic method of control. Veterinariya (Moscow), 7, 63-65. (In Russian)

Splittstoesser, C. M., Kawanishi, C. Y. and Tashiro, H. 1978. Infection of the European chafer, Amphimallon majalis by Bacillus popilliae: Light and electron microscope observations. J. Invertebr. Pathol., 31, 84-90.

Stairs, G. R. 1964. Changes in the susceptibility of Galleria mellonella (Linnaeus) larvae to nuclear-polyhedrosis virus following blockage of the phagocytes with Indian ink. J. Insect Pathol., 6, 373-386.

Stairs, G. R. 1965. Dosage-mortality response of Galleria mellonella (Linnaeus) to a nuclear-polyhedrosis virus. J. Invertebr. Pathol., 7, 5-9.

Steinhaus, E. A. 1949. Principles of Insect Pathology, McGraw-Hill, N.Y. 757 pp.

Sutter, G. R., Rothenbuhler, W. C. and Raun, E. S. 1968. Resistance to American foulbrood in honeybees. VII. Growth of resistant and susceptible larvae. J. Invertebr. Pathol., 12, 25-28.

Uzigawa, K. and Aruga, H. 1965. On the selection of resistant strains to the infectious flacherie virus in the silkworm Bombyx mori L.. J. Sericult. Sci. Japan, 35, 23-27. (In Japanese)

Van Herrewege, J. 1969. La toxine thermostable de Bacillus thuringiensis var. thuringiensis Berliner (Serotype 1): Effects sur Drosophila melanogaster Meigen. These Univ. Lyon. 90 pp.

Vansulin, S. A. 1974. On the resistance of Culex pipiens molestus Forsk. to Boverin. Parazitologiya, 8, 274-275.

Watanabe, H. 1965. Resistance to peroral infection by the cytoplasmic-polyhedrosis virus in the silkworm, Bombyx mori (Linnaeus). J. Invertebr. Pathol., 7, 257-258.

Watanabe, H. 1966a. Relative virulence of polyhedrosis viruses and host-resistance in the silkworm Bombyx mori L. (Lepidoptera: Bombycidae). Appl. Ent. Zool., 1, 139-144.

Watanabe, H. 1966b. Genetic resistance to peroral infection with the cytoplasmic-polyhedrosis virus in the silkworm, Bombyx mori L. J. Sericult. Sci. Japan, 35, 27-31. (In Japanese)

Watanabe, H. 1966c. Probit analysis for determining the resistance of the silkworm, Bombyx mori L., to the infection by cytoplasmic polyhedrosis virus. J. Sericult. Sci. Japan, 35, 289-295. (In Japanese)

Watanabe, H. 1966d. Some aspects of the mechanism of resistance to peroral infection by cytoplasmic-polyhedrosis virus in the silkworm, Bombyx mori. L. J. Sericult. Sci. Japan, 35, 411-417. (In Japanese).

Watanabe, H. 1967. Development of resistance in the silkworm, Bombyx mori, to peroral infection of a cytoplasmic-polyhedrosis virus. J. Invertebr. Pathol., 9, 474-479.

Watanabe, H. 1971. Resistance of the silkworm to cytoplasmic-polyhedrosis virus. In The Cytoplasmic-Polyhedrosis Virus of the Silkworm, H. Aruga and Y. Tanada, eds., Univ. of Tokyo Press. pp. 169-184.

Watanabe, H. and Takamiya, K. 1976. Susceptibility of the silkworm larvae, Bombyx mori, reared under different light conditions to polyhedrosis viruses. J. Sericult. Sci. Japan, 45, 403-406. (In Japanese)

Watanabe, H. and Maeda, S. 1978. Genetic resistance to peroral infection with a densonucleosis virus in the silkworm, Bombyx mori. J. Sericult. Sci. Japan, 47, 209-214. (In Japanese)

Watanabe, H., Tanaka, S. and Simizu, T. 1974. Interstrain difference in the resistance of the silkworm, Bombyx mori to a flacherie and a cytoplasmic-polyhedrosis virus. J. Sericult. Sci. Japan, 43, 98-100. (In Japanese

Weiser, J. 1969. Immunity of insects to protozoa. In Immunity to Parasitic Animals, G. J. Jackson, R. Herman and I. Singer, eds., Appleton-Century-Crofts, N.Y. pp. 129-147.

Whitlock, V. H. 1977. Failure of a strain of Heliothis armigera (Hübn) (Noctuidae: Lepidoptera) to develop resistance to a nuclear polyhedrosis virus and a granulosis virus. J. Ent. Soc. S. Afr., 40, 251-253.

Wilson, B. H. and Burns, E. C. 1968. Induction of resistance to Bacillus thuringiensis in a laboratory strain of house flies. J. Econ. Entomol., 61, 1747-1748.

Woodard, D. B. and Fukuda, T. 1977. Laboratory resistance of the mosquito Anopheles quadrimaculatus to the mermithid nematode Diximermis peterseni. Mosquito News, 37, 192-195.

Yamaguchi, K. 1979. Natural recovery of the fall webworm, Hyphantria cunea to infection by a cytoplasmic-polyhedrosis virus of the silkworm, Bombyx mori. J. Invertebr. Pathol., 33, 126-128.

Yamvrias, C. 1962. Contribution a l'etude du mode d'action de Bacillus thuringiensis Berliner vis-a-vis de la Teigne de la Farine Anagasta (Ephestia) Kuehniella Zeller (Lepidoptera). Entomophaga, 7, 101-159.

Yokoyama, T. 1979. Silkworm selection and hybridization. In Genetics in Relation to Insect Management, M. A. Hoy and J. J. McKelvey Jr., eds., Rockefeller Found. pp. 71-83.

AUTHOR INDEX*

*Senior authors of literature cited are listed; full citations appear on page numbers given in italics.

SUBJECT INDEX